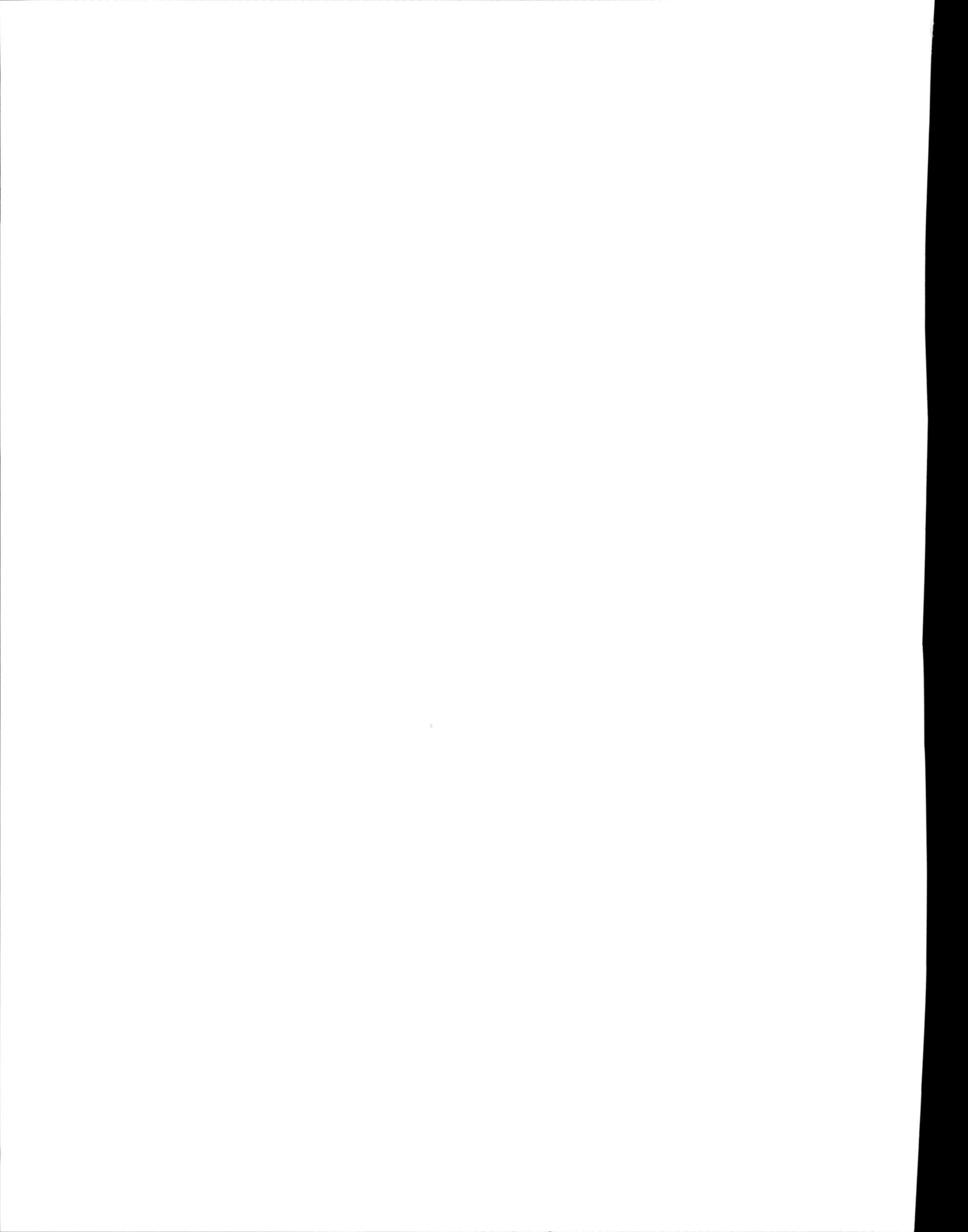

Poultry Science: Breeding, Rearing and Management of Animals

Poultry Science: Breeding, Rearing and Management of Animals

Editor: Carlos Hassey

R CALLISTO REFERENCE

www.callistoreference.com

Callisto Reference,
118-35 Queens Blvd., Suite 400,
Forest Hills, NY 11375, USA

Visit us on the World Wide Web at:
www.callistoreference.com

© Callisto Reference, 2018

ISBN: 978-1-63239-908-3 (Hardback)

Cataloging-in-Publication Data

Poultry science : breeding, rearing and management of animals / edited by Carlos Hassey.
   p. cm.
Includes bibliographical references and index.
ISBN 978-1-63239-908-3
1. Poultry. 2. Animal culture. 3. Livestock. I. Hassey, Carlos.
SF487 .P68 2018
636.5--dc23

Table of Contents

Preface

Poultry science involves the rearing of domestic birds such as chickens, ducks, geese, etc. for food. Standardized shelter construction as well as provision of nutritional feed are responsibilities that have to be undertaken by those rearing poultry birds. Technology that is implemented in this sector may be mechanized or manual. This book attempts to assist those with a goal of delving into the field of poultry science and management. It elucidates new techniques and their applications in a multidisciplinary approach. With state-of-the-art inputs by acclaimed experts of this field, this book targets students and professionals.

After months of intensive research and writing, this book is the end result of all who devoted their time and efforts in the initiation and progress of this book. It will surely be a source of reference in enhancing the required knowledge of the new developments in the area. During the course of developing this book, certain measures such as accuracy, authenticity and research focused analytical studies were given preference in order to produce a comprehensive book in the area of study.

This book would not have been possible without the efforts of the authors and the publisher. I extend my sincere thanks to them. Secondly, I express my gratitude to my family and well-wishers. And most importantly, I thank my students for constantly expressing their willingness and curiosity in enhancing their knowledge in the field, which encourages me to take up further research projects for the advancement of the area.

Editor

Hypoxia-Induced miR-15a Promotes Mesenchymal Ablation and Adaptation to Hypoxia during Lung Development in Chicken

Rui Hao[1], Xiaoxiang Hu[1], Changxin Wu[2], Ning Li[1,3]*

[1] State Key Laboratory for Agrobiotechnology, China Agricultural University, Beijing, P. R. China, [2] College of Animal Science and Technology, China Agricultural University, Beijing, P. R. China, [3] College of Animal Science, Yunnan Agricultural University, Kunming, P. R. China

Abstract

The lungs undergo changes that are adaptive for high elevation in certain animal species. In chickens, animals bred at high elevations (e.g., Tibet chickens) are better able to hatch and survive under high-altitude conditions. In addition, lowland chicken breeds undergo physiological effects and suffer greater mortality when they are exposed to hypoxic conditions during embryonic development. Although these physiological effects have been noted, the mechanisms that are responsible for hypoxia-induced changes in lung development and function are not known. Here we have examined the role of a particular microRNA (miRNA) in the regulation of lung development under hypoxic conditions. When chicks were incubated in low oxygen (hypoxia), miR-15a was significantly increased in embryonic lung tissue. The expression level of miR-15a in hypoxic Tibet chicken embryos increased and remained relatively high at embryonic day (E)16–20, whereas in normal chickens, expression increased and peaked at E19–20, at which time the cross-current gas exchange system (CCGS) is developing. *Bcl-2* was a translationally repressed target of miR-15a in these chickens. miR-16, a cluster and family member of miR-15a, was detected but did not participate in the posttranscriptional regulation of *bcl-2*. Around E19, the hypoxia-induced decrease in Bcl-2 protein resulted in apoptosis in the mesenchyme around the migrating tubes, which led to an expansion and migration of the tubes that would become the air capillary network and the CCGS. Thus, interfering with miR-15a expression in lung tissue may be a novel therapeutic strategy for hypoxia insults and altitude adaptation.

Editor: Joseph Najbauer, University of Fécs Medical School, Hungary

Funding: This work was funded by The High-Tech Research and Development Program of China (2011AA100301) and Yunnan Flexible Talents Development Program (2010C1118). The funders had no role in study design, data collection and analysis, decision to publish, or preparation of the manuscript.

Competing Interests: The authors have declared that no competing interests exist.

* E-mail: ninglcau@cau.edu.cn

Introduction

The cross-current gas exchange system (CCGS) and blood-gas barrier (BGB) morphology are two important factors that determine the avian lung air-diffusing capacity [1]. Both play a critical role in maintaining embryo homeostasis under low-oxygen conditions (hypoxia). Specific processes that take place during embryonic lung development and structures in animals that live at high altitudes are the basis for highly efficient pulmonary ventilation, which leads to altitude adaptation [2,3]. For plain chickens at low altitudes, inadequate oxygen exchange results in hypoxia syndrome and is lethal [4–6]. Therefore, determining the particular altitude adaptation characteristics of high-altitude chickens is important. Improving ventilation and protecting organs is an important strategy against the high death rate in normal animals raised at high altitudes, and studies such as these may suggest therapeutic targets for hypoxia syndrome. However, the molecular mechanisms of chicken lung development and hypoxia adaptation at high altitude have not been well defined.

For *in vitro* incubation, chickens are often used to examine the interaction between embryonic development and environmental conditions. In contrast to mammals, in chicks the mechanisms that underlie the development of their lungs involve a complex morphology and the generation of a cross-current system. Simple lung buds begin to develop at embryonic day (E4), but a complex gas-exchanging lung does not develop until E20–21, when the chick hatches from the shell. Pulmonary ventilation is an important factor for later-stage embryos. Makanya and colleagues have found that during chicken embryo development, the atrium, infundibulate, and air capillaries (ACs) form in turn from E8 to E19. During development, the ACs branch and anastomose with their neighboring cognates during E16–19. This process is a direct result of epithelial cell attenuation and canalization of the ACs [7]. Increasing evidence shows that these processes are accompanied by mesenchyme reduction via apoptosis. Lung mesenchyme degradation brings the ACs and blood capillaries into close proximity with each other [8].

Areas of high altitude, such as in Tibet, are regarded as "restricted areas of life". The Tibetan chicken, an indigenous chicken bred at high altitudes, is well known for its high hatchability under conditions of reduced oxygen. The hatchability of lowland breeds is approximately zero under the hypoxia that is simulated at an altitude of ≥4000 m, whereas the natural hatchability of Tibetan chicks can reach 70–80% [4–6]. An oxygen-deprived environment limits embryonic development and leads to organ paramorphia [9]. Hypoxia stress adaptation is a characteristic of high-altitude animals. But lowland animals also show low-oxygen–induced adaptation during embryonic development. Maina and Kalenga reported that in rat lungs under hypoxic conditions, the mean thickness of the BGB is half or less than of its thickness under normal conditions [10,11]. Non-homogeneous effects of hypoxia in internal organs have been observed in avian and mammalian species during development [12,13]. Specific weights (organ (lung) weight/body weight) are obviously decreased in size from E19 in plain chickens under hypoxia, whereas other organs tend to increase in weight [13]. Thus, hypoxia induces a lower density lung structure. However, the mechanisms of hypoxia-induced embryonic lung development and function remain unclear.

MicroRNAs (miRNAs) are non-coding RNA molecules of 20–24 nucleotides long that regulate the expression of other genes by inhibiting translation or cleaving complementary target mRNAs [14,15]. miRNA functions include the regulation of cell proliferation, differentiation, apoptosis, maintenance of stem cell pluripotency, tumor genesis, and others [16–20]. Several miRNAs regulate proapoptotic and antiapoptotic genes. miR-15a and miR-16, two members of the miR-15a/16 cluster, play a role in proapoptosis regulation by inhibiting the translation of the antiapoptotic protein Bcl-2 via binding to the 3′-untranslated region (3′-UTR) of bcl-2 mRNA [21]. These two miRNAs do not, however, always work in the same way [22]. Recent reports have focused on the influence of miRNAs on homeostasis and, in particular, on the repression of certain genes during hypoxia [23]. Because miRNAs have the ability to modify gene expression rapidly and reversibly, they are ideal mediators for sensing and responding to hypoxic stress. Their ability to modify regulatory pathways can affect the ability of an organism to survive under and adapt to hypoxic conditions [23]. Therefore, during embryonic development, miRNA-mediated regulatory circuits may provide flexible and conditional alternatives to embryonic development. In this study, we identified the influence of long-term hypoxia stress on lung development and showed that the miR-15a, downregulation of HIF-1, was responsive to the oxygen concentration and induced mesenchymal ablation through direct inhibition of the antiapoptotic gene chicken bcl-2 by binding to a unique target region. The chick embryo dying may start from an imbalance in homeostasis that begins in the lung. The inhibition of miR-15a or activation of HIF-1 or Bcl-2 may prevent hypoxia-induced lung damage and reduce chick embryo death.

Materials and Methods

Ethics statement

All animal work was conducted according to the guidelines for the care and use of experimental animals established by the Ministry of Science and Technology of the People's Republic of China (Approval number: 2006-398). The chicken incubation and chicken tissue samples collection procedure were approved by the Animal Welfare Committee of China Agricultural University (Permit Number: XK622). All efforts were made to minimize suffering.

Animals

The White Leghorn chickens eggs were obtained from the Chicken Farm of China Agriculture University. The Tibet chicken breed was obtained from Nyingchi Tibet and feeding in the Chicken Farm of China Agriculture University. White leghorn chicken eggs and Tibet (hT) chickens were divided into four groups: hypoxic White Leghorn (hW) chickens and normal White Leghorn (nW) chickens, hypoxic Tibet (hT) chickens and normal Tibet (nT) chickens. The hT group and hW group were incubated

under plateau-simulating conditions (37°C; 14% O_2; that is the similar condition at the 3200 m AMSL (Above Mean Sea Level)). The other two groups (nT, nW) were incubated under normal condition as control (37°C; 21% O_2). Because of the different hatchability, the starting incubation amount of eggs for each group was with big difference. We used 300, 550, 270 and 300 eggs for nW, hW, nT and hT samples incubation. Finally, the amount of nW, hW, nT, hT samples (eggs/chicks) were 209, 163, 193 and 192, respectively. The different organ samples (brain, lung, heart and liver) were all from these eggs/chicks. We want to make the best value for our egg/chick samples.

Northern blot analysis

Total RNA was isolated from different tissues from adult chickens and chicken embryos using TRIzol (Invitrogen, Carlsbad, CA, USA). Total RNA (20 μg) was fractionated using a 15% denaturing polyacrylamide/8M urea gel. The RNA was then transferred to a GeneScreen Plus membrane (Perkin–Elmer, Waltham, Massachusetts, USA). The 5′ ends of the DNA probes were labeled with [^{32}P]ATP (Amersham, Pittsburgh, PA, USA) using T4 polynucleotide kinase (New England BioLabs, Ipswich, Massachusetts, USA). Hybridization and washing were performed as described [24]. Hybridization signals were detected using a Phosphor Screen (Molecular Dynamics, Sunnyvale, CA, USA).

RNA purification and real-time PCR

Total RNA was isolated with Trizol reagent (Invitrogen) according to the manufacturer's protocol. After treatment with DNase I (RNase-free; TAKARA), oligo(dT) was used to synthesize single-stranded cDNA from 2 μg total RNA using M-MLV reverse transcriptase (Invitrogen). miRNA-specific reverse transcription (RT) stem-loop primers were as follows:

5sRT: 5′-CTCAACTGGTGTCGTGGAGTCGGCAATT-CAGTTGAGAAGCCTAC-3′;

16RT: 5′-CTCAACTGGTGTCGTGGAGTCGGCAATT-CAGTTGAGCACCAATA -3′;

15aRT: 5′-CTCAACTGGTGTCGTGGAGTCGGCAATT-CAGTTGAGACAAACC-3′;

144R: 5′-CTCAACTGGTGTCGTGGAGTCGGCAATT-CAGTTGAGGAGTACA-3′.

miRNA was amplified using SYBR Green (Applied Biosystems, Carlsbad, CA, US) and the TaqMan Universal PCR Master Mix Kit (Applied Biosystems) with the 7900HT Fast Real-Time PCR System (Applied Biosystems). Primers and the TaqMan probe were designed using Primer Select software (Primer Express v2.0, Applied Biosystems). Real-time primer sequences were as follows:

GAPDH forward primer: 5′-CGATCTGAACTACATGGTT-TACATGTT-3′;

GAPDH reverse primer: 5′-CCCGTTCTCAGCCTTGACA-3′;

bcl-2 forward primer: 5′-AGCGTCAACCGGGAGATGT-3′;

bcl-2 reverse primer: 5′-GCATCCCATCCTCCGTTGT-3′;

HIF-1 forward primer: 5′-CAGGTACAA GAGCAACCAA CCA-3′;

HIF-1 reverse primer: 5′-TGGATAAT GACATGGCTA ATGAATTC-3′

HIF-1 probe: 5′-FAM-AGTTCACCTGAGCCC-MGB-3′;

universal reverse primer: 5′-CTCAACTGGTGTCGTG-GAGTC-3′

miR-144: 5′-CGCGGCTACAGTATAGATGATG-3′

miR-15a: 5′-GCTGG TAGCAGCACATAATGG-3′

miR-16: 5′- GCGGAGTAGCAGCACGTAAA-3′

RN5S: 5′- GCTCTGGAATACCGGGTGCTGT-3′ (Table S1 and S2).

Western blot analysis

Equal amounts of protein were resolved by SDS-PAGE and transferred to polyvinylidene difluoride membranes. Membranes were incubated with Bcl-2 (C-2) antibody (1:500; Santa Cruz Biotechnology, Santa Cruz, CA, US) or HIF-1α (C-19) antibody (1:500; Santa Cruz Biotechnology). Primary antibody binding was visualized with horseradish peroxidase–conjugated secondary antibody and detected with enhanced chemiluminescence (Beyotime, Jangsu, China). Secondary antibodies were goat anti–mouse IgG (H+L) (1:10000; Santa Cruz Biotechnology) and rabbit anti–goat IgG (1:10000; Santa Cruz Biotechnology). For a loading control, membranes were reprobed with primary antibody against actin (2Q1055; 1:5000; Santa Cruz Biotechnology).

Immunohistochemical staining

Paraffin-embedded tissues were sectioned (4 µm thick) and mounted on glass slides. They were then dewaxed followed by antigen retrieval in 0.01M sodium citrate at 95°C for 10 min. Immunohistochemistry was done using mouse monoclonal anti–Bcl-2 (C-2) (IgG1; 1:100; Santa Cruz Biotechnology) in blocking reagent at 25°C for 20 min. The primary antibody was omitted for negative controls. The slides were then treated consecutively with horseradish peroxidase–conjugated goat anti–mouse IgG (1:200; Chemicon International, Inc., Temecula, CA, USA) and incubated for 1 h at room temperature. Slides were then incubated using a DAB substrate kit (Roche Applied Science, Indianapolis city, IN, USA), and the color reaction was allowed to develop for 5–10 min. After washing, slides were coverslipped.

In situ hybridization

In situ hybridization was performed as described [25]. Hybridization was done using the 5′-digoxigenin–labeled miR-CURY LNA microRNA Detection Probes anti-gga-miR-16 and anti-gga-miR-15a (EXIQON, Vedbaek, Denmark). The 5′-digoxigenin–labeled Scramble-miR Probe (EXIQON) was used as a negative control.

TUNEL staining

TUNEL staining was performed using the in situ Colorimetric TUNEL Apoptosis Assay Kit (Beyotime), according to the manufacturer's manual. Briefly, incubated the frozen sections in the cold 0.1% Triton X-100 for 2 min. Then moved the slides into the 0.3% H_2O_2 in methanol for 2 min at RT. Washed 3 times with PBS. That was followed by incubation with TdT enzyme solution for 60 min at 37°C. The reaction was terminated by incubation in a stop/wash buffer for 10 min at 37°C. Then used the Streptavidin-HRP for coloration. Incubated the sections in the Streptavidin-HRP working solution for 30 min at RT, following with the colour reagent incubation for 30 min

Cell culture

The human embryonic kidney 293T cells (American Type Culture Collection, Rockville, MD) were cultured at 37°C in 5% CO_2 in DMEM supplemented with 10% fetal bovine serum (HyClone, Logan, UT, US) and antibiotics.

Plasmid transfection and luciferase assay

The chicken bcl-2 3′-UTR containing a miR-15a–binding site and a miR-16–binding site was cloned into the psiCHICK-2 vector (Promega, Madison, WI, US) using the following primers:

forward primer: 5′-CTCGAGAGTCACCCAGTTTATCGT-3′;

reverse primer: 5′-CTCGAGGATTCTTCCGCTTCGTCA-3′.(Table S3)

The synthesized fragments containing the reverse complement sequences of gga-miR-15a and gga-miR-16 were also individually cloned into the psiCHICK-2 vector and used as a positive control. The psiCHICK-2 vector was transfected into the human embryonic kidney 293T cells with miR-15a/16 mimics (Applied Biosystems) or a miR-15a/16 inhibitor (Applied Biosystems) or a negative control (Applied Biosystems). The concentration of mimic is 10 pmol. The relative proportion of mimic and inhibitor was 1:10. The firefly and Renilla luciferase activity of each transfection was determined with the Dual-Glo Luciferase Assay System (Promega) 44 h post-transfection.

Results

The expression of miR-15a and miR-144 is induced by hypoxia stress in chicken embryo

Xu and colleagues [26] published that the expression levels of both miR-15a and miR-144 are higher in the lung tissue. Our studies confirmed the result and the tissue special expression. In later stages of chick embryo development, relatively strong expression of miR-15a and miR-144 was detected in lung tissues with Northern blotting, whereas expression was weak in other organs (Fig. 1A). To determine whether both miR-15a and miR-144 are highly expressed throughout chicken embryonic development or just a temporally variable expression pattern, we compared the miR-15a and miR-144 expression profiles in lung tissues. We set the hatcher to simulate the plateau condition (AMSL is 3200 m, 37°C, 14% O_2, PO_2 is 14.093 kPa) and incubated both the Tibet chicken eggs and White Leghorn chicken eggs in it to simulate the hT and hW groups. The nT and nW chicken groups are the eggs incubated in normal plain condition (37°C, 21% O_2, PO_2 is 100 kPa). Lung samples were obtained from E13 to the third day after hatching (d3). Real-time PCR showed no difference in miR-15a and miR-144 expression from E13 to E17 in the four groups (data not shown). At E18, the hT group showed significant upregulation of miR-15a that remained relatively high and stable through d3 (Fig. 1B). miR-15a in hW showed continuously increasing expression from E18 and a peak at E20 and the hatch out day (d1) (Fig. 1B). Expression of miR-15a in nW chickens nearly coincided with that of nT chickens except for a relative peak value for nW at E19 (Fig. 1B). As a control, miR-15a in nT chickens showed a relatively smooth line, indicating stable, low expression of miR-15a (Fig. 1B). A comparison of interclass differences showed that at E18, expression of miR-15a was higher in lung tissues from hT than in the lung tissues from the other three groups. At E19, the time at which the respiratory CCGS develops, miR-15a was also upregulated in hW and nT chickens. Comparing the expression levels at E20, we found that expression of miR-15a was relatively low in both tissues under normal conditions (nT, nW chickens), and relatively high in the hW and hT sample which incubated under plateau simulating condition (Fig. 1B). Quantitative analysis of miR-144 revealed no significant changes over time (data not shown). As a cluster and family member, miR-16 was reported to have the same function as miR-15a [21]. To determine whether the peak value of miR-15a expression was lung specific or was a systemic reaction to hypoxia and whether expression of miR-15a was the same as that of miR-16, we performed real-time PCR for miR-15a and miR-16 using chick embryo lung, heart, brain and liver at E16 and E19. hW showed remarkable upregulation of miR-15a at E19 as compared with that at E16 in the embryonic lung (Fig. 1C). In the heart and brain, miR-15a expression in hW and hT was also increased from

E16 to E19, but the range was not as large as that in the lung (Fig. 1C). In the liver, no differences in expression between the two embryonic stages or among the experimental groups were identified (Fig. 1C). We identified a small increase in miR-16 from E16 to E19 in embryonic lung in hW (Fig. 1C). However, there were no clear changes in miR-16 in the other groups. Thus, miR-16 also appeared to be affected by hypoxia at a time when the respiratory gas exchange was developing. Thus, miR-15a expression was sensitive to oxygen concentration, and the strongest response was detected in lung tissue. We also found differences in hypoxia-induced expression of miR-15a and miR-16 between high-altitude and plain chicks. The period of development of pulmonary ventilation is crucial for development of lung organ function. Although miR-16 belongs to the same cluster and family as miR-15a [21], they were differentially sensitive to hypoxia. miR-144 also appeared responsive to hypoxia (data not shown), but it is not involved in temporal lung development.

Hypoxia stress upregulates HIF-1α and Bcl-2 protein

Hypoxia-inducible transcription factor 1 alpha (HIF-1α) is either anti-apoptotic or pro-apoptotic, according to the cell type and experimental design. Because severe or prolonged hypoxia induces apoptosis [27], HIF-1 can be used to monitor this hypoxia-induced response. To explore the reaction of the lung to hypoxia, we detected HIF-1α transcripts using quantitative PCR and HIF-1α protein using western blotting (Fig. 2A, C). Hypoxia-induced HIF-1α mRNA expression in hW increased from E16 to E19, whereas HIF-1α in hT kept higher (from E18 to E19), until E20 the trend reverse. In the normoxia experimental groups, nW and nT chickens, HIF-1α mRNA expression showed a slight upward trend, but there was no difference between these two groups (Fig. 2A). These changes were further confirmed with Western blotting, which showed that HIF-1α protein was induced by hypoxia stress (Fig. 2C). Interestingly, HIF-1α protein in nW chickens was upregulated compared to nT chickens, especially at E19 (Fig. 2C). During normoxia, HIF-1α expression can be regulated at the translational level [28]. These results suggest that nW also suffered acute hypoxia stress at E19.

Bcl-2 is downstream of HIF-1α in the HIF-1α–induced anti-apoptotic pathway [29]. Because hypoxic cell death is prevented by anti-apoptotic proteins such as Bcl-2 [30], we investigated bcl-2 mRNA expression during the whole embryo development and identified no remarkable variation in bcl-2 mRNA (Fig. 2B, and data not shown). However, Bcl-2 protein was strongly decreased in E19 hW lung samples as compared with its expression at E16 and E18 (Fig. 2C). Moreover, in other groups, Bcl-2 protein levels were decreased to different degrees in association with the degree of hypoxic stress (the time length) (Fig. 2C). Thus, the regulation of Bcl-2 protein levels may not be at the transcriptional level but rather at the translational level. We note that miR-15a expression was increased at E19 during hypoxia (Fig. 1B).

bcl-2 is a target gene of miR-15a but not miR-16 in chicken lung tissues under hypoxia

miR-15a and miR-16 negatively regulate bcl-2 by directly binding to a particular sequence in the 3′-UTR of bcl-2 and inhibiting its translation [21,31]. This repression is sufficient to induce apoptosis [32]. Our observation of hypoxia-induced miR-15a expression (Fig. 1B) and reduced Bcl-2 protein levels at E19 (Fig. 2C) indicated an inverse relationship between miR-15a and Bcl-2 protein expression, which suggested a causative role for miR-15a in the downregulation of bcl-2. However, as a cluster member of miR-15a, miR-16 was identified as hypoxia insensitive. To further verify whether miR-15a or miR-16 repressed chicken bcl-2

by binding directly to the predicted binding site in its 3′-UTR as shown in human cells [31], we analyzed the complementation of the chicken bcl-2 mRNA (NM_205339.1) and chicken miR-15a/16 using four different prediction algorithms: TargetScan [33], RNA22 [34], DIANA [35,36], and PicTar [37]. Interestingly, human bcl-2 mRNA (NM_000633) has one target located at nt 2529–2536 of its 3′-UTR [21] that is not conserved with the chicken bcl-2 3′-UTR. Chicken bcl-2 was predicted to contain a miR-15a/16: bcl-2 site located at nt 369–390 of its 3′-UTR (Fig. 3A). The RNA22/PicTar results showed that the target site sequence complemented only with chicken miR-15a, and the TargetScan result suggested that both the chicken miR-15a and miR-16 would work (data not shown). We cloned the entire chicken bcl-2 3′-UTR, containing the miR-15a/16 binding site, and inserted it into the multiple cloning sites of the psiCHICK-2 vector. We transfected this luciferase reporter vector into 293T cells. We also co-transfected these cells with chicken miR-15a mimic, mimic and inhibitor, or mimic control (miRNA mimics are small, chemically modified double-stranded RNAs that mimic endogenous miRNAs and enable miRNA functional analysis by up-regulation of miRNA activity. miRNA inhibitors are small, chemically modified single-stranded RNA molecules designed to specifically bind to and inhibit endogenous miRNA molecules and enable miRNA functional analysis by down-regulation of miRNA activity. The controls are small control molecules). We performed the same experiment with chicken miR-16. Chicken miR-15a decreased luciferase activity of the reporter vector containing the 3′-UTR sequence, and the level of luciferase activity recovered in the presence of the chicken miR-15a inhibitor. However, no change was observed with the chicken miR-16 mimic (Fig. 3C). Together, the results suggested that chicken miR-15a decreased chicken bcl-2 translation by directly acting on a miR-15a–specific response element in the 3′-UTR of chicken bcl-2 mRNA and that this effect may be different in chicken bcl-2 as compared with that in human bcl-2. Chicken miR-16, another member of the cluster, did not affect bcl-2 regulation.

Hypoxia-induced miR-15a overexpression promotes the formation of CCGS and a thin BGB

At E19, the lung begins to function, but the development of the complex CCGS is still occurring. To further understand a possible role for hypoxia-induced miR-15a in the development of a functional lung in vivo, we immunostained sections from hT, hW, nT, and nW chicken lungs for Bcl-2 protein and performed terminal deoxynucleotidyl transferase nick-end labeling (TUNEL) staining.

Hypoxia upregulated Bcl-2 protein expression at E18 (Fig. 4G, H). The strongest staining was of hW chicken lung sections at E18 (Fig. 4H). Staining of hT and nW lung sections at E18 were similar. Their staining were weaker than what detected of the hW sections (Fig. 4F, G). No staining was observed in nT chickens at E18 (Fig. 4E). As expected, the staining was localized in the mesenchyma around the ACs of the chicken lung (Fig. 4F', G', H'). No special staining signal was detected in other sections (Fig. 4A, B, C, D).

Apoptotic cells were observed at E19, except for in nT chickens (Fig. 5K). TUNEL staining was darker in hW lung sections at E19 (Fig. 5N) and in hT lung sections at E19 (Fig. 5M) as compared with staining in nW lung sections at E19 (Fig. 5L). Sections viewed at higher magnification showed that apoptotic cells were located in the mesenchyme around the migrating epithelial tubes, including capillaries (Fig. 5L', M', N'). Interesting, at E19, no significant difference of Bcl-2 protein expression was detected in all the groups (Fig. 4K–N). Taken together, these results suggested that

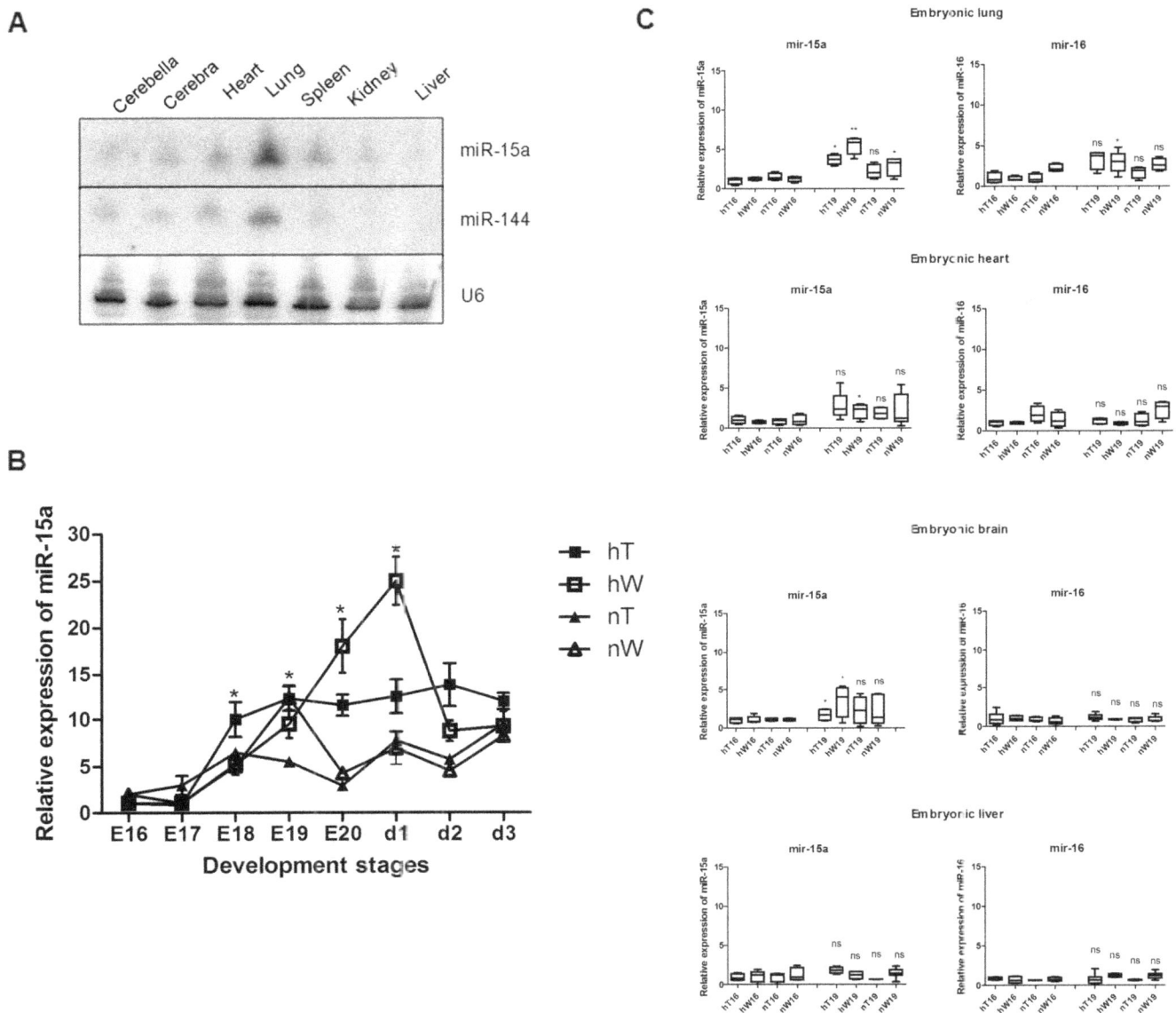

Figure 1. Expression of miRNAs in chicken lung tissues. (A) Total RNA from tissues from E19 chicken embryos was blotted with probes for miR-15a, miR-144 and U6 (loading control). miR-15a and miR-144 were identified in late-stage embryonic lung tissue. In each sample, the total RNA was mixed with samples from 9 chickens. (B) Quantitative expression analysis of miR-15a showed hypoxia-related expression that was affected by both the species and environmental conditions. The hW chicken group was most sensitive relative to other three groups. In the nW chicken group, there was a response at E19. At E19 of the nW chicken group, the expression of miR-15a remained relatively high in the hT chicken group compared with the nT, nW chicken groups and was largely unchanged in the nT chicken group through the whole embryo stages. Data are expressed as the mean ± SEM for each group. (C) Quantitative expression analysis of miR-15a and miR-16 in the embryonic lung, heart, brain and liver at E16 and E19 tissues and were expressed as the mean ± SEM for each group. Under hypoxia stress, miR-15a was more highly expressed at E19 than at E16 in the brain, heart and lung for the hW group and in the lung and brain for the hT group. miR-16 showed a weak response to stress in the embryonic lung (hW). Result statistically different are indicated with an asterisk/s (* $P<0.05$; ** $P<0.01$; ns : not significant). E16-20 = embryonic d13-20, respectively. d1, d2, d3 = the 1st day after hatching, the 2nd day after hatching, the 3rd day after hatching.

the lung appeared resistant to apoptosis at E18 which is likely the result of upregulation of Bcl-2. However, at E19, the time of further lung development, the tissues became more sensitive to the oxygen concentration, and apoptosis was induced by the hypoxia-induced upregulation of miR-15a. Of interest, the tube density in hW at E19 was also higher than that in sections from those other two groups. That indicated adaptive lung development which has also been reported in mammals living as altitude adaptation [3,38]. These responses to hypoxia stress indicate an oxygen-

concentration dose effect process in the chick lung. The apoptotic cells were located in the mesenchyme around the migrating tubes. These mesenchymal cells are in direct contact with the air and are the targets of apoptosis, which may lead to expansion and migration of tubes, implying precise temporospatial regulation. During the subsequent CCGS establishment when the ACs are branching and anastomosing with their neighboring cognates, mesenchymal cell apoptosis participates in the approximation of ACs and blood capillaries (BCs), to establish a thin BGB.

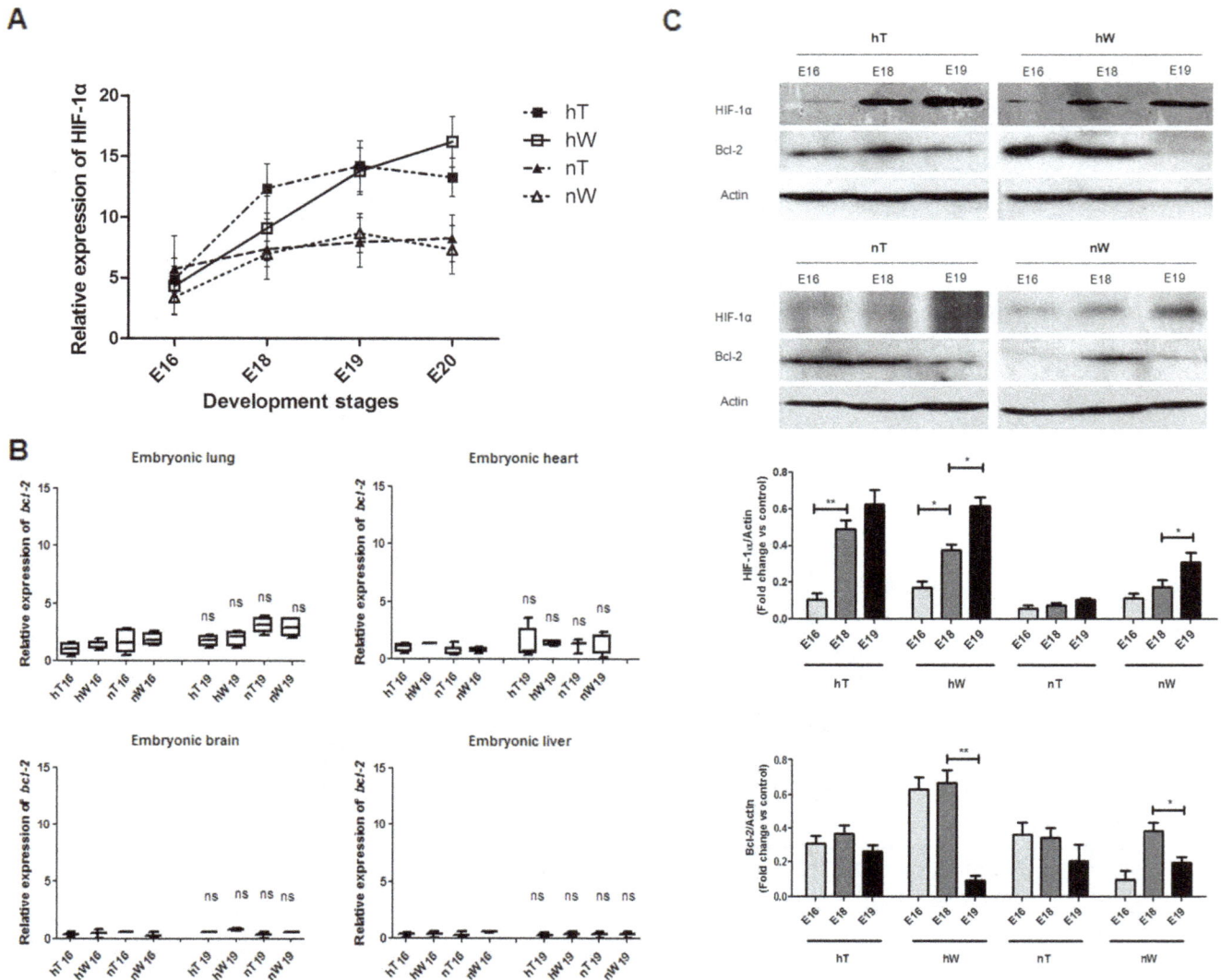

Figure 2. HIF-1α and Bcl-2 expression in lung tissue under hypoxia and normoxia conditions. Data were obtained from hT, hW, nT, and nW chicken tissues at E16, E18, E19, and E20. (A) Quantitative expression analysis of HIF-1α mRNA in embryonic lung at E16, E18, E19, and E20. The stress reaction in hW was robust as compared with the smooth response in hT. Data are expressed as the mean ± SEM for each group. (B) Quantitative expression analysis of *bcl-2* mRNA in the embryonic lung, heart, brain and liver at E16 and E19. There were no changes in the expression of *bcl-2* mRNA between the E16 and E19 in each kind of embryonic tissues. Data are expressed as the mean ± SEM for each group. (C) Analysis of HIF-1α and Bcl-2 expression at the protein level in lung tissue (shown in western blot and densitometry value). HIF-1α protein expression increased from E16 to E19, however in different level in hT, hW, nT, and nW group. Bcl-2 protein did show different levels of expression across different time points and different groups. (* $P<0.05$; ** $P<0.01$; ns = not significant).

Discussion

Adaptation to hypoxia of the Tibet chicken appears to involve numerous changes at the organism's level. These chicken have bigger hearts and lungs, higher hemoglobin oxygen transfer efficiency and other adaptation characteristics [13,39–42]. Such adaptive changes may counteract the reduced oxygen uptake and mitigate tissue injury due to hypoxia. There are two oxygen sensitive periods during the chick embryo development, the embryo day 0–4 (early stage) and the respiratory gas exchange (around the E19, the late stage) [43]. Increased oxygen tension in the early and late stages of development results in increased hatchability, while during while the middle stage of development increased oxygen causes little or no change [44–46].

The major finding of our study is that, during the late stage of chick embryo development, lung mesenchymal cells become sensitive to the oxygen concentration when the lung assumes the gas-exchange function from the allantois at E19. Hypoxia stress induces proapoptotic chicken miR-15a expression and subsequently inhibits the translation of the antiapoptotic protein Bcl-2, resulting in mesenchyme ablation. Chicken miR-16, a cluster and family member of chicken miR-15a, does not affect the hypoxia-induced pathway. In vivo, mesenchymal cell death was seen around the tubes, and thus, apoptosis led to expansion and migration of new tubes. The mesenchyme ablation led to the establishment of CCGS and a thin BGB. Thus, BGB formation was associated with the oxygen concentration. The distance between ACs and BCs will also be affected by this process. Because differences exist among species, long-term hypoxia stress in normal

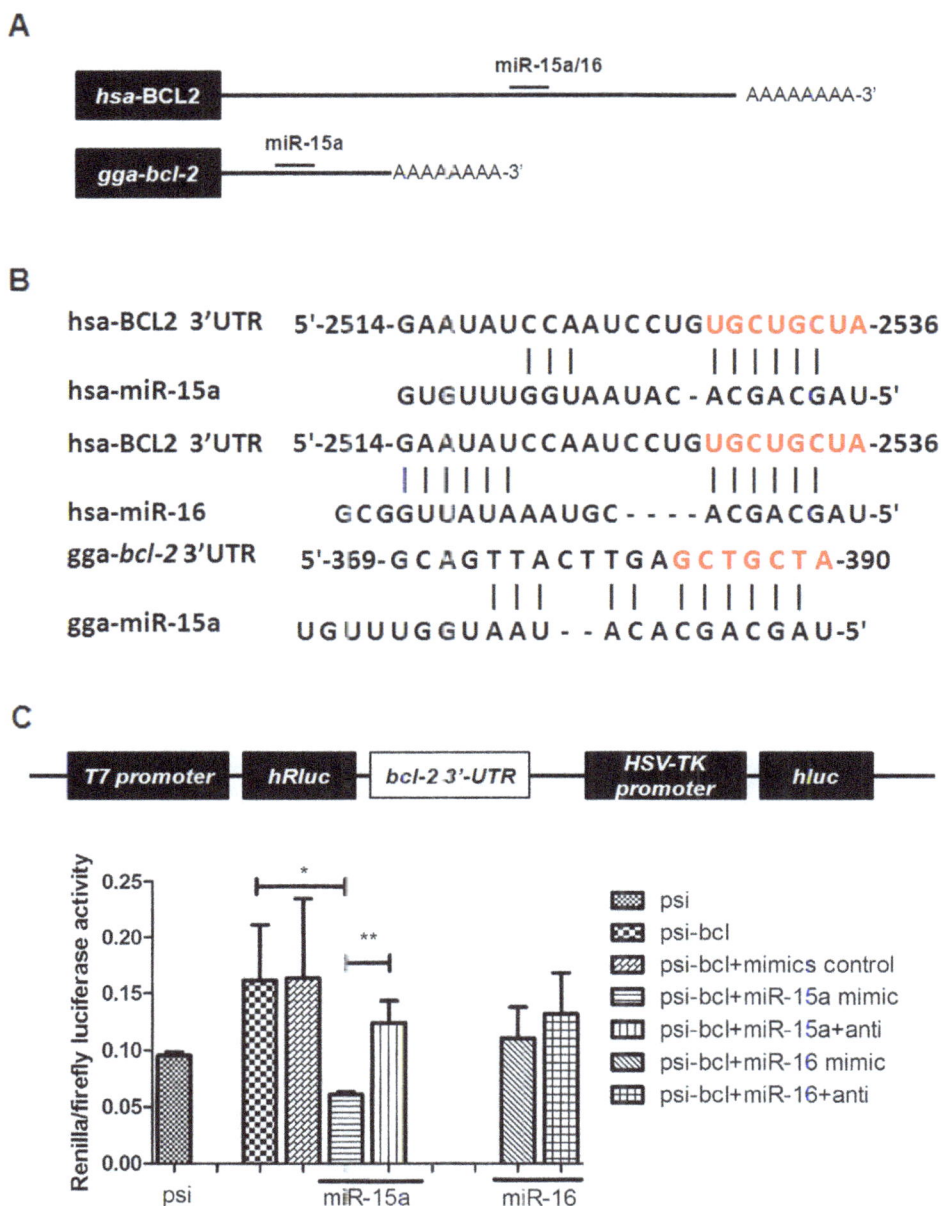

Figure 3. bcl-2 is a target gene of miR-15a, but not miR-16 in chicken lung. (A) miR-15a regulates the translation of *bcl-2* mRNA through a sequence that is not conserved with human sequence. miR-15a and the binding site in the gga-*bcl-2* 3′-UTR are shown, but miR-16 shows no target site in this part of the sequence. (B) miR-15a/16 have a consistent target site in human *bcl-2* 3′-UTR. The miR-15a binding site in the *bcl-2* 3′-UTR sequence mediates translation repress on by miR-15a. (C) The luciferase reporter vector contains two luciferase cDNAs, Renilla luciferase (hRluc) and firefly luciferase (hluc). The *bcl-2* 3′-UTF was fused to the hRluc cDNA downstream sequence. In co-transfected cells, the miR-15a mimic decreased the expression of hRluc and miR-15a mimic inhibitor rescued hRluc activity; no differences were seen for miR-16. Data are expressed as the mean ± SEM. * $P<0.05$, ** $P<0.01$.

chicks is lethal. The lung is the most sensitive organ at later incubation times, and an imbalance between organ ablation and body development will be lethal (Fig. 6A). To our knowledge, our experiments have provided the first evidence that establishment of a thin BGB can be mediated by the oxygen concentration during the sensitive period of lung development. Development of a thin BGB requires precise temporospatial regulation and is tightly correlated with the incubation conditions. The response to stress controls mesenchymal cell death via a miRNA-mediated mechanism. We also show that of the two cluster and family members,

only chicken miR-15a is responsive to hypoxia stress and participates in downstream regulation.

The ability to sense and respond to hypoxia is of fundamental importance during embryonic development as well as environmental adaption. Dysregulation of oxygen homeostasis is an important cause of embryo death in hypoxic conditions. An organism's oxygen homeostasis begins with the lung gas-exchange system. Among the extant vertebrate taxa, the avian lung is reported to be the most efficient and complex gas exchanger and is a CCGS [47,48]. Microscopy was used to identify the functional development of the avian lung in 2006 and showed establishment

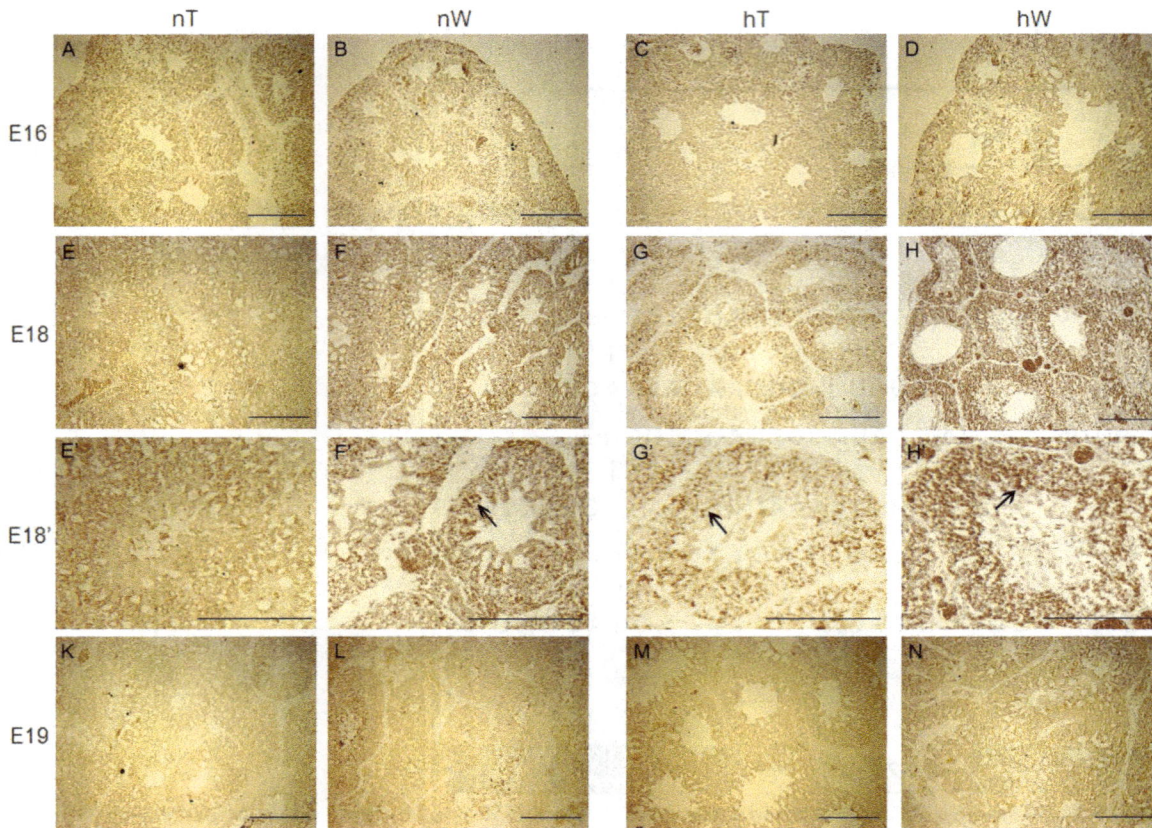

Figure 4. Immunohistochemical staining (dark brown coloring) for Bcl-2 protein in lung specimens from chickens incubated under normal and hypoxic conditions. Bcl-2 protein expression in lung samples from nW, hT, and hW gradually increased moderately from E16 to E18 before the onset of lung functioning (E–H), and disappeared at E19. By E18 in nW lungs, weaker Bcl-2 expression was observed relative to hT (G) and hW (H) chicken groups. In nT chicken sections, Bcl-2 protein staining was nearly unchanged throughout the developmental stages examined (A, E, K). To see details of the E18 lung structure, higher-magnification photomicrographs were taken. Bcl-2 staining was detected in the mesenchyme (arrows) around the ACs and not in the infundibula or atrias of chicken lung (F', G', H')". In the hW section at E18, Bcl-2 staining was strong (arrow in H'), but weaker in hT and nW at this stage (arrows in G', F'). Scale bar = 200 μm.

of a thin BGB by the time of hatching [8,49–52]. However, the lung of various domestic chicks begins to develop at E18 [8]. On later days (around E19), the mesenchymal cells of chick lung immediately surrounding the migrating epithelial tubes become the initial targets of apoptosis. In this study, we report for the first time that at E19, initiation of pulmonary ventilation, which brings air into the lung ACs, was induced immediately and promoted establishment of an oxygen concentration–dependent CCGS. The induced establishment of the CCGS was implemented by induced mesenchymal apoptosis, resulting in expansion and migration of tubes. The observation that the lung organ/body weight ratio of normal chickens was lower than in hypoxic conditions indirectly reflects the hypoxia-induced disintegration of mesenchymal cells [13]. Adaptive lung development has also been reported in mammals living at higher altitudes [38]. E19 is the peak of mortality for embryos growing in both normal and hypoxic conditions [53]. Our data showed relative hypoxic stress at E19 even when embryos are incubated in normal conditions. We suggest that when an acute injury disrupts developmental homeostatic regulation, hypoxia may result in the of the affected animal.

miRNAs were discovered nearly two decades ago, and their importance was highlighted when they were identified as conserved in higher eukaryotes. Recent reports have focused on the influence of miRNAs on homeostatic and adaptive development, and in particular, response to hypoxia. An ideal mediator of hypoxic stress should be able to modify gene expression both rapidly and reversibly, a widely accepted characteristic of miRNAs [23]. Therefore, miRNAs may be a molecular rheostat that facilitates survival and adaptation to hypoxic conditions by tuning and switching regulatory circuits on and off. Various miRNAs appear to regulate hypoxia responses in a variety of different organisms, cell types, and disease states [22,54–57]. More than 90 hypoxia-regulated miRNAs (HRMs) have been described, and many regulate certain cell types only. However, as a general characteristic, these HRMs respond to hypoxic stress rapidly (less than 48 h). These hypoxic stress responses are regulated through different pathways. miR-107 is directly regulated by p53, which is one of the most frequently mutated genes in human cancers [58–60]. HIF-1 was identified as target gene of miR-107 both in vitro and in vivo, suggesting a p53-miR-107-HIF-1 hypoxia response pathway [60]. miR-206 was also proved as a direct regulator of HIF-1alpha. In cultured hypoxic pulmonary artery smooth muscle cells (PASMCs), miR-206 promotes the Hypoxia-induced pulmonary hypertension though the HIF-1a/Fhl-1 pathway [61]. HIF-1 and microRNAs build a network of hypoxia response, HIF-1 was reported as regulator of miR-9, it induced upregulation of miR-9 in PASMCs. As a phenotypic switch, knockdown of miR-9 was

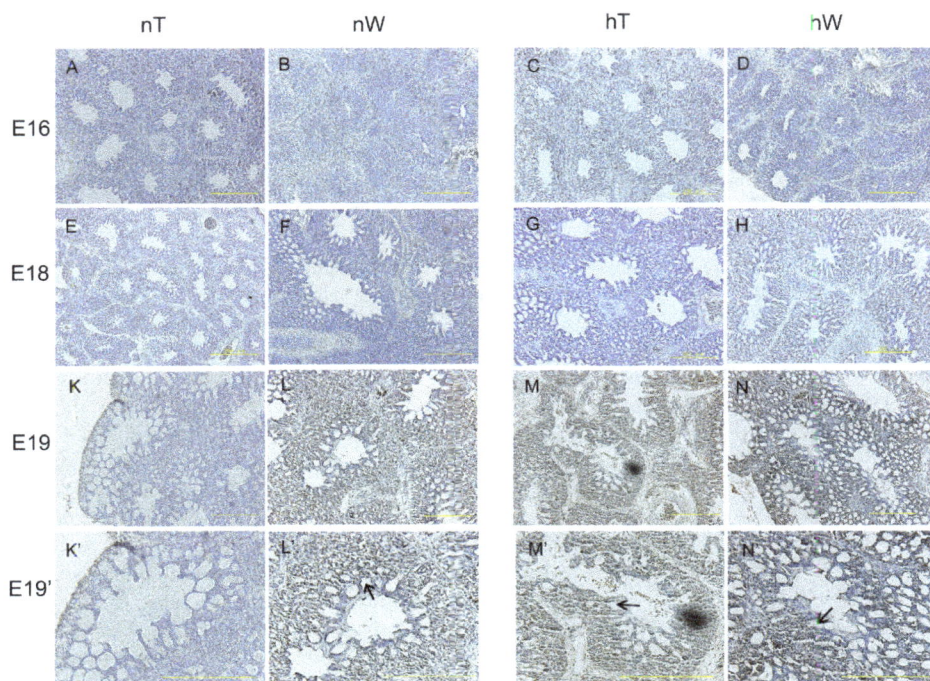

Figure 5. Analysis of apoptosis in lung specimens from chickens incubated under normal and hypoxic conditions. From E16 to E18, no TUNEL staining was identified (A–H). At E19, apoptotic cells were localized in the mesenchyme surrounding the atrias and infundibula of the chicken lungs (L, M, N). There was no obvious staining in nT chicken at E19 (K). At higher magnification, staining was clearly seen in the regions between ACs and not in the parabronchi, atrias, or infundibula (arrows in L', M', N'). hW staining at E19 (arrow in N') was clearly darker than that observed in hT or nW lung sections at this stage (arrows in M', N'). The tube density in hW at E19 was also higher than that in sections from those other two groups. Scale bar = 200 μm.

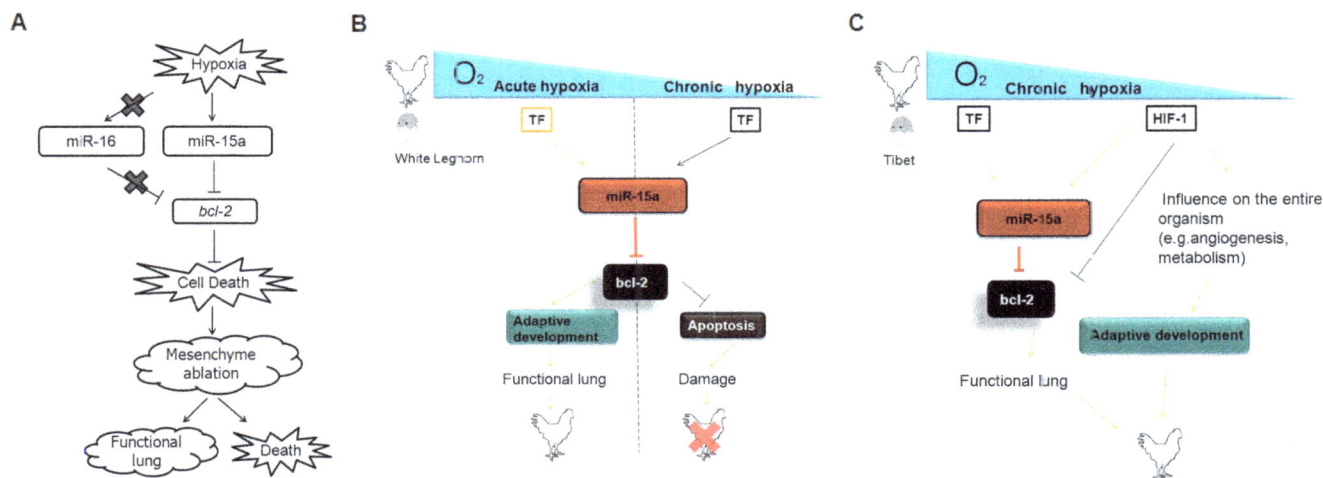

Figure 6. Schematic representation of hypoxia-induced gga-miR-15a control of mesenchymal cell death/ablation during functional lung establishment and adaptive development. (A) In chicken lung, hypoxia stress stimulates gga-miR-15a expression, and miR-15a may bind to the 3′-UTR of gga-bcl-2 and inhibit its antiapoptotic activity through posttranscriptional gene silencing. Mesenchymal apoptosis then occurs, which leads to a functional lung with a thinner EGB or an acute reaction resulting in death in different chicken species. gga-miR-16 shows no response to hypoxia, and there is no target site for gga-miR-16 in the 3′-UTR region of gga-bcl-2. (B) Mesenchymal cell death/ablation indeed appeared necessary for the formation of a thinner BGB and is promoted by hypoxia (the oxygen concentration). Particular transcriptional factors (TF) may respond to the hypoxia stress and induce gga-miR-15a expression. For plain chickens, the induced apoptosis would be helpful for the formation of a thinner BGB, but it could be lethal if they are exposed to low oxygen concentrations for a long time. (C) However, for high-altitude chickens, apoptosis is genetically initiated to form a functional BGB, and it is under control of the whole organism (C). Acute hypoxia: related hypoxia condition induced by the CCGS occurring around E19. Chronic hypoxia: long-term altitude condition with low oxygen concentration.

followed by hypoxia exposure attenuation in PASMCs proliferation [62]. miR-210 and miR-373 were discovered in HeLa and MCF-7 cells. Both miRNAs contain hypoxia-responsive elements (HREs) in their promoter regions, suggesting direct regulation by HIF-1 [63]. Most hypoxic injury results in cell apoptosis. Hayashita and Frankel reported that miR-21 and the miR-17-92 cluster function as antiapoptotic miRNAs in various cancers [64,65]. In contrast, Cimmino et al. showed that miR-15a and miR-16 promote apoptosis by posttranscriptional gene silencing of bcl-2 [21]. In this study, our data suggested that the expression of miR-15a, but not miR-16, is sensitive to the oxygen concentration and was significantly increased in lung mesenchymal cells in chicken. This results in select apoptosis in mesenchymal cells around the migrating tubes through reduction of Bcl-2 levels, as shown in vitro and vivo. Although miR-15a and miR-16 belong to the same cluster and miRNA family, Yin et al. also reported that miR-15a plays a causative role in the regulation of apoptosis by directly targeting bcl-2 during ischemic vascular injury [22]. Together, our experimental data extend these early findings and further show that chicken miR-15a can repress bcl-2 translation by directly binding to the 3′-UTR of bcl-2 with a different target site sequence. To our knowledge, our work is the first to document that chicken miR-15a, but not miR-16, regulates apoptosis by directly targeting chicken bcl-2 at a specific target region during later chick lung CCGS development.

Kang et al. [66] reported that the miR-15a/16-1 cluster is a homo-cluster. Because the homologous miRNA in the homo-cluster has almost the same proximal targets or genes in the same pathway, homo-cluster miRNAs show a direct regulatory coordination involving one step. Thus, there is no intermediate regulator between the miRNAs and their targets; direct coordinated regulation has the advantage of accuracy and quickness. Hua and colleagues identified a group of regulatory miRNAs, including miR-16, which regulates the expression of vascular endothelial growth factor [67]. miR-16 may function during BC network formation, which is perpendicular to the long axis of ACs. miR-16 shows unequal expression among tissues [68]. In this study, we suppose that in chick, members of the miR-15a/16-1 cluster respond to hypoxia through different target genes and pathways. We showed relatively high expression of miR-144 in chicken embryo lungs. Analysis of miR-144 showed a lower response to hypoxia. miR-144 expression has been reported to be limited to the blood islands at stage 8 [69], and it is crucial for erythroid homeostasis [70]. miR-144 also modulates oxidative stress tolerance and is associated with anemia severity in sickle cell disease [71]. Thus, we suggest that miR-144 may be responsive to hypoxia but may have no direct relationship with lung development.

Lung organogenesis is a field that has attracted many investigators. Since "molecular embryology" was reviewed, more challenges have emerged in developmental lung biology. miRNAs play an important role not only in cell proliferation and tumorigenesis but also in apoptosis that is associated with organism and organ development. Yang et al. identified 167 differentially expressed miRNAs during rat lung organogenesis, and Dong reported 117 miRNAs that are dynamically and temporally regulated during mouse lung organogenesis. Mujahid et al. found that, during the important window in lung development, 25 miRNAs showed different expression profiles across gestation, and 37 miRNAs changed between males and females [72–74]. Transcription factors connect the extracellular environment to the signaling network in cells. HIF-1 is a widely recognized hypoxic stress transcription factor [23]. In a recent report, HIF-1 was shown to function in lung branching morphogenesis and epithelial maturation. Functional loss of HIF-1 results in neonatal lethality and severe respiratory distress. Interestingly, some abnormalities were detected in the endothelial barrier, which indicates a disorder in development of the epithelial and mesenchyme [75]. By controlling both O_2 delivery and utilization, HIF-1 functions as a important regulator of O_2 homeostasis. Under tissue hypoxia or ischemia, HIF-1 protect the heart against injury by inducing angiogenic growth factors, which stimulate vascular remodeling to increase blood flow following [76]. Intrestingly, for the High altitude adaptive species, during the Tibet chicken embryo lung development, HIF-1 and other factors induce complex CCGS and a thin BGB which is necessary for the High altitude adaptive lung function. In the present study, we detected upregulation of both HIF-1 and miR-15a with a similar expression profile during E16–E19. We hypothesize that the induced ablation of mesenchymal cells may involve HIF-1 as a hypoxia sensor, which then indirectly decreases Bcl-2 protein levels by inducing miR-15a expression. However, more studies are needed to elucidate the details. Other transcription factors are likely to participate in this regulatory network. The sensor of oxygen concentration is important here, because apoptosis occurs only around the ACs and not at the top to the atria. In the Tibet chicken, induced apoptosis is necessary for establishment of functional lungs containing complex CCGS and a thin BGB. However, when the induced apoptosis is uncontrolled and organ homeostasis is lost, normal chickens will die for pressure overload (Fig. 6B, C). Other molecules have been reported as miR-15a regulators in different cell types or tissues, and some are hypoxia regulated. miR-709 responds to apoptosis stimuli and prevents expression of miR-15a/16 in mouse [77]. Myc and HDAC3 are regarded as miR-15a/16 regulators in mantle cells and other non-Hodgkin B-cell lymphoma [78,79]. HERΔ16, a clinically important oncogenic isoform of HER2, is another regulator of miR-15a/16 in breast tumors [80]. In human tumors, E2F1 is a positive regulator of miR-15a and miR-16, but only miR-15a inhibits expression of cyclin E [81]. In a mouse model of ischemia-induced cerebral injury, PPARδ (peroxisome proliferator-activated receptorδ) was identified as a regulator of miR-15a that inhibits the induction of apoptosis by preventing miR-15a expression [22]. Intrestingly, three isoforms of PPARs (α, γ and δ) were all documented direct downstream of miRs [22,82,83]. PPRAα were identified had specific functional variants between highland Tibetan populations and normal [85]. However, although PPRAα is in the Gene Ontology biological process 'response to hypoxia', but it is not even in the HIF pathway [84]. So, additional studies will be needed to further understand the interaction between miRNAs, gene functional variation and conditional development.

In this study, we identified the influence of long-term hypoxia stress on lung development. In conclusion, the mechanisms of hypoxia-induced, miR-15a–mediated CCGS and thin BGB establishment are important for understanding adaptational development of chick lung at particular stages. HIF-1 may regulate miR-15a in this setting. We also showed that only miR-15a and not miR-16 was responsive to the oxygen concentration and induced mesenchymal ablation through direct inhibition of the antiapoptotic gene chicken bcl-2 by binding to a unique target region. The stage at which embryos die may results from an imbalance in homeostasis that begins in the lung. Therefore, either inhibition of miR-15a or activation of HIF-1 or Bcl-2 may potentially be therapeutic options for preventing hypoxia-induced lung damage and may reduce death at particular stages of chick embryo development.

Supporting Information

Table S1   Reverse transcription primers and run method.

Table S2   Real-time PCR primers and run method.

Table S3   *bcl-2* 3′-UTR cloning primers.

References

1. Powell FL (1982) Diffusion in avian lungs. Fed Proc 41: 2131–2133.
2. Tang JR, Seedorf GJ, Muehlethaler V, Walker DL, Markham NE, et al. (2010) Moderate postnatal hyperoxia accelerates lung growth and attenuates pulmonary hypertension in infant rats after exposure to intra-amniotic endotoxin. Am J Physiol Lung Cell Mol Physiol 299: L735–748.
3. Durmowicz AG, Hofmeister S, Kadyraliev TK, Aldashev AA, Stenmark KR (1993) Functional and structural adaptation of the yak pulmonary circulation to residence at high altitude. J Appl Physiol 74: 2276–2285.
4. Visschedijk AH, Ar A, Rahn H, Piiper J (1980) The independent effects of atmospheric pressure and oxygen partial pressure on gas exchange of the chicken embryo. Respir Physiol 39: 33–44
5. Visschedijk AH, Tazawa H, Piiper J (1985) Variability of shell conductance and gas exchange of chicken eggs. Respir Physiol 59: 339–345.
6. Visschedijk AH (1985) Gas exchange and hatchability of chicken eggs incubated at simulated high altitude. J Appl Physiol 58: 416–418.
7. Makanya AN, El-Darawish Y, Kavoi BM, Djonov V (2011) Spatial and functional relationships between air conduits and blood capillaries in the pulmonary gas exchange tissue of adult and developing chickens. Microsc Res Tech 74: 159–169.
8. Maina JN (2004) Morphogenesis of the laminated, tripartite cytoarchitectural design of the blood-gas barrier of the avian lung: a systematic electron microscopic study on the domestic fowl, Gallus gallus variant domesticus. Tissue Cell 36: 129–139.
9. Tuder RM, Yun JH, Bhunia A, Fijalkowska I (2007) Hypoxia and chronic lung disease. J Mol Med (Berl) 85: 1317–1324.
10. Maina JN, King AS, Settle G (1989) An allometric study of pulmonary morphometric parameters in birds, with mammalian comparisons. Philos Trans R Soc Lond B Biol Sci 326: 1–57.
11. Kalenga M, Tschanz SA, Burri PH (1995) Protein deficiency and the growing rat lung. II. Morphometric analysis and morphology. Pediatr Res 37: 789–795.
12. Mortola JP (2001) Respiratory Physiology of Newborn Mammals: A Comparative Perspective The Johns Hopkins University Press, Baltimore, Maryland.
13. Azzam MA, Mortola JP (2007) Organ growth in chicken embryos during hypoxia: implications on organ "sparing" and "catch-up growth". Respir Physiol Neurobiol 159: 155–162.
14. Bartel DP (2004) MicroRNAs: genomics, biogenesis, mechanism, and function. cell 116: 281–297.
15. Filipowicz W, Bhattacharyya SN, Sonenberg N (2008) Mechanisms of post-transcriptional regulation by microRNAs: are the answers in sight? Nat Rev Genet 9: 102–114.
16. Jovanovic M, Hengartner MO (2006) miRNAs and apoptosis: RNAs to die for. Oncogene 25: 6176–6187.
17. Schickel R, Boyerinas B, Park SM, Peter ME (2008) MicroRNAs: key players in the immune system, differentiation, tumorigenesis and cell death. Oncogene 27: 5959–5974.
18. Barroso-Deljesus A, Lucena-Aguilar G, Sanchez L, Ligero G, Gutierrez-Aranda I, et al. (2011) The Nodal inhibitor Lefty is negatively modulated by the microRNA miR-302 in human embryonic stem cells. FASEB J.
19. Foshay KM, Gallicano GI (2007) Small RNAs, big potential: the role of MicroRNAs in stem cell function. Curr Stem Cell Res Ther 2: 264–271.
20. Gunaratne PH (2009) Embryonic stem cell microRNAs defining factors in induced pluripotent (iPS) and cancer (CSC) stem cells. Curr Stem Cell Res Ther 4: 168–177.
21. Cimmino A, Calin GA, Fabbri M, Iorio MV Ferracin M, et al. (2005) miR-15 and miR-16 induce apoptosis by targeting BCL2. Proc Natl Acad Sci U S A 102: 13944–13949.
22. Yin KJ, Deng Z, Hamblin M, Xiang Y, Huang H, et al. (2010) Peroxisome proliferator-activated receptor delta regulation of miR-15a in ischemia-induced cerebral vascular endothelial injury. J Neurosci 30: 6398–6408.
23. Pocock R (2011) Invited review: decoding the microRNA response to hypoxia. Pflugers Arch 461: 307–315.
24. Lee RC, Ambros V (2001) An extensive class of small RNAs in Caenorhabditis elegans. science 294: 862–864.
25. Obernosterer G, Martinez J, Alenius M (2007) Locked nucleic acid-based in situ detection of microRNAs in mouse tissue sections. Nat Protoc 2: 1508–1514.
26. Xu H, Wang X, Du Z, Li N (2006) Identification of microRNAs from different tissues of chicken embryo and adult chicken. FEBS Lett 580: 3610–3616.
27. Piret JP, Mottet D, Raes M, Michiels C (2002) Is HIF-1alpha a pro- or an anti-apoptotic protein? Biochem Pharmacol 64: 889–892.
28. Greijer AE, van der Wall E (2004) The role of hypoxia inducible factor 1 (HIF-1) in hypoxia induced apoptosis. J Clin Pathol 57: 1009–1014.
29. Cory S, Adams JM (2002) The Bcl2 family: regulators of the cellular life-or-death switch. Nat Rev Cancer 2: 647–656.
30. Shimizu S, Eguchi Y, Kosaka H, Kamiike W, Matsuda H, et al. (1995) Prevention of hypoxia-induced cell death by Bcl-2 and Bcl-xL. nature 374: 811–813.
31. Calin GA, Cimmino A, Fabbri M, Ferracin M, Wojcik SE, et al. (2008) MiR-15a and miR-16-1 cluster functions in human leukemia. Proc Natl Acad Sci U S A 105: 5166–5171.
32. Calin GA, Croce CM (2006) Genomics of chronic lymphocytic leukemia microRNAs as new players with clinical significance. Semin Oncol 33: 167–173.
33. Grimson A, Farh KK, Johnston WK, Garrett-Engele P, Lim LP, et al. (2007) MicroRNA targeting specificity in mammals: determinants beyond seed pairing. Mol Cell 27: 91–105.
34. Schuster P, Fontana W, Stadler PF, Hofacker IL (1994) From sequences to shapes and back: a case study in RNA secondary structures. Proc Biol Sci 255: 279–284.
35. Maragkakis M, Alexiou P, Papadopoulos GL, Reczko M, Dalamagas T, et al. (2009) Accurate microRNA target prediction correlates with protein repression levels. BMC Bioinformatics 10: 295.
36. Maragkakis M, Reczko M, Simossis VA, Alexiou P, Papadopoulos GL, et al. (2009) DIANA-microT web server: elucidating microRNA functions through target prediction. Nucleic Acids Res 37: W273–276.
37. Krek A, Grun D, Poy MN, Wolf R, Rosenberg L, et al. (2005) Combinatorial microRNA target predictions. Nat Genet 37: 495–500.
38. Lalthantluanga R, Wiesner H, Braunitzer G (1985) Studies on yak hemoglobin (Bos grunniens, Bovidae): structural basis for high intrinsic oxygen affinity? Biol Chem Hoppe Seyler 366: 63–68.
39. Zhang H, Burggren WW (2012) Hypoxic level and duration differentially affect embryonic organ system development of the chicken (Gallus gallus). Poult Sci 91: 3191–3201.
40. Wang XY, He Y, Li JY, Bao HG, Wu C (2013) Association of a missense nucleotide polymorphism in the MT-ND2 gene with mitochondrial reactive oxygen species production in the Tibet chicken embryo incubated in normoxia or simulated hypoxia. Anim Genet 44: 472–475.
41. Bao HG, Wang XY, Li JY, Wu CX (2011) Comparison of effects of hypoxia on glutathione and activities of related enzymes in livers of Tibet chicken and Silky chicken. Poult Sci 90: 648–652.
42. Bao HG, Zhao CJ, Li JY, Zhang H, Wu C (2007) A comparison of mitochondrial respiratory function of Tibet chicken and Silky chicken embryonic brain. Poult Sci 86: 2210–2215.
43. ZHANG H, Wu C (2006) Influences of Oxygen on Embryonic Mortality and Hatchability of Chicken Eggs. Acta Veterinaria et Zootechnica Sinica 37(2): 112–116.
44. Christensen VL, Bagley LG (1988) Improved hatchability of turkey eggs at high altitudes due to added oxygen and increased incubation temperature. Poult Sci 67: 956–960.
45. Christensen VL, Bagley LG (1989) Efficacy of fertilization in artificially inseminated turkey hens. Poult Sci 68: 724–729.
46. Christensen VL, Bagley RA (1984) Vital gas exchange and hatchability of turkey eggs at high altitude. Poult Sci 63: 1350–1356.
47. Duncker HR (1971) The lung air sac system of birds. A contribution to the functional anatomy of the respiratory apparatus. Ergeb Anat Entwicklungsgesch 45: 7–171.
48. King A (1989) Form and function in birds. London: Academic Press.
49. Gallagher BC (1986) Basal laminar thinning in branching morphogenesis of the chick lung as demonstrated by lectin probes. J Embryol Exp Morphol 94: 173–188.
50. Gallagher BC (1986) Branching morphogenesis in the avian lung: electron microscopic studies using cationic dyes. J Embryol Exp Morphol 94: 189–205.

Acknowledgments

We thank our lab members for the discussion. Thanks to Dr. Xiaobo Wang for advice on Q-PCR and Northern blot methods. We are also grateful to Dr. Haigang Bao and Dr. Ming Tang for advice on egg hatching experiments and Jinjin Mao for help with in making figures and assistance with immunohistochemical staining.

Author Contributions

Conceived and designed the experiments: RH XH CW NL. Performed the experiments: RH. Analyzed the data: RH. Contributed reagents/materials/analysis tools: RH XH CW. Wrote the paper: RH.

51. Maina JN (2003) A systematic study of the development of the airway (bronchial) system of the avian lung from days 3 to 26 of embryogenesis: a transmission electron microscopic study on the domestic fowl, Gallus gallus variant domesticus. Tissue Cell 35: 375–391.

52. Maina JN (2003) Developmental dynamics of the bronchial (airway) and air sac systems of the avian respiratory system from day 3 to day 26 of life: a scanning electron microscopic study of the domestic fowl, Gallus gallus variant domesticus. Anat Embryol (Berl) 207: 119–134.

53. Liu C, Zhang LF, Song ML, Bao HG, Zhao CJ, et al. (2009) Highly efficient dissociation of oxygen from hemoglobin in Tibetan chicken embryos compared with lowland chicken embryos incubated in hypoxia. Poult Sci 88: 2689–2694.

54. Camps C, Buffa FM, Colella S, Moore J, Sotiriou C, et al. (2008) hsa-miR-210 Is induced by hypoxia and is an independent prognostic factor in breast cancer. Clin Cancer Res 14: 1340–1348.

55. Fasanaro P, D'Alessandra Y, Di Stefano V, Melchionna R, Romani S, et al. (2008) MicroRNA-210 modulates endothelial cell response to hypoxia and inhibits the receptor tyrosine kinase ligand Ephrin-A3. J Biol Chem 283: 15878–15883.

56. Hebert C, Norris K, Scheper MA, Nikitakis N, Sauk JJ (2007) High mobility group A2 is a target for miRNA-98 in head and neck squamous cell carcinoma. Mol Cancer 6: 5.

57. Kulshreshtha R, Ferracin M, Wojcik SE, Garzon R, Alder H, et al. (2007) A microRNA signature of hypoxia. Mol Cell Biol 27: 1859–1867.

58. Oren M (2003) Decision making by p53: life, death and cancer. Cell Death Differ 10: 431–442.

59. Vogelstein B, Kinzler KW (2004) Cancer genes and the pathways they control. Nat Med 10: 789–799.

60. Yamakuchi M, Lotterman CD, Bao C, Hruban RH, Karim B, et al. (2010) P53-induced microRNA-107 inhibits HIF-1 and tumor angiogenesis. Proc Natl Acad Sci U S A 107: 6334–6339.

61. Yue J, Guan J, Wang X, Zhang L, Yang Z, et al. (2013) MicroRNA-206 is involved in hypoxia-induced pulmonary hypertension through targeting of the HIF-1alpha/Fhl-1 pathway. Lab Invest 93: 748–759.

62. Shan F, Li J, Huang QY (2014) HIF-1 Alpha-Induced Up-Regulation of miR-9 Contributes to Phenotypic Modulation in Pulmonary Artery Smooth Muscle Cells During Hypoxia. J Cell Physiol.

63. Crosby ME, Kulshreshtha R, Ivan M, Glazer PM (2009) MicroRNA regulation of DNA repair gene expression in hypoxic stress. Cancer Res 69: 1221–1229.

64. Hayashita Y, Osada H, Tatematsu Y, Yamada H, Yanagisawa K, et al. (2005) A polycistronic microRNA cluster, miR-17-92, is overexpressed in human lung cancers and enhances cell proliferation. Cancer Res 65: 9628–9632.

65. Frankel LB, Christoffersen NR, Jacobsen A, Lindow M, Krogh A, et al. (2008) Programmed cell death 4 (PDCD4) is an important functional target of the microRNA miR-21 in breast cancer cells. J Biol Chem 283: 1026–1033.

66. Wang J, Haubrock M, Cao KM, Hua X, Zhang CY, et al. (2011) Regulatory coordination of clustered microRNAs based on microRNA-transcription factor regulatory network. BMC Syst Biol 5: 199.

67. Hua Z, Lv Q, Ye W, Wong CK, Cai G, et al. (2006) MiRNA-directed regulation of VEGF and other angiogenic factors under hypoxia. PLoS One 1: e116.

68. Yue J, Tigyi G (2010) Conservation of miR-15a/16-1 and miR-15b/16-2 clusters. Mamm Genome 21: 88–94.

69. Darnell DK, Kaur S, Stanislaw S, Konieczka JH, Yatskievych TA, et al. (2006) MicroRNA expression during chick embryo development. Dev Dyn 235: 3156–3165.

70. Rasmussen KD, Simmini S, Abreu-Goodger C, Bartonicek N, Di Giacomo M, et al. (2010) The miR-144/451 locus is required for erythroid homeostasis. J Exp Med 207: 1351–1358.

71. Sangokoya C, Telen MJ, Chi JT (2010) microRNA miR-144 modulates oxidative stress tolerance and associates with anemia severity in sickle cell disease. blood 116: 4338–4348.

72. Yang Y, Kai G, Pu XD, Qing K, Guo XR, et al. (2012) Expression profile of microRNAs in fetal lung development of Sprague-Dawley rats. Int J Mol Med 29: 393–402.

73. Dong J, Jiang G, Asmann YW, Tomaszek S, Jen J, et al. (2010) MicroRNA networks in mouse lung organogenesis. PLoS One 5: e10854.

74. Mujahid S, Logvinenko T, Volpe MV, Nielsen HC (2013) miRNA regulated pathways in late stage murine lung development. BMC Dev Biol 13: 13.

75. Bridges JP, Lin S, Ikegami M, Shannon JM (2012) Conditional hypoxia inducible factor-1alpha induction in embryonic pulmonary epithelium impairs maturation and augments lymphangiogenesis. Dev Biol 362: 24–41.

76. Semenza GL (2014) Hypoxia-inducible factor 1 and cardiovascular disease. Annu Rev Physiol 76: 39–56.

77. Tang R, Li L, Zhu D, Hou D, Cao T, et al. (2012) Mouse miRNA-709 directly regulates miRNA-15a/16-1 biogenesis at the posttranscriptional level in the nucleus: evidence for a microRNA hierarchy system. Cell Res 22: 504–515.

78. Zhang X, Chen X, Lin J, Lwin T, Wright G, et al. (2011) Myc represses miR-15a/miR-16-1 expression through recruitment of HDAC3 in mantle cell and other non-Hodgkin B-cell lymphomas. Oncogene.

79. Wu G, Yu F, Xiao Z, Xu K, Xu J, et al. (2011) Hepatitis B virus X protein downregulates expression of the miR-16 family in malignant hepatocytes in vitro. Br J Cancer 105: 146–153.

80. Cittelly DM, Das PM, Salvo VA, Fonseca JP, Burow ME, et al. (2010) Oncogenic HER2{Delta}16 suppresses miR-15a/16 and deregulates BCL-2 to promote endocrine resistance of breast tumors. Carcinogenesis 31: 2049–2057.

81. Ofir M, Hacohen D, Ginsberg D (2011) MiR-15 and miR-16 are direct transcriptional targets of E2F1 that limit E2F-induced proliferation by targeting cyclin E. Mol Cancer Res 9: 440–447.

82. Lin Q, Gao Z, Alarcon RM, Ye J, Yun Z (2009) A role of miR-27 in the regulation of adipogenesis. FEBS J 276: 2348–2358.

83. Iliopoulos D, Malizos KN, Oikonomou P, Tsezou A (2008) Integrative microRNA and proteomic approaches identify novel osteoarthritis genes and their collaborative metabolic and inflammatory networks. PLoS One 3: e3740.

84. Ashburner M, Ball CA, Blake JA, Botstein D, Butler H, et al. (2000) Gene ontology: tool for the unification of biology. The Gene Ontology Consortium. Nat Genet 25: 25–29.

85. Scheinfeldt LB, Tishkoff SA (2010) Living the high life: high-altitude adaptation. Genome Biol 11: 133.

2

Behavioural and Physiological Effects of Finely Balanced Decision-Making in Chickens

Anna C. Davies[1]*, Christine J. Nicol[1], Mia E. Persson[2], Andrew N. Radford[3]

1 School of Clinical Veterinary Science, University of Bristol, Bristol, United Kingdom, **2** Avian Behavioural Genomics and Physiology Group, Linköping University, Linköping, Sweden, **3** School of Biological Sciences, University of Bristol, Bristol, United Kingdom

Abstract

In humans, more difficult decisions result in behavioural and physiological changes suggestive of increased arousal, but little is known about the effect of decision difficulty in other species. A difficult decision can have a number of characteristics; we aimed to monitor how finely balanced decisions, compared to unbalanced ones, affected the behaviour and physiology of chickens. An unbalanced decision was one in which the two options were of unequal net value (1 (Q1) vs. 6 (Q6) pieces of sweetcorn with no cost associated with either option); a finely balanced decision was one in which the options were of equal net value (i.e. hens were "indifferent" to both options). To identify hens' indifference, a titration procedure was used in which a cost (electromagnetic weight on an access door) was applied to the Q6 option, to find the individual point at which hens chose this option approximately equally to Q1 via a non-weighted door. We then compared behavioural and physiological indicators of arousal (head movements, latency to choose, heart-rate variability and surface body temperature) when chickens made decisions that were unbalanced or finely balanced. Significant physiological (heart-rate variability) and behavioural (latency to pen) differences were found between the finely balanced and balanced conditions, but these were likely to be artefacts of the greater time and effort required to push through the weighted doors. No other behavioural and physiological measures were significantly different between the decision categories. We suggest that more information is needed on when best to monitor likely changes in arousal during decision-making and that future studies should consider decisions defined as difficult in other ways.

Editor: Elsa Addessi, CNR, Italy

Funding: This study was funded by the BBSRC (Grant reference: BB/F016662/1). The funders had no role in study design, data collection and analysis, decision to publish, or preparation of the manuscript.

Competing Interests: The authors have declared that no competing interests exist.

* Email: Anna.c.davies@bristol.ac.uk

Introduction

Humans and non-human animals are faced with decisions in all aspects of their lives [1,2]. The process of decision-making in humans is associated with changes in behaviour and physiology, including galvanic skin responses and heart-rate (HR) [3], indicative of increased arousal. Moreover, more difficult decisions lead to increases in such response parameters e.g. [4–7]. Although behavioural and physiological processes have been studied during decision-making in chickens [8], little is known about whether decisions of greater difficulty result in increased arousal in non-human species. Procedures involving decision-making are used widely in animal research, such as when testing preferences for environmental or social resources [9–11]. For example, choice tests have been used to determine chickens' preferred dustbathing substrate [12] and lighting [13]. Since the results of these tests can have practical, political and welfare implications, any procedural influences must be considered as part of an overall interpretation [14,15].

A complicating factor is that decisions may be "difficult" in more than one way. A difficult decision might, for instance, involve finely balanced options (e.g. two options with the same net

value: [16]), have the risk of a critical outcome (e.g. a risk of predation: [17]), have an ambiguous outcome (e.g. insufficient information available: [18]), vary in more than one dimension (e.g. cost, motivation and resource type: [19]) or require processing of a large amount of information (e.g. numerous options available: [20]). When options are finely balanced (e.g. the net benefit of accessing a large quantity of food with an associated cost is similar to accessing a small quantity of food with no associated cost), individuals are likely to become indifferent to the outcome, choosing both options at approximately equal frequency. When individuals are close to indifference, decisions may not be solved by simple prioritisation (i.e. choosing the alternative with the highest probability-weighted utility) e.g. [21,22]. Significantly more cognitive effort is therefore invested in finely balanced problems [23] and such decision-making can result in behavioural indicators of frustration and anxiety [24]. In the present study we focussed on comparing the physiology and behaviour of chickens making decisions in which options were either finely balanced or unbalanced.

An unbalanced decision was one in which two options were of unequal net value (i.e. a small quantity of food vs. a large quantity of food). The first part of our study used a titration methodology, in which weighted push-doors were used, to obtain individually-determined points of indifference (i.e. a finely balanced decision). The second part of our study compared the behaviour (head movements and latency to choose) and physiology (surface body temperature and heart-rate variability (HRV)) of individual chickens when making finely balanced and unbalanced decisions. Head movement increases and surface body temperature decreases have been measured in chickens in response to both aversive stimuli (head movements: [25]; surface body temperature: [26–28]) and to a signalled palatable reward (head movements: [8,29]; surface body temperature: [30]). Decreases in HRV, which provides more information on the source of cardiac stimulation (i.e. parasympathetic or sympathetic) than HR [31], occur in response to acutely stressful situations in birds [32–34].

We predicted that around the time of finely balanced decision-making, chickens would be more stressed, hence their HRV and surface body temperature would decrease (from levels at the start of the test) compared with an unbalanced alternative. We also predicted that hens would show increased arousal during decision-making (make more head movements) and have increased latencies to choose, due to increased cognitive demand when decisions were finely balanced.

Methods

Ethics Statement

All work was conducted under UK Home Office licence (30/2779) and had University ethical approval. We also conducted the study in compliance with ASAB ethical guidelines. The hens were rehomed to small responsible free-range holdings after the study.

Animals, Housing and Husbandry

Sixteen Columbian Black Tail laying-hens were obtained at approximately 16–18 weeks old from a commercial pullet rearer in Devon. They were transported in poultry crates in a well-ventilated van for approximately 2 h prior to delivery. On arrival all birds were weighed, health-checked, treated with preventative red mite powder (diatomaceous earth: Oak Tree Poultry) and individually tagged for identification using numbered leg rings. They were housed in groups of four, in four out of eight available pens (0.96×1.2 m, 2 m high) in the same room (home room). During weekly cleaning, each group of four birds was switched to the opposite pen within the same room to avoid a housing side-bias. All birds were checked at least once per day and were weighed weekly.

Hens in all pens were fed *ad libitum* feed (Farmgate Layers Mash, BOCM Pauls, Ipswich, Suffolk, UK) via two external feed troughs (total length: 0.77 m), reached by an opening in the pen 0.15 m from floor level. Water was provided via a hanging drinker which was placed in the back corner of each pen (0.2 m high). A nest box (0.39×0.38×0.47 m) and a round perch (0.25 m high, stretching the width of the pen) were also provided. Wood shavings were used as bedding at a depth of 5–10 cm. The room temperature was kept at 19–22°C and the lighting schedule was 12 L: 12 D (light period 7 am–7 pm).

Experimental room and Procedure

The experimental room contained two pens (one on each side of the room) of the same size as the home room, but which contained no feeders, nest boxes or perches. The experimental pens could be joined by a Perspex tunnel (1.79×0.24 m, 0.47 m high), which formed part of a T-maze apparatus. The experimental room was separated from the home room by solid wooden doors and a corridor, providing an area where hens could be tested away from

the noise of conspecifics. Within the experimental room, a CCTV camera was attached to the ceiling above the test apparatus, which was connected to a computer on one side of the room. Another computer for electrocardiogram (ECG) monitoring was set-up on the other side of the room.

Aim 1 – Establishing Finely Balanced and Unbalanced Decisions

The first aim of this study was to devise a protocol for identifying finely balanced and unbalanced decisions in individual hens. This was achieved using the T-maze test apparatus described above, with incorporated push-doors at each end of the Perspex tunnel. Each push-door could be attached to an electromagnet to alter the force required to push through the door (figure 1). The first step to achieving this aim was to habituate and train the hens. This was followed by a titration procedure.

After a five day settling period, habituation to the ECG monitor, training to establish associations between feed bowls and reward quantity, and habituation to the T-maze test procedure (including push-doors) began. Habituation and training criteria needed to be satisfied at each stage before individuals could progress. Habituation and training took 4–5 weeks, depending on individual progression.

ECG recording habituation. We made sure birds were familiar with wearing a harness and an ECG monitor for later phases of the study. ECG was recorded as in [8] using non-invasive remote telemetric units [35] and ECG cables contained within a harness. Hens were gradually habituated to wearing the harness, initially using a simplified harness (with no monitor) for a short period of time. As habituation criteria were met (that hens were able to walk and behave normally in their home environment without moving backwards or stopping excessively), the amount of time wearing the harness was gradually increased in 15 min increments, to a maximum time of 6 h. The ECG cables were then added to the harness and finally the monitor (weighing approx. 100 g). Harness habituation was conducted in both the home and experimental rooms, including within the T-maze.

Training associations between feed bowl characteristics and food reward. The hens were individually trained to discriminate between two stimuli (feed bowls), each associated with a different quantity of food (either 1 (Q1) or 6 (Q6) pieces of sweetcorn). Sweetcorn was used instead of a more motivating stimulus like mealworms [36] because we previously found that HR increased considerably in anticipation of a mealworm reward [8] and we wanted to focus on arousal caused by decision-making. During preliminary studies, hens were motivated to eat the sweetcorn reward. We chose to manipulate food quantity rather than quality as we wanted to avoid individual differences in preference for different food types.

The two feed bowls used for each hen were of different size (90 or 140 mm diameter) and colour (green or blue) to aid discrimination of the reward quantity (Q1 and Q6). There were four possible combinations of how size and colour were allocated to reward quantity (figure 2), and bowl combinations were systematically allocated so that one hen from each pen was trained to each. Once hens approached both bowls without stopping or hesitating, discrimination training continued in the T-maze along with habituation to other aspects of the apparatus.

T-Maze Habituation. Initial habituation to the T-maze apparatus was conducted in home groups by connecting opposite pens in the home room with a tunnel. Three 3-h group sessions were sufficient to ensure all hens in each group had walked through the tunnel without showing fearful behaviour (stopping or hesitating whilst walking). Habituation then continued individually

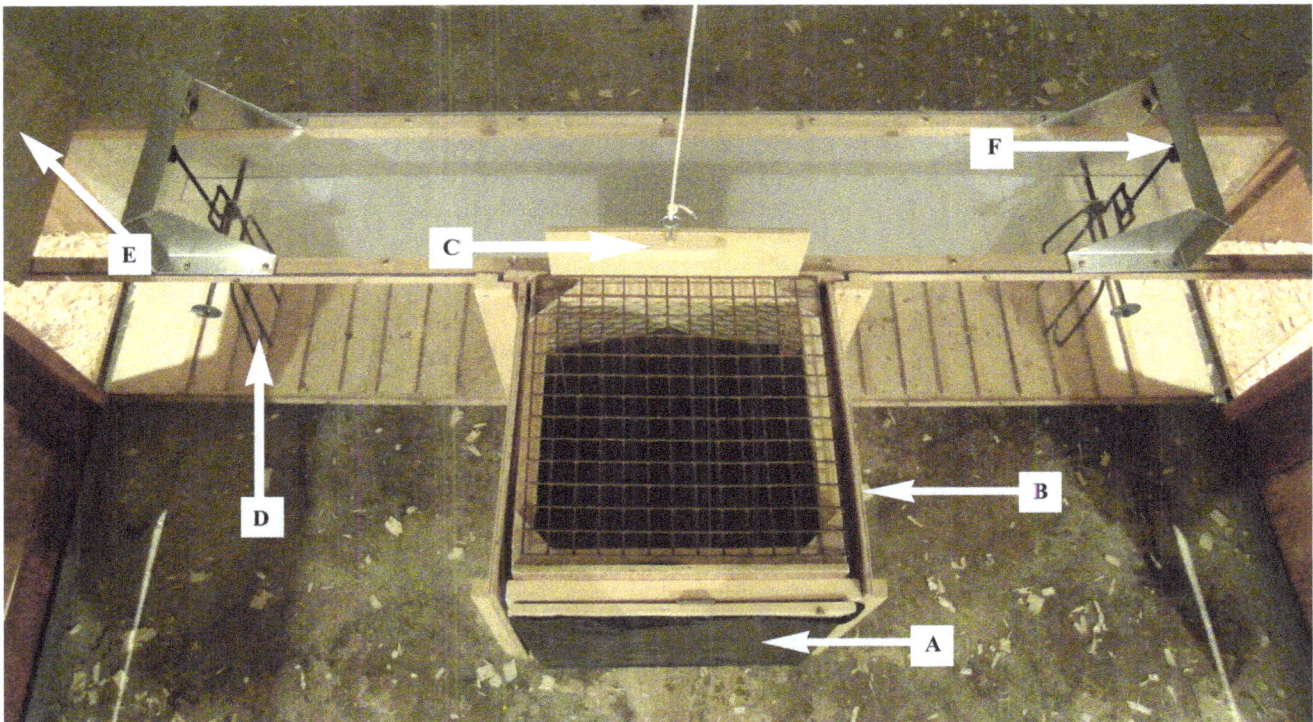

Figure 1. T-maze test apparatus consisting of a Perspex tunnel with push-doors and an attached wooden start-box. The tunnel connects the two pens in the experimental room. **A** indicates the rear of the start-box which was made of black plastic. **B** indicates the wooden side-doors which were removed to reveal wire mesh, through which the feed bowls could be viewed. **C** indicates the tunnel-door, which was raised using a pulley mechanism to allow access to the tunnel. **D** marks the push-doors which were located at either end of the tunnel. **E** marks the pen-door which was placed in the pen entrance once the hen entered the pen, to prevent her from re-entering the tunnel. **F** marks the electromagnet which could be altered to change the force required to open the push-door.

in the experimental room until hens became accustomed to leaving the start-box and walking through the tunnel with the push-doors (figure 1) fixed open (approximately seven sessions). A tunnel-door suspended by a pulley mechanism was then added to the start-box, to prevent hens from entering the tunnel for a short period at the start of each test (to allow for an initial assessment of behaviour and physiology during the testing phase). To begin with, hens were confined within the start-box for 5 s, but this was gradually increased (when they showed no escape attempts or excessive vocalisations) to 30 s (this took approximately five sessions).

Once accustomed to an initial confinement period, habituation to side-door removal began. The removable wooden side-doors of the start-box were kept in place at the start of each test to prevent hens from viewing the conditioned stimuli (feed bowls placed at both pen entrances). After a 10 s period within the start-box, side-doors were sequentially removed (to ensure that hens were looking at the food bowls on either side of the T-maze) and replaced, then both were removed simultaneously (viewing period). The tunnel-door was then raised and birds were allowed 10 s to leave the start-box (this took approximately six sessions).

When hens had successfully reached criteria in all other aspects of the T-maze procedure, the push-doors (with no weight applied) were introduced. It took approximately eight sessions to train all hens (using sweetcorn) to push through the doors to access the feed bowls. During habituation, a mixture of unidirectional and free-choice trials was conducted and both side-door removal order and the location of each reward quantity (Q1 or Q6) relative to the start-box were systematically varied to ensure equally balanced training. By the end of the training period, hens were well habituated to handling and to all other aspects of the T-maze (including the start-box), having completed approximately 50 individual training trials. If side biases became evident during habituation, additional training was conducted.

Prior to discrimination training in the T-maze, hens were food deprived for an individually-determined period of time to standardise hunger motivation. We selected the minimum food

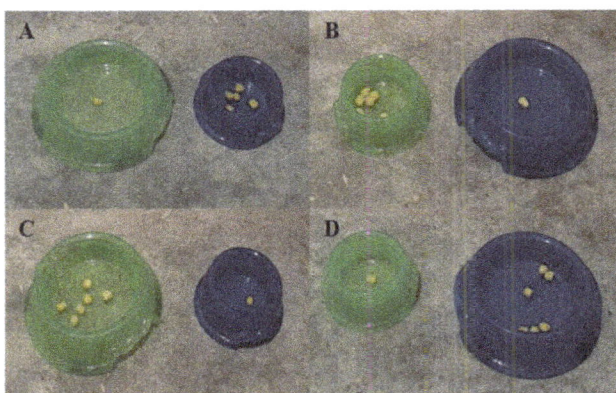

Figure 2. Different coloured and sized feed bowls were used to aid discrimination learning. The feed bowls used were counterbalanced for colour and size, with four chickens being trained on each of the four possible combinations (**A–D**).

deprivation period (tested during the training phase) that ensured each hen would perform multiple (up to 8) consecutive tests. The required period of food deprivation varied between hens, ranging from 80 to 390 min. As we were using a within-subjects design, the inter-hen range of food deprivation times should not influence treatment-based findings.

When the doors were unweighted and hens consistently chose the Q6 option in the T-maze (at least 90% of the time over consecutive 12 sessions), they were deemed able to discriminate between the two stimuli and this was defined as an unbalanced decision. The mean percentage of Q6 choices made across the 12 consecutive sessions was 97%.

Titration Phase. The aim of the titration phase was to detect the point at which the two options (Q1 and Q6) had the same net value (i.e. a finely balanced decision). This was done by gradually increasing the force required for hens to push through the Q6 push-door (Q1 door force was always zero), to find the point at which Q6 and Q1 were selected with approximately equal frequency (25–75% of choices) across 12 consecutive sessions. One hen's balance point fell outside the acceptable threshold and the hen was excluded from further testing. A finely balanced decision was defined for the remaining 15 hens – the mean percentage of Q6 choices made across the 12 consecutive sessions was 49% – with the electromagnetic force that had to be applied to the Q6 door varying greatly between individuals (figure 3). The door force at the point of indifference can be seen in figure 4.

Aim 2 – Behavioural and Physiological Responses to Balanced and Unbalanced Decisions

Once the unbalanced and finely balanced decisions had been individually defined, each hen underwent 10 consecutive days of testing during which aspects of their behaviour and physiology were monitored. On each day (which involved either an unbalanced or a finely balanced decision), hens were food deprived as for training, before being given two unidirectional (forced) trials (one to either option) followed by one free-choice test (when the door weights remained as in the forced trials). The forced trials allowed hens to 'assess' whether the free-choice test

would be unbalanced or finely balanced. The test protocol is outlined in figure 5. Prior to starting each free-choice test, the ECG monitor and a stopwatch were activated simultaneously. The test commenced when hens were placed in the start-box and confined for 10 s with the wooden side-doors in place, so that initial (start-box) physiological measures could be taken without conditioned stimuli (feed bowls) being visible. The side-doors were sequentially removed from the start-box for 5 s and replaced, and then both side-doors were simultaneously removed for a 10 s viewing period. The tunnel-door was then raised, allowing access to the push-doors ('the push-door period'). Once hens had accessed their chosen feed bowl via the push-door, the pen-door was closed and they were confined within the pen for 90 s (to allow consumption of their reward: 'the pen period'). If a hen failed to enter the tunnel within the first 60 s of each test, she was gently encouraged into the tunnel and the tunnel-door was replaced. A maximum time of 300 s was given for hens to enter a pen, after which time the test was stopped and the hen was removed from the T-maze.

Only one free-choice test was given each day to ensure that hunger motivation was kept as constant as possible throughout the experiment. Tests were carried out on each individual at the same time of day across the testing phase. Five tests were conducted per hen for both finely balanced and unbalanced decisions during the testing period. The order in which each hen experienced each decision category was systematically alternated, with no more than two consecutive tests from either decision category. The side of the T-maze where Q1 and Q6 were presented, the order of unidirectional trials and the order of side-door removal were also systematically alternated using the same criteria. Side biases were investigated and checked as the experiment progressed. The unidirectional trials given ensured that hens visited both sides of the T-Maze.

For each free-choice test, the following measures were taken during the push-door period: the latency to the push-door, the number of switches between push-doors and the number of attempts made to enter the pen. The side of the T-maze which hens first tried to enter was recorded as their 'first push' and their

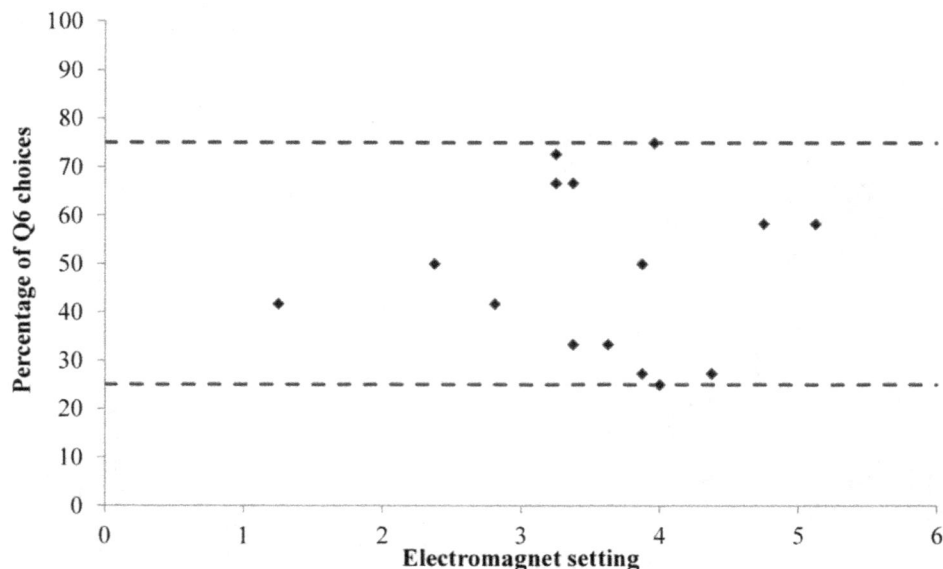

Figure 3. The electromagnet setting at which individual hens chose Q6 within our defined threshold of 25–75% (dashed line) of choices across 12 consecutive trials, measured at the end of the titration phase.

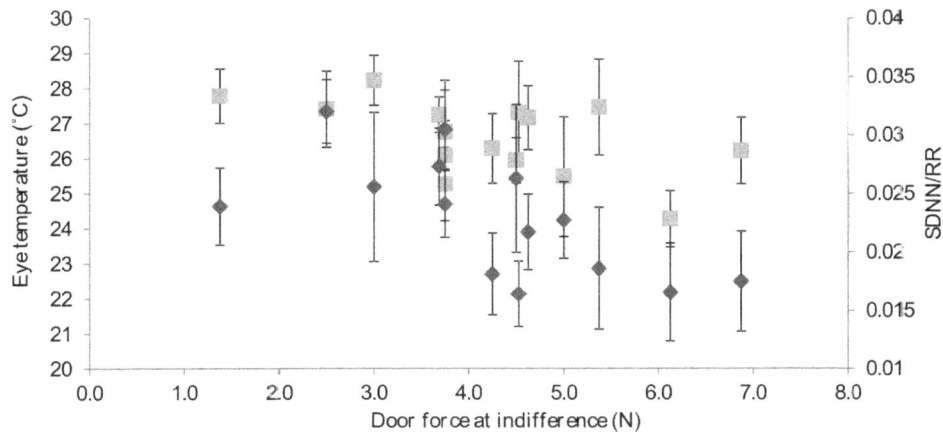

Figure 4. The association between the door force (in Newtons) at indifference and mean ±1 SE SDNN/RR (dark grey) and eye temperature (light grey) during the viewing period when making finely balanced decisions.

'ultimate choice' was also noted. The latency to the push-door and latency to the pen were both recorded as outlined in figure 5. A CCTV camera was fixed above the start-box and video was continuously recorded using WebCCTV software. The videos were analysed using The Observer Software version 10.0 to measure the number of head movements during the 10 s viewing period.

Throughout each free-choice test, the ECG was recorded continuously onto a micro-SD flash card which was inserted into the monitor. The monitor communicated with a base unit (attached to a computer via USB connection) and was controlled using RVC Telemetry Software version 1.5. HRV data were extracted using Spike 2 Software (version 6) from three 10 s periods: start-box, viewing and pen. As we were primarily interested in whether more difficult decisions induced arousal, we extracted two measures of HRV which are influenced by both branches of the autonomic nervous system [37] – mean interbeat interval (RR) and the quotient of the standard deviation of interbeat intervals and RR (i.e. the coefficient of variance (SDNN/RR)). Measures of HRV were taken at the start of each test (start-box period) to check for individual differences and to account for the influence of handling on arousal.

Surface body temperature was also recorded during the start-box and viewing periods using a thermal video camera (FLIR SC305), and was extracted using FLIR ResearchIR Software version 1.2 SP2. The rear surface of the wooden start-box was made from black plastic (through which we were able to detect heat radiation from the chicken), allowing us to record surface body temperature without hens being able to view the experimenter. The emissivity of the black plastic was calculated to be 0.292. A clear image of the side of the head was taken from each video to obtain the eye and maximum head temperature (hottest part of exposed skin). Video was recorded at 3 frames/second so it was usually possible to obtain clear images even if hens were moving.

Statistical Analysis

The data were analysed using *IBM SPSS Statistics 19*. Fifteen hens were given 10 free-choice tests each. HRV data were not collected from one individual as she did not behave normally when wearing the monitor during training. This hen did not wear the monitor during testing, hence we did not collect HRV data from her reducing the HRV sample to 14 individuals. Temperature, head movement and latency data were collected for all 15

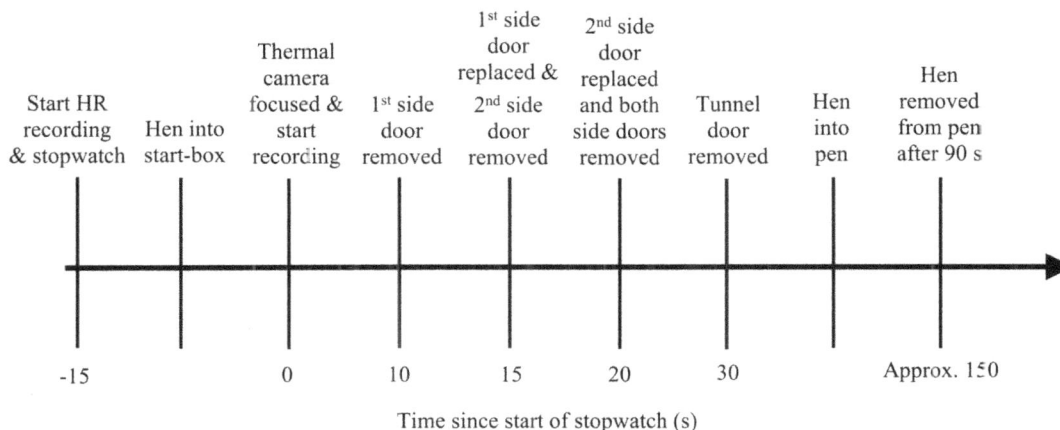

Figure 5. The test procedure during the testing phase. A stopwatch was used to monitor the precise time at which each event occurred so that accurate start-box and viewing period HRV and temperatures could be extracted from the data files.

individuals. Prior to conducting analyses we investigated the possibility of interactions by visual, inspection of plotted results for all measured variables and found no evidence of such interactions. For each set of data, the assumptions of parametric testing were checked and data were transformed if possible, then analysed using paired-samples t-tests. Where transformations were not possible or unsatisfactory, Wilcoxon tests were used. For analyses, a mean of each measure (% Q6 choices, RR intervals, SDNN/RR, head movements, maximum head and eye temperature, latency to first push and pen, number of switches made, number of attempts to enter the pen) was taken from the five repeats to each hen, to look for differences between finely balanced and unbalanced decisions during the relevant periods (start-box, viewing, push-door, pen). Unless otherwise stated, means ± SE are presented (in figures or text) for parametric data and the median, interquartile range (IQR) and range are presented for non-parametric data as appropriate. A measure of effect size is given alongside significant results. As means were used for the analyses, the range of individual hen standard deviations is given where appropriate.

Results

Aim 1 – Establishing Finely-Balanced and Unbalanced Decisions

To validate the titration procedure, the percentage of Q6 choices made during the testing phase was calculated. Overall, significantly more Q6 choices were made in the unbalanced condition (median: 100; IQR: 60–100; range: 60–100%) than in the finely balanced condition (median: 60; IQR: 40–80; range: 0–100%; Wilcoxon signed rank test: $z = 2.83$, $n = 15$, $p < 0.01$, $r = 0.52$).

Aim 2 – Behavioural and Physiological Responses to Balanced and Unbalanced Decisions

Start-box period. There were no significant differences between treatments in RR intervals (range of individual standard deviation: finely balanced $= 3.5$–22.4 ms, unbalanced $= 4.3$–10.7 ms; paired samples t-test: $t_{13} = 0.43$, $p = 0.68$, figure 6a) or in SDNN/RR (range of individual standard deviation: finely balanced $= 0.004$–0.03, unbalanced $= 0.004$–0.03; paired samples t-test: $t_{13} = 0.72$, $p = 0.49$, figure 6b). There were also no significant differences in start-box maximum head temperature (range of individual standard deviation: finely balanced $= 0.9$–$5.2°$C, unbalanced $= 1.2$–$4.1°$C; paired samples t-test: $t_{14} = 0.91$, $p = 0.38$, figure 7a) or eye temperature (range of individual standard deviation: finely balanced $= 1.3$–$3.8°$C, unbalanced $= 1.4$–$4.2°$C; paired samples t-test: $t_{14} = 0.71$, $p = 0.49$, figure 7b) between finely balanced and unbalanced decisions.

Viewing period. During the viewing period, there was no significant difference in RR between finely balanced and unbalanced decisions (range of individual standard deviation: finely balanced $= 2.4$–15.0 ms, unbalanced $= 2.7$–13.8 ms; paired samples t-test: $t_{13} = 0.28$, $p = 0.79$, figure 6a). However, there was a strong trend for a lower SDNN/RR when making a finely balanced decision compared to an unbalanced one (range of individual standard deviation: finely balanced $= 0.006$–0.01, unbalanced $= 0.002$–0.05; $t_{13} = 2.10$, $p = 0.055$, figure 6b). There was a significant negative correlation between the door weight at indifference and SDNN/RR during the viewing period (Pearson correlation coefficient $= -0.68$, $n = 14$, $p = 0.008$, figure 4). There were no significant differences between treatments in the number of head movements (range of individual standard deviation: finely balanced $= 0.7$–3.8, unbalanced $= 0.5$–2.9; $t_{14} = 0.26$, $p = 0.80$)

or in the maximum head (range of individual standard deviation: finely balanced $= 1.3$–$4.2°$C, unbalanced $= 1.2$–$3.7°$C; $t_{14} = 0.65$, $p = 0.52$, figure 7a) and eye temperatures (range of individual standard deviation: finely balanced $= 1.1$–$3.8°$C, unbalanced $= 1.5$–$3.7°$C; $t_{14} = 0.81$, $p = 0.43$, figure 7b). There was also a significant negative correlation between the door weight at indifference and the eye temperature during the viewing period (Pearson correlation coefficient $= 0.55$, $n = 15$, $p = 0.033$, figure 4).

Push-door period. Latency to the first push on a door was not significantly influenced by experimental condition (range of individual standard deviation: finely balanced $= 1.0$–36.5 s, unbalanced $= 0.4$–50.1 s; paired-samples t-test: $t_{14} = 0.08$, $p = 0.94$). Hens did make significantly more switches between push-doors (finely balanced: median: 0.2; range: 0–0.6; unbalanced: median: 0; range: 0–0; Wilcoxon signed rank test: $z = 2.59$, $n = 15$, $p = 0.01$, $r = 0.47$) and significantly more attempts to enter the pen (finely balanced: median: 1.6; range: 1–6.2; unbalanced: median: 1; range: 1–1.4; $z = 3.06$, $n = 15$, $p < 0.01$, $r = 0.56$, figure 8a) in the finely balanced treatment compared to when the decision was unbalanced.

Pen period. Hens took significantly longer to reach the pen when making finely balanced decisions compared to unbalanced ones (range of individual standard deviation: finely balanced 1.6–56.5 s, unbalanced $= 0.5$–68.5 s; paired samples t-test: $t_{14} = 2.73$, $p = 0.02$, eta squared $= 0.35$, figure 8b). Once in the pen, the RR interval was significantly shorter (range of individual standard deviation: finely balanced $= 3.0$–20.8 ms, unbalanced $= 1.9$–9.9 ms; $t_{13} = 3.22$, $p = 0.007$, eta squared $= 0.46$, figure 6a) and the SDNN/RR was significantly lower (range of individual standard deviation: finely balanced $= 0.001$–0.05, unbalanced $= 0.005$–0.03; $t_{13} = 2.46$, $p = 0.029$, eta squared $= 0.34$, figure 6b) when hens had made finely balanced decisions compared to unbalanced ones.

Discussion

Our titration methodology, based on the idea that two options are substitutable when they are of equal net value [16], was successful in determining the cost (the force needed to open the access door) required for hens to choose a larger food reward at an equivalent frequency to a smaller food reward. The electromagnetic force required to generate finely balanced choices varied between individuals, highlighting the importance of conducting this protocol on an individual hen basis. Finely balanced decisions between two substitutable options are likely to be inherently more difficult than unbalanced decisions [16,23,24], so our method allowed subsequent testing of behavioural and physiological correlates of this aspect of decision difficulty in hens. The titration method we have developed could also prove useful in examining the effect of this aspect of decision difficulty in other species.

Three significant physiological and behavioural differences were found between the finely balanced and unbalanced conditions, but all of these occurred outside the 'viewing period' (when we expected hens to make their decision and that arousal caused by decision-making would be detectable) after hens had left the start-box. Measures were taken during the push-door and pen periods to monitor the effect of the decision consequence on arousal. Under finely balanced conditions, hens took longer to reach the pen and there were differences in the two measures of HRV: mean RR interval was significantly shorter and the coefficient of variance (SDNN/RR) was significantly lower when hens had made finely balanced decisions. These differences are likely to be artefacts of the additional time and effort caused by hens having to push against the weighted door, of them switching more often

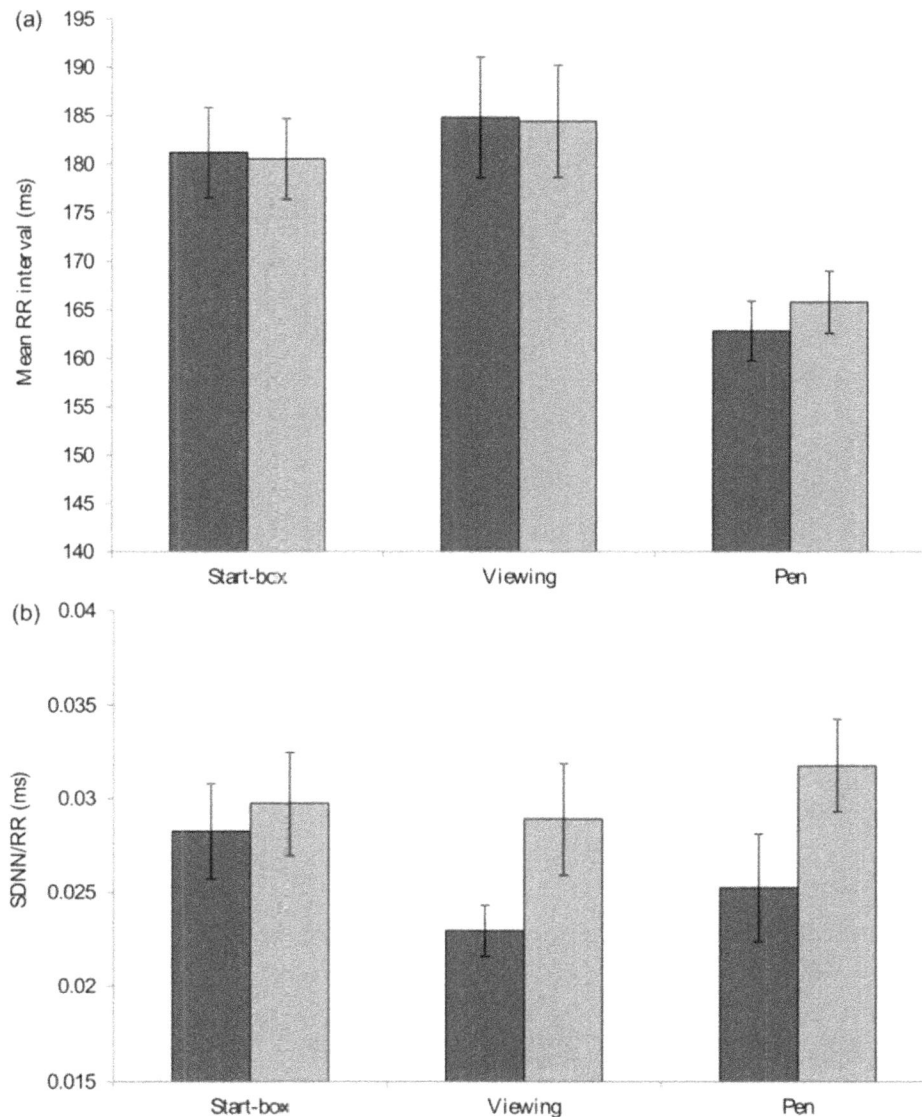

Figure 6. Mean ±1 SE (a) RR interval, (b) coefficient of variance (SDNN/RR), during each period within the test when making finely balanced (dark grey) and unbalanced (light grey) decisions.

between push-doors, and of additional attempts to enter the pen in the finely balanced condition. We had predicted that decreases in measures of HRV in the pen and longer latencies to pen in the finely balanced condition would be accompanied by a decrease in HRV during the viewing period and an increased latency to the first push, if affected by the difficulty of the decision. Although the coefficient of variance was influenced by experimental condition, the difference was not quite statistically significant, and there were no treatment-related differences in latency to the first push and the RR interval, suggesting that the observed differences in pen measures were unlikely to be caused by the difficulty of decision-making. We did find however, that the door weight at indifference was negatively correlated with some measures of arousal during the viewing period (eye temperature and the coefficient of variance), suggesting that hens pushing heavier doors may have anticipated the extra weight.

We found no other significant influences of experimental condition on the eye and maximum head temperature change or on the number of head movements made during decision-making. This could be interpreted in several ways. First, although the perceived difficulty of a decision affects human behaviour and physiology [4–7], it might not result in the same responses around the time of decision-making in other animals or more specifically in birds. However, previous research has shown that some non-human animals, such as dolphins, gorillas and honey bees, behave similarly to humans during other types of difficult tasks [18,38,39]. The key difference may therefore relate to the manner in which a decision is defined as difficult. In the aforementioned studies, the type of difficult decisions examined tends to involve a degree of uncertainty, for example as a result of ambiguous cues. And in the studies of human arousal during decision-making, difficult decisions often involved a degree of risk [5,6,7]. These are factors that we are currently examining in separate work, but here we explored behavioural and physiological effects of an aspect of decision difficulty that has not previously been examined in other species (i.e. a decision between two options of equal net value).

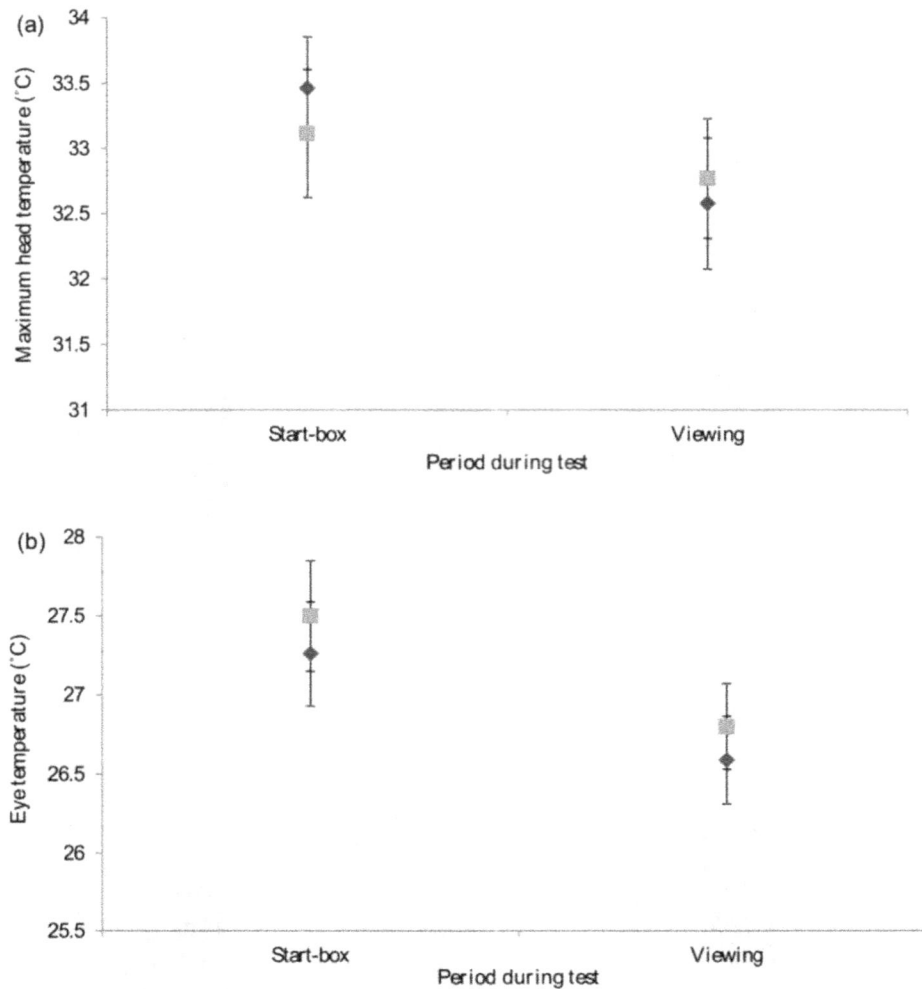

Figure 7. Mean ±1 SE (a) maximum head temperature, (b) eye temperature, during the start-box and viewing periods when making finely balanced (dark grey) and unbalanced (light grey) decisions.

One reason why the birds were not aroused when making finely balanced decisions may have been because they adopted a simple non-arousing strategy to deal with the situation. When humans must make decisions but are indifferent to the outcomes, they can adopt strategies such as choosing randomly, consistently choosing one option (status quo maintenance) or alternating between options e.g. [40–42]. Although potentially costly in terms of time [23], the use of non-evaluative decision strategies may be less arousing than evaluative decision-making in humans, though to our knowledge this has not been tested. Possibly our finely balanced decisions were difficult, but not stressful enough to result in physiological arousal.

One additional possibility is that the overall method and the behavioural and physiological measures we took were not sensitive enough to detect changes during the relevant period. However, we have previously found strongly significant changes in these measures in our related studies with chickens [8] and others have found differences in these measures in anticipation of and in response to positive and negative stimuli [25,27,29,30]. Often during human decision-making experiments, antecedents and consequences of decision-making are monitored, rather than directly measuring changes during the process, due to the difficulty of identifying the exact point at which decisions are made. For the

purpose of our current experiment, we considered that the birds' decision period most likely coincided with the 10-s viewing period, when the hens could view options but not yet access them. It is possible that hens did not make their choice during our defined decision period, or that changes in arousal occurred later than we had predicted. The coefficient of variance (HRV) showed a strong trend ($p = 0.055$) towards a significant difference between our experimental conditions which suggests that some decreases in HRV occurred during the viewing period. It is possible that a time-lag between decision-making and the onset of physiological arousal exists.

Although the titration methodology we developed was successful in identifying finely balanced and unbalanced decisions in individual chickens, to progress work in this area more information is needed on when best to monitor changes in arousal during decision-making. Additionally, future work could aim to identify whether such finely balanced decisions result in behavioural and physiological arousal in other species, and to explore alternative definitions of difficult decisions. As experimental protocols involving decision-making are used widely in animal welfare research [9–11], it is essential that all procedural influences are taken into consideration for an accurate interpretation of results.

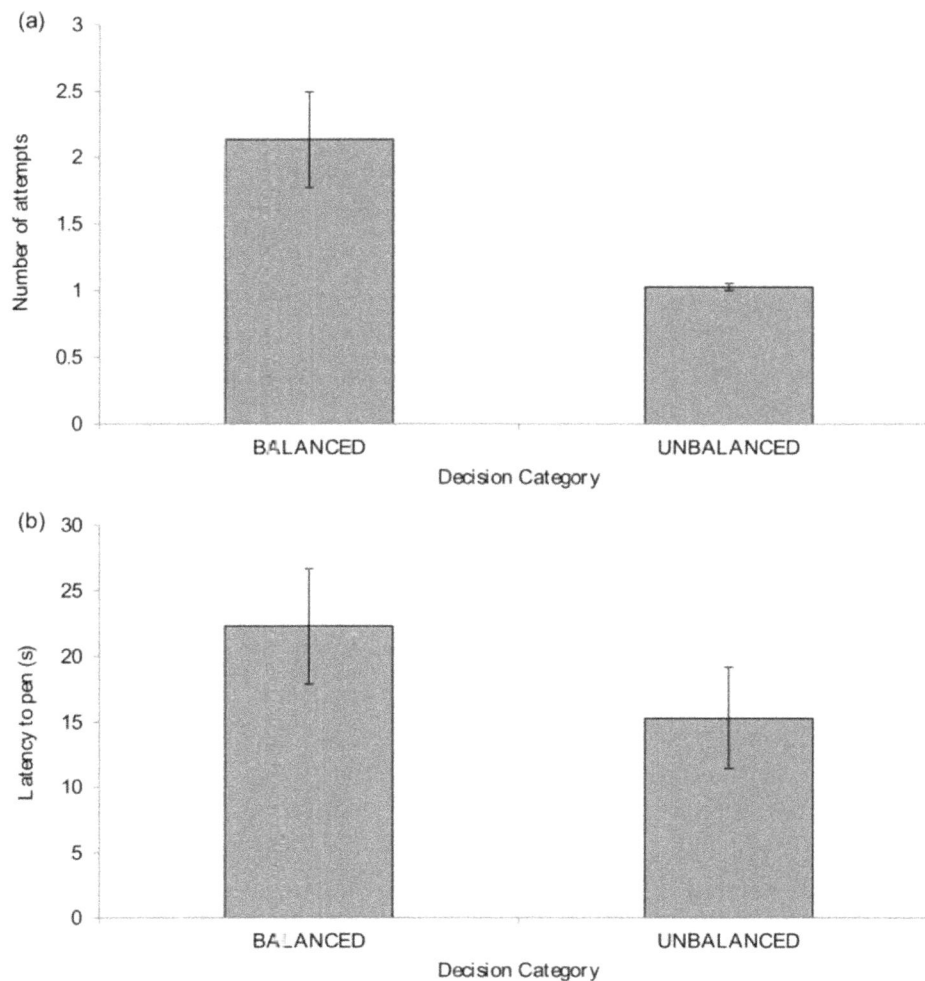

Figure 8. Mean ±1 SE (a) number of attempts made, (b) latency to pen.

Acknowledgments

We would like to thank Professor Mike Mendl for helpful discussions and comments on the manuscript and the Animal Services Unit at the University of Bristol for caring for the hens.

Author Contributions

Conceived and designed the experiments: AD CN MP AR. Performed the experiments: AD MP. Analyzed the data: AD. Contributed reagents/materials/analysis tools: AD CN AR. Wrote the paper: AD CN AR.

References

1. Saaty (1988) What is the analytic hierarchy process? In: Mitra G, Greenberg HJ, Lootsma FA, Rijckaert MJ, Zimmerman HJ, editors. Mathematical models for decision support. NATO ASI Series Vol 48. Springer Berlin Heidelberg. pp.109–121.

2. McFarland DJ (1977) Decision making in animals. Nature 269: 15–21.

3. Crone EA, Somsen RJM, Van Beek B, Van Der Molen MW (2004) Heart rate and skin conductance analysis of antecendents and consequences of decision-making. Psychophysiology 41: 531–540.

4. Gerard HB (1967) Choice difficulty, dissonance and the decision sequence. J Pers 35: 91–108.

5. Mann L, Janis IL, Chaplin R (1969) Effecs of anticipation on predecisional processes. J Pers Soc Psychol 11: 10–16.

6. Critchley HD, Mathias J, Dolan RJ (2001) Neural activity in the human brain relating to uncertainty and arousal during anticipation. Neuron 29: 537–545.

7. Van Harreveld F, Rutjens BT, Rotteveel M, Nordgren LF, van der Pligt J (2009) Ambivalence and decisional conflict as a cause of psychological discomfort: feeling tense before jumping off the fence. J Exp Soc Psychol 45: 167–173.

8. Davies AC, Radford AN, Nicol CJ (2014) Behavioural and physiological expression of arousal during decision-making in laying hens. Physiol Behav 123: 93–99.

9. Kirkden RD, Pajor EA (2006) Using preference, motivation and aversion tests to ask scientific questions about animals' feelings. Appl Anim Behav Sci 100: 29–47.

10. Jensen MB, Pedersen LJ (2008) Using motivation tests to assess ethological needs and preferences. Appl Anim Behav Sci 113: 340–356.

11. Fraser D, Nicol CJ (2011) Preference and motivation research. In: Appleby MC, Mench JA, Olsson IAS, Hughes BO, editors. Animal Welfare 2nd Edition. CABI. pp.183–199.

12. Shields SJ, Garner JP, Mench JA (2004) Dustbathing by broiler chickens: a comparison of preference for four different substrates. Appl Anim Behav Sci 87: 69–82.

13. Widowski TM, Keeling LJ, Duncan IJH (1992) The preferences of hens for compact fluorescent over incandescent lighting. Can J Anim Sci 72: 203–211.

14. Bateson M (2004) Mechanisms of decision-making and the interpretation of choice tests. Anim Welf 13: 115–120.

15. Browne WJ, Caplen G, Statham P, Nicol CJ (2011) Mild environmental aversion is detected by a discrete-choice preference testing method but not by a free-access method. Appl Anim Behav Sci 134: 152–163.

16. Sørensen DB, Ladewig J, Ersbøl AK, Matthews L (2004) Using the cross point of demand functions to assess animal priorities. Anim Behav 68: 949–955.

17. Lima SL, Dill LM (1990) Behavioural decisions made under the risk of predation: a review and prospectus. Can J Zool 68: 619–640.

18. Perry CJ, Barron AB (2013) Honey bees selectively avoid difficult choices. Proc Natl Acad Sci USA 110: 19155–19159.

19. Cooper JJ, Mason GJ (2000) Increasing costs of access to resources cause re-scheduling of behaviour in American mink (*Mustela vison*): implications for the assessment of behavioural priorities. Appl Anim Behav Sci 66: 135–151.

20. Churchland AK, Kiani R, Shadlen MN (2008) Decision-making with multiple alternatives. Nat Neurosci 11: 693–702.

21. Manski CF (1975) Maximum score estimation of the stochastic utility model of choice. J Econom 3: 205–228.

22. Varian HR (1990) Goodness-of-fit in optimizing models. J Econom 46: 125–140.

23. Moffatt PG (2005) Stochastic choice and the allocation of cognitive effort. Exp Econ 8: 369–388.

24. Ventricelli M, Focaroli V, De Petrillo F, Macchitella L, Paglieri F, Addessi E (2013) How capuchin monkeys (Cebus apella) behaviourally cope with increasing delay in a self-control task. Behav Process 100: 146–152.

25. Zimmerman PH, Buijs SAF, Bolhuis JE, Keeling LJ (2011) Behaviour of domestic fowl in anticipation of positive and negative stimuli. Anim Behav 81: 569–577.

26. Cabanac M, Aizawa S (2000) Fever and tachycardia in a bird (*Gallus domesticus*) after simple handling. Physiol Behav 69: 541–545.

27. Edgar JL, Lowe JC, Paul ES, Nicol CJ (2011) Avian maternal response to chick distress. Proc R Soc B 278: 3129–3134.

28. Edgar JL, Nicol CJ, Pugh CA, Paul ES (2013) Surface temperature changes in repsonse to handling in domestic chickens. Physiol Behav 119: 195–200.

29. Moe RO, Nordgreen J, Janczak AM, Spruijt BM, Zanella AJ, et al. (2009) Trace classical conditioning as an approach to the study of reward-related behaviour in laying hens: A methodological study. Appl Anim Behav Sci 121: 171–178.

30. Moe RO, Stubsjøen SM, Bohlin J, Flø A, Bakken M (2012) Peripheral temperature drop in response to anticipation and consumption of a signaled palatable reward in laying hens (*Gallus domesticus*). Physiol Behav 106: 527–533.

31. Von Borell E, Langbein J, Després G, Hansen S, Leterrier C, et al. (2007) Heart rate variability as a measure of autonomic regulation of cardiac activity for assessing stress and welfare in farm animals – A review. Physiol Behav 92: 293–316.

32. Kjaer JB, Jørgensen H (2011) Herat-rate variability in domestic chicken lines genetically selected on feather pecking behaviour. Genes Brain Behav 10: 747–755.

33. Cyr NE, Dickens MJ, Romero ML (2009) Heart-rate and heart-rate variabilirty responses to acute and chronic stress in a wild-caught passerine bird. Physiol Biochem Zool 82: 332–344.

34. Korte SM, Ruesnik W, Blokhuis HJ (1999) Heart-rate variability during manual restraint in chicks from high- and low-feather pecking lines of laying hens. Physiol Behav 65: 649–652.

35. Lowe JC, Abeyesinghe SM, Demmers TGM, Wathes CM, McKeegan DEF (2007) A novel telemetric logging system for recording physiological signals in unrestrained animals. Comput Electron Agr 57: 74–79.

36. Bruce EH, Prescott NB, Wathes CM (2003) Preferred food rewards for laying hens in behavioural experiments. Br Poult Sci 44: 345–349.

37. Task Force of the European Society of Cardiology, North American Society of Pacing and Electrophysiology. (1996). Heart rate variability: standards of measurement, physiological interpretation and clinical use. Circulation 93: 1043–1065.

38. Smith DJ, Schull J, Strote J, McGee K, Egnor R, et al. (1995) The uncertain response in the bottlenosed dolphin (*Tursiops truncatus*). J Exp Psychol 124: 391–408.

39. Suda-King C, Bania AE, Stromberg EE, Subiaul F (2013) Gorillas' use of the escape response in object choice memory tests. Animal Cognition 16: 65–84.

40. Mandler M (2005) Incomplete preferences and rational intransitivity of choice. Games Econ Behav 50: 255–277.

41. Eliaz K, Ok EA (2006) Indifference or indecisiveness? Choice theoretic foundations of incomplete preferences. Games Econ Behav 56: 61–86.

42. Danan E (2010) Randomization vs. Selection: How to choose in the absence of preference? Manage Sci 56: 503–518.

3

The *Staurotypus* Turtles and Aves Share the Same Origin of Sex Chromosomes but Evolved Different Types of Heterogametic Sex Determination

Taiki Kawagoshi[1], Yoshinobu Uno[1], Chizuko Nishida[2], Yoichi Matsuda[1,3]*

1 Laboratory of Animal Genetics, Department of Applied Molecular Biosciences, Graduate School of Bioagricultural Sciences, Nagoya University, Nagoya, Japan, **2** Department of Natural History Sciences, Faculty of Science, Hokkaido University, Sapporo, Japan, **3** Avian Bioscience Research Center, Graduate School of Bioagricultural Sciences, Nagoya University, Nagoya, Japan

Abstract

Reptiles have a wide diversity of sex-determining mechanisms and types of sex chromosomes. Turtles exhibit temperature-dependent sex determination and genotypic sex determination, with male heterogametic (XX/XY) and female heterogametic (ZZ/ZW) sex chromosomes. Identification of sex chromosomes in many turtle species and their comparative genomic analysis are of great significance to understand the evolutionary processes of sex determination and sex chromosome differentiation in Testudines. The Mexican giant musk turtle (*Staurotypus triporcatus*, Kinosternidae, Testudines) and the giant musk turtle (*Staurotypus salvinii*) have heteromorphic XY sex chromosomes with a low degree of morphological differentiation; however, their origin and linkage group are still unknown. Cross-species chromosome painting with chromosome-specific DNA from Chinese soft-shelled turtle (*Pelodiscus sinensis*) revealed that the X and Y chromosomes of *S. triporcatus* have homology with *P. sinensis* chromosome 6, which corresponds to the chicken Z chromosome. We cloned cDNA fragments of *S. triporcatus* homologs of 16 chicken Z-linked genes and mapped them to *S. triporcatus* and *S. salvinii* chromosomes using fluorescence in situ hybridization. Sixteen genes were localized to the X and Y long arms in the same order in both species. The orders were also almost the same as those of the ostrich (*Struthio camelus*) Z chromosome, which retains the primitive state of the avian ancestral Z chromosome. These results strongly suggest that the X and Y chromosomes of *Staurotypus* turtles are at a very early stage of sex chromosome differentiation, and that these chromosomes and the avian ZW chromosomes share the same origin. Nonetheless, the turtles and birds acquired different systems of heterogametic sex determination during their evolution.

Editor: Roscoe Stanyon, University of Florence, Italy

Funding: This research was supported by Grants-in-Aid for Scientific Research on Innovative Areas (No. 23113004) and Scientific Research (B) (No. 22370081) from the Ministry of Education, Culture, Sports, Science, and Technology (MEXT), Japan. The funders had no role in study design, data collection and analysis, decision to publish, or preparation of the manuscript.

Competing Interests: The authors have declared that no competing interests exist.

* Email: yoimatsu@agr.nagoya-u.ac.jp

Introduction

The constitutions of sex chromosome and sex-determination systems of reptiles are extraordinarily diverse. Reptiles exhibit both genotypic sex determination (GSD)– in which the sex of offspring is determined by a sex-determining gene on the sex chromosome – and temperature-dependent sex determination (TSD)– in which the sex ratio depends on the incubation temperature of embryos. GSD systems are found in all snakes, many lizards, and a small number of turtles [1]. Almost all snakes exhibit female heterogamety (ZZ/ZW), whereas lizards and turtles with GSD exhibit both male heterogamety (XX/XY) and female heterogamety [2]. In lizards, sex chromosomes have been identified for 181 species, with 115 species shown to exhibit male heterogamety and 66 shown to exhibit female heterogamety [3]. Given that the distribution of XY and ZW species shows no clear phylogenetic segregation [3–6], it seems likely that the sex chromosomes differentiated independently in each lineage. On the other hand,

whereas 18 turtle species from the order Testudines exhibit GSD [7], differentiated sex chromosomes have been identified for only nine such species (six XX/XY species and three ZZ/ZW species) [2,7–12]. Among them, the ZW sex chromosomes of Chinese soft-shelled turtle (*Pelodiscus sinensis*, Trionychidae) have conserved linkage homology with chicken chromosome 15 [11], whereas the XY sex chromosomes of the black marsh turtle (*Siebenrockiella crassicollis*, Geoemydidae) share linkage homology with chicken chromosome 5 [12]. These results suggest that the sex chromosomes of birds and turtles differentiated from different autosomal pairs of the common ancestor of Archosauromorpha, which diverged 250–270 million years ago (MYA) [13–15]. The group Archosauromorpha contains Archosauria (diapsid amniotes whose living representatives consist of birds and crocodilians) and all other saurians that are closer to Archosauria than they are to Lepidosauria (including tuataras, lizards, snakes and amphisbaenia). However, the origins of the sex chromosomes of the other seven GSD turtle species with differentiated sex chromosomes are still unknown. Identification of the linkage groups of the sex chromosomes and their homologies in other reptilian and avian species will improve our understanding of the evolutionary mechanisms that drive the genetic determination of sex and the differentiation of sex chromosomes in extant vertebrates.

The Mexican giant musk turtle (*Staurotypus triporcatus*, Kinosternidae) and the giant musk turtle (*Staurotypus salvinii*) inhabit the region from eastern and southern North America to Argentina and have heteromorphic X and Y sex chromosomes [16,17]. The X and Y chromosomes were only slightly different in terms of the sizes of the short arms and secondary constrictions in the two species, as determined by conventional Giemsa staining. Neither the structural differences between the X and Y chromosomes at the molecular level nor their linkage groups have been determined. The present study involved comparative mapping of functional genes for the X and Y chromosomes of *S. triporcatus* and *S. salvinii* in order to elucidate the origin and evolution of the sex chromosomes of *Staurotypus* turtles. The homology of the X chromosomes of *Staurotypus* turtles with the chicken Z chromosome was found by cross-species hybridization with chromosome paints of Chinese soft-shelled turtle (*Pelodiscus sinensis*); therefore, we isolated *S. triporcatus* homologs of 16 chicken Z-linked genes and mapped them to chromosomes of *S. triporcatus* and *S. salvinii*. Comparison of the cytogenetic maps of the X chromosomes of these two turtle species with that of the Z chromosome of the ostrich (*Struthio camelus*), which is one of the most primitive extant avian species and retains the ancestral type of avian Z chromosomes, sheds light on the differentiation of the X and Y chromosomes of *Staurotypus* turtles and the evolution of sex chromosomes in Testudines.

Materials and Methods

Cell culture and chromosome preparation

For each of *S. triporcatus* and *S. salvinii*, a male that had been bred in captivity was purchased and used for this study. After intra-peritoneal injection of a fatal dose of pentobarbital, the heart, lung, and mesentery were removed and used for cell culture at 26°C in a humidified atmosphere of 5% CO_2 in air. Animal care and all experimental procedures were approved by the Animal Experiment Committee, Graduate School of Bioagricultural Sciences, Nagoya University (approval no. 2010052401), and the experiments were conducted according to the Regulations on Animal Experiments in Nagoya University. Cell culturing and chromosome preparation were performed as described previously [12]. Fibroblasts of the ostrich used in our previous study [18] were recovered from liquid nitrogen and subsequently cultured for chromosome preparation. For gene mapping by fluorescence in situ hybridization (FISH), replication banding was performed to identify each chromosome precisely, as described previously [12,19]. The fibroblast cell cultures were treated with BrdU (12 µg/ml) (Sigma-Aldrich) at the late replication stage for 12 h, including 45 min of colcemid treatment, and chromosome preparations were made using an air-drying method. The cultured cells of the ostrich were harvested after 6 h of treatment with BrdU (25 µg/ml) under conditions of 39°C with 5% CO_2 in air. After staining the slides with Hoechst 33258 (1 µg/ml) for 10 min, replication bands were obtained by heating them at 65°C for 3 min and exposing them to UV light at 65°C for an additional 6.5 min. The slides were kept at −80°C until use.

C-banding

To examine the chromosomal distribution of constitutive heterochromatin in *S. triporcatus* and *S. salvinii*, C-banding was performed by the standard barium hydroxide/saline/Giemsa method [20] with slight modification; chromosome slides were treated with 0.2N HCl at room temperature for 5 min and then 5% Ba $(OH)_2$ at 50°C for 2 min.

Chromosome painting

Cross-species chromosome painting with chromosome-specific DNA probes of *P. sinensis* was performed for *S. triporcatus*. The *P. sinensis* chromosome paints were prepared and provided by Fengtang Yang and Patricia O'Brien, both from the Department of Veterinary Medicine, Cambridge University, UK. Chromosome painting was performed as described previously [12,21]. One microgram of DNA probe was labeled with biotin-16-dUTP (Roche Diagnostics) using a nick translation kit (Roche Diagnostics). After pre-hybridization for 15 min at 37°C, hybridization was carried out at 37°C for five days. After hybridization, the slide was washed, incubated with fluorescein-conjugated avidin (Roche Diagnostics), and stained with 0.75 µg/ml propidium iodide (PI).

Molecular cloning of *S. triporcatus* and ostrich homologs of chicken genes

Testis and brain of *S. triporcatus* and testis of the ostrich were homogenized and lysed with TRIzol Reagent (Life Technologies), and total RNA was extracted following the manufacturer's instructions. Testis tissues of the ostrich used in our previous study [18] were recovered from liquid nitrogen. Molecular cloning of *S. triporcatus* and ostrich homologs of the chicken Z-linked genes was performed by reverse transcription polymerase chain reaction (RT-PCR) using the PCR primers shown in Table S1. The nucleotide sequences of cDNA fragments were determined and compared as described previously [22].

FISH mapping

FISH was performed for chromosomal localization of the 18S–28S ribosomal RNA (rRNA) genes and cDNA fragments of functional genes as described by Kawagoshi et al. [11] and Matsuda and Chapman [19]. After FISH of the rRNA genes, Ag-NOR staining was performed to visualize nucleolar organizing regions (NORs) on the same metaphase spreads following Howell and Black [23]. For chromosome mapping of functional genes, 250 ng of cDNA fragments were labeled with biotin-16-dUTP (Roche Diagnostics) by nick translation. After hybridization, the probe DNA was hybridized with goat anti-biotin antibody (Vector Laboratories), stained with Alexa Fluor 488 rabbit anti-goat IgG (H+L) conjugate (Life Technologies-Molecular Probes), and then counter-stained with 0.75 µg/ml PI.

Results

Karyotypes of *S. triporcatus* and *S. salvinii*

Twenty Giemsa-stained metaphase spreads of *S. triporcatus* and 18 metaphase spreads of *S. salvinii* were examined for karyotyping. The chromosome numbers were $2n = 54$ in all metaphase spreads of both species, as reported previously [16]. Karyotypes of both species consisted of four pairs of large chromosomes including sex chromosomes (chromosomes 1–3 and X and Y chromosomes), seven pairs of medium-sized and/or small chromosomes (chromosomes 4–10), and 16 pairs of indistinguishable microchromosomes (Figure 1). The sex chromosomes were morphologically differentiated: whereas the X chromosomes were acrocentric in *S. triporcatus* and subtelocentric in *S. salvinii*, with a secondary constriction on the long arm near the centromere, the Y chromosomes were both acrocentric; and the size of the

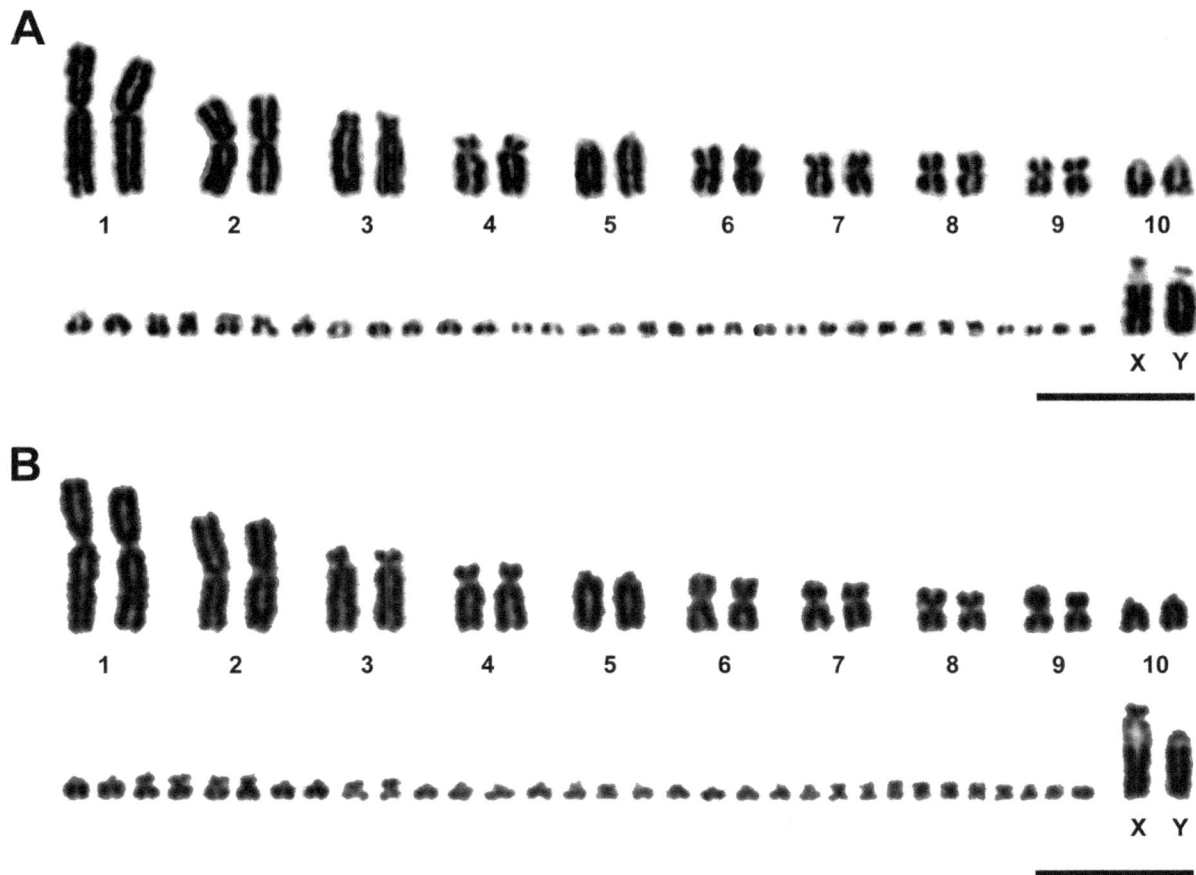

Figure 1. Giemsa-stained karyotypes of male *S. triporcatus* **and** *S. salvinii.* (A) *S. triporcatus.* (B) *S. salvinii.* The X and Y chromosomes have large and small secondary constrictions, respectively. Scale bars = 10 μm.

secondary constriction was larger in the X chromosomes than in the Y chromosomes.

C-positive heterochromatin blocks were observed in the centromeric regions of almost all autosomes and the telomeric regions of several pairs of autosomes in both species (Figures 2A, B). Chromosomal regions surrounding the secondary constrictions on the X and Y chromosomes were heterochromatized and showed C-positive bands in both species (Figures 2C, D).

Chromosomal locations of the 18S-28S rRNA genes and NORs in *S. triporcatus* and *S. salvinii*

FISH signals of the 18S–28S rRNA genes were detected in the secondary constrictions of the X and Y chromosomes, one of the copies of chromosome 2, and a pair of microchromosomes in *S. triporcatus* (Figure 3A). In *S. salvinii*, signals were detected only in the secondary constrictions of the X and Y chromosomes (Figure 3D). There was a remarkable difference in the size of hybridization signals between the X and Y chromosomes in both species, which corresponded to the difference in the size of secondary constrictions. NORs were detected in the secondary constrictions of the X and Y chromosomes in both species using Ag-NOR staining, whereas no NORs were found for chromosome 2 and a pair of microchromosomes in *S. triporcatus* (Figures 3C, F), in which small FISH signals of rRNA genes were observed (Figure 3A).

Chromosome homology of the *S. triporcatus* X chromosome with the chicken Z chromosome

Hybridization of the chromosome 6 paint of *P. sinensis* to the X and Y chromosomes of *S. triporcatus* (Figure 4) indicated that the *S. triporcatus* X and Y sex chromosomes are a counterpart of *P. sinensis* chromosome 6, which is homologous to the chicken Z chromosome [24,25].

Chromosomal locations of *S. triporcatus* homologs of chicken Z-linked genes

On the basis of the result that *S. triporcatus* X and Y sex chromosomes are homologous to the chicken Z chromosome, we cloned *S. triporcatus* homologs of 16 chicken Z-linked genes: *ACO1, ATP5A1, CHD1, DMRT1, FER, GHR, HMGCR, KIF2A, NARS, NFIB, NTRK2, RNF20, RPS6, SPIN, TMOD,* and *VCP*. Nucleotide sequence identities in the equivalent regions of cDNA fragments of these 16 genes between *S. triporcatus* and chicken ranged from 77.7% to 94.4% (Table 1). Hoechst-stained bands obtained by the replication banding method enabled precise determination of the subchromosomal locations of the genes (Figure 5). For FISH mapping, 25–30 metaphase spreads were observed for each gene. The hybridization efficiency ranged from 20% to 36% on the X chromosome, and from 23% to 38% on the Y chromosome. Sixteen homologs of chicken Z-linked genes were all localized to the long arm of *S. triporcatus* X and Y chromosomes in the same order (Figure 6).

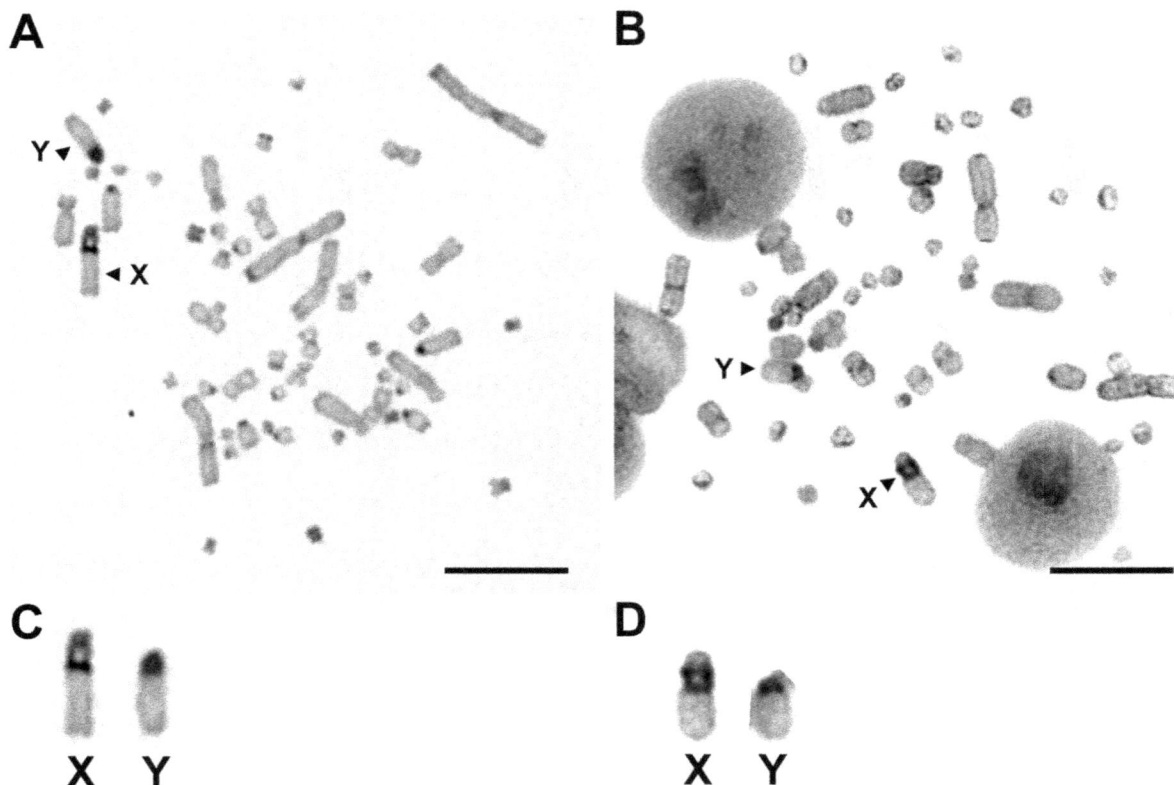

Figure 2. C-banded metaphase spreads of male *S. triporcatus* and *S. salvinii*. (A) *S. triporcatus*. (B) *S. salvinii*. (C, D) Enlarged photographs of the X and Y chromosomes of *S. triporcatus* (C) and *S. salvinii* (D). Scale bars = 10 μm.

Comparison between the *S. triporcatus* X chromosome and the ostrich Z chromosome

We cloned ostrich homologs of eight chicken Z-linked genes, *ACO1*, *FER*, *HMGCR*, *KIF2A*, *NARS*, *NFIB*, *RNF20*, and *VCP*, by RT-PCR using the PCR primers shown in Table S1 and mapped them to ostrich chromosomes by FISH (Figure S1). Although *ACO1* (*IREBP1*) was previously mapped to the ostrich Z chromosome [26,27], we cloned a cDNA fragment of this gene and mapped it to determine its precise location on the ostrich Z chromosome. We also mapped *DMRT1* to ostrich chromosomes using the cDNA fragments isolated in our previous study [18]. We then constructed a cytogenetic map of the ostrich Z and W chromosomes with 16 functional genes by adding seven ostrich Z-linked genes (*ATP5A1*, *CHD1*, *GHR*, *NTRK2*, *RPS6*, *SPIN*, and *TMOD*), which were cloned and mapped in our previous studies (Figure S2) [18,27]. Nucleotide sequence identities in the equivalent regions of cDNA fragments of 16 genes ranged from 79.6% to 94.4% between *S. triporcatus* and the ostrich (Table 2). In general, the identities of nucleotide sequences were higher in 14 genes than in those between *S. triporcatus* and chicken; exceptions were for *NFIB* and *VCP*, for which the nucleotide sequence identities did not differ (Tables 1 and 2). Eleven genes (*RPS6*, *NTRK2*, *SPIN*, *FER*, *CHD1*, *HMGCR*, *KIF2A*, *GHR*, *ATP5A1*, *NARS*, and *VCP*) were localized to the ostrich Z and W chromosomes in the same order, whereas five genes (*TMOD*, *ACO1*, *RNF20*, *DMRT1*, and *NFIB*) were not mapped to the W chromosome (Figure S2). This indicated that the proximal region of the ostrich Z chromosome that contained these five genes had been deleted in the W chromosome. The order of 16 genes on the ostrich Z chromosome was almost the same as those on the X and

Y chromosomes of *S. triporcatus* (Figure 6), although the precise order among several genes located close together was not determined.

Comparison of the XY chromosomes between *S. triporcatus* and *S. salvinii*

Sixteen genes were also all localized to the X and Y chromosomes of *S. salvinii*, and their locations and orders completely matched those of *S. triporcatus* (Figures S3 and S4). The hybridization efficiency ranged from 23% to 38% for 25–30 metaphase spreads.

Discussion

The origin and evolutionary process of the X and Y sex chromosomes of *S. triporcatus* and *S. salvinii* were investigated using cross-species chromosome painting and chromosome mapping of cDNA clones of sex-linked genes isolated from *S. triporcatus*. Cross-species chromosome painting revealed that the X and Y chromosomes of *S. triporcatus* are homologous to *P. sinensis* chromosome 6, which corresponds to the chicken Z chromosome [24,25]. The homology with the chicken Z chromosome has been also reported for the red-eared slider (*Trachemys scripta elegans*) chromosome 6 and Nile crocodile (*Crocodylus niloticus*) chromosome 6 [28]; however, the homology of these chromosomes with *P. sinensis* chromosome 6 is still not known.

S. triporcatus homologs of 16 chicken Z-linked genes were all shown to be localized to the long arm of the X and Y chromosomes of *S. triporcatus* and *S. salvinii* in the same order.

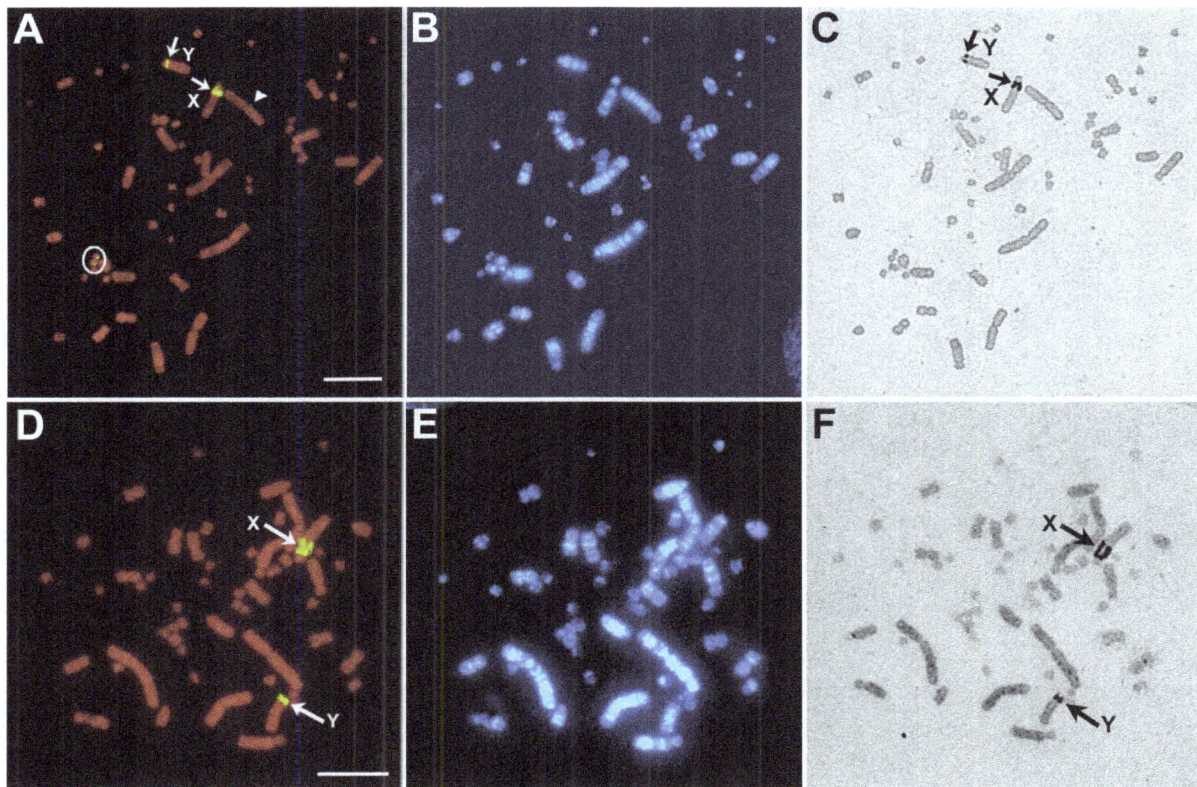

Figure 3. Chromosomal distribution of the 18S-28S rRNA genes and NORs on metaphase spreads of male *S. triporcatus* and *S. salvinii*. (A–C) *S. triporcatus.* (D–F) *S. salvinii.* FISH signals of the 18S–28S rRNA genes were localized to the secondary constrictions of the X and Y chromosomes (indicated by arrows), one of the copies of chromosome 2 (an arrowhead), and a pair of microchromosomes (a circle) in *S. triporcatus* (A), and the secondary constrictions of the X and Y chromosomes in *S. salvinii* (D). Ag-stained NORs were also distributed in the secondary constrictions of the X and Y chromosomes in *S. triporcatus* (C) and *S. salvinii* (F). However, no NORs were detected on chromosome 2 and a pair of microchromosomes in *S. triporcatus*, where the FISH signals of the rRNA genes were detected. (B, E) Hoechst-stained patterns of the same PI-stained metaphase spreads (A) and (D), respectively. Scale bars = 10 μm.

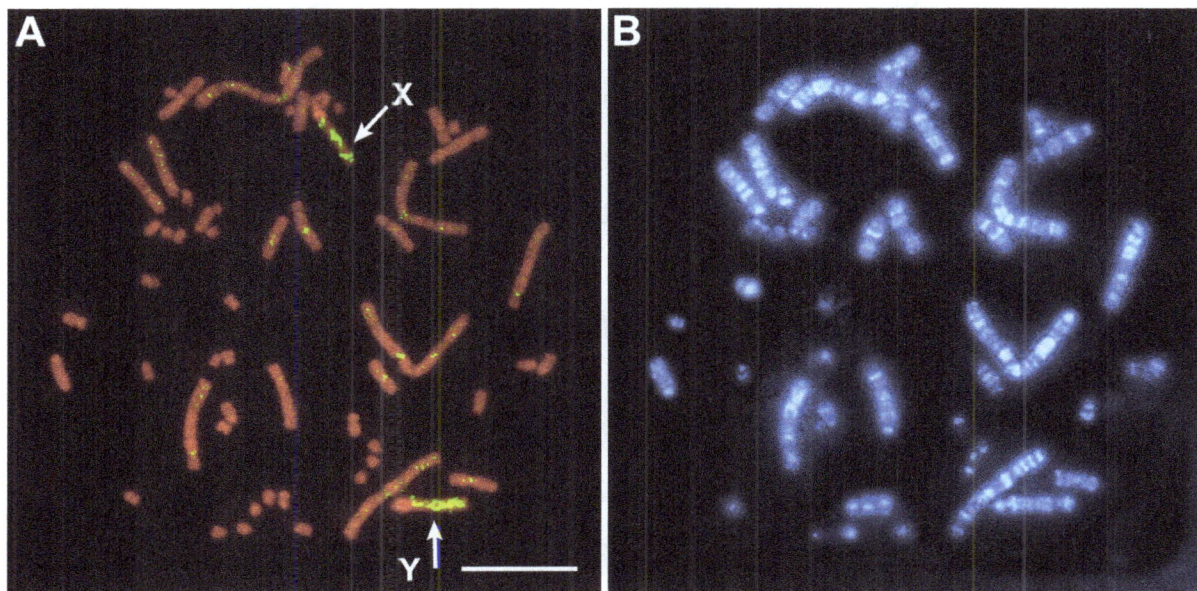

Figure 4. Chromosome painting with chromosome 6-specific DNA probe of *P. sinensis* to metaphase spread of male *S. triporcatus*. (A) The probe painted the X and Y chromosomes on PI-stained metaphase spread of *S. triporcatus* (indicated by arrows). (B) Hoechst-stained pattern of the same metaphase spread as in (A). Scale bar = 10 μm.

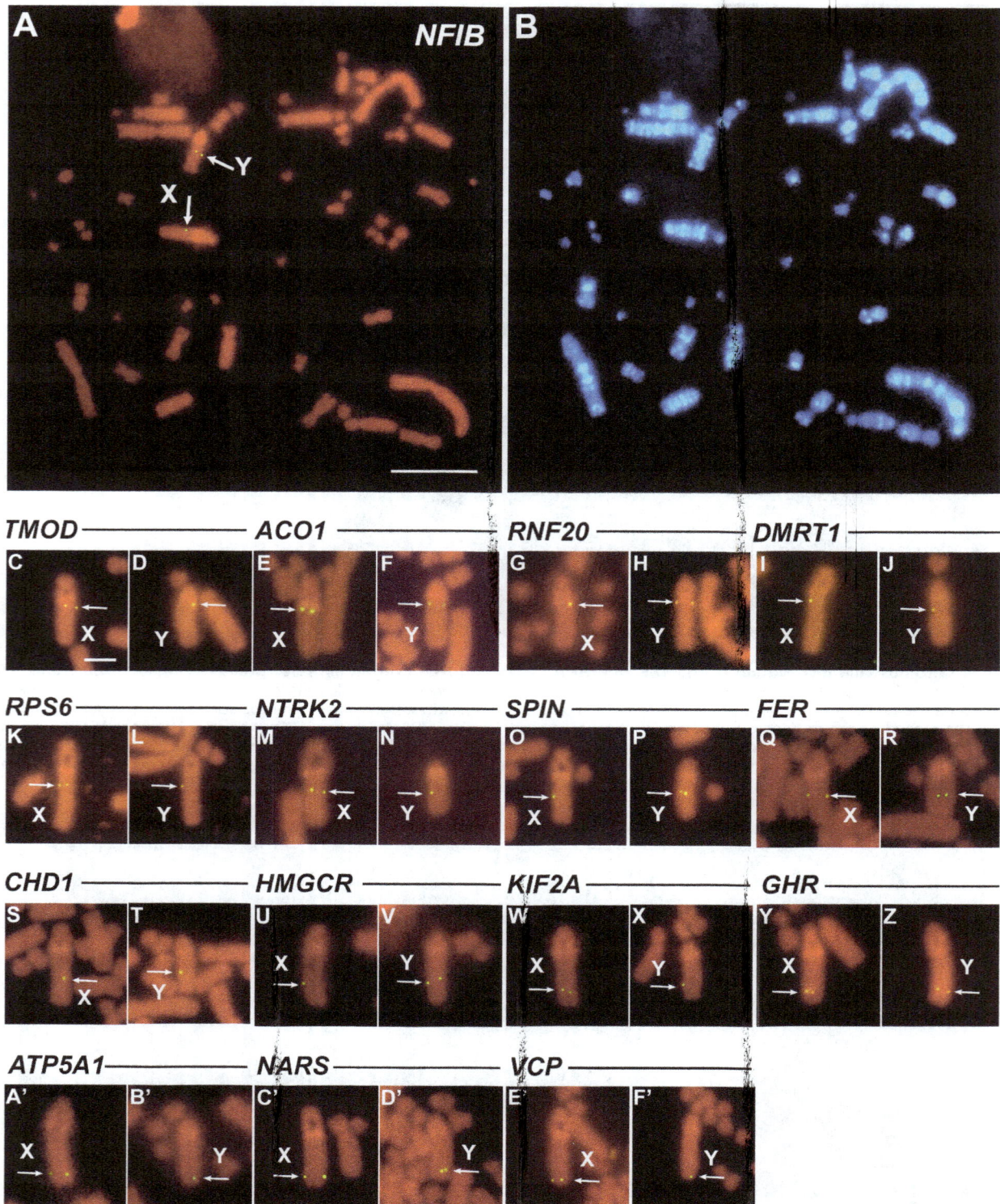

Figure 5. Chromosomal locations of *S. triporcatus* homologs of 16 chicken Z-linked genes in male *S. triporcatus*. (A, B) FISH pattern of *NFIB* on PI-stained metaphase spread (A) and Hoechst-stained pattern of the same metaphase spread (B). (C–Z, A'–F') FISH signals of *TMOD* (C, D), *ACO1* (E, F), *RNF20* (G, H), *DMRT1* (I, J), *RPS6* (K, L), *NTRK2* (M, N), *SPIN* (O, P), *FER* (Q, R), *CHD1* (S, T), *HMGCR* (U, V), *KIF2A* (W, X), *GHR* (Y, Z), *ATP5A1* (A', B'), *NARS* (C', D'), and *VCP* (E', F') on PI-stained X and Y chromosomes. Arrows indicate the hybridization signals of the genes. Scale bars represent 10 μm (A, B) and 2.5 μm (C–Z, A'–F').

These results suggest that the XY sex chromosomes of *Staurotypus* turtles share the same origin as avian ZW sex chromosomes; however; *Staurotypus* turtles and birds acquired different types of heterogametic sex-determination system during their evolution,

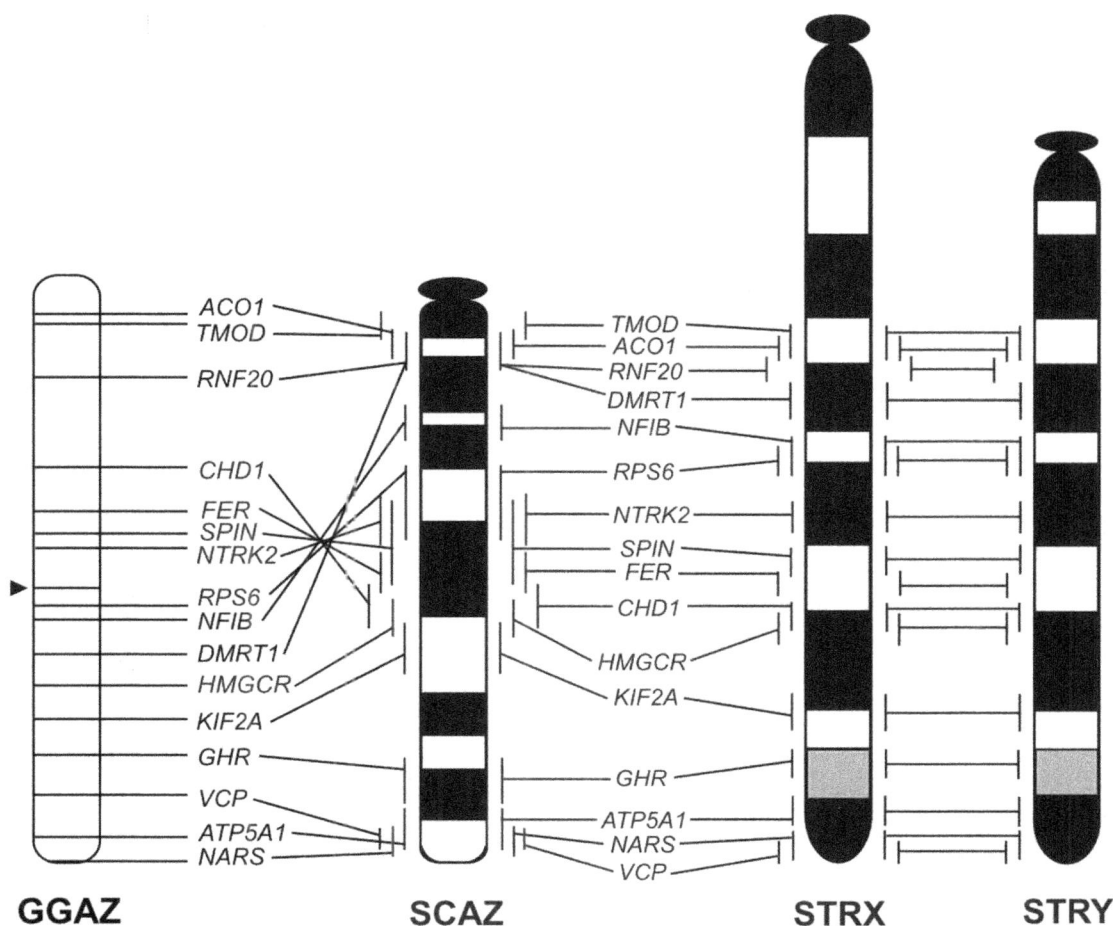

Figure 6. Comparative cytogenetic maps of 16 functional genes on the chicken Z chromosome (GGAZ), the ostrich Z chromosome (SCAZ), and the X and Y chromosomes of *S. triporcatus* (STRX and STRY, respectively). The gene order of 16 genes on the chicken Z chromosome was taken from the Ensembl Chicken Genome Browser (http://www.ensembl.org/Gallus_gallus). The chicken Z chromosome is inverted to facilitate comparison of the gene order. Arrowhead represents the location of the centromere.

and the X and Y chromosomes of *S. triporcatus* and *S. salvinii* are at a very early stage of differentiation. The only structural difference between the X and Y chromosomes in *S. triporcatus* was in the vicinity of the secondary constriction near the centromere, where meiotic recombination would have been suppressed. In *S. salvinii*, in addition to the difference in the size of the secondary constriction, the X and Y chromosomes were morphologically different: the X was subtelocentric, whereas the Y was acrocentric. The cessation of meiotic recombination very likely accounts for the difference in the copy number of the 18S–28S rRNA genes: this might have resulted from either a decrease in the copy number on the Y chromosome and/or amplification on the X chromosome. Alternatively, Sites et al. [17] suggested that the *S. salvinii* X chromosome was evolutionarily derived from the translocation of the NOR followed by the addition of a heterochromatic short arm onto the X, which occurred in one of the homomorphic proto-sex chromosomes, and the Y has remained unchanged. However, the initial step of sex chromosome differentiation in *Staurotypus* turtles remains unknown because the morphology of the homomorphic proto-sex chromosomes has not yet been identified.

The order of 16 genes on the *S. triporcatus* X chromosome was nearly identical to that of the ostrich Z chromosome, which bears the primitive gene order of avian sex chromosomes [27,29]

(Figure 6). This result suggests that the X chromosomes of *S. triporcatus* and *S. salvinii* and the ostrich Z chromosome are derived from the same autosomal pair of the common ancestor, and that the primitive gene order has been retained in both lineages independently since the time when Archosauromorpha diverged from the common ancestor of sauropsids 250–270 MYA [13–15]. In the chicken Z chromosome, the order was inverted in a region across the centromere, where seven genes (*DMRT1*, *NFIB*, *RPS6*, *NTRK2*, *SPIN*, *FER*, and *CHD1*) are contained, compared with those in the ostrich Z chromosome and the X chromosomes of two *Staurotypus* species. Moreover, the order of *DMRT1–NFIB–RPS6–NTRK2–SPIN–FER–CHD1* is probably the same as those of the ostrich Z and *Staurotypus* X chromosomes, although the location of the centromere differed (Figure 6). This result leads us to predict that a large paracentric inversion occurred at the breakpoints between *RNF20* and *DMRT1* and between *CHD1* and *HMGCR* in the ancestral acrocentric Z chromosome, and that subsequent repositioning of the centromere led to the metacentric chicken Z chromosome. Our previous studies revealed that whereas the ZW sex chromosomes of *P. sinensis* have homology with chicken chromosome 15, the XY chromosomes of *S. crassicollis* are homologous to chicken chromosome 5 [11,12]. These results indicate that the sex chromosomes of these three turtle species

Table 1. The cDNA fragments of *S. triporcatus* (STR) homologs of chicken Z-linked genes and nucleotide sequence identities between *S. triporcatus* and chicken (*Gallus gallus*, GGA) cDNA fragments.

Gene[a]	Length of cDNA fragment (bp)	Identity (%) between STR and GGA[b]	Accession number of *S. triporcatus* homolog
ACO1	1135	83.1 (943/1135)	AB747261
ATP5A1	1102	86.8 (956/1102)	AB747262, AB747263
CHD1	893	88.8 (793/893)	AB747264
DMRT1	684	81.2 (553/681)	AB747265
FER	760	91.4 (695/760)	AB747266
GHR	898	77.7 (698/898)	AB747267
HMGCR	1077	84.4 (909/1077)	AB747268
KIF2A	664	93.2 (619/664)	AB747269
NARS	1083	85.0 (921/1083)	AB747271
NFIB	820	94.4 (774/820)	AB747272
NTRK2	554	89.7 (497/554)	AB747273
RNF20	1159	84.9 (984/1159)	AB747274
RPS6	658	86.4 (569/658)	AB747275
SPIN	628	93.3 (586/628)	AB747276
TMOD	1007	82.2 (828/1007)	AB747277
VCP	995	90.1 (897/995)	AB747278

[a]*ACO1*, aconitase 1, soluble; *ATP5A1*, ATP synthase, H$^+$ transporting, mitochondrial F1 complex, alpha subunit, isoform 1, cardiac muscle; *CHD1*, chromodomain helicase DNA binding protein 1; *DMRT1*, *doublesex* and *mab-3* related transcription factor 1; *FER*, (fps/fes related) tyrosine kinase; *GHR*, growth hormone receptor; *HMGCR*, 3-hydroxy-3-methylglutaryl-CoA reductase; *KIF2A*, kinesin heavy chain member 2A; *NARS*, asparaginyl-tRNA synthetase; *NFIB*, nuclear factor I/B; *NTRK2*, neurotrophic tyrosine kinase receptor, type 2; *RNF20*, ring finger protein 20, E3 ubiquitin protein ligase; *RPS6*, ribosomal protein S6; *SPIN*, spindlin; *TMOD*, tropomodulin 1; *VCP*, valosin containing protein.
[b]The number in parenthesis indicates the number of identical bases/the number of bases in the overlapped region between cDNA fragments of two species.

differentiated independently from different autosomal pairs of the common ancestor in each lineage. This suggests great diversity of sex chromosomal origins and a considerable level of plasticity of sex determination in Testudines. Such diversity of sex chromosomal origins within the same order has also been found in squamate reptiles [22,30–33]. The homology of the micro-X sex chromosome of the green anole lizard (*Anolis carolinensis*) to chicken chromosome 15 [31] indicates that *A. carolinensis* and *P. sinensis* happen to share the same origin of sex chromosomes. However, it remains unclear whether the gene order of the sex chromosomes has been conserved.

The family Kinosternidae is composed of two subfamilies, Staurotypinae and Kinosterninae [34]; however, molecular phylogenetic analysis has indicated that these two clades show monophyly within the family [35]. Staurotypinae comprises only three species: the narrow-bridged musk turtle (*Claudius angustatus*), *S. triporcatus*, and *S. salvinii*. These three species have similar karyotypes with 2n = 54. *C. angustatus* also exhibits GSD; however, this species has no heteromorphic sex chromosomes [16,36]. The karyotypes of Kinosterninae species differ from those of Staurotypinae in terms of the diploid chromosome number (2n = 56), and no GSD species have been reported in this subfamily [2,16,37]. These observations collectively suggest that TSD was probably the primitive state in Kinosternidae and that GSD arose in the lineage of Staurotypinae; it thus seems likely that *Staurotypus* and *Claudius* share the ancestral XY sex chromosome system for this group but that *Claudius* remains at a more primitive stage of differentiation or that *Claudius* sex chromosomes are more recently derived than those in *Staurotypus*. The level of homology of the sex chromosomes between *C. angustatus* and the other two *Staurotypus* species remains unknown;

therefore, identification of the *C. angustatus* sex chromosomes and their linkage groups are needed to clarify the ancestral form of sex chromosomes and the initial step of sex chromosome differentiation in Staurotypinae.

S. triporcatus and *S. salvinii* are the second case of reptilian species for which sex chromosomes were found to have the same origin as the avian Z sex chromosome. The first case is the Hokou gecko (*Gekko hokouensis*), in which six chicken Z-linked genes (*ACO1*, *ATP5A1*, *CHD1*, *DMRT1*, *GHR*, and *RPS6*) were all mapped to the Z chromosome in the same order as that of the ostrich Z chromosome [22,27]. In *G. hokouensis*, the W homolog of *DMRT1* was located in the pericentromeric region where multiple rearrangements including a pericentric inversion occurred. Consequently, recombination should have been suppressed between the Z and W chromosomes. This suggests that functional divergence may have occurred in the W homolog. *DMRT1* is a strong candidate of the sex-determining gene in birds, which is deleted in the chicken W chromosome and also in the W chromosomes of paleognathous birds, emu (*Dromaius novaehollandiae*), double-wattled cassowary (*Casuarius casuarius*), and ostrich [18,38], and is considered to be involved in testis determination by twofold gene dosage in ZZ males [39,40]. By contrast, in the African clawed frog (*Xenopus laevis*), a paralog of *DMRT1* located only on the W chromosome, *DM-W*, was identified as the ovary-determinant gene [41]. In *S. triporcatus* and *S. salvinii*, the X and Y homologs of *DMRT1* were mapped near the secondary constrictions where the X and Y chromosomes might be structurally differentiated. However, the male-specific region on the Y chromosome, which is involved in male sex determination, is still unknown because no intra-chromosomal rearrangement, partial deletion of the Y chromosome, and/or

Table 2. The cDNA fragments of ostrich (*S. camelus*, SCA) homologs of chicken Z-linked genes and nucleotide sequence identities among *S. triporcatus* (STR), ostich and chicken (*Gallus gallus*, GGA) cDNA fragments.

Gene[a]	Length of cDNA fragment (bp)	Identity (%) between STR and SCA[b]	Identity (%) between SCA and GGA[b]	Accession number of ostrich homolog
ACO1	1133	83.6 (948/1133)	91.7 (1039/1133)	AB755561
ATP5A1	990	88.1 (873/990)	92.5 (916/990)	AB254864[c], AB254866[c]
CHD1	874	89.4 (780/872)	92.1 (805/874)	AB254867[c]
DMRT1	1262	87.1 (420/482)	88.3 (575/651)	AB536738[d]
FER	761	92.5 (703/760)	94.3 (718/761)	AB747279
GHR	832	79.6 (653/820)	86.8 (712/820)	AB254871[c]
HMGCR	1074	85.7 (920/1074)	91.6 (984/1074)	AB747280
KIF2A	666	93.8 (623/664)	95.8 (637/665)	AB747281
NARS	1085	86.0 (931/1083)	91.7 (994/1084)	AB747283
NFIB	820	94.4 (774/820)	95.2 (781/820)	AB747284
NTRK2	500	90.8 (454/500)	94.8 (474/500)	AB254873[c]
RNF20	1171	86.4 (999/1156)	91.8 (1076/1171)	AB747285
RPS6	612	87.3 (534/612)	93.8 (574/612)	AB254876[c]
SPIN	580	94.3 (547/580)	97.8 (567/580)	AB254878[c]
TMOD	901	83.9 (756/901)	90.1 (812/901)	AB254879[c]
VCP	995	90.0 (896/995)	93.4 (929/995)	AB747356

[a]ACO1, aconitase 1, soluble; ATP5A1, ATP synthase, H⁺ transporting, mitochondrial F1 complex, alpha subunit, isoform 1, cardiac muscle; CHD1, chromodomain helicase DNA binding protein 1; DMRT1, *doublesex* and mab-3 related transcription factor 1; FER, (fps/fes related) tyrosine kinase; GHR, growth hormone receptor; HMGCR, 3-hydroxy-3-methylglutaryl-CoA reductase; KIF2A, kinesin heavy chain member 2A; NARS, asparaginyl-tRNA synthetase; NFIB, nuclear factor I/B; NTRK2, neurotrophic tyrosine kinase receptor, type 2; RNF20, ring finger protein 20, E3 ubiquitin protein ligase; RPS6, ribosomal protein S6; SPIN, spindlin; TMOD, tropomodulin 1; VCP, valosin containing protein.
[b]The number in parenthesis indicates the number of identical bases/the number of bases in the overlapped region between cDNA fragments of two species.
[c]The nucleotide sequences were obtained from Tsuda et al. [27].
[d]The nucleotide sequence was obtained from Ishijima et al. [18].

structurally differentiated Y-linked gene has yet been found. Hence, another molecular cytogenetic approach is needed to identify the critical sex-determining region in these species.

Supporting Information

Figure S1　Chromosomal locations of ostrich homologs of nine chicken Z-linked genes in female ostrich. (A, B) FISH pattern of *NARS* on PI-stained metaphase spread (A) and Hoechst-stained pattern of the same metaphase spread (B). (C–N) FISH signals of *ACO1* (C), *RNF20* (D), *DMRT1* (E), *NFIB* (F), *FER* (G), *HMGCR* (I), *KIF2A* (K), and *VCP* (M) on PI-stained Z chromosomes, and FISH signals of *FER* (H), *HMGCR* (J), *KIF2A* (L), and *VCP* (N) on PI-stained W chromosomes. No signals of *ACO1*, *RNF20*, *DMRT1*, and *NFIB* were detected on the W chromosomes. Arrows indicate the hybridization signals of the genes. Scale bars represent 10 μm (A, B) and 2.5 μm (C–N).

Figure S2　Comparative cytogenetic maps of 16 functional genes on the Z chromosome (SCAZ) and W chromosome (SCAW) of the ostrich (*S. camelus*, SCA). The chromosomal locations of seven genes (*TMOD*, *RPS6*, *NTRK2*, *SPIN*, *CHD1*, *GHR*, and *ATP5A1*) written in red were taken from our previous report [27].

Figure S3　Chromosomal locations of *S. salvinii* homologs of 16 chicken Z-linked genes in male *S. salvinii*. (A, B) FISH pattern of *VCP* on PI-stained metaphase spread (A) and Hoechst-stained pattern of the same metaphase spread (B). (C–Z,

A'–E') FISH signals of *TMOD* (C, D), *ACO1* (E, F), *RNF20* (G, H), *DMRT1* (I, J), *NFIB* (K, L), *RPS6* (M, N), *NTRK2* (O, P), *SPIN* (Q, R), *FER* (S, T), *CHD1* (U), *HMGCR* (V, W), *KIF2A* (X, Y), *GHR* (Z, A'), *ATP5A1* (B', C'), and *NARS* (D', E') on PI-stained X and Y chromosomes. Arrows indicate the hybridization signals of the genes. Scale bars represent 10 μm (A, B) and 2.5 μm (C–Z, A'–E').

Figure S4　Comparative cytogenetic maps of 16 functional genes on the X and Y chromosomes of *S. triporcatus* (STRX and STRY) and *S. salvinii* (SSAX and SSAY).

Table S1　Degenerate oligonucleotide primers used for molecular cloning of *S. triporcatus* homologs of 16 chicken Z-linked genes.

Acknowledgments

We are grateful to Fengtang Yang and Patricia O'Brien (both from the Department of Veterinary Medicine, Cambridge University, UK) for providing chromosome-specific DNA probes of *P. sinensis*.

Author Contributions

Conceived and designed the experiments: TK YM. Performed the experiments: TK YU CN. Analyzed the data: TK YU CN YM. Contributed reagents/materials/analysis tools: TK CN. Contributed to the writing of the manuscript: TK YM.

References

1. Valenzuela N, Lance V (eds.) (2004) Temperature-dependent sex determination in vertebrates. Washington: Smithsonian Books.

2. Olmo E, Signorino G (2005) Chromorep: a reptile chromosomes database. Available: http://chromorep.univpm.it/.

3. Ezaz T, Sarre SD, O'Meally D, Graves JAM, Georges A (2009) Sex chromosome evolution in lizards: independent origins and rapid transitions. Cytogenet Genome Res 127: 249–260.

4. Janzen FJ, Phillips PC (2006) Exploring the evolution of environmental sex determination, especially in reptiles. J Evol Biol 19: 1775–1784.

5. Organ CL, Janes DE (2008) Evolution of sex chromosomes in Sauropsida. Integr Comp Biol 48: 512–519.

6. Pokorná M, Kratochvíl L (2009) Phylogeny of sex-determining mechanisms in squamate reptiles: are sex chromosomes an evolutionary trap? Zool J Linn Soc 156: 168–183.

7. Badenhorst D, Stanyon R, Engstrom T, Valenzuela N (2013) A ZZ/ZW microchromosome system in the spiny softshell turtle, *Apalone spinifera*, reveals an intriguing sex chromosome conservation in Trionychidae. Chromosome Res 21: 137–147.

8. Ezaz T, Valenzuela N, Grützner F, Miura I, Georges A, et al. (2006) An XX/XY sex microchromosome system in a freshwater turtle, *Chelodina longicollis* (Testudines: Chelidae) with genetic sex determination. Chromosome Res 14: 139–150.

9. Martinez PA, Ezaz T, Valenzuela N, Georges A, Graves JAM (2008) An XX/XY heteromorphic sex chromosome system in the Australian chelid turtle *Emydura macquarii*: a new piece in the puzzle of sex chromosome evolution in turtles. Chromosome Res 16: 815–825.

10. Kawai A, Nishida-Umehara C, Ishijima J, Tsuda Y, Ota H, et al. (2007) Different origins of bird and reptile sex chromosomes inferred from comparative mapping of chicken Z-linked genes. Cytogenet Genome Res 117: 92–102.

11. Kawagoshi T, Uno Y, Matsubara K, Matsuda Y, Nishida C (2009) The ZW micro-sex chromosomes of the Chinese soft-shelled turtle (*Pelodiscus sinensis*, Trionychidae, Testudines) have the same origin as chicken chromosome 15. Cytogenet Genome Res 125: 125–31.

12. Kawagoshi T, Nishida C, Matsuda Y (2012) The origin and differentiation process of X and Y chromosomes of the black marsh turtle (*Siebenrockiella crassicollis*, Geoemydidae, Testudines). Chromosome Res 20: 95–110.

13. Kumazawa Y, Nishida M (1999) Complete mitochondrial DNA sequences of the green turtle and blue-tailed mole skink: statistical evidence for Archosaurian affinity of turtles. Mol Biol Evol 16: 784–792.

14. Rest JS, Ast JC, Austin CC, Waddell PJ, Tibbetts EA, et al. (2003) Molecular systematics of primary reptilian lineages and the tuatara mitochondrial genome. Mol Phylogent Evol 29: 289–297.

15. Wang Z, Pascual-Anaya J, Zadissa A, Li W, Niimura Y, et al. (2013) The draft genomes of soft-shell turtle and green sea turtle yield insights into the development and evolution of the turtle-specific body plan. Nat Genet 45: 701–706.

16. Bull JJ, Moon RG, Legler JM (1974) Male heterogamety in kinosternid turtles (genus *Staurotypus*). Cytogenet Cell Genet 13: 419–425.

17. Sites JW Jr, Bickham JW, Haiduk MW (1979) Derived X chromosome in the turtle genus *Staurotypus*. Science 206: 1410–1412.

18. Ishijima J, Uno Y, Nishida C, Matsuda Y (2014) Genomic structures of the kW1 loci on the Z and W chromosomes in ratite birds: structural changes at an early stage of W chromosome differentiation. Cytogenet Genome Res 142: 255–267.

19. Matsuda Y, Chapman VM (1995) Application of fluorescence *in situ* hybridization in genome analysis of the mouse. Electrophoresis 16: 261–272.

20. Sumner AT (1972) A simple technique for demonstrating centromeric heterochromatin. Exp Cell Res 75: 304–306.

21. Matsubara K, Nishida-Umehara C, Kuroiwa A, Tsuchiya K, Matsuda Y (2003) Identification of chromosome rearrangements between the laboratory mouse (*Mus musculus*) and the Indian spiny mouse (*Mus platythrix*) by comparative FISH analysis. Chromosome Res 11: 57–64.

22. Kawai A, Ishijima J, Nishida C, Kosaka A, Ota H, et al. (2009) The ZW sex chromosomes of *Gekko hokouensis* (Gekkonidae, Squamata) represent highly conserved homology with those of avian species. Chromosoma 118: 43–51.

23. Howell WM, Black DA (1980) Controlled silver-staining of nucleolus organizer regions with a protective colloidal developer: a 1-step method. Experientia 36: 1014–1015.

24. Matsuda Y, Nishida-Umehara C, Tarui H, Kuroiwa A, Yamada K, et al. (2005) Highly conserved linkage homology between birds and turtles: Birds and turtle chromosomes are precise counterparts of each other. Chromosome Res 13: 601–615.

25. Uno Y, Nishida C, Tarui H, Ishishita S, Takagi C, et al. (2012) Inference of the protokaryotypes of amniotes and tetrapods and the evolutionary processes of microchromosomes from comparative gene mapping. PLoS One 7: e53027.

26. Ogawa A, Murata K, Mizuno S (1998) The location of Z- and W-linked marker genes and sequence on the homomorphic sex chromosomes of the ostrich and the emu. Proc Natl Acad Sci USA 95: 4415–4418.

27. Tsuda Y, Nishida-Umehara C, Ishijima J, Yamada K, Matsuda Y (2007) Comparison of the Z and W sex chromosomal architectures in elegant crested tinamou (*Eudromia elegans*) and ostrich (*Struthio camelus*) and the process of sex chromosome differentiation in palaeognathous birds. Chromosoma 116: 159–173.

28. Kasai F, O'Brien PC, Martin S, Ferguson-Smith MA (2012) Extensive homology of chicken macrochromosomes in the karyotypes of *Trachemys scripta elegans* and *Crocodylus niloticus* revealed by chromosome painting despite long divergence times. Cytogenet Genome Res 136: 303–307.

29. Nishida-Umehara C, Tsuda Y, Ishijima J, Ando J, Fujiwara A, et al. (2007) The molecular basis of chromosome orthologies and sex chromosomal differentiation in palaeognathous birds. Chromosome Res 15: 721–734.

30. Matsubara K, Tarui H, Toriba M, Yamada K, Nishida-Umehara C, et al. (2006) Evidence for different origin of sex chromosomes in snakes, birds, and mammals and step-wise differentiation of snake sex chromosomes. Proc Natl Acad Sci USA 103: 18190–18195.

31. Alföldi J, Di Palma F, Grabherr M, Williams C, Kong L, et al. (2011) The genome of the green anole lizard and a comparative analysis with birds and mammals. Nature 31: 587–591.

32. O'Meally D, Ezaz T, Georges A, Sarre SD, Graves JAM (2012) Are some chromosomes particularly good at sex? Insights from amniotes. Chromosome Res 20: 7–19.

33. Young MJ, O'Meally D, Sarre SD, Georges A, Ezaz T (2013). Molecular cytogenetic map of the central bearded dragon, *Pogona vitticeps* (Squamata: Agamidae). Chromosome Res 21: 361–374.

34. Vitt LJ, Caldwell JP (2013) Herpetology: An introductory biology of amphibians and reptiles. Waltham: Academic Press, 776 p.

35. Iverson JB, Le M, Ingram C (2013) Molecular phylogenetics of the mud and musk turtle family Kinosternidae. Mol Phylogenet Evol 69: 929–939.

36. Vogt RC, Flores-Villela O (1992) Effects of incubation temperature on sex determination in a community of neotropical freshwater turtles in southern Mexico. Herpetologica 48: 265–270.

37. Valenzuela N, Adams DC (2011) Chromosome number and sex determination coevolve in turtles. Evolution 65: 1808–1813.

38. Shetty S, Kirby P, Zarkower D, Graves JAM (2002) DMRT1 in a ratite bird: evidence for a role in sex determination and discovery of a putative regulatory element. Cytogenet Genome Res 99: 245–251.

39. Nanda I, Zend-Ajusch E, Shan Z, Grützner F, Schartl M, et al. (2000) Conserved synteny between the chicken Z sex chromosome and human chromosome 9 includes the male regulatory gene *DMRT1*: a comparative (re)view on avian sex determination. Cytogenet Cell Genet 89: 67–78.

40. Smith CA, Roeszler KN, Ohnesorg T, Cummins DM, Farlie PG, et al. (2009) The avian Z-linked gene *DMRT1* is required for male sex determination in the chicken. Nature 461: 267–271.

41. Yoshimoto S, Okada E, Umemoto H, Tamura K, Uno Y, et al. (2008) A W-linked DM-domain gene, DM-W, participates in primary ovary development in *Xenopus laevis*. Proc Natl Acad Sci USA 105: 2469–2474.

4

Development of Retroviral Vectors for Tissue-Restricted Expression in Chicken Embryonic Gonads

Luke S. Lambeth[1,3]*, Thomas Ohnesorg[1], David M. Cummins[4], Andrew H. Sinclair[1,2,3], Craig A. Smith[1,2,3]

1 Murdoch Childrens Research Institute, Royal Children's Hospital, Melbourne, VIC, Australia, 2 Department of Paediatrics, The University of Melbourne, Melbourne, VIC, Australia, 3 Poultry Cooperative Research Centre, Armidale, NSW, Australia, 4 CSIRO Animal, Food and Health Sciences, Australian Animal Health Laboratory, Geelong, VIC, Australia

Abstract

The chicken embryo has long been a useful model organism for studying development, including sex determination and gonadal differentiation. However, manipulating gene expression specifically in the embryonic avian gonad has been difficult. The viral vector RCASBP can be readily used for embryo-wide transgene expression; however global mis-expression using this method can cause deleterious off-target effects and embryo-lethality. In an attempt to develop vectors for the over-expression of sequences in chicken embryonic urogenital tissues, the viral vector RCANBP was engineered to contain predicted promoter sequences of gonadal-expressed genes. Several promoters were analysed and it was found that although the *SF1* promoter produced a tissue-restricted expression pattern that was highest in the mesonephros and liver, it was also higher in the gonads compared to the rest of the body. The location of EGFP expression from the *SF1* promoter overlapped with several key gonad-expressed sex development genes; however expression was generally low-level and was not seen in all gonadal cells. To further validate this sequence the key testis determinant DMRT1 was over-expressed in female embryos, which due to insufficient levels had no effect on gonad development. The female gene aromatase was then over-expressed in male embryos, which disrupted the testis pathway as demonstrated by a reduction in AMH protein. Taken together, although these data showed that the *SF1* promoter can be used for functional studies *in ovo*, a stronger promoter sequence would likely be required for the functional analysis of gonad genes that require high-level expression.

Editor: Dmitry I. Nurminsky, University of Maryland School of Medicine, United States of America

Funding: This work was supported by an Australian Research Council (ARC) Future Fellowship awarded to CAS and The National Health and Medical Research Council (Project Grant #1031214) awarded to TO. No additional external funding was received for this study. The funders had no role in study design, data collection and analysis, decision to publish, or preparation of the manuscript.

Competing Interests: This research was conducted within the Poultry CRC. There are no patents, products in development or marketed products to declare.

* Email: luke.lambeth@mcri.edu.au

Introduction

One of the main strengths of using chickens and other avian species for studies in development is the ability to manipulate an embryo that develops outside the maternal body. Numerous techniques can be used to track and manipulate factors involved in embryonic development and differentiation. These include the transplantation of tissues, implantation of beads soaked with growth factors, and the introduction of recombinant DNA by electroporation and retroviruses [1].

There are three principal methods for gene transfer into the developing chicken embryo; transfection, viral infection and electroporation. *In ovo* electroporation has been widely described and allows for controlled gene expression both temporally and spatially. Various areas of the embryo can be specifically targeted by electroporating different tissues and stages of development, including the midbrain [2], the somites [3], the retina [4] and the spinal cord [5,6], and the forelimb mesoderm [7]. Although the targeting of intermediate mesoderm or primordial gonads by electroporation has been described, this method is generally not very well established. Electroporation of viral vectors has been used to achieve ectopic PITX2 expression in the gonads of developing chicken embryos [8], and more recently the developing left gonad was targeted for overexpression of DMRT1 [9]. Another approach to targeting gonadal expression has exploited the transfection of migrating primordial germ cells. Lipofection of recombinant Tol2 transposon and transposase plasmids into very early stage (14HH) chicken embryos, results in effective integration in primordial germ cells, which subsequently migrate to the developing gonads and deliver GFP reporter expression [10].

The most widely used retroviral system used in avian developmental studies is RCAS (Replication Competent ALV LTR with a Splice acceptor), which is a modified version of an avian Rous sarcoma virus [11]. A cDNA copy can be inserted downstream of the viral *env* gene which is transcribed by a promoter within one of the viral long terminal repeats (LTRs) and subsequently spliced. Infection with RCAS permits sustained mis-expression of inserted sequences following the stable integration of the viral DNA into the host genome. Infection at early developmental stages or of highly proliferative cell populations can result in large areas of the embryo being infected. Indeed, injection of an RCASBP (a modified RCAS with Bryan RSV Polymerase) encoding EGFP into the blastoderms of susceptible eggs results in embryo-wide EGFP expression, including the urogenital systems of both male and female embryos [12]. Despite the considerable potential for experimental analysis of candidate genes using this method, embryo-wide overexpression also could induce unwanted off-target effects. Many of the key factors involved in sex development are transcription factors, such as the male up-regulated genes DMRT1 and SOX9, and the female up-regulated gene FOXL2. Embryo-wide over-expression of these genes is expected to be lethal, and indeed we have found this to be the case when RCASBP was used to deliver DMRT1 or FOXL2 (CA Smith, unpublished data). In contrast, knockdown of DMRT1 in chicken embryos was achieved using a U6 promoter to express shRNAs from RCASBP, resulting in feminization of male gonads [13]. Although the shRNA was expressed throughout the embryo, the urogenital system restricted expression pattern of DMRT1 meant that this protein was suppressed in these tissues only, and no overt off-targeting effects were noted.

The viral vector RCANBP is derived from RCASBP, however it lacks the splice acceptor site downstream of the *env* gene. This vector thereby permits the introduction of exogenous internal promoter sequences to direct transgene expression instead of the viral LTR. Infection of RCANBP that contains EGFP under control of the human cytomegalovirus (CMV) promoter results in widespread EGFP expression in E8.5 embryos [14]. Although this expression is strong, it is in restricted regions including the retinal-pigmented epithelium, liver, and proliferating zones in developing bones. In contrast, EGFP expression from RCASBP/EGFP is generally more widespread throughout the embryo [14].

The manipulation of genes in embryonic chicken gonads is of interest to the field of sex determination and differentiation. In the chicken and all other birds, a ZZ male: ZW female sex chromosome system exists, the inheritance of which, determines sex. The exact molecular mechanisms leading to sexual differentiation however, are still not fully understood. Gonad sexual differentiation appears to be highly conserved and many of the important signaling factors involved in ovarian or testis development in mammals are also implicated in birds [15]. Therefore, gonad-specific expression of key sex development factors would help to further advance studies in chicken sex determination. To address this, we have identified and compared the ability of several gonad-expressed gene promoters to drive reporter expression in the embryonic urogenital system of chicken embryos.

Results

Promoter characterisation

To develop retroviral vectors for tissue-specific expression in chicken embryonic gonads, promoter regions of several genes that are expressed in embryonic gonads were characterised. These genes included: Wilm's tumor suppressor *(WT1)*, Steroidogenic factor 1 *(SF1, NR5A1)*, Anti-Müllerian hormone *(AMH) and* aromatase *(CYP19A1)*. The DNA sequences of the regions located directly upstream of each open reading frame were obtained from the UCSC Genome Browser from the Chicken May 2006 (WUGSC 2.1/galGal3) Assembly.

For each putative promoter region, the sequence was analysed for the presence of potential regulatory elements and transcription factor binding sites (Figure 1). An analysis of the putative chicken *SF1* promoter was previously described, which reported the identification of several promoter elements and its activity *in vitro* [16]. In the current study, a 424 nt region directly upstream of the chicken *SF1* coding sequence that included 125 nt of sequence downstream of the predicted TSS (accession AB018710) was cloned and verified. Binding elements including a TATA-box, GC-box, CCAAT-box and an E-box were identified as described previously [16]. For the aromatase promoter *(AROMp)*, a 947 nt region upstream of the aromatase coding sequence was cloned that

included 39 nt of sequence downstream of the TSS in ovary [17]. Analysis of this sequence revealed a TATA box from −28 to −21 and an SF1 consensus-binding site from −133 to −125. The SF1 binding site was the same sequence as the SFRE consensus sequence [18] and is very similar to those found in the mouse 3β-Hydroxysteroid dehydrogenase 1 (HSD1) and Cyp17 promoters [19]. The chicken *AMH* promoter was previously characterised and analysed for the presence of potential SOX9 binding sites [20]. In this study, a 303 nt region directly upstream of the coding sequence was cloned that included 45 nt of sequence downstream of the TSS. Consistent with the study by Oreal et al., features within the *AMHp* region included a TATA box and an estrogen responsive element (ERE), as well as a consensus SF1 binding site (5′-TCAAGGCCA-3′). To isolate a putative chicken *WT1* promoter sequence *(WT1p)*, a region of 594 nt directly upstream of the predicted *WT1* coding sequence was cloned. Like the human *WT1* promoter [21], this sequence did not have a TATA-box or a CCAAT-box, and we were not able to identify any typical gonad transcription factor consensus binding sites. In addition to the chicken promoters, the mouse *SF1* promoter was also used to drive EGFP expression from RCANBP. This sequence was previously described and tested in mice, where it produces strong and specific expression in mouse gonads [22]. To act as a positive control for EGFP expression from RCANBP using an internal promoter, the well characterised and widely used Simian virus 40 *(SV40)* promoter was also included.

Promoter validation in ovo

To assess the ability of each of the promoter sequences to express a reporter gene in chicken embryos, RCANBP vectors were generated that contained each of the promoter sequences upstream of EGFP. High titre RCANBP viral stocks for each vector were generated and used to infect blastoderm stage embryos. The expression of EGFP was then monitored in E7.5 embryos by wholemount fluorescent microscopy. For each embryo, the expression of EGFP was first analysed for the whole embryo of both sexes. To test for gonad-restricted expression, the urogenital systems (mesonephros and gonads) were revealed by removing the viscera (Figure 2). Non-injected negative controls showed only background levels of fluorescence, whereas very strong EGFP was detected in the urogenital system for *SV40p* control in both sexes. In addition to the urogenital systems, EGFP expression in the *SV40p* infected embryos was evident throughout the entire embryo at high levels, indicating that as expected, this promoter exhibited ubiquitous transcriptional activity. For embryos infected with the RCANBP viruses encoding the various gonad factor promoters, a variety of EGFP expression patterns were observed. For *WT1p*, a low level of embryo-wide EGFP was evident that did not show any increase in the urogenital system. EGFP expression from *SF1p* was at low levels throughout the embryo, except for the liver and mesonephros, which both showed high levels of expression in both male and female (Figure 2B). For *AMHp*, EGFP expression was only evident in the urogenital system, as the rest of the embryo appeared to be negative. The level in the urogenital system however, was far lower than *SF1p*. The *aromatase* promoter produced strong EGFP expression throughout the entire embryo. The level of expression was very consistent across all tissues, including the urogenital system, and therefore did not show any tissue-specificity. Surprisingly, the mouse *SF1p* produced very weak expression overall, with no detectable expression in the gonads and a moderate level of expression in the liver (data not shown).

Taken together, these data showed that each of the promoters tested produced varying activities in early stage chicken embryos.

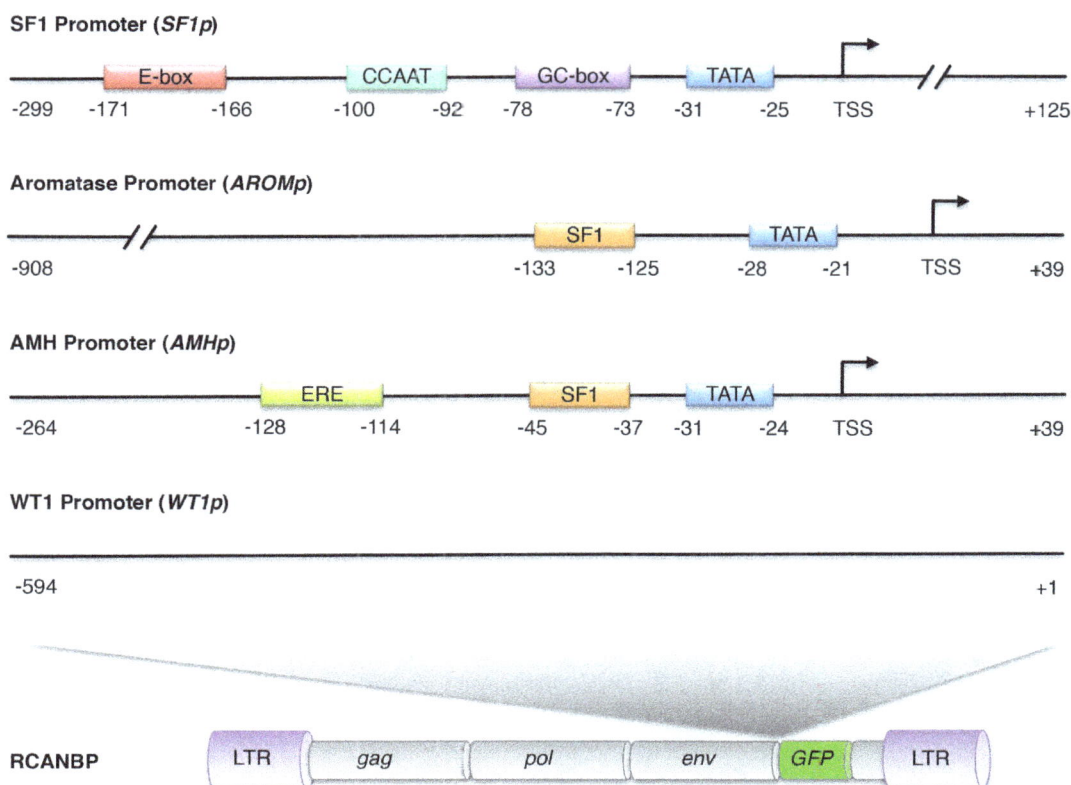

Figure 1. Schematic representation of putative gonad promoter sequences. All numbers shown are relative to the transcriptional start site (TSS) for each putative promoter sequence. The *SF1p* contains several promoter elements that have been described previously [16]. Both *aromatase* and *AMH* promoters contain TATA boxes and consensus SF1 binding sites. The *AMH* promoter also contains an estrogen responsive element (ERE). The *WT1* promoter is TATA-less and no other binding elements were identified. All promoter sequences were cloned into the RCANBP viral vector directly upstream of the EGFP open reading frame.

Since *AMHp* and *WT1p* showed only weak urogenital expression they were not pursued any further in this study. Although the *aromatase* promoter produced high-level EGFP expression, since it was not restricted to the urogenital system, it was also not pursued any further. The *SF1* promoter produced the most potentially useful expression pattern, as levels of EGFP were higher in the mesonephros and potentially the gonads compared to the rest of the embryo (Figure 2B). To further analyse the extent of its gonad-specific activity, the expression of EGFP was analysed by immunostaining. Embryos infected with RCANBP-SF1p-EGFP were dissected at E7.5 and gonad tissues were compared to samples of forelimb, which provided a representation of the rest of the embryo (Figure 3). Expression in the gonads of both sexes was evident, and although the forelimbs did show some immunoreactive EGFP expression, it was at greatly reduced levels compared to the gonads (Figure 3). P27 staining for the presence of viral epitope confirmed that the virus was present in each of the tissues.

The cellular location of EGFP expression from *SF1p* was then analysed in the context of other key sex development genes. The location of EGFP protein was compared to DMRT1, SOX9, aromatase and FOXL2 in the gonads of E7.5 embryos injected with RCANP-SF1p-EGFP. Analysis of the overlayed images of SOX9 and DMRT1 with EGFP revealed that in some cells EGFP was co-expressed (white arrows) with these proteins. Although, as EGFP is expressed in the cytoplasm and DMRT1 and SOX9 are expressed in the nucleus of cells in the cords of developing male gonads, these proteins did not co-localise (Figure 4). SOX9 is usually absent in developing female gonads, however, DMRT1 is

expressed in female germ cells, which are located in the outer gonadal cortex at E7.5 (Figure 4A). The high power image of the cortex showed that although EGFP was expressed in some cells within the cortex, it did not appear to be expressed in any of the DMRT1 positive cells (i.e., germ cells, which are known to silence RCAS/RCAN viruses [12]).

Female pathway genes aromatase and FOXL2 were then analysed for localisation with *SF1p* expressed EGFP (Figure 4B). Aromatase showed strong cytoplasmic expression in the female medulla and closer analysis showed that EGFP co-localised with aromatase in some cells as indicated by the orange colouring (white arrows). In contrast, FOXL2 showed nuclear expression primarily in cells within the medulla, and when overlayed with EGFP, it was apparent that both of these proteins were in some cases present in the same cells (white arrows).

To more closely analyse the relationship between EGFP and germ cells, the left and right gonads of both male and females were stained for the germ cell marker chicken vasa homologue (CVH). The left gonad of females characteristically shows predominant germ cell localisation within the thickened cortex, whereas the right gonad, and both the left and right gonads of males, show scattered germ cells localised throughout the medulla (Figure 5). When overlayed with EGFP, it was clear that in the left and right gonads of both sexes, germ cells in the cortex and medulla did not have any EGFP expression. Taken together with the DMRT1 staining in the female left gonad, these data suggest that *SF1p* does not express EGFP in germ cells when delivered from RCANBP. It does, however appear to show expression patterns that overlap

A

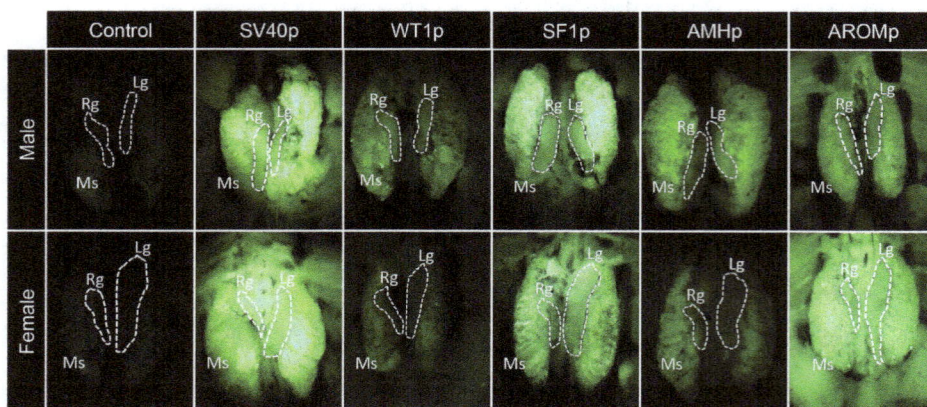

B

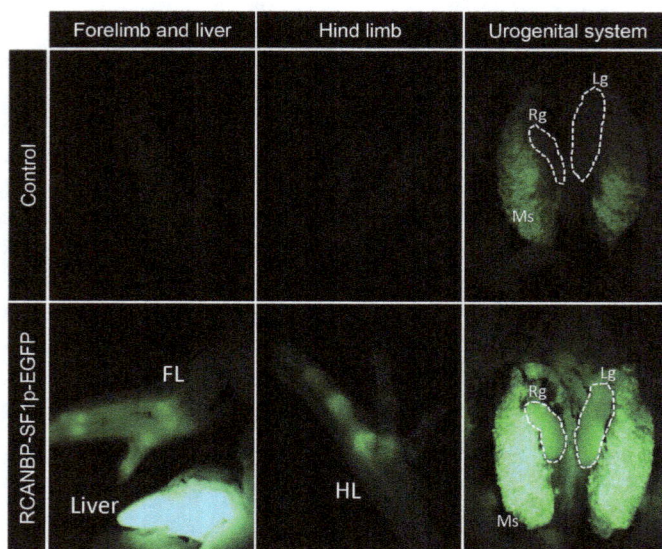

Figure 2. Wholemount fluorescent microscopy of novel gonad promoter expressed EGFP. Tissues from E7.5 embryos infected with RCANBP viruses containing *SV40* (*SV40p*), *WT1* (*WT1p*), *SF1* (*SF1p*), *AMH* (*AMHp*) and *aromatase* (*AROMp*) promoters. Dashed white lines delineate the left (Lg) and right (Rg) gonads, which sit on top of the mesonephros (Ms). A: Strong EGFP expression was evident for *SV40p* and *AROMp*, however, this was not confined to the urogenital systems. *WT1p* and *AMHp* produced low-level expression in the urogenital systems. EGFP expressed from *SF1p* was moderate in the urogenital system, and included the gonads. B: RCANBP-SF1p-EGFP infected E7.5 embryo; in addition to EGFP expression in the urogenital system, embryos also showed EGFP expression in the liver, forelimb (FL) and hind limb (HL).

with several key sex pathway genes in the somatic cells of both sexes.

SF1p-mediated over-expression of DMRT1

DMRT1 is normally expressed at higher levels in males compared to females and has been shown to be critical for testis development [9,13]. We have previously attempted DMRT1 over-expression from RCASBP, which produces embryo-wide delivery via the viral LTR promoter, but found that it induced early stage embryo-lethality [13]. To test if *SF1p* could be used to over-express a testis pathway gene in female gonads, the *DMRT1* open reading frame was cloned downstream of this promoter sequence in

RCANBP (called RCANBP-DMRT1). Embryos infected with RCANBP-DMRT1 at the blastoderm stage showed no signs of increased mortality or developmental abnormalities. Immuno-staining showed that in female gonads DMRT1 was expressed at higher levels compared to the control female (Figure 6). However, this over-expression was not as high as that seen in the male control. In the RCANBP-DMRT1 infected male there did not appear to be any observable increase in the level of DMRT1 expression and none was observed outside the cords. However, we have previously seen that despite robust levels of RCASBP mediated over-expression of aromatase in both male and female gonads, no expression was observed in cell types that do not

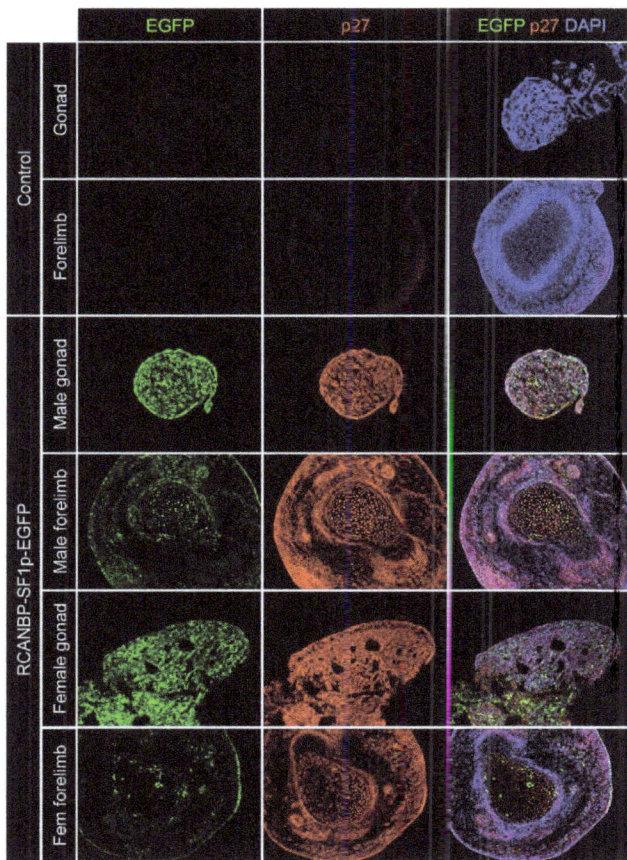

Figure 3. Tissue restricted EGFP expression from the *SF1* promoter. Immunostaining for EGFP (green) and the RCANBP viral antigen p27 (red), in RCANBP-SF1p-EGFP infected E7.5 embryos (magnification 10×). Control (male) gonad and forelimb tissues are negative for EGFP and p27. For both male and female embryos infected with the *SF1p* vector, EGFP expression is higher in gonad tissues compared to the forelimb.

normally express this protein [23]. To analyse the effect of DMRT1 over-expression in female gonads, the expression of the male gene SOX9 and the female gene aromatase were analysed by immunostaining (Figure 6). Both the control and RCANBP-DMRT1 infected males showed robust SOX9 expression, while in the control female and in the RCANBP-DMRT1 infected female no SOX9 was detected. Strong aromatase expression in the control and the RCANBP-DMRT1 infected females was also evident, however, the RCANBP-DMRT1 infected male had no ectopic expression of this protein. These data indicated that despite increased levels of DMRT1 expression in female gonads, this was not sufficient to masculinize female gonads.

SF1p-mediated over-expression of aromatase

To test the ability of *SF1p* to over-express a gene involved in ovarian development in male gonads, the *aromatase* open reading frame was inserted downstream of *SF1p* in RCANBP (RCANBP-SF1p-Arom). Aromatase is normally expressed in a female-specific manner. We have previously reported that its global over-expression causes male-to-female gonadal sex reversal, which included the down-regulation of key testis genes and up-regulation of ovarian development genes [23]. The gonads of RCANBP-SF1p-Arom infected E7.5 embryos were analysed by immuno-

staining. The male left gonad had elevated aromatase expression compared to the control male, however this was much lower than the control female (Figure 7A). The control female had a characteristic thickened outer cortex and the male had defined cord structures and lacked a cortex region. Like the control male, the RCANBP-SF1p-Arom infected male also had cord structures and lacked a thickened outer cortex. The levels of the key male protein AMH were then analysed in RCANBP-Arom infected males (Figure 7B). Typical AMH expression was evident in the control male gonads, with strong staining throughout the cords. In contrast, in the RCANBP-Arom infected male AMH was reduced and its expression pattern was disrupted from its normal pattern. These data show that *SF1p* could be used to over-express a key female sex development factor in male embryos and was able to disrupt normal testis development.

Discussion

In an effort to advance the methods available for studying sex development in avian species, we have characterized and compared several promoters for their activity in chicken embryonic gonads. Unlike other model systems such as the mouse, the production of transgenic animals for over-expression and knock-down studies is not yet routine for avian species. Therefore, the use of various techniques for the introduction of DNA vectors into live embryos is the most practical approach for mis-expression of genes.

Several gonad-specific expression systems have been described for the mouse, however these usually make use of large genomic fragments of genes involved in sex development to act as promoters for tissue specific expression. A major limitation on the use of the RCASBP and RCANBP for the delivery of recombinant sequences is a restriction on insert size of about 2.5 kb. Longer sequences may produce non-replication-competent viruses and thus lower titers [24–26]. The size of promoter sequences tested in this study was therefore limited to less than 1 kb and only the minimum predicted region required for activity was preferably used (which would then allow pairing an ORF of at least 1.5 kb in size). For *SF1p*, since the chicken minimal promoter was already validated *in vitro*, a fragment size of about 400 nt was used as this sequence was found to show similar activity to those tested up to about 1 kb [16]. Based on the analysis of the human *WT1* promoter [21], a fragment of about 600 nt was selected as this size provided the best expression. For *AMHp* a shorter sequence of about 300 nt was used, as it contained predicted binding elements that might confer activity.

Using these gonad sequences to transcribe EGFP *in ovo* from RCANBP provided the first indication of promoter activity. Although the expression of reporter genes from internal promoters in RCANBP has been previously reported *in vitro* [26] and in chicken embryos [14], activity in the gonads has not been described. Since strong EGFP expression from the viral LTR promoter can be seen in developing gonads from RCASBP [12], it was clear that this virus can effectively target this tissue. In the current study, infection with RCANBP containing the *SV40* promoter, embryos also showed strong EGFP expression through-out the entire embryo. Importantly, this included the urogenital system. This experiment provided a positive control and showed that expression from RCANBP in chicken gonads using an internal promoter was achievable.

It was anticipated that reporter expression driven from the various gonadal promoters should at least in part reflect some of the endogenous expression patterns of those genes. SF1 is expressed endogenously in the gonads of both sexes prior to and

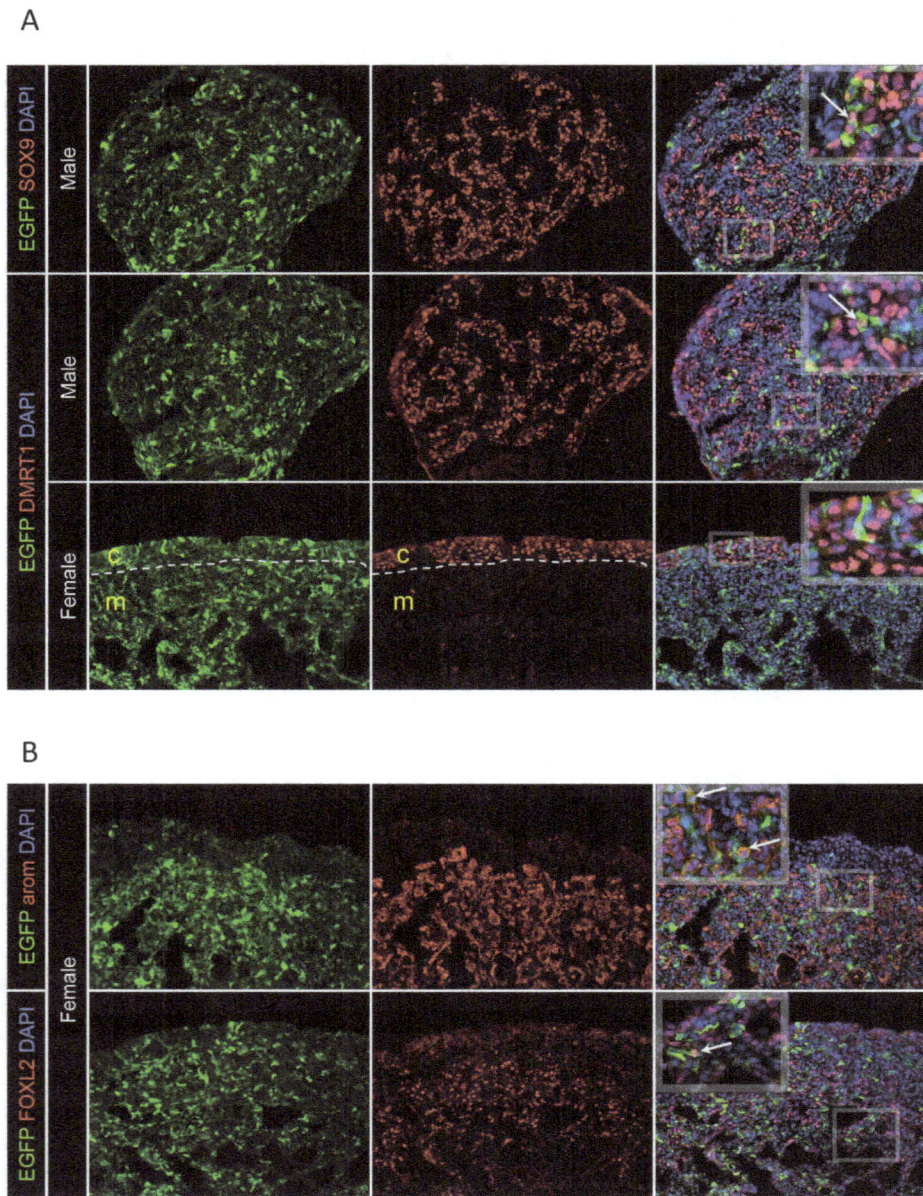

Figure 4. Cellular location of *SF1* promoter expressed EGFP compared to key gonad factors. Immunostaining for EGFP (green) and key testis and ovarian developmental proteins (red) in RCANBP-SF1p-EGFP infected E7.5 embryos (magnification 20×). A: Male genes: In male gonads, co-staining of EGFP with DMRT1 or SOX9 showed that both proteins were expressed in some cells simultaneously (white arrows). However, in female gonads DMRT1 expression did not overlap with EGFP. B: Female genes: In female gonads, co-staining of EGFP with aromatase or FOXL2 showed that both proteins were expressed in some cells simultaneously (white arrows).

during gonadal sex differentiation, but becomes female up-regulated as development proceeds [27]. Similarly, WT1 is expressed in the gonads of both sexes, but also in developing kidneys [27,28], while AMH is only expressed in male embryonic gonads [29], and aromatase is entirely female-specific in embryonic gonads [30,31]. However, the relevant core promoter fragments of these genes did not faithfully reflect the endogenous expression of these genes. EGFP reporter expression in E7.5 embryos varied from very low activity of the *AMHp* sequence to embryo-wide strong expression from the *AROMp* sequence. Endogenous aromatase is expressed female-specifically, so it was also interesting to note that the *AROMp* sequence provided strong embryo-wide EGFP expression in both sexes.

Taken together, these data therefore suggest that transcription of the native transcripts of these promoters requires additional up and/or downstream sequences or other structural features to achieve their normal patterns of expression. Taking the core promoter fragments out of context for testing in RCANBP clearly showed that most of these sequences could drive expression in non-gonadal sites, and thereby suggests that repressors or insulators were absent from the promoter regions used. Conversely, low-level reporter expression from promoters such as *AMHp* indicated that additional sequences are required for robust gonad-restricted expression. A relevant example of this requirement is demonstrated by a BAC transgene containing the mouse *SF1* gene 5'-flanking sequences. A 47 kb region can direct EGFP expression

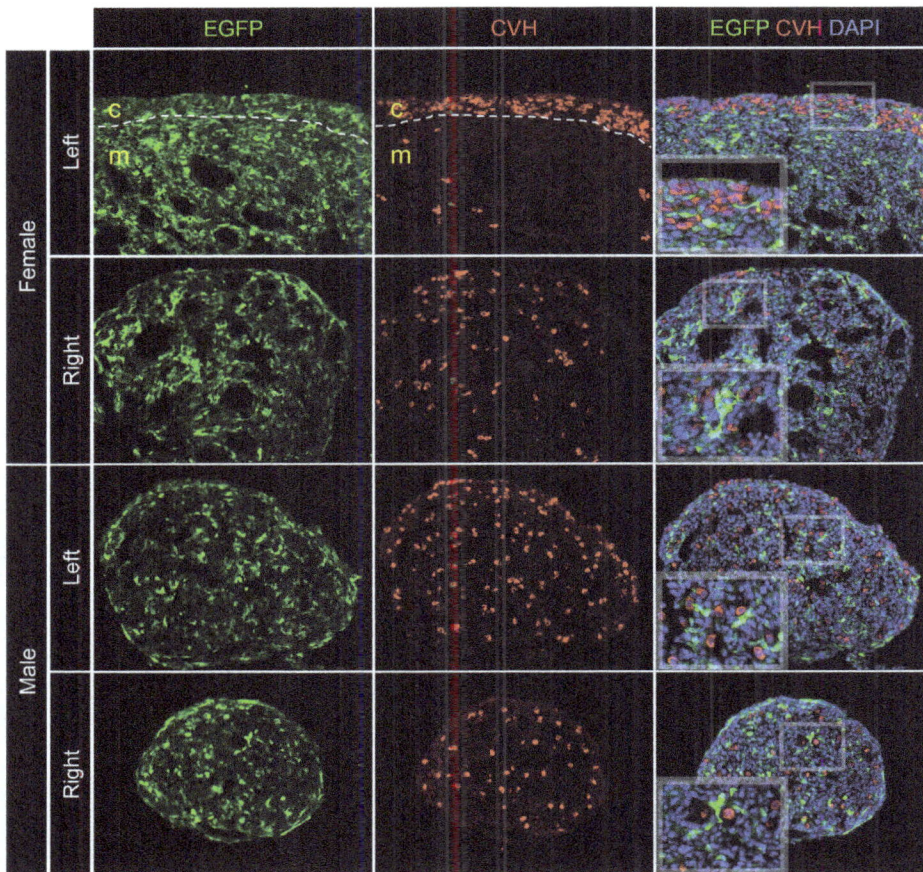

Figure 5. Cellular location of *SF1* promoter expressed EGFP compared to germ cells. Immunostaining for EGFP (green) and CVH (red) in RCANBP-SF1p-EGFP infected E7.5 embryos (magnification 20×). The expression of EGFP from *SF1p* did not overlap with any cells that were positive for the germ cell marker CVH.

to the gonads, adrenal cortex, spleen and ventromedial hypothalamic nucleus in mice [32], but this fragment lacks important sequences that are present in a longer 111 kb version of the same region that can direct additional expression in the hypothalamus and pituitary [33].

The initial analysis suggested that *SF1p* showed the best potential for future application as a tissue-restricted promoter sequence. In particular, after the liver and mesonephros, the gonad showed the highest EGFP expression levels when using this promoter in RCANBP. There was however, low-level expression elsewhere in the embryo, particularly in the fore and hindlimbs (Figure 2B). Further analysis of *SF1p* expressed EGFP in gonads by immunostaining showed that compared to the forelimb (used to represent a non-gonadal tissue showing that showed some expression), the expression was higher in the gonads of both sexes compared to forelimbs. These data suggested that *SF1p* was a potentially an appropriate candidate for tissue restricted expression of sequences of interest.

Immunostaining gonads for EGFP expression along with various other gonad development genes provided a more in depth analysis of *SF1p* activity. Since EGFP was expressed in some of the same cells that were also expressing DMRT1, SOX9, FOXL2 and aromatase, *SF1p* could potentially be used to transcribe sequences in cell lineages that express key sex-determining genes. Staining for the germ cell marker CVH showed that EGFP was not expressed in the germ cells of neither male nor female gonads. This finding is

consistent with previous observations that although RCAS-based viruses can infect germ cells, their transcriptional activity appears to be silenced, at least in embryos [12]. Therefore it does not necessarily reflect on the ability of *SF1p* to deliver expression in germ cells, as a non-RCAS vector might produce different results. It was also evident that EGFP expression from this promoter showed variegated expression in gonadal tissues (Figures 3, 4 and 5). The effect of viral integration site on transgene expression may at least in part account for this, especially considering that Rous Sarcoma Virus shows numerous insertion sites when infected either as a virus particle [34] or by DNA transfection [35], showing no apparent preference for specific integration sites.

A critical test for the potential use of *SF1p* for studies in sex development was to over-express genes involved in gonadal sex differentiation. To this end, the key testis development gene DMRT1 and the female-specific gene aromatase were tested. Previously, global DMRT1 knockdown in male embryos resulted in gonad feminization and ovarian development, which included the up-regulation of aromatase and down-regulation of SOX9 [13]. We have previously used RCASBP to globally over-express DMRT1 in chicken embryos, however infection with this virus induces early-stage embryo-lethality [13]. This was not surprising considering that it encodes a transcription factor regulating cell fate decisions [36,37]. Recently, we reported that site-specific electroporation of RCASBP encoding DMRT1 into female gonads was able to avoid embryonic toxicity and activate testis

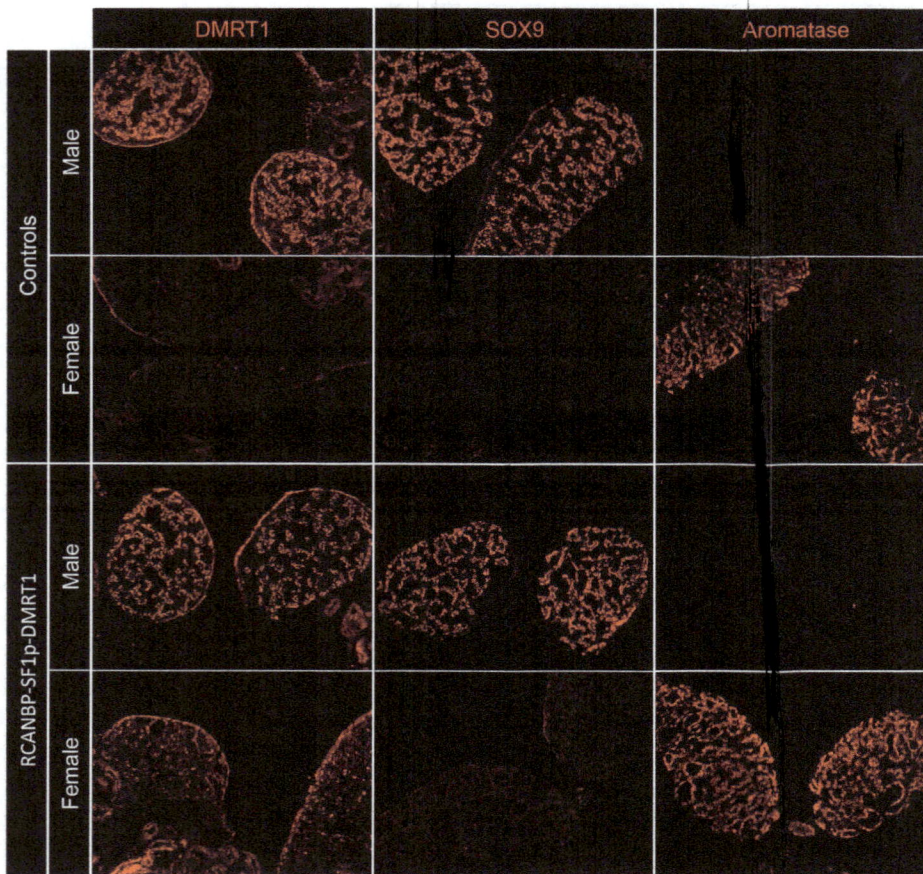

Figure 6. Over-expression of DMRT1 from *SF1p* in embryonic gonads. Immunostaining for key testis and ovarian developmental genes (red) in RCANBP-SF1p-EGFP infected E7.5 embryos (magnification 10×). Low-level over-expression of DMRT1 was evident in the gonads of RCANBP-SF1p-DMRT1 infected embryos compared to the control female. This over-expression did not cause any change in the expression of SOX9 in the female or aromatase in the male.

pathway genes [9], however this method required a great deal of optimization, technical expertise and viral infection was limited to the left gonad. In the study reported here, embryos infected with RCANBP-DMRT1 showed normal survival rates, suggesting that embryo-wide expression was avoided. Given that DMRT1 over-expression was detected in female gonads (Figure 6), any non-gonadal expression was not at sufficient levels to induce embryo lethality at the time points analysed. Despite the elevated DMRT1 expression in female gonads, there did not appear to be any effect on the other sex development genes analysed. This was likely due to insufficient levels of DMRT1 being over-expressed (i.e. to match the level of a normal male), given that robust levels of DMRT1 over-expression in female gonads was required to induce male pathways genes, disrupt cortex formation and reduce aromatase expression [9]. It is therefore likely that a promoter stronger than *SF1p* would be required to match the level of expression required to effect gonad development. This notion is consistent with the hypothesis that a sufficient level of DMRT1 expression is required to initiate the testis developmental pathway [38].

Over-expression of aromatase in genetic male embryos from *SF1p* provided an example of the application of this sequence to over-express a female gene that is usually absent in male gonads. Aromatase over-expression in male embryos has been shown to override testis development and induce the ovarian program of

gonad differentiation [23]. In the current study, it was found that despite only very modest aromatase over-expression in male embryos from *SF1p*, this was enough to disrupt the expression key male marker AMH. Although this phenotype was mild compared to that seen using the stronger viral LTR promoter of RCASBP [23], it did show that this promoter sequence can be used to over-express a gene to induce a gonad phenotype. The lack of effect of DMRT1 over-expression and positive effect of aromatase expression might be at least in part due to differences in structure and functions of these proteins. DMRT1 is a transcription factor that is expressed in the gonads of both sexes, and its level of expression most likely determines its activity in each of these tissues. In contrast, aromatase is an entirely female specific enzyme that can have a graded or transient effect when it is expressed at varying levels in male gonads, resulting in the formation ovotestis [23].

In summary, a number of potential gonadal promoters were tested for their ability to drive gene expression specifically in the embryonic urogenital system. *SF1p* offered a tissue-restricted expression pattern with high-level EGFP expression in the liver and mesonephros, moderate levels in the gonad, and lower levels elsewhere in the embryo. This sequence was capable of expressing EGFP in some of the same cells as key male and female gonad genes and was used to over-express the key female gene aromatase to induce a phenotype. It was also evident that this promoter would only be suitable for applications where a low or moderate

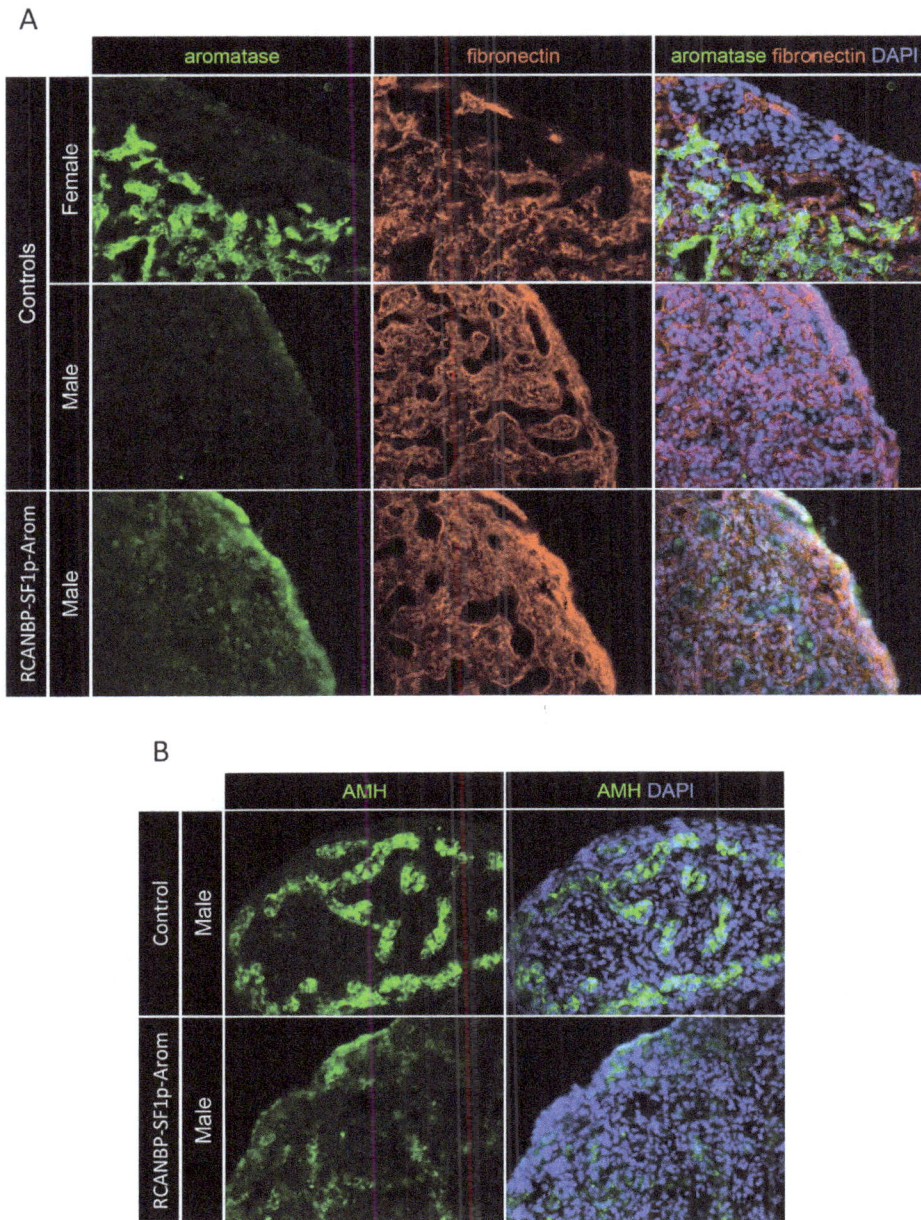

Figure 7. Over-expression of aromatase from *SF1p* in embryonic gonads. A: Immunostaining for aromatase (green), fibronectin (red) and DAPI (blue) in RCANBP-SF1p-EGFP infected E7.5 embryos (magnification 40×). Low-level aromatase over-expression was evident in the male left gonad compared to the control male. B: Immunostaining for key male protein AMH (green) and DAPI (blue) in RCANBP-SF1p-EGFP infected E7.5 embryos (magnification 40×). The over-expression of aromatase in the male gonad caused a reduction and disruption of AMH expression.

level of over-expression is required. It will be interesting to see if SF1p can be employed for the expression of high efficacy short hairpin RNAs for RNA interference, especially since the use of weaker promoters *in vivo* are able to avoid cytotoxic effects seen with some stronger promoters [39]. In addition, there is the potential for linking *SF1p* with enhancer sequences, particularly with regions that might show gonad-specificity such as TESCO [40] and WT1 [21]. This study provides a novel method of achieving tissue restricted expression in embryonic chicken gonads and will be of use in the field of sex determination and gonadal development.

Materials and Methods

Ethics statement

All experiments were carried out with respect for the principles of laboratory animal care and were consistent with the *Australian Code of Practice for the Care and Use of Animals for Scientific Purposes, 7TH Edition 2004* and the *Prevention of Cruelty to Animals Act, Victoria 1986.* This included official approval from the Murdoch Childrens Research Institute Animal Ethics Committee (AEC # A627).

Vector construction and virus preparation

Prior to insertion into RCANBP vectors, all promoters were first cloned upstream of EGFP in the transfer vector pSLAX. For the *SV40* promoter, the EGFP ORF was cloned from pEGFP-N1 (Clonetech) on *Nco*I and *Xba*I and ligated into pGL3-Promoter (Promega) digested with the same enzymes. This removed the luciferase ORF from pGL3-Promoter and relaced it with EGFP under the control of the SV40 promoter (*SV40p*). This vector was then digested with *Bgl*II and *Bam*HI and ligated into pSLAX digested with the same enzymes. For the other promoter sequences, a pSLAX vector encoding EGFP was first generated. The EGFP ORF was PCR amplified from pEGFP-N1 using a forward primer with an introduced *Hind*III site and a reverse primer with *Bam*HI, and then cloned into pSLAX on the same sites. All promoters were then amplified using forward primers with introduced *Eco*RI sites or *Nco*I sites and cloned into pSLAX with EGFP using the same enzymes. Primer sequences for *SF1p* were; forward 5′-ACCTCCCCGCAGTTCCTCTCTCCTGG-3′ and reverse 5′-AAGTACTCACCTCGATGCGGC-3′; *WT1* promoter: forward 5′-CATGCCATGGGAATTCATCTGCTC-GCAGACCTTCTG-3′ and reverse 5′-CATGCCATGGGAA-TTCGCTAGCGAGCTCAGATCTGGGAGATATATTATAC-CACAGTAACCTC-3′; *AMH* promoter: forward 5′-TCAC-CATGGTCTAGATTCCACCCTCCTCTCCAA-3′ and reverse 5′-TCACCATGGTGTTCTGCTGCACCCACAG-3′, *aromatase* promoter: forward 5′-ATATAGAATTCGCGAGACCAAATC-ACAAGATAAA-3′ and reverse 5′- ATATAGAATTCGCAGG-CTGGTGAAGTAGTTCAGTG-3′; mouse *SF1* promoter: forward 5′-TCACCATGGTCTAGACACACCCTTAGCCCAGC-AGTC-3′ and reverse 5′-TCACCATGGTCCCAGGCCTCAG-GTAGGGCA-3′. The promoter/EGFP cassettes were then excised from pSLAX using *Cla*I and ligated into RCANBP/A proviral DNA that had also been digested with *Cla*I. All constructs were verified by DNA sequencing. The viral DNA was then transfected into chicken fibroblastic DF1 cells using Lipofectamine 2000 (Invitrogen) and propagated for approximately 2 weeks. Recombinant virus was harvested from culture medium, concentrated by ultracentrifugation and titered as previously described [12].

Embryo manipulation

High titer virus (approximately 10^8 Infectious Units/mL) was injected into day 0 blastoderms, and eggs were sealed with parafilm and incubated at 37.5°C. Embryos were harvested on embryonic day 7.5 (HH32), as at this time point the first physical differences in male and female gonad development can be observed. These experiments involved at least 50 embryos for each experiment and at least 5 embryos per sex for each experimental condition were analysed. Immunostaining tissues with p27 antibody, which detects a viral epitope, confirmed RCANBP infection.

Immunofluorescence

Tissues were fixed for 15 minutes in 4% PFA/PBS at room temperature, prior to processing for tissue section immunofluorescence, as described previously [41]. At least 5 embryos per time point and/or treatment were examined. Briefly, 10 μm sections were cut on a cryostat, permeablised in PBS 1% Triton X-100 and blocked in PBS 2% BSA for 1 hour. Primary antibodies were either raised in-house (rabbit anti-chicken aromatase (1:5000), rabbit anti-chicken DMRT1 (1:5000), rabbit anti-chicken vasa homologue (CVH) (1:6000), rabbit anti-chicken FOXL2 (1:6000), or were obtained commercially (rabbit anti-p27 (1:1000); Charles River Services, goat anti-AMH (1:1000); Millipore, rabbit anti-mouse SOX9 (1:6000); Santa Cruz, mouse anti-fibronectin (1:500); Serotec, goat anti-GFP (1:500). Alexa-fluor secondary antibodies were used (donkey or goat anti-rabbit, mouse or goat 488 or 594; Molecular Probes). Sections were counterstained with DAPI.

PCR sexing

Infected and control embryos were dissected at indicated time points. For genetic sexing of embryos, a small piece of limb tissue was digested in PCR compatible Proteinase K buffer and the genomic DNA was used for rapid PCR sexing [42]. By this method, only females show a W-linked (female-specific) *Xho*I band. Amplification of 18S rRNA in both sexes served as an internal control.

Acknowledgments

The authors thank collaborators Tim Doran, Jim McKay, Scott Tyack, Ton Schat, Kelly Roeszler, Yu Cao and Katie Ayers for their input and guidance throughout this study. This research was partly conducted within the Poultry CRC, established and supported under the Australian Government's Cooperative Research Centres Program, and was supported by the Victorian Government's Operational Infrastructure Support Program.

Author Contributions

Conceived and designed the experiments: LSL TO DMC AHS CAS. Performed the experiments: LSL TO DMC CAS. Analyzed the data: LSL TO CAS. Contributed reagents/materials/analysis tools: LSL TO CAS. Wrote the paper: LSL CAS.

References

1. Davey MG, Tickle C (2007) The chicken as a model for embryonic development. Cytogenet Genome Res 117: 231–239. doi:10.1159/000103184

2. Farley EK, Gale E, Chambers D, Li M (2011) Effects of in ovo electroporation on endogenous gene expression: genome-wide analysis. Neural Dev 6: 17. doi:10.1186/1749-8104-6-17

3. Scaal M, Gros J, Lesbros C, Marcelle C (2004) In ovo electroporation of avian somites. Dev Dyn 229: 643–650. doi:10.1002/dvdy.10433

4. Islam MM, Doh ST, Cai L (2012) In ovo electroporation in embryonic chick retina. J Vis Exp. doi:10.3791/3792

5. Kolle G, Jansen A, Yamada T, Little M (2003) In ovo electroporation of Crim1 in the developing chick spinal cord. Dev Dyn 226: 107–111. doi:10.1002/dvdy.10204

6. Wilson NH, Stoeckli ET (2011) Cell type specific, traceable gene silencing for functional gene analysis during vertebrate neural development. Nucleic Acids Research 39: e133. doi:10.1093/nar/gkr628

7. Swartz M, Eberhart J, Mastick GS, Krull CE (2001) Sparking new frontiers: using in vivo electroporation for genetic manipulations. Dev Biol 233: 13–21. doi:10.1006/dbio.2001.0181

8. Guioli S, Lovell-Badge R (2007) PITX2 controls asymmetric gonadal development in both sexes of the chick and can rescue the degeneration of the right ovary. Development 134: 4199–4208. doi:10.1242/dev.010249

9. Lambeth L, Raymond CS, Roeszler KN, Kuroiwa A, Nakata T, et al. (2014) Over-expression of DMRT1 induces the male pathway in embryonic chicken gonads. Dev Biol. doi:10.1016/j.ydbio.2014.02.012

10. Tyack SG, Jenkins KA, O'Neil TE, Wise TG, Morris KR, et al. (2013) A new method for producing transgenic birds via direct in vivo transfection of primordial germ cells. Transgenic Res 22: 1257–1264. doi:10.1007/s11248-013-9727-2

11. Hughes SH, Greenhouse JJ, Petropoulos CJ, Sutrave P (1987) Adaptor plasmids simplify the insertion of foreign DNA into helper-independent retroviral vectors. Journal of Virology 61: 3004–3012.

12. Smith CA, Roeszler KN, Sinclair AH (2009) Robust and ubiquitous GFP expression in a single generation of chicken embryos using the avian retroviral vector, RCASBP. Differentiation 77: 473–482. doi:10.1016/j.diff.2009.02.001

13. Smith CA, Roeszler KN, Ohnesorg T, Cummins DM, Farlie PG, et al. (2009) The avian Z-linked gene DMRT1 is required for male sex determination in the chicken. Nature 461: 267–271. doi:10.1038/nature08298

14. Sato N, Matsuda K, Sakuma C, Foster DN, Oppenheim RW, et al. (2002) Regulated gene expression in the chicken embryo by using replication-competent retroviral vectors. Journal of Virology 76: 1980–1985.

15. Chue J, Smith CA (2011) Sex determination and sexual differentiation in the avian model. FEBS Journal 278: 1027–1034. doi 10.1111/j.1742-4658.2011.08032.x

16. Kudo T, Sutou S (1999) Structural characterization of the chicken SF-1/Ad4BP gene. Gene 231: 33–40.

17. Matsumine H, Herbst MA, Ou SH, Wilson JD, McPhaul MJ (1991) Aromatase mRNA in the extragonadal tissues of chickens with the henny-feathering trait is derived from a distinctive promoter structure that contains a segment of a retroviral long terminal repeat. Functional organization of the Sebright, Leghorn, and Campine aromatase genes. J Biol Chem 266: 19900–19907.

18. Horard B, Vanacker J-M (2003) Estrogen receptor-related receptors: orphan receptors desperately seeking a ligand. J Mol Endocrinol 31: 349–357.

19. Busygina TV, Ignatieva EV, Osadchuk AV (2003) Consensus sequence of transcription factor SF-1 binding site and putative binding site in the 5′ flanking regions of genes encoding mouse steroidogenic enzymes 3betaHSDI and Cyp17. Biochemistry Mosc 68: 377–384.

20. Oreal E, Pieau C, Mattei MG, Josso N, Picard JY, et al. (1998) Early expression of AMH in chicken embryonic gonads precedes testicular SOX9 expression. Dev Dyn 212: 522–532. doi:10.1002/(SICI)1097-0177(199808)212:4<522::AID-AJA5>3.0.CO;2-J.

21. Fraizer GC, Wu YJ, Hewitt SM, Maity T, Ton CC, et al. (1994) Transcriptional regulation of the human Wilms' tumor gene (WT1). Cell type-specific enhancer and promiscuous promoter. J Biol Chem 269: 8892–8900.

22. Wilhelm D, Englert C (2002) The Wilms tumor suppressor WT1 regulates early gonad development by activation of Sf1. Genes Dev 16: 1839–1851. doi:10.1101/gad.220102

23. Lambeth LS, Cummins DM, Doran TJ, Sinclair AH (2013) Overexpression of aromatase alone is sufficient for ovarian development in genetically male chicken embryos. PLoS ONE. doi:10.1371/journal.pone.0068362

24. Boerkoel CF, Federspiel MJ, Salter DW, Payne W, Crittenden LB, et al. (1993) A new defective retroviral vector system based on the Bryan strain of Rous sarcoma virus. Virology 195: 669–679. doi:10.1006/viro.1993.1418

25. Fisher GH, Orsulic S, Holland E, Hively WP, Li Y, et al. (1999) Development of a flexible and specific gene delivery system for production of murine tumor models. Oncogene 18: 5253–5260. doi:10.1038/sj.onc.1203087

26. Petropoulos CJ, Hughes SH (1991) Replication-competent retrovirus vectors for the transfer and expression of gene cassettes in avian cells. Journal of Virology 65: 3728–3737.

27. Smith CA, Smith MJ, Sinclair AH (1999) Gene expression during gonadogenesis in the chicken embryo. Gene 234: 395–402. doi:10.1016/S0378-1119(99)00179-1.

28. Carmona R, Gonzalez-Iriarte M, Perez-Pomares JM, Munoz-Chapuli R (2001) Localization of the Wilm's tumour protein WT1 in avian embryos. Cell and Tissue Research 303: 173–186.

29. Nishikimi H, Kansaku N, Saito N, Usami M, Ohno Y, et al. (2000) Sex differentiation and mRNA expression of P450c17, P450arom and AMH in gonads of the chicken. Mol Reprod Dev 55: 20–30. doi:10.1002/(SICI)1098-2795(200001)55:1<20::AID-MRD4>3.0.CO;2-E.

30. Smith CA, Andrews JE, Sinclair AH (1997) Gonadal sex differentiation in chicken embryos: expression of estrogen receptor and aromatase genes. J Steroid Biochem Mol Biol 60: 295–302.

31. Andrews JE, Smith CA, Sinclair AH (1997) Sites of estrogen receptor and aromatase expression in the chicken embryo. Gen Comp Endocrinol 108: 182–190. doi:10.1006/gcen.1997.6978.

32. Stallings NR, Hanley NA, Majdic G, Zhao L, Bakke M, et al. (2002) Development of a transgenic green fluorescent protein lineage marker for steroidogenic factor 1. Molecular Endocrinology 16: 2360–2370.

33. Shima Y, Zubair M, Ishihara S, Shinohara Y, Oka S, et al. (2005) Ventromedial hypothalamic nucleus-specific enhancer of Ad4BP/SF-1 gene. Molecular Endocrinology 19: 2812–2823. doi:10.1210/me.2004-0431

34. Lerner TL, Skalka AM, Hanafusa H (1981) Integration of Rous sarcoma virus DNA into chicken embryo fibroblasts: no preferred proviral acceptor site in the DNA of clones of singly infected transformed chicken cells. Journal of Virology 40: 421–430.

35. Copeland NG, Jenkins NA, Cooper GM (1981) Integration of Rous sarcoma virus DNA during transfection. Cell 23: 51–60.

36. Krentz AD, Murphy MW, Kim S, Cook MS, Capel B, et al. (2009) The DM domain protein DMRT1 is a dose-sensitive regulator of fetal germ cell proliferation and pluripotency. Proc Natl Acad Sci USA 106: 22323–22328. doi:10.1073/pnas.0905431106

37. Matson CK, Murphy MW, Griswold MD, Yoshida S, Bardwell VJ, et al. (2010) The mammalian doublesex homolog DMRT1 is a transcriptional gatekeeper that controls the mitosis versus meiosis decision in male germ cells. Dev Cell 19: 612–624. doi:10.1016/j.devcel.2010.09.010.

38. Smith CA, Sinclair AH (2004) Sex determination: insights from the chicken. Bioessays 26: 120–132. doi:10.1002/bies.10400

39. Giering JC, Grimm D, Storm TA, Kay MA (2008) Expression of shRNA from a tissue-specific pol II promoter is an effective and safe RNAi therapeutic. Mol Ther 16: 1630–1636. doi:10.1038/mt.2008.144

40. Sekido R, Lovell-Badge R (2008) Sex determination involves synergistic action of SRY and SF1 on a specific Sox9 enhancer. Nature 453: 930–934. doi:10.1038/nature06944

41. Smith CA, Katz M, Sinclair AH (2003) DMRT1 is upregulated in the gonads during female-to-male sex reversal in ZW chicken embryos. Biol Reprod 68: 560–570.

42. Clinton M, Haines L, Belloir B, McBride D (2001) Sexing chick embryos: a rapid and simple protocol. Br Poult Sci 42: 134–138. doi:10.1080/713655025

5

Two Polymorphisms Facilitate Differences in Plasticity between Two Chicken Major Histocompatibility Complex Class I Proteins

Alistair Bailey[1,2], **Andy van Hateren**[1,2], **Tim Elliott**[1,2]*, **Jörn M. Werner**[1,3]*

1 Institute for Life Sciences, University of Southampton, Southampton, United Kingdom, 2 Cancer Sciences Unit, Faculty of Medicine, University of Southampton, Southampton, United Kingdom, 3 Centre for Biological Sciences, Faculty of Natural & Environmental Sciences, University of Southampton, Southampton, United Kingdom

Abstract

Major histocompatibility complex class I molecules (MHC I) present peptides to cytotoxic T-cells at the surface of almost all nucleated cells. The function of MHC I molecules is to select high affinity peptides from a large intracellular pool and they are assisted in this process by co-factor molecules, notably tapasin. In contrast to mammals, MHC homozygous chickens express a single MHC I gene locus, termed BF2, which is hypothesised to have co-evolved with the highly polymorphic tapasin within stable haplotypes. The BF2 molecules of the B15 and B19 haplotypes have recently been shown to differ in their interactions with tapasin and in their peptide selection properties. This study investigated whether these observations might be explained by differences in the protein plasticity that is encoded into the MHC I structure by primary sequence polymorphisms. Furthermore, we aimed to demonstrate the utility of a complimentary modelling approach to the understanding of complex experimental data. Combining mechanistic molecular dynamics simulations and the primary sequence based technique of statistical coupling analysis, we show how two of the eight polymorphisms between BF2*15:01 and BF2*19:01 facilitate differences in plasticity. We show that BF2*15:01 is intrinsically more plastic than BF2*19:01, exploring more conformations in the absence of peptide. We identify a protein sector of contiguous residues connecting the membrane bound α_3 domain and the heavy chain peptide binding site. This sector contains two of the eight polymorphic residues. One is residue 22 in the peptide binding domain and the other 220 is in the α_3 domain, a putative tapasin binding site. These observations are in correspondence with the experimentally observed functional differences of these molecules and suggest a mechanism for how modulation of MHC I plasticity by tapasin catalyses peptide selection allosterically.

Editor: Junwen Wang, The University of Hong Kong, Hong Kong

Funding: This work was supported by a MRC/Microsoft Research UK studentship and centenary award and CRUK (Programme number CRUK/A10601). The funders had no role in study design, data collection and analysis, decision to publish, or preparation of the manuscript.

Competing Interests: The authors have declared that no competing interests exist.

* E-mail: T.J.Elliott@soton.ac.uk (TE); J.M.Werner@soton.ac.uk (JMW)

Introduction

Major histocompatibility complex class I molecules (MHC I) select peptides for presentation to CD8+ cytotoxic T-cells at the surface of almost all nucleated cells. This MHC I antigen processing and presentation system is a key mechanism in the surveillance and recognition by the immune system of diseased, infected or cancerous cells. Yet understanding how the peptide selection process determines the intensity and specificity of the cytotoxic T-cell response to pathogens remains one of the most important unsolved problems in immunology [1]. Peptides are primarily, but not always, derived from degraded proteins and defective ribosomal products inside the cell and are loaded onto MHC I molecules within the endoplasmic reticulum. As part of this peptide loading complex [2–5], MHC I associates with several proteins, most notably the co-factor molecule tapasin, the molecule that most helps MHC I select high affinity peptides [6–9]. It is via tapasin that MHC I co-locates with the transporter associated with antigen presentation (TAP) [10] that supplies peptides from the cytosol. MHC class I molecules have a common tertiary structure (Figure 1) consisting of a heavy chain formed of

α_1–α_2 peptide binding domain and the membrane bound α_3 domain with a non-covalently bound monomorphic β_2-microglobulin light chain (β_2m). Peptides usually of 8–10 amino acids in length bind into the groove formed between the α_1 and α_2 helices.

In humans the major histocompatibility complex is a large genomic region spanning approximately 3.5 mega base pairs of DNA nucleotides [11]. It contains genes encoding three classical MHC I alleles that are co-dominantly expressed and are highly polymorphic. The exact reasons for MHC I gene diversity is still unknown, but these genes appear to be at least in part subject to negative frequency dependent, balancing selection processes [12]. That is to say that there is a drive to maintain multiple MHC I alleles, specifically rare alleles, which survive perhaps due to their fitness advantage in presenting pathogen derived peptides. In the human MHC region, the genes for tapasin and TAP are distant from the MHC I genes, have few alleles and exhibit little sequence diversity and have no known functional distinctions. Thus, although in humans TAP favours peptides with hydrophobic C-termini, it has a broad transport specificity [13] and the majority of the specificity for selection of peptide from the available pool is encoded into the MHC I molecule. Likewise tapasin enhances the

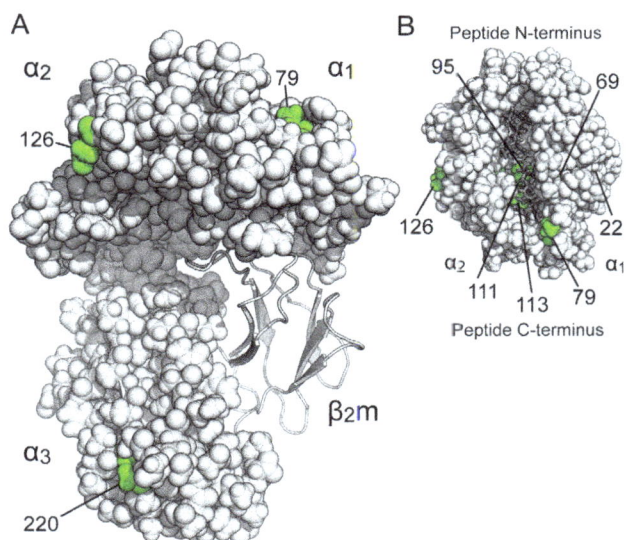

Figure 1. The structure and polymorphisms of chicken MHC Class I alleles BF2*15:01 and BF2*19:01. A) The structure of the lumenal domain of a chicken MHC Class I molecule. A space filling representation of the heavy chain is shown, formed of α_1– α_2 peptide binding domain and the membrane proximal α_3 domain, creating a complex with a non-covalently bound β_2m light chain shown as a ribbon representation. B) The peptide is shown as a stick representation in grey, non-covalently bound into the groove formed between the α_1 and α_2 helices. The sites of the polymorphic residues between BF2*15:01 and BF2*19:01 indicated in green, with the location of residue 22 indicated in the peptide binding domain below the α_1 helix.

peptide selection function for all MHC I alleles. The recently characterised TAPBPR molecule may also play a role in the peptide selection process [14].

In contrast to most mammals, chickens have a compact major histocompatibility complex spanning only about 92 kilo base pairs [15]. This contains a single dominantly expressed MHC I gene closely located with the tapasin and TAP genes that are rarely disrupted by recombination events [15,16]. In chickens the proximity of MHC I, tapasin and TAP genes and the absence of recombination are hypothesised to have led to a diverse set of co-evolving haplotypes with a high degree of allelic polymorphism and sequence diversity of MHC I, tapasin and TAP genes [17,18]. For certain haplotypes, the peptide specificity of TAP appears to complement the peptide binding motif of the MHC I molecule [19], and tapasin provides complementary enhanced peptide selection functionality, supporting the co-evolution hypothesis.

The chicken haplotypes B19 and B15 express MHC I proteins BF2*19:01 and BF2*15:01 respectively which share a similar peptide binding specificity [20,21], but differ by only eight amino acids in their primary sequences [18]. As shown in Figure 1, seven of these eight polymorphic residues are in the α_1– α_2 peptide binding domain, with the eighth polymorphic residue in the membrane bound α_3 domain. The peptide binding domain residues 79 and 126 and α_3 domain residue 220 are located on the protein surface, whilst the other polymorphic positions are buried and not immediately accessible. Residues 126 and 220 are located on the surface that is the putative tapasin facing side of MHC I in mammals [22,23].

We have recently described differences in the abilities of BF2*15:01 and BF2*19:01 to select peptide in the presence and absence of both their complementary and mismatched tapasin

[24]. This work showed that there are intrinsic differences in the abilities of these MHC I molecules to select high affinity peptides and that the complementary tapasin allele best enhances their selection capabilities *in vivo*. Notably, BF2*15:01 was less dependent on tapasin for exchange of low affinity peptides for high affinity peptides than BF2*19:01. Furthermore, this work identified position α_3 domain residue 220 as relevant to tapasin function and the intrinsic peptide selection properties of these molecules.

Our recent work examining two human HLA–B*44 alleles, that differ by a single amino acid, concluded that it was differences in protein plasticity, the intrinsic ability of the molecule to change shape, that determined the relative dependence on tapasin of these molecules for high affinity peptide selection [unpublished data]. We therefore hypothesised that differences in protein plasticity may also explain the functional differences observed for BF2*15:01 and BF2*19:01. We sought to characterise how the polymorphisms between these molecules could alter the plasticity of the MHC I structure in order to rationalise the observed functional differences at the structural level.

We have combined two modelling approaches: mechanistic molecular dynamics simulations [25–28] of BF2*15:01 and BF2*19:01 and the sequence analysis method of statistical coupling analysis [29,30]. The aim of this first approach was to use molecular dynamics as a computational microscope [31] to examine whether differences in plasticity arise from the polymorphisms between BF2*15:01 and BF2*19:01 as described by their protein dynamics. These dynamics were quantified in terms of sites of local flexibility identified in each MHC I structure, the global motions of each molecule and their relative abilities to explore the conformational space. The aim of the second approach was to identify evolutionarily conserved primary sequence positions in MHC I forming networks of residues, called protein sectors, which are physically connected in the tertiary heavy chain structure. The overarching aim was to identify which heavy chain residues are most likely to encode differences in MHC I plasticity and therefore the biological properties of BF2*15:01 and BF2*19:01. Collectively, this created a framework in which to interpret the experimentally observed differences of these MHC I molecules in their intrinsic peptide selection ability and tapasin dependence in terms of protein dynamics, which are reflected in the evolutionary history of their primary sequences.

Results

Briefly, homology models of BF2*15:01 and BF2*19:01 both with and without the 8-mer peptide KRLIGKRY were created using MODELLER [32,33] based upon the crystal structure of BF2*21 PDB ID: 3BEV [34], and assessed using SWISS-MODEL [35–37]. This peptide binds to both molecules with equal affinity [24]. Three independent molecular dynamics simulations of 150 nanoseconds (ns) for each structure were performed using the GROMACS package [25]. Having discarded the first 10ns to remove any effects of the system reaching equilibrium, the final 140ns of each simulation were concatenated to generate a 420ns trajectory for each structure of BF2*15:01 and BF2*19:01 in the peptide bound and peptide free state. Residue numbering of these proteins as presented here is for the lumenal domain of the MHC I molecule as found in all PDB structure files i.e. 274 heavy chain residues starting at position 1 after the signal peptide. Further details of the methodology and quality assessment are supplied in materials and methods and supporting information (SI).

Quantification of the flexibility of BF2*15:01 and BF2*19:01 identifies local differences in protein plasticity

To quantify protein plasticity in terms of per residue flexibility during the molecular dynamics simulation we used the conformational angle φ. This is the dihedral angle of internal rotation in the main chain of a protein created by rotation around the N-Cα bond. For each 5 picosecond trajectory snapshot, 84,000 structures over 420ns, the φ angle was measured for each residue in the MHC I complex. The standard deviation of the φ angles therefore quantifies the extent to which each residue varies from their average conformation over the trajectory and thus indicates which regions of the protein are flexible and which are not.

Examination of the peptide bound BF2*15:01 and BF2*19:01 molecules φ angle standard deviations revealed that they displayed a similar degree of flexibility (Figure 2A,C). Most residues had φ angle standard deviation below 25°; hence we examined more closely those residues displaying flexibility greater than this threshold. There were 17 and 15 residues in BF2*15:01 and BF2*19:01 respectively, with φ angle standard deviations greater than 25°. Unsurprisingly, the unstructured loop regions exhibit the greatest flexibility, such as residues 53 and 90 on loops preceding and following the α_1 helix respectively. Both molecules have φ angle standard deviations of about 25° in the α_2 helix around residues 146 and 150 and flanking the peptide N-terminus binding site around residues 168 and 170. Residues around positions 190 and 260 in these hairpin turns in the α_3 domain indicate that they are highly flexible in both molecules in the peptide bound state.

In contrast, on removal of the peptide there was a marked difference in plasticity between the alleles. We observed an increase in the φ angle standard deviation greater than 25° from 17 to 25 sites on BF2*15:01 (Figure 2A), but only an increase from 15 to 19 sites on BF2*19:01 (Figure 2C). In the BF2*15:01 peptide binding domain, the increased flexibility of residues 132 and 146 in the α_2 helix flanking the peptide C-terminus binding site suggests these residues might create hinge points about which the helix could rotate (Figure 2B). There was also increased flexibility in α_2 helix flanking the peptide N-terminus binding site around residues 168 and 170 and in the α_1 helix around residue 72. In the α_3 domain there is a decrease in the φ angle standard deviation around the 191 hairpin and a large increase in the loop containing residues 220–225. This is a region of MHC I that is a putative tapasin binding site (Figure 2B).

For BF2*19:01, upon removal of the peptide, the largest increases in flexibility were observed in the residues around 191 hairpin in the α_3 domain, in contrast to BF2*15:01. In common with BF2*15:01 there was an increase in flexibility around residue 222 in the putative tapasin binding loop in the α_3 domain (Figure 2C,D). These changes coincided with the flexibility of the peptide binding domain of BF2*19:01 remaining broadly as in the peptide bound state, whereas BF2*15:01 became more plastic on removal of the peptide.

Overall, this measure of local flexibility suggested that the BF2*15:01 heavy chain has a more intrinsically plastic structure than the BF2*19:01 heavy chain, and so next we looked for further evidence of this difference in the global dynamics of these proteins.

Identification of the global motions of BF2*15:01 and BF2*19:01 and conformational exploration by Principal Component Analysis

To understand how local flexibility impacts upon plasticity in terms of the global dynamics of these MHC I molecules we used Principal Component Analysis (PCA). PCA aims to identify the modes of motion corresponding to the directions along which the covariance of backbone atomic motions during the simulation are maximised. This is to say that in contrast to examining residues individually, as with the conformational angle analysis, we identify the collective motions of the atoms in MHC I and rank them according to their contribution to the overall motion i.e. how principal each component is. The underlying assumptions are that the dynamics of MHC I are best expressed in terms of a few modes containing large variances and that these are relevant to function [38,39]. Here PCA provides us with three pieces of information about the dynamics of BF2*15:01 and BF2*19:01: 1) PCA identifies and quantifies which collective atomic motions most contribute to the overall motion of the molecule during simulation. 2) We can project the dominant collective motions onto the MHC I structure to observe their quality and compare them between molecules. 3) Having identified the dominant motions, we can also examine how many different conformations are explored by these collective motions, and how frequently each conformation occurs. Further details of PCA analysis are provided in the materials and methods. It is important to note that one would not expect the φ and PCA analyses to be directly correlated on a residue by residue basis. Indeed, the rationale for using the different analyses presented is to try and build a detailed picture of the dynamics from different perspectives. For example, a residue that shows great local flexibility may not necessarily undergo large amplitude motions; rather it may be a residue whose flexibility facilitates other residues to undergo large amplitude motions. Lack of direct correlation between local flexibility and the amplitude of motion for a given residue is therefore not unexpected.

Therefore, we first calculated the variance contributed by each individual principal component (PC), and the percentage of total variance accounted for by the PCs cumulatively (Figure 3A). It is clear that the first 50 PCs are sufficient to describe almost all of the backbone atomic motions in all simulations. For BF2*15:01 the first two principal components account for about 35% of the total variance in the peptide bound state with nearly 30% contained within the dominant PC1 mode. This increases to about 45% in the peptide free state, corresponding with nearly a doubling of the actual backbone variance of these modes. For BF2*19:01 the first two principal components also account for about 35% of the total variance in the peptide bound state, with about 25% contained within the dominant PC1 mode. The contribution of the first two PCs to the total variance falls to below 30% on the removal of the peptide, with nearly a halving of the variance contributed by PC1. This suggests that both molecules have a similar plasticity as quantified by PCA in the native peptide bound state, but display contrasting degrees of plasticity in the non-native peptide free state in correspondence with the overall differences observed between the molecules in the φ angle standard deviation analysis.

To then examine these similarities and differences, qualitatively as well as quantitatively, the top two principal components were visualised as porcupine plots showing the direction and magnitude of the motion of each backbone atom along PC1 and PC2 (Figure 3B, Movies S1). In the peptide bound state, the magnitudes of the atomic fluctuations are similar for both BF2*15:01 and BF2*19:01 along both the modes PC1 and PC2. However, qualitatively they are different. The BF2*19:01 heavy chain domains have a twisting mode for PC1 whilst BF2*15:01 displays a swinging motion between heavy chain domains. Both molecules show twisting dynamic between domains for the heavy chain PC2 mode.

On removal of the peptide both molecules display the same quality of heavy chain motions for PC1 and PC2, but the amplitudes of the motions are greater for BF2*15:01 than BF2*19:01. Dominant mode PC1 describes an opening and

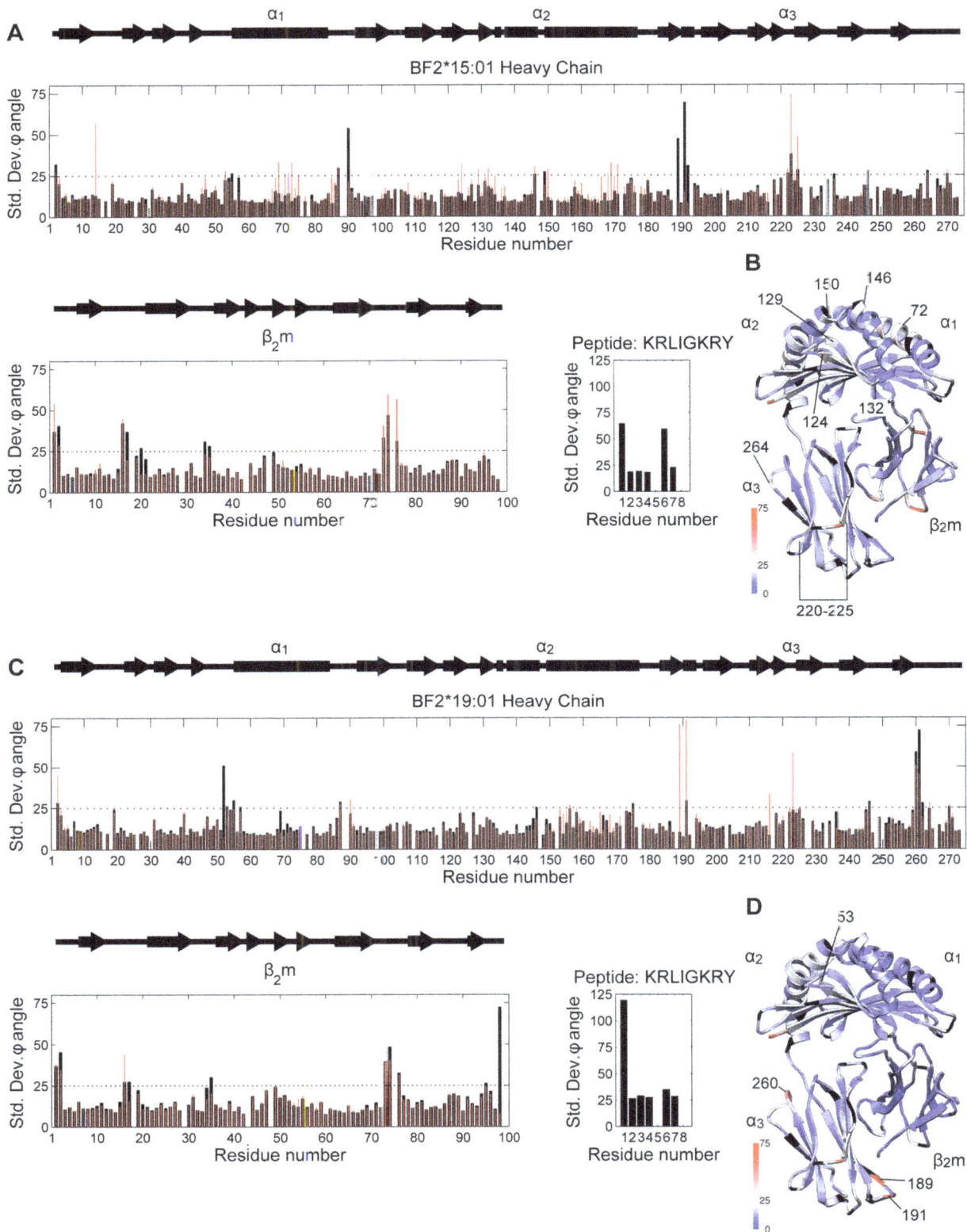

Figure 2. Quantification of the flexibility of MHC I by conformational φ angle standard deviation. A) and C) The standard deviation of the internal angle of rotation φ measuring the rotation around N-Cα bond of each residue of BF2*15:01 and BF2*19:01 from 420ns of molecular dynamics simulation in the peptide bound and peptide free states. Peptide bound measurements are shown as black bars and peptide free as red bars. B) and D) Ribbon representations of BF2*15:01 and BF2*19:01 with the peptide free simulations φ angle standard deviations mapped as increasing from blue to white to red, with annotations on the BF2*15:01 heavy chain. Glycine residues are coloured black.

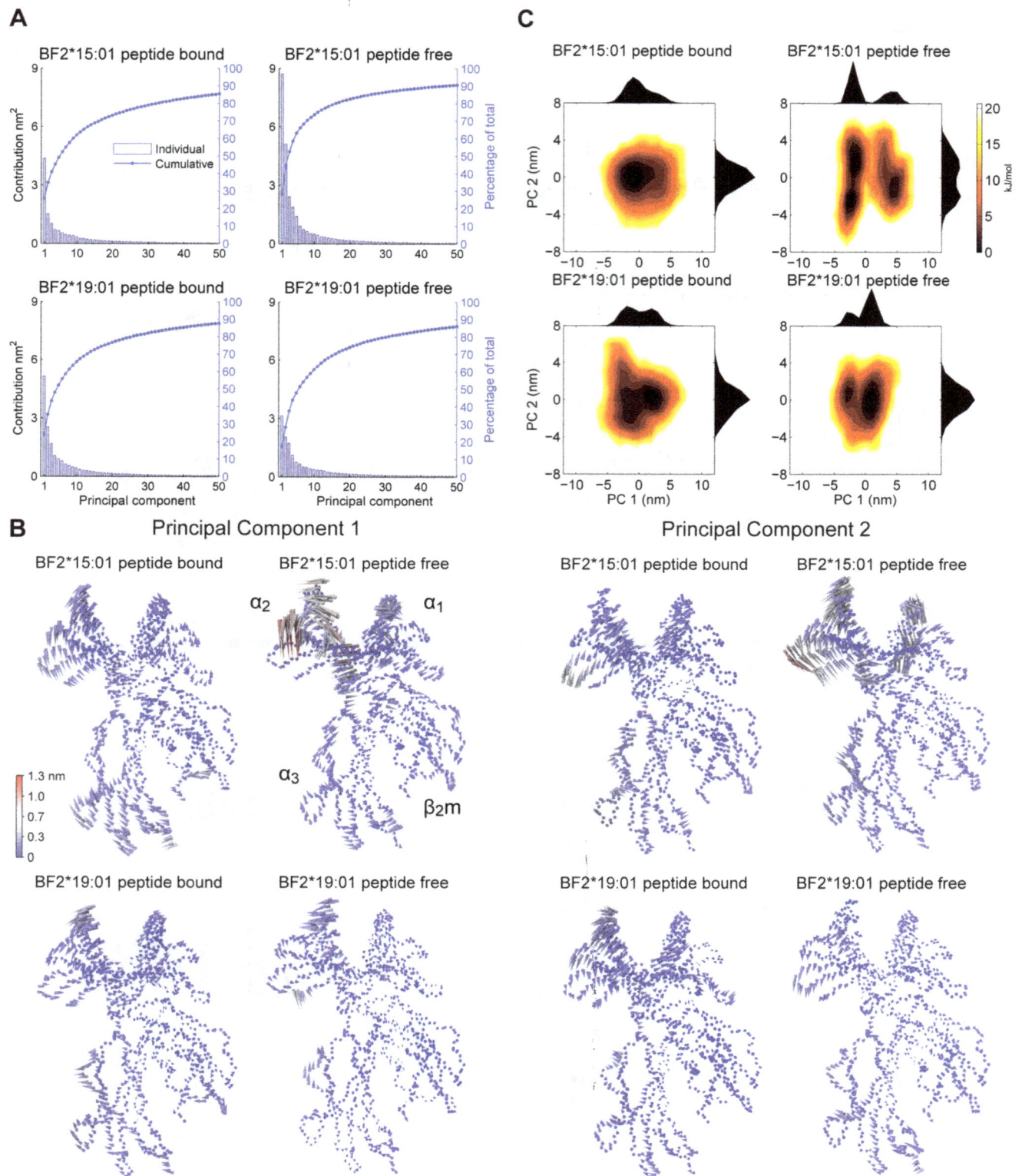

Figure 3. The global dynamics of MHC I identified by Principal Component Analysis. For each 420ns molecular dynamics simulation of BF2*15:01 and BF2*19:01 PCA was performed using a common peptide free backbone structure. A) Contributions of the first 50 PCs to the total variance of the backbone atomic motions. B) Porcupine plots indicate the magnitude and direction of motion for each backbone atom along PC1 and 2 in both the peptide bound and peptide free states. The magnitude between extremes is indicated by the colour bar. C) Gibbs free energy landscapes are generated from the principal coordinates of PC1 and PC2 and transformed by treatment as a Boltzmann ensemble. Individual probability densities for PC1 and PC2 are plotted on the outside adjacent axes.

closing of the helices flanking the peptide binding groove corresponding with a twisting between heavy chain domains centred about the domain-domain linker region. This suggests that the motions of the heavy chain domains are dynamically coupled [40]. In other words large domain-domain motions appear to correspond with large conformational changes in the peptide binding domain and vice versa. The peptide free PC2 mode describes a combined rocking and twisting motion between peptide binding groove and heavy chain domains. For BF2*15:01 the PC1 motion is pronounced with the greatest amplitudes occurring in the α_2 helix between residues 134 and 150 that flank the peptide C-terminus binding site. In contrast, BF2*19:01 PC1 shows a reduction in the amplitude of domain-domain motions as compared with the peptide bound state. For PC2 BF2*15:01 again has much greater amplitudes than BF2*19:01 with the largest motions in the peptide binding domain helices.

To examine the extent to which these molecules are actually able to explore the different conformations indicated in the porcupine plots, the simulation trajectories were treated as Boltzmann ensembles and plotted as Gibbs free energy landscapes (Figure 3C). This was done by calculating a transformation of the joint probability distribution of the coordinates of the top two dominant modes. This information about how BF2*15:01 and BF2*19:01 explore their conformational landscapes further indicates any differences in plasticity by indicating how likely they are to populate different conformations. In the peptide bound state both molecules inhabit a single energy minimum indicating the stability of the peptide bound state. Thus although we observe the possibility of various conformations in the porcupine plots (Figure 3B), these landscapes suggest that in fact peptide bound MHC I mostly inhabits a single conformation and infrequently visits other states.

On the removal of peptide from BF2*15:01, the molecule explores a larger region of the energy landscape and populates three local energy minima. The probability distribution broadens along PC2 and forms a two peaked distribution along PC1 with a dominant and sub-dominant peak separated by a large energy barrier. In one half of the landscape is a minima corresponding with the sub-dominant PC1 probability distribution peak. On the other half of the landscape, there are two distinct energy minima corresponding with the dominant PC1 probability distribution peak, separated by a small energy barrier along the PC2 axis. In contrast to BF2*15:01, removal of the peptide from BF2*19:01 leads to the appearance of two closely located energy minima, but with an increased energy barrier between these states and less of the landscape explored overall.

In summary, PCA indicates that in absence of peptide BF2*15:01 is more able to explore different conformations, consistent with the suggestion from the φ angle analysis that the BF2*15:01 molecule is more plastic than BF2*19:01 (Figure 3C). When peptide is present both molecules display similar dynamics and plasticity, also consistent with the φ angle analysis. Furthermore the difference in the plasticity between BF2*15:01 and BF2*19:01 in the absence of peptide manifests itself as a greater intrinsic ability of the BF2*15:01 molecule to explore peptide binding groove conformations, and that these motions are coupled to the motions of the α_3 domain (Figure 3B).

Statistical coupling analysis identifies a protein sector spanning the peptide binding domain and the α_3 domain in chicken MHC class I

Finally we sought to identify positions in the primary amino acid sequences of BF2*15:01 and BF2*19:01 that could form networks of residues through which the observed dynamics and differences in plasticity act. To this end we used the previously described technique of statistical coupling analysis (SCA) [29,30,41–43] to identify non-random correlations between sequence positions of a multiple sequence alignment (MSA), generated using BF2*15:01 heavy chain as the query sequence. To briefly summarise, using a MSA of 141 sequences (MSA S1), the positional conservation of each heavy chain residue was calculated (Figure 4A). This conservational weighting was then used to calculate a matrix of correlations between all pairs of MSA positions to quantify the evolutionary history of each pair of sequence positions in the alignment. Eigenvalue decomposition of this positional correlation matrix identified the top six statistically significant eigenmodes describing weighted groups of positions (Figure S1). Using Independent Component Analysis (ICA) as previously described [30], we transformed these eigenmodes and projected the heavy chain positions along three maximally independent axes (Figure S2). One of these directions, IC2, identified a group of heavy chain residues that was used to define a single MHC I heavy chain protein sector (Figure S2). Further details are provided in the materials and methods and in Refs [29,30,41–43].

Here SCA identified a protein sector that creates a contiguous network of 85 residues, constituting approximately 31% of heavy chain residues, connecting the peptide binding domain and the α_3 domain of the chicken MHC I heavy chain (Figure 4, Sector S1). The sector connects residues along the α_1 helix (Figure 4A–D) and passes across the peptide binding groove via residue 96 at the interaction site of β_2m (Figure 4C) to connect with the α_2 helix. Connection to the α_3 domain is through the domain linker residues 177, 178 and 179 (Figure 4A–C) and the sector continues almost to the region of the heavy chain that would ordinarily connect to the transmembrane domain (Figure 4A–C). Importantly, this protein sector also includes many of the residues identified as sites of local flexibility in φ angle standard deviation analysis, such as 71,129 and 150 in the peptide binding domain and α_3 domain residues 191, 224 and 264 (Figure 1B and 4). This sector is also consistent with the notion that MHC I heavy chain domains are dynamically coupled as suggested from the PCA analysis.

Of the eight polymorphic residues that differ between BF2*15:01 and BF2*19:01, coloured green in Figure 4, two are identified as sector residues, coloured blue in Figure 4. These polymorphic residues are position 22 in the peptide binding groove just below the α_1 helix and position 220 in the α_3 domain (Figure 4A–C). We have recently shown that when amino acid position 220 is swapped between BF2*15:01 and BF2*19:01, their ability to exchange low affinity peptides for high affinity peptide *in vitro* is altered [24]. This observed allosteric effect on peptide editing may be explained at the molecular level by this sector spanning the peptide binding domain and the the α_3 domain. Interestingly, whilst mutation of position 220 had a detrimental effect on this measure of intrinsic peptide selection ability for both alleles, mutation of non-sector position 126 did not. Furthermore, position 220 is known to reside in, or near, a putative tapasin interaction site [44,45] and the polymorphism at position 220 influenced the extent to which each allele benefited from presence of tapasin when these assays were repeated. Alongside the dynamics simulation data, these observations suggest that by interacting with position 220 on MHC I, tapasin exerts an allosteric influence on the peptide binding domain via the protein sector we have identified.

These observations give the first indication of how we might connect the protein dynamics to the observed functional differences of BF2*15:01 and BF2*19:01. Here we have identified

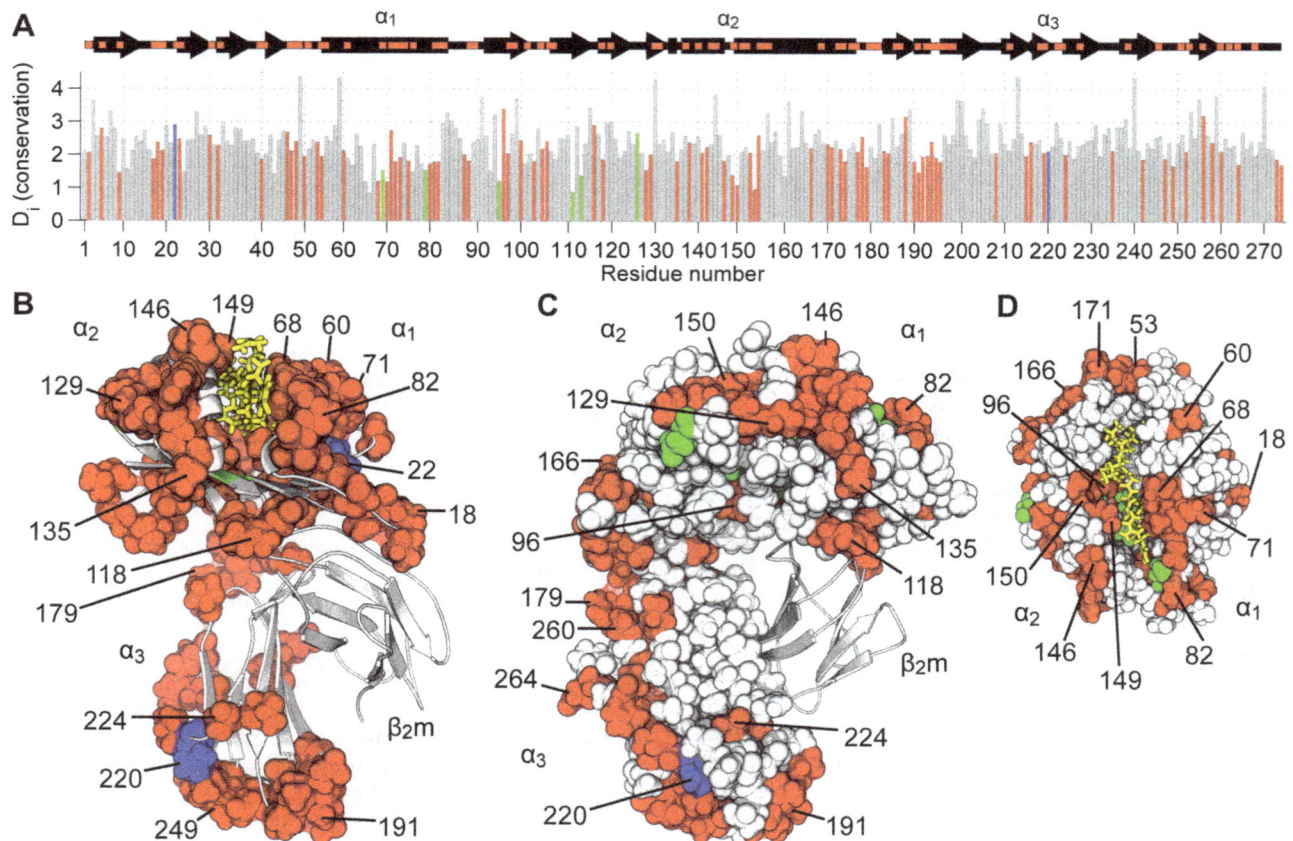

Figure 4. Identification of a protein sector in chicken MHC I. Statistical coupling analysis (SCA) was carried out on a multiple sequence alignment (MSA) of 141 sequences obtained from a similarity search querying the BF2*15:01 heavy chain as described in [30]. A) The degree of conservation of each heavy chain residue i in the MSA is computed as the Kullback-Leibler relative entropy D_i. Bigger bars indicate greater conservation. The 85 protein sector residues are in red, 6 polymorphic residues between BF2*15:01 and BF2*19:01 are in green and the 2 residues that are both polymorphic and part of the protein sector are in blue. All other residues are in grey. B) Protein sector residues are mapped as spheres onto a ribbon representation of the BF2*15:01 structure. Colours as (A), with the peptide as yellow sticks. C) and D) Space filling representations of the MHC I heavy chain, coloured as (B). The contiguous network of residues forming a protein sector comprises of 31% of heavy chain residues.

a sector in MHC I of 85 sequence positions that have been evolutionarily conserved and containing just two of the polymorphic residues. These polymorphisms at positions 22 and 220 correspond with the differences in their intrinsic plasticity indicated by the changes in dynamic coupling between BF2*15:01 and BF2*19:01 seen during the peptide free molecular dynamics simulations. This sector therefore indicates which residues through which dynamic coupling may occur in MHC I and suggests a link between polymorphisms and changes in protein dynamics.

Discussion

The aim of the study presented here was to examine the protein dynamics of two chicken MHC I molecules in the light of the co-evolving haplotype hypothesis [19] and experimentally observed differences in the intrinsic peptide selection abilities of these alleles [24]. The co-evolving haplotype hypothesis proposes that in chickens, unlike in mammals, the genes for MHC I, tapasin and TAP have evolved together with optimal function resulting from alleles of these proteins encoded from the same haplotype that share complementary functions. Recently, this hypothesis was tested experimentally using the same MHC I molecules presented in this study, BF2*15:01 and BF2*19:01 [24] which proposed a

mechanistic basis for the co-evolution of chicken tapasin and MHC I molecules. These experiments demonstrated that the mismatching of MHC I and tapasin molecules from these haplotypes impaired the maturation of MHC I *in vivo*. Secondly, as described in the previous section, differences in the intrinsic peptide selection properties of BF2*15:01 and BF2*19:01 in the absence of tapasin *in vitro* were also observed. Importantly, position 220 in the α_3 domain was shown to influence these intrinsic properties as well as tapasin function.

Both BF2*15:01 and BF2*19:01 have a similar specificity for the peptides that they can bind [20,21], so the hypothesis we examined here was whether the differences observed *in vitro* and *in vivo* could be a consequence of differences in protein dynamics arising from the polymorphisms. Furthermore we wanted to test the utility of computational models in deepening our understanding of complex biological data and provide a rational framework for future investigations.

The work presented here relies entirely upon models, so the initial limitation we must address is:

1. The reliability and limitations of homology models and molecular dynamics simulations

Our analysis of the plasticity of BF2*15:01 and BF2*19:01 made use of both homology models of these molecules and molecular dynamics simulations. The limitations on these modelling methods are firstly the quality of the models generated; the timescales available and the approximations made in the treatment of the molecules in the molecular dynamics simulations.

Assessment of the quality of the homology models using SWISS-MODEL [35] indicates they exhibit a similar degree of quality to the X-ray structure on which they are based (Figure S3). Similarly the molecular dynamics simulations have been assessed for stability of the trajectories using time block analysis of the root mean square fluctuations of the average atom positions (Figure S4). The force field used here [28,46] demonstrated good agreement between simulated folding events and experimentally observed structures in recent investigations into protein folding. This is important as a further complication is the removal of the peptide which creates non-physical structures that cannot be directly compared against experimental observation. So whilst these simulations may not be a true representation of reality, they are consistent with the behaviour of proteins where more direct comparison has been possible. Lastly, here the simulations explore ~0.5 μs timescale which is below the >1 μs timescale on which we might expect large dynamic events to occur [47], however they do provide an indication of what protein dynamics are possible.

2. BF2*15:01 and BF2*19:01 exhibit differences in protein plasticity

Analysis of the local flexibility of BF2*15:01 and BF2*19:01 using φ angle standard deviations and the global dynamics using PCA indicates that these molecules do differ in their intrinsic plasticity. BF2*15:01 appears to be an intrinsically more plastic protein than BF2*19:01 in the absence of peptide. On removal of the peptide we observe a greater increase in the number of local sites of flexibility for BF2*15:01 than BF2*19:01 (Figure 2). This proposal is supported by the PCA analysis of the global dynamics. Here we observe that the molecules display similar dynamics in the peptide bound state both quantitatively and qualitatively and mostly occupied a single conformational state. This is consistent with the observation of a stable conformation for MHC I seen in crystallographic structures. On removal of the peptide, BF2*15:01 explores multiple conformational minima (Figure 3C) corresponding with an increase in the ability to explore a range of peptide binding groove conformations correlated with domain-domain motions (Figure 3B). Conversely, peptide free BF2*19:01 displays a decrease in the global motions described by the top two modes and explores less of the energy landscape than peptide free BF2*15:01.

As previously described [38] there are several assumptions and limitations to PCA analysis: One is the assumption that the dynamics most important for protein function are described by the first few principal components which contain the largest variances. Another is that this is a linear analysis and therefore it neglects motions the may be spread across several components. However there are many examples in the literature supporting the notion that the top principal modes do contain functional motions, such as for T4 lysozyme [48].

Whilst the peptide free states represent non-physical structures, the notion of conformational intermediates for MHC I has been hypothesized on the basis of several pieces of circumstantial experimental evidence [49–51]. These simulations do not refute that proposal and further inform how they might arise via the plasticity encoded in the primary sequences of these alleles.

3. Modulation of protein plasticity by MHC I polymorphisms

Analysis of the MHC I heavy chain using SCA revealed an allosteric protein sector, connecting residues in both helices of the peptide binding domain, passing across the base of the ligand binding site at a site of interaction with β2m, and down through the domain linker into the α3 domain. Importantly, the sector contains only two of the eight polymorphisms that exist between BF2*15:01 and BF2*19:01. One sector polymorphism is in the peptide binding domain beneath the α1 helix at position 22 and the other in the α3 domain at position 220 (Figure 4). These observations suggest a role for these residues in modulating differences in protein plasticity when compared to the molecular dynamics simulations. The molecular dynamics simulations indicate that these polymorphisms are capable of modulating the amplitude of domain coupled backbone motions (Figure 2B), but do not inform us as to their relative importance. The protein sector identifies a possible network through which the dynamics may act and two residues that may modulate the differences between BF2*15:01 and BF2*19:01. What is most striking is that this analysis identifies position 220 as a sector residue. This was identified in the *in vitro* experiments as influencing the ability of tapasin to enhance peptide dissociation, a critical component of MHC I peptide selection. Exchange of position 220 between BF2*15:01 and BF2*19:01 influenced the magnitude of tapasin function, suggesting a possible mechanism for how tapasin might modulate the peptide binding groove conformation via the α3 domain and catalyse the selection of high affinity peptides by MHC I. The significance of position 22 in the peptide binding domain is not obvious, but intriguing. Our speculation is that a more intrinsically plastic molecule, such as we propose for BF2*15:01, is better able to sample the conformational space and more quickly select a high affinity peptide than a less plastic molecule and is therefore less reliant on tapasin to access the relevant conformations, and vice versa. Alternative approaches to sector identification using conformational angles from molecular dynamics trajectories have been demonstrated [52,53]. In this approach, the measurement of the mutual information between the side chain conformations, rotameric states, of each residue in the molecular dynamics trajectory looks for evidence of local or distant coupling between residues. The mutual information for two residues measures the extent to which the rotameric state of one residue depends upon the rotameric state of the other. Thus, as has been shown for β-lactamase [54], a mutual information matrix can be constructed for all the residues in a protein to indicate dynamically coupled residues that may form a sector. Robust measurement of mutual information of rotameric states using the approach described in [54] requires a large ensemble of structures far greater than what is presented here, but this would be an important line of investigation for future work in terms of cross-validation of the SCA.

4. Evidence that protein plasticity and protein dynamics are evolutionarily conserved features of MHC I

Our identification of a protein sector indicates how a subset of conserved heavy chain residues connect to form a contiguous network throughout the MHC I heavy chain. Alongside the molecular dynamics data, this suggests the possibility that protein dynamics are a conserved feature of MHC I molecules. SCA has previously been used to identify an allosteric sector in Hsp70 proteins [30]. This showed how two functional protein domains could be coupled, and thus conserved, through a network of connecting residues. The implication of this finding was that there exists a dynamic mechanism in Hsp70 molecules operating through this sector. Our work supports the notion that protein dynamics are conserved features of proteins that are encoded into the primary sequence and which underpin biological function. This is most apparent from the observation that BF2*15:01 and BF2*19:01 have similar protein dynamics in the peptide bound state, that is after they have achieved their function, but different dynamics in the peptide free state, prior to peptide selection.

These observations, alongside our recent work with human HLA-B44 alleles [unpublished data], have led us to believe that protein plasticity is an important determinant of MHC class I function. Moreover, this work focuses and complements experimental investigations into the mechanism of peptide selection by MHC I by reducing the target for future investigation from 274 heavy chain residues to 85. Our expectation is that understanding the mechanism by which protein plasticity manifests intrinsically and is modulated by co-factor molecules such as tapasin will be a key part of understanding the peptide selection process determining the immune response in other species, including humans. We foresee establishing techniques such as molecular dynamics and SCA as fully integrated tools in the investigative process as a means of accelerating these developments.

Materials and Methods

Homology modelling and molecular dynamics simulations

The starting conformation of BF2*15:01 and BF2*19:01 were derived from the experimentally determined structure from the RSCB Protein Databank of BF2*B21, PDB id: 3BEV [34]. These homology models were created used MODELLER [32,33]. Quality assurance of these models was performed with SWISS-MODEL incorporating PROCHECK [35–37].

The GROMACS version 4.5.3 [25] molecular dynamics package was used for the all atom simulations. The simulations used the Amber99SB-ILDN [28] force field and TIP3P [55] explicit water molecules using the Simple Point Charge water system [56], and Sodium counter ions were added to neutralise the charge of the system. The protein structures were placed in rhombic dodecahedron shaped box centred at 1.5 nm from the edge with periodic boundary conditions. Covalent bond lengths were constrained using the P-LINCS algorithm [57] and the water angles were constrained using the SETTLE algorithm [58] allowing an integration time step of 2 fs to be used. Nosé-Hoover temperature coupling [59,60] and Parinello-Rhaman pressure coupling [61,62] used a time constant of 0.5 ps with reference baths of 300 Kelvin and 1 bar respectively to maintain the average thermodynamic properties of the protein and solvent comprising the system. Electrostatic interactions use a cut-off of 1 nm with the interactions beyond this cut-off treated using the particle mesh Ewald method [63]. Van der Waals forces used a cut-off of 1 nm. The neighbour list is updated every five steps. Each system initially underwent an energy minimization over 1000 steps of 2 fs to relax the structure and remove the forces from the systems that were introduced by the protonation of the molecule and addition of solvent. This was followed by a 5 ns equilibration of the water surround the protein with the protein atoms restrained using a randomly generated initial starting velocity. Full production runs were performed with the position restraints released. To analyse conformational dynamics, concatenated trajectories of 420 ns were created from three independent repeats of 150 ns, with the first 10 ns of each simulation discarded. Quality assurance and post processing, including PCA, was performed using a combination of the suite of utilities provided with GROMACS. Additional post processing tasks were performed using MATLAB™ and bespoke UNIX awk scripts. Visualisation of the protein structures and molecular dynamics trajectories was performed using the VMD [64] and USCF Chimera [65] packages.

Principal Component Analysis

Principal component analysis is performed as follows using the GROMACS g_covar and the g_anaeig utilities:

A mass weighted variance-covariance matrix is built using the backbone atoms. This is a symmetric $3N \times 3N$ matrix comprising of the fluctuation of the atom positions with coordinates x as a function of the trajectory t such that:

$$C = <(x(t) - <x>).(x(t) - <x>^T>$$ (1)

where $<>$ indicates the conformational ensemble average. This matrix C therefore contains as elements, for each atom pair, the difference between the mean product of their atomic positions and the product of their mean atom positions i.e. the difference between their average position as a pair and the product of their individual average positions. Atom pairs moving together in the same direction give rise to positive covariances and pairs moving in the opposite direction give rise to negative covariances. Non-correlated atoms give near zero covariances. The variance for each atom is contained on the main diagonal.

With reference to the covariance matrix C generated by equation 1, Eigen decomposition of matrix C is performed using the eigenvector matrix P, its inverse P^{-1} and the diagonal matrix D which has the corresponding eigenvalues on the diagonal, such that:

$$C = PDP^{-1}$$ (2)

The eigenvalues along the diagonal in D represent the mean square fluctuations for each eigenvector in C (the columns of P) and therefore indicate how much each eigenvector contributes to the total fluctuation. The eigenvectors are sorted according to size of the eigenvalues. Projection of the data onto the first eigenvector transforms the data into a new coordinate system with the greatest variance residing on this first coordinate. This coordinate is called the first principal component and the first eigenvector is also known as the first principal mode. This projection can be done for each principal mode μ_i to yield the principal components as a function of the trajectory $p_i(t)$ as follows:

$$p_i(t) = \mu_i.(x(t) - <x>)$$ (3)

The variance of principal component $<p^2_i>$ is equal to its corresponding eigenvalue in D. The projections can then be transformed back into Cartesian coordinates, $x'_i(t)$ for visualisation by rearranging equation 3 such that a linear equation describes the coordinates x as a function of the trajectory in terms of the principal coordinates and the average ensemble coordinates $<x>$:

$$x'_i(t) = p_i(t).\mu_i + <x>$$ (4)

Statistical Coupling Analysis

Statistical coupling analysis (SCA) was carried out on a multiple sequence alignment (MSA) of 141 sequences generated using BF2*15:01 heavy chain as the query sequence for a PSI-BLAST [66] search of the non-redundant UniProtKB/SwissProt sequences database (July 2013), using the BLOSUM62 scoring matrix and an expectation threshold of 0.0001 Automatic alignment of the search results was performed with Clustal Omega [67,68] followed by manual alignment using SeaView [69].

SCA was performed using a process described in [30,41] and implemented in the SCA 5.0 toolbox for MATLAB™ available from the Ranganathan laboratory website: http://systems.swmed.edu/rr_lab/sca.html. The sector was defined by empirical fitting of the Students t-distribution to a histogram of the positional weights along IC2 with a cumulative probability density cut-off in the tail of the distribution of 85%. No mechanistic basis is implied by the use of this distribution and the choice of cut-off is that which produces a sector comprising a contiguous network of residues in the tertiary structure. These top 15% of positions in the IC2 distribution represent about 31% of the heavy chain residues, consistent with previous definitions of protein sectors [41].

Supporting Information

Figure S1   Histograms of the eigenvalues of the SCA positional correlation matrix. Histograms of the eigenvalues from decomposition of the positional correlation matrix for the BF2 MHC I heavy chain multiple sequence alignment in blue. The top six eigenmodes are indicated with arrows. Eigenvalues generated from decomposition of 100 randomized alignments are shown in red.

Figure S2   Identification of the protein sector by Independent Component Analysis. A plot of the top three independent components generated by transformation of the top six eigenmodes of the SCA matrix using Independent Component Analysis as previously described in [29,41] to test for the existence of quasi-independent sectors. The identified protein sector is indicated along IC2 in red. The polymorphisms between BF2*15:01 and BF2*19:01 are shown in green. Two pusedo-sectors identified by the ICA are indicated in cyan and magenta. These putative sectors were discarded as they are not contiguous in the tertiary structure and most residues are close to zero.

Figure S3   Ramachandran plots comparing the homology models of BF2*15:01 and BF2*19:01 to the template

structure of BF2*21. Ramachandran plots indicating the conformational φ and ψ angles to assess the quality of the homology models of A) BF2*15:01 and B) BF2*19:01 in comparison to the crystallographic template structure of C) BF2*21 were generated using SWISS-MODEL incorporating PROCHECK [34–37].

Figure S4   Time block assessment of the stability of the molecular dynamics simulations of BF2*15:01 and BF2*19:01. Each plot shows the Root Mean Square Fluctuation (RMSF) of the atoms from their average position during each 10 nanosecond time block of each molecular dynamics simulation trajectory as an indication of the overall stability of each simulation and between simulations.

MSA S1   The multiple sequence alignment used for the SCA in fasta format.

Sector S1   The protein sector identified by SCA for BF2 heavy chain using PDB residue numbering.

Movies S1   Animations of molecular dynamics simulations of BF2*15:01 and BF2*19:01 projected onto the first two principal components. The peptide is removed so that a common structure is used for the projections. The magnitude of the motions for BF2*15:01 peptide free PC1 has created the appearance of a broken molecule. This is an artefact of the rendering.

Acknowledgments

The authors would like to thank Jim Kaufman for providing feedback during the preparation of this manuscript and Syma Khalid and Thomas Piggott for their assistance and guidance in the use of GROMACS. All the molecular dynamic simulations were performed using the IRIDIS High Performance Computing Facility, and we acknowledge the associated support services at the University of Southampton in the completion of this work.

Author Contributions

Conceived and designed the experiments: AB AvH TE JMW. Performed the experiments: AB. Analyzed the data: AB. Wrote the paper: AB AvH TE JMW.

References

1. Neefjes J, Jongsma ML, Paul P, Bakke O (201 Towards a systems understanding of MHC class I and MHC class II antigen presentation. Nat Rev Immunol 11: 823–836.

2. Elliott T (1997) How does TAP associate with MHC class I molecules? Immunology Today 18: 375–379.

3. Suh WK, Mitchell EK, Yang Y, Peterson PA, Waneck GL, et al. (1996) MHC class I molecules form ternary complexes with calnexin and TAP and undergo peptide-regulated interaction with TAP via their extracellular domains. J Exp Med 184: 337–348.

4. Lewis JW, Elliott T (1998) Evidence for successive peptide binding and quality control stages during MHC class I assembly. Curr Biol 8: 717–720.

5. Cresswell P, Bangia N, Dick T, Diedrich G (1999) The nature of the MHC class I peptide loading complex. Immunol Rev 172: 21–28.

6. Ortmann B, Copeman J, Lehner PJ, Sadasivan B, Herberg JA, et al. (1997) A Critical Role for Tapasin in the Assembly and Function of Multimeric MHC Class I-TAP Complexes. Science 277: 1306–1309.

7. Sadasivan B, Lehner PJ, Ortmann B, Spies T, Cresswell P (1996) Roles for calreticulin and a novel glycoprotein, tapasin, in the interaction of MHC class I molecules with TAP. Immunity 5: 103–114.

8. Williams AP, Peh CA, Purcell AW, McCluskey J, Elliott T (2002) Optimization of the MHC class I peptide cargo is dependent on tapasin. Immunity 16: 509–520.

9. Howarth M, Williams A, Tolstrup AB, Elliott T (2004) Tapasin enhances MHC class I peptide presentation according to peptide half-life. Proc Natl Acad Sci U S A 101: 11737–11742.

10. Androlewicz MJ, Ortmann B, van Endert PM, Spies T, Cresswell P (1994) Characteristics of peptide and major histocompatibility complex class I/beta 2-microglobulin binding to the transporters associated with antigen processing (TAP1 and TAP2). Proceedings of the National Academy of Sciences of the United States of America 91: 12716–12720.

11. Horton R, Wilming L, Rand V, Lovering RC, Bruford EA, et al. (2004) Gene map of the extended human MHC. Nat Rev Genet 5: 889–899.

12. Aguilar A, Roemer G, Debenham S, Binns M, Garcelon D, et al. (2004) High MHC diversity maintained by balancing selection in an otherwise genetically monomorphic mammal. Proc Natl Acad Sci U S A 101: 3490–3494.

13. van Endert PM, Riganelli D, Greco G, Fleischhauer K, Sidney J, et al. (1995) The peptide-binding motif for the human transporter associated with antigen processing. J Exp Med 182: 1883–1895.

14. Boyle LH, Hermann C, Boname JM, Porter KM, Patel PA, et al. (2013) Tapasin-related protein TAPBPR is an additional component of the MHC class I presentation pathway. Proc Natl Acad Sci U S A.

15. Kaufman J, Milne S, Gobel TW, Walker BA, Jacob JP, et al. (1999) The chicken B locus is a minimal essential major histocompatibility complex. Nature 401: 923–925.

16. Wong GK, Liu B, Wang J, Zhang Y, Yang X, et al. (2004) A genetic variation map for chicken with 2.8 million single-nucleotide polymorphisms. Nature 432: 717–722.

17. Kaufman J (1999) Co-evolving genes in MHC haplotypes: the "rule" for nonmammalian vertebrates? Immunogenetics 50: 228–236.

18. Shaw I, Powell TJ, Marston DA, Baker K, van Hateren A, et al. (2007) Different evolutionary histories of the two classical class I genes BF1 and BF2 illustrate drift and selection within the stable MHC haplotypes of chickens. Journal of Immunology 178: 5744–5752.

19. Walker BA, Hunt LG, Sowa AK, Skjodt K, Gobel TW, et al. (2011) The dominantly expressed class I molecule of the chicken MHC is explained by coevolution with the polymorphic peptide transporter (TAP) genes. Proc Natl Acad Sci U S A 108: 8396–8401.

20. Wallny HJ, Avila D, Hunt LG, Powell TJ, Riegert P, et al. (2006) Peptide motifs of the single dominantly expressed class I molecule explain the striking MHC-determined response to Rous sarcoma virus in chickens. Proc Natl Acad Sci U S A 103: 1434–1439.

21. Kaufman J, Volk H, Wallny HJ (1995) A "minimal essential Mhc" and an "unrecognized Mhc": two extremes in selection for polymorphism. Immunol Rev 143: 63–88.

22. Lewis JW, Neisig A, Neefjes J, Elliott T (1996) Point mutations in the alpha 2 domain of HLA-A2.1 define a functionally relevant interaction with TAP. Curr Biol 6: 873–883.

23. Peace-Brewer AL, Tussey LG, Matsui M, Li G, Quinn DG, et al. (1996) A point mutation in HLA-A*0201 results in failure to bind the TAP complex and to present virus-derived peptides to CTL. Immunity 4: 505–514.

24. van Hateren A, Carter R, Bailey A, Kontouli N, Williams AP, et al. (2013) A mechanistic basis for the co-evolution of chicken tapasin and major histocompatibility complex class I (MHC I) proteins. J Biol Chem 288: 32797–32808.

25. Van Der Spoel D, Lindahl E, Hess B, Groenhof G, Mark AE, et al. (2005) GROMACS: fast, flexible, and free. J Comput Chem 26: 1701–1718.

26. Hess B (2009) GROMACS 4: Algorithms for highly efficient, load-balanced, and scalable molecular simulation. Abstracts of Papers of the American Chemical Society 237.

27. Pronk S, Pall S, Schulz R, Larsson P, Bjelkmar P, et al. (2013) GROMACS 4.5: a high-throughput and highly parallel open source molecular simulation toolkit. Bioinformatics 29: 845–854.

28. Lindorff-Larsen K, Piana S, Palmo K, Maragakis P, Klepeis JL, et al. (2010) Improved side-chain torsion potentials for the Amber ff99SB protein force field. Proteins 78: 1950–1958.

29. Lockless SW, Ranganathan R (1999) Evolutionarily conserved pathways of energetic connectivity in protein families. Science 286: 295–299.

30. Smock RG, Rivoire O, Russ WP, Swain JF, Leibler S, et al. (2010) An interdomain sector mediating allostery in Hsp70 molecular chaperones. Mol Syst Biol 6: 414.

31. Dror RO, Dirks RM, Grossman JP, Xu H, Shaw DE (2012) Biomolecular Simulation: A Computational Microscope for Molecular Biology. Annual Review of Biophysics 41: 429–452.

32. Sali A, Blundell TL (1993) Comparative protein modelling by satisfaction of spatial restraints. J Mol Biol 234: 779–815.

33. Eswar N, Webb B, Marti-Renom MA, Madhusudhan MS, Eramian D, et al. (2007) Comparative protein structure modeling using MODELLER. Curr Protoc Protein Sci Chapter 2: Unit 2 9.

34. Koch M, Camp S, Collen T, Avila D, Salomonsen J, et al. (2007) Structures of an MHC class I molecule from B21 chickens illustrate promiscuous peptide binding. Immunity 27: 885–899.

35. Arnold K, Bordoli L, Kopp J, Schwede T (2006) The SWISS-MODEL workspace: a web-based environment for protein structure homology modelling. Bioinformatics 22: 195–201.

36. Laskowski RA, Macarthur MW, Moss DS, Thornton JM (1993) Procheck – a Program to Check the Stereochemical Quality of Protein Structures. Journal of Applied Crystallography 26: 283–291.

37. Benkert P, Biasini M, Schwede T (2011) Toward the estimation of the absolute quality of individual protein structure models. Bioinformatics 27: 343–350.

38. Amadei A, Linssen AB, Berendsen HJ (1993) Essential dynamics of proteins. Proteins 17: 412–425.

39. Amadei A, Linssen AB, de Groot BL, van Aalten DM, Berendsen HJ (1996) An efficient method for sampling the essential subspace of proteins. J Biomol Struct Dyn 13: 615–625.

40. Cooper A, Dryden DT (1984) Allostery without conformational change. A plausible model. Eur Biophys J 11: 103–109.

41. Halabi N, Rivoire O, Leibler S, Ranganathan R (2009) Protein sectors: evolutionary units of three-dimensional structure. Cell 138: 774–786.

42. Reynolds KA, McLaughlin RN, Ranganathan R (2011) Hot spots for allosteric regulation on protein surfaces. Cell 147: 1564–1575.

43. McLaughlin RN Jr, Poelwijk FJ, Raman A, Gosal WS, Ranganathan R (2012) The spatial architecture of protein function and adaptation. Nature 491: 138–142.

44. Van Hateren A, James E, Bailey A, Phillips A, Dalchau N, et al. (2010) The cell biology of major histocompatibility complex class I assembly: towards a molecular understanding. Tissue Antigens 76: 259–275.

45. Simone LC, Georgesen CJ, Simone PD, Wang X, Solheim JC (2012) Productive association between MHC class I and tapasin requires the tapasin transmembrane/cytosolic region and the tapasin C-terminal Ig-like domain. Mol Immunol 49: 628–639.

46. Lindorff-Larsen K, Piana S, Dror RO, Shaw DE (2011) How fast-folding proteins fold. Science 334: 517–520.

47. Henzler-Wildman K, Kern D (2007) Dynamic personalities of proteins. Nature 450: 964–972.

48. de Groot BL, Hayward S, van Aalten DM, Amadei A, Berendsen HJ (1998) Domain motions in bacteriophage T4 lysozyme: a comparison between molecular dynamics and crystallographic data. Proteins 31: 116–127.

49. Elliott T, Elvin J, Cerundolo V, Allen H, Townsend A (1992) Structural requirements for the peptide-induced conformational change of free major histocompatibility complex class I heavy chains. European Journal of Immunology 22: 2085–2091.

50. Bouvier M, Wiley DC (1998) Structural characterization of a soluble and partially folded class I major histocompatibility heavy chain/beta 2m heterodimer. Nat Struct Biol 5: 377–384.

51. Springer S, Doring K, Skipper JC, Townsend AR, Cerundolo V (1998) Fast association rates suggest a conformational change in the MHC class I molecule H-2Db upon peptide binding. Biochemistry 37: 3001–3012.

52. McClendon CL, Friedland G, Mobley DL, Amirkhani H, Jacobson MP (2009) Quantifying Correlations Between Allosteric Sites in Thermodynamic Ensembles. J Chem Theory Comput 5: 2486–2502.

53. Dubay KH, Bothma JP, Geissler PL (2011) Long-range intra-protein communication can be transmitted by correlated side-chain fluctuations alone. PLoS Comput Biol 7: e1002168.

54. Bowman GR, Geissler PL (2012) Equilibrium fluctuations of a single folded protein reveal a multitude of potential cryptic allosteric sites. Proc Natl Acad Sci U S A 109: 11681–11686.

55. Jorgensen WL, Chandrasekhar J, Madura JD, Impey RW, Klein ML (1983) Comparison of Simple Potential Functions for Simulating Liquid Water. Journal of Chemical Physics 79: 926–935.

56. Berendsen HJC, Postma JPM, Van Gunsteren WF, Hermans J (1981) Interaction models for water in relation to protein hydration; Pullman B, editor: Reidel. 331–342 p.

57. Hess B, Kutzner C, van der Spoel D, Lindahl E (2008) GROMACS 4: Algorithms for highly efficient, load-balanced, and scalable molecular simulation. Journal of Chemical Theory and Computation 4: 435–447.

58. Miyamoto S, Kollman PA (1992) Settle – an Analytical Version of the Shake and Rattle Algorithm for Rigid Water Models. Journal of Computational Chemistry 13: 952–962.

59. Nosé S (1984) A molecular dynamics method for simulations in the canonical ensemble. Molecular Physics 52: 255–268.

60. Hoover WG (1985) Canonical dynamics: Equilibrium phase-space distributions. Phys Rev A 31: 1695–1697.

61. Parrinello M, Rahman A (1981) Polymorphic transitions in single crystals: A new molecular dynamics method. Journal of Applied Physics 52: 7182–7190.

62. Nosé S, Klein ML (1983) Constant pressure molecular dynamics for molecular systems. Molecular Physics 50: 1055–1076.

63. Essmann U, Perera L, Berkowitz ML, Darden T, Lee H, et al. (1995) A Smooth Particle Mesh Ewald Method. Journal of Chemical Physics 103: 8577–8593.

64. Humphrey W, Dalke A, Schulten K (1996) VMD: Visual molecular dynamics. Journal of Molecular Graphics 14: 33–&.

65. Pettersen EF, Goddard TD, Huang CC, Couch GS, Greenblatt DM, et al. (2004) UCSF chimera – A visualization system for exploratory research and analysis. Journal of Computational Chemistry 25: 1605–1612.

66. Altschul SF, Madden TL, Schaffer AA, Zhang J, Zhang Z, et al. (1997) Gapped BLAST and PSI-BLAST: a new generation of protein database search programs. Nucleic Acids Res 25: 3389–3402.

67. Sievers F, Wilm A, Dineen D, Gibson TJ, Karplus K, et al. (2011) Fast, scalable generation of high-quality protein multiple sequence alignments using Clustal Omega. Mol Syst Biol 7: 539.

68. Goujon M, McWilliam H, Li W, Valentin F, Squizzato S, et al. (2010) A new bioinformatics analysis tools framework at EMBL-EBI. Nucleic Acids Res 38: W695–699.

69. Gouy M, Guindon S, Gascuel O (2010) SeaView version 4: A multiplatform graphical user interface for sequence alignment and phylogenetic tree building. Mol Biol Evol 27: 221–224.

6

Cell Cycle Modulation by Marek's Disease Virus: The Tegument Protein VP22 Triggers S-Phase Arrest and DNA Damage in Proliferating Cells

Laëtitia Trapp-Fragnet[1]*, **Djihad Bencherit**[1], **Danièle Chabanne-Vautherot**[1], **Yves Le Vern**[2], **Sylvie Remy**[1], **Elisa Boutet-Robinet**[3,4], **Gladys Mirey**[3,4], **Jean-François Vautherot**[1], **Caroline Denesvre**[1]

1 INRA, UMR1282 Infectiologie et Santé Publique, Equipe Biologie des Virus Aviaires, Nouzilly, France, 2 INRA, UMR1282 Infectiologie et Santé Publique, Laboratoire de Cytométrie, Nouzilly, France, 3 INRA, UMR 1331, Toxalim, Research Centre in Food Toxicology, Toulouse, France, 4 University of Toulouse, UPS, UMR1331, Toxalim, Toulouse, France

Abstract

Marek's disease is one of the most common viral diseases of poultry affecting chicken flocks worldwide. The disease is caused by an alphaherpesvirus, the Marek's disease virus (MDV), and is characterized by the rapid onset of multifocal aggressive T-cell lymphoma in the chicken host. Although several viral oncogenes have been identified, the detailed mechanisms underlying MDV-induced lymphomagenesis are still poorly understood. Many viruses modulate cell cycle progression to enhance their replication and persistence in the host cell, in the case of some oncogenic viruses ultimately leading to cellular transformation and oncogenesis. In the present study, we found that MDV, like other viruses, is able to subvert the cell cycle progression by triggering the proliferation of low proliferating chicken cells and a subsequent delay of the cell cycle progression into S-phase. We further identified the tegument protein VP22 (pUL49) as a major MDV-encoded cell cycle regulator, as its vector-driven overexpression in cells lead to a dramatic cell cycle arrest in S-phase. This striking functional feature of VP22 appears to depend on its ability to associate with histones in the nucleus. Finally, we established that VP22 expression triggers the induction of massive and severe DNA damages in cells, which might cause the observed intra S-phase arrest. Taken together, our results provide the first evidence for a hitherto unknown function of the VP22 tegument protein in herpesviral reprogramming of the cell cycle of the host cell and its potential implication in the generation of DNA damages.

Editor: Bruce W. Banfield, Queen's University, Canada

Funding: This study was supported by the "Ligue contre le Cancer du Grand-Ouest", comities of "Indre et Loire" and "Vendée" (http://www.ligue-cancer.net). The "Cluster en Infectiologie de la Région Centre" (Http://www.infectiologie-regioncentre.fr) supported the technical collaboration between the "Biologie des Virus Aviaires" laboratory and the Research Center in Food Toxicology. The funders had no role in study design, data collection and analysis, decision to publish, or preparation of the manuscript.

Competing Interests: The authors have declared that no competing interests exist.

* E-mail: fragnet@tours.inra.fr

Introduction

Gallid herpesvirus 2 (GaHV-2), more frequently referred to as Marek's disease virus (MDV), is an alphaherpesvirus (type species of the genus Mardivirus) and the causative agent of a highly infectious lymphoproliferative disease termed Marek's disease (MD) affecting many birds in the *Phasianidae* family. Despite global vaccination campaigns that are effective to prevent disease development, MDV field strains continue to spread in poultry and appear to evolve towards increased virulence. The dissemination of MDV in poultry is mediated by infectious viral particles associated with dander and feather debris [1,2]. With the exception of the feather follicle epithelium, the site where free infectious viral particles are shed, the virus remains strictly cell-associated and progression of the infection is restricted to viral cell-to cell spread [3]. The MDV particle is composed of a 180-kbp double-strand DNA genome packaged in an icosaedric capsid surrounded by a tegument layer, which insures the morphological and functional continuity between the capsid and the host cell derived viral envelope. By homology with other alphaherpes-

viruses, a number of viral proteins composing the tegument have been identified, including a major tegument protein, VP22 (pUL49), various trans-activators and two protein kinases (pUL13 and pUS3). The UL49-encoded VP22 protein is abundantly expressed in infected cells and is essential for MDV replication [4,5,6]. VP22 is a specific tegument protein of alphaherpesviruses and conserved among this subfamily. To date, the absolute requirement of the UL49 gene for viral replication was initially demonstrated for MDV [5] and afterwards for Varicella Zoster virus (VZV) [7]. The deletion of VP22 in other alphaherpesviruses including Herpes Simplex virus 1 (HSV-1), Pseudorabies virus (PRV), Bovine herpesvirus 1 (BoV-1) still allows viral replication, even though viral spread is reduced in some cell types [8,9,10,11,12]. While its role in virus infection remains unclear, it was demonstrated for HSV-1 that VP22 interacts with and recruits various viral proteins, such as the trans-activators ICP0, ICP4 and viral glycoproteins composing the infectious virions [9,10,13]. Furthermore, VP22 was shown to interact with cellular proteins involved in the organization of microtubules and nucleosome assembly [14,15]. The VP22 protein encoded by

MDV shares common functional features with VP22 encoded by other alphaherpesviruses [5,16]. It was previously shown that MDV-VP22 shows both a cytoplasmic and nuclear location in infected cells and accumulates in the nucleus upon overexpression in cells [4]. Moreover, MDV-VP22 exhibits a strong affinity to DNA, especially heterochromatin, and to microtubules [4,17]. We previously demonstrated the role of VP22 in MDV cell-to-cell spread, which could explain the necessity of VP22 in MDV replication [16,18]. It was recently shown that recombinant MDV viruses expressing VP22 with a C or N-terminal GFP-tag are highly attenuated in vivo suggesting that VP22 might play a role in MDV-induced lymphomagenesis [6,19]. However, the precise role of VP22 in MDV replication and MD pathogenesis remains unclear. Notably, the functional significance of the VP22 nuclear distribution is still unknown, even if previous reports on VP22 encoded by alphaherpesviruses evoke a possible regulatory function of VP22 within nuclei [17,20,21,22].

Virus infection frequently results in the disturbance of key cellular processes within the host cell. The subversion of cell cycle pathways is a well-established mechanism by which viruses create the most suitable environment for their replication. Especially, the induction of S-phase is either mandatory or at least advantageous for lytic replication of a number of viruses. The eminent role of cellular factors from the DNA synthesis machinery in viral replication was demonstrated for viruses from different families such as the Flaviviridae, Retroviridae, Parvoviridae, and Polyomaviridae [23,24,25,26,27,28]. In contrast, herpesviruses encode their own DNA polymerase and accessory proteins, and thus theoretically do not require an S-phase environment to support their replication (reviewed in [29,30]). Nevertheless, several studies have demonstrated the importance of the S-phase in the life cycle of VZV and Epstein-Barr virus (EBV) [31,32]. For EBV, S-phase cyclin-dependent kinase activity is essential for the expression of immediate early and early viral proteins and is thus required for viral replication [31]. Vice versa, EBV lytic replication is able to provide a S-phase-like cellular environment by modulating DNA damage pathways [33]. The impact of the S-phase environment on the viral life cycle is not restricted to lytic viral replication but is also involved in the episomal genome maintenance during viral latency or reactivation processes, as was recently shown for EBV and the Kaposi's Sarcoma-associated herpesvirus (KSHV) [34,35]. Strikingly, infections with oncogenic viruses (e.g., SV40, HPV, HTLV-1, EBV) are often associated with S-phase deregulation and genomic instability, preferentially occurring during this critical phase of the cell cycle [24,27,33,36].

In relation with cell cycle delay, DNA damage signaling is often triggered upon viral infections (reviewed in [37,38]). Particularly, DNA damage response (DDR) pathways are preferential targets of herpesviruses, including HSV-1, EBV, KSHV, human cytomegalovirus and murine gamma-herpesvirus 68 [33,39,40,41,42,43]. The role of DDR in herpesviruses life cycle is complex. On the one hand, recent evidence suggests that DDR acts as an efficient antiviral response [44]. On the other hand, DDR modulation can be beneficial for herpesviruses by facilitating viral replication, viral genome processing or latency establishment [33,34,40,42,45]. Moreover, during the course of cellular infections with large DNA tumour viruses, such as human papillomaviruses (HPV) or gammaherpesviruses (e.g. EBV and KSHV), the generation of DNA damage and/or activation of DDR were found to be associated with genomic instability which in turns can participate to virus-induced tumorigenesis (reviewed in [46,47,48]).

In the present study, we set out to elucidate an important aspect of MDV-host cell interaction by analyzing the impact of MDV and virus-encoded proteins on the regulation of the cell cycle. We demonstrate that MDV lytic infection activates the proliferation of chicken primary skin cells concomitant with a delay in S-phase. By studying the effects of transient vector-driven overexpression in a proliferating chicken cell line, we identified the VP22 tegument protein as a potent cell cycle modulator encoded by MDV. A comparative experimental approach employing VP22 variants with a C- or N-terminal eGFP-tag allowed us to show that an unmodified C-terminus of VP22 is required to elicit the observed S-phase arrest. Moreover, the cell cycle regulating activity of VP22 relies on its ability to be associated with chromatin in the nucleus. In order to define the mechanisms underlying the drastic S-phase arrest observed in VP22 expressing cells, we investigated the impact of VP22 expression on DNA integrity. Strikingly, we found that the DNA of cells expressing this viral protein showed significant DNA damage, as was assessed by comet assay. Together, these data provide new insights into the interaction of MDV with the host cell during lytic replication and pinpoint to a novel powerful function of VP22 that may help to better understand the pre-eminent role of VP22 in MDV replication and more generally in the life cycle of the virus.

Materials and Methods

Cell Culture and Viruses

Chicken Embryo Skin Cells (CESC) were prepared from 12-days-old chicken embryos (LD1 Brown Leghorn chicken strain) and maintained in culture as previously described [4]. This procedure was carried out in strict compliance with the French legislation for animal experiments and ethics stating that the use of embryos from oviparous species before the last third of their normal development (i.e. before day 14 for chicken embryos) is not submitted to regulation (Art. R.214-88). Thus, the preparation of CESC from 12-days-old chicken embryos does not require the permission of governmental or local authorities. Embryos were sacrificed by opening the eggshell, cervical dislocation and immersion in William's Medium E (Lonza) supplemented with collagenase as described by Dorange et al., 2000. The chicken hepatocellular carcinoma cell line LMH was cultured on gelatin-coated flasks in William's Medium E (Lonza) supplemented with 2 mM glutamine and 10% fetal bovine serum (FBS) at 37°C in a 5% CO_2 atmosphere. As positive controls for DNA damage analyses, LMH cells were treated for 24 h with 1.5 μM etoposide, a topoisomerase IIα inhibitor potent inducer of DNA double strand breaks.

Recombinant viruses were generated from the avirulent MDV-BAC20 strain cloned as bacterial artificial chromosome (BAC) [49]. The recEGFPVP22 recombinant virus harboring the UL49 gene fused at its 5′ end with the eGFP gene was previously described [16].

Parental BAC20 and recEGFPVP22 viruses were produced after transfection of BAC-DNA into CESC as previously reported [16]. Infections were performed by co-culture of 7×10^6 fresh CESC in a 100-mm diameter plate with infected cells at a ratio of 10^4 PFU/plate.

Plasmids

The pcDNA3-UL49 and pcDNA3-UL48 plasmids encoding the wild-type (wt) VP22 and VP16 tegument proteins of the RB-1B oncogenic MDV-RB-1B strain, have been previously described [4]. Two plasmids harboring the VP22 protein cloned in frame with the enhanced green fluorescent protein (eGFP) were used: (i) the peGFP-UL49, encoding a VP22 tagged with eGFP at its N-terminal extremity [18] and (ii) the pUL49-eGFP in which VP22 is tagged with eGFP at its C-terminus. The latter construct was generated by PCR amplification of the UL49-eGFP fragment from the purified rUL49-eGFP BAC-DNA kindly provided by B. Kaufer (Institut für Virologie, Freie Universität Berlin, Germany) [19]. The primer pairs used for amplification were UL49FCL-BamHI/eGFPendNotI (Table 1). The PCR product was inserted into the PCR2.1 TOPO TA cloning vector (Invitrogen) and the BamHI/NotI fragment was then sub-cloned into the peGFP-N1 vector (BD Biosciences, Clontech) where the internal eGFP cassette was previously removed by BglII/NotI enzymatic cleavage. The pGE109 plasmid harbouring the UL49 gene encoded by HSV-1 was kindly provided by G. Elliott [50]. The HSV1-UL49 gene was cloned in frame with eGFP at the Bgl II site in the peGFP-C1 vector (BD Biosciences, Clontech). The VZV-ORF63 encoding the VZV orthologue of UL49 was amplified from pcDNA63wt (kindly provided by C. Sadzot-Delvaux) with the primer pair 5FUL49VZVXhoI/3RUL49VZVBamHI (Table 1) [51]. The PCR product was TA-cloned into the pGEMT-easy cloning vector (Promega) and subsequently sub-cloned in fusion with eGFP in the peGFP-C1 vector at the Xho I and Bam HI sites.

The genes encoding UL37 (pUL37), UL54 (encoding the ICP27 trans-activator), and the two viral kinases UL13 (pUL13) and US3 (pUS3) were amplified from RB-1B genomic DNA with the primer pairs UL37F/UL37R; UL13F/UL13R; UL54F/UL54R, and US3F/US3R, respectively (Table 1). Amplification products were inserted into the pGEMT-easy cloning vector (Promega). The UL37, UL13, and US3 genes were sub-cloned under control of the cytomegalovirus immediate early promoter into the pcDNA3.1 vector (Invitrogen) at the NotI site, and the UL54 gene was cloned into the pcDNA3.1 vector at the EcoRV site. All intermediate and final constructs were verified by sequencing (Eurofins, MWG Operon).

Transient Expression

The different eukaryotic expression vectors were transfected into CESC or LMH by using Lipofectamine 2000, according to the manufacturer's instructions (Invitrogen). Briefly, cells at 80% of confluency plated on 60-mm dishes were rinsed twice with OptiMEM (Fischer Scientific) and were transfected with 5 μg of the plasmid of interest. After 6 h of incubation at 37°C, the transfection mix was removed and serum complemented fresh medium was added. Cells were harvested 24 h or 48 h after transfection for further analysis. Each transfection was performed in triplicate.

Cell Cycle Analysis

At the time points indicated, 1.10^6 vector-transfected or infected cells were trypsinized and washed twice in phosphate-buffered saline (PBS) prior to fixation with 70% ethanol at 4°C for 24 h. Cells were then washed twice in cold PBS and incubated in PBS containing 500 μg/ml Ribonuclease A (Sigma-Aldrich) at 37°C for 1 h. After filtration through a 30-μm pore size membrane, cells were stained with 10 μg/ml propidium iodide (Invitrogen) for 15 min in the dark. Flow cytometry analysis was performed using a MoFlo high-speed cell sorter (Beckman Coulter, Fort Collins, CO, USA) equipped with a solid-state laser operating at 488 nm and 100 mW. Cellular DNA content was analyzed with a 740 nm long-pass filter. Doublets were discarded on the basis of combination of pulse width and area/peak fluorescence. eGFP autofluorescence was detected with a 530/40 nm band-pass filter and the cell cycle distribution was specifically analyzed for eGFP-positive versus eGFP-negative cells. Cell cycle profiles were analyzed with the MultiCycle AV software (Phoenix Flow Systems, California, USA).

Reverse Transcription-Polymerase Chain Reaction (RT-PCR) and Real Time Quantitative RT-PCR (qRT-PCR)

Total RNA was extracted from 10^6 cells with Trizol according to the manufacturer's instructions (Sigma-Aldrich). RNAs were treated with RNAse-free RQ1 DNAse (Promega, France) and RNA concentration was measured with a NanoDrop spectrophotometer. One μg of each total RNA preparation was reverse transcribed using 100 μg/mL oligo(dT) primers (Promega) and M-MLV reverse transcriptase according to the manufacturer's recommendations (Promega).

The expression of the different cellular genes involved in cell cycle regulation was analyzed by qPCR. Amplification of the cDNA by qPCR (CFX96 Touch Real-Time PCR Detection System; Bio-Rad) was performed in triplicate, using 200 ng of

Table 1. Primers used for the cloning of the genes of interest in eukaryotic expression vectors.

Primer names	Primer sequences
UL49FCLBamHI[1]	5'GG**GGATCC**CATGGGGGATTCTGAAAGGCGG3'
eGFPendNotI[2]	5'AGGCGGCCGCTTACTTGTACAGCTCGTCCATG3'
5FUL49VZV XhoI[3]	5'GG**CTCGAG**CTTTGGCATCTTCCGACGGGTGA3'
3RUL49VZVBamHI[1]	5'GG**GGATCC**CTATTTTCGCGTATCAGTTC3'
UL37F	5'AGGCCTATGTCTGCCGTAACGACCGA3'
UL37R	5'CTGCAGTTATGCATTATCACCGTTTG3'
UL13F	5'CCAGATCTATGGATACTGAATCAAAAAAC3'
UL13R	5'CCGAATTCCTAGTTCCATAACAACAAATC3'
UL54F	5'AGGCCTATGTCTGTAGATGCATTCTCTC3'
UL54R	5'AGGCCTCTAGAATTACATACCAAACAGAGTATTGCAG3'
US3F	5'CCAGATCTATCTCTTCGAGTCCGGAGGC3'
US3R	5'CCGAATTCTTACATATGAGCGGCAGTTATC3'

[1]BamHI restriction site is indicated in bold.
[2]NotI restriction site is underlined.
[3]XhoI restriction site is indicated in bold underlined.

cDNA, 7.5 µl 2×iQ Supermix SYBR green (Bio-Rad), 1 µl ultrapure water (Sigma-Aldrich) and 0.75 µl of each specific primer (10 µM) selected according to the EST data deposited in Genbank (described in Table 2). The PCR program consisted of a 5 min activation step at 95°C, followed by 39 cycles of 95°C for 10 s and 60°C for 10 s. Expression of the chicken glyceraldehyde phosphate dehydrogenase (GAPDH) was used for the normalization of all target gene mRNAs to enable cross-comparisons among the samples. The relative changes in gene expression were determined by the 2(−ΔΔCT) method.

The expression of MDV genes (ICP4, UL13, US3, UL49, UL37, UL54 and UL48) was assessed by RT-PCR performed with 100 ng of the synthesized cDNA prepared from LMH or CESC cells transfected with the corresponding expression vector and 10 µM of specific primers (Table 2). The GAPDH gene was used as internal control. Specific PCR products were resolved by agarose (2%) gel electrophoresis.

Immunofluorescence Microscopy

At 24 h or 48 h post-transfection, cells grown on glass coverslips were fixed with 4% paraformaldehyde (PFA) for 20 min at room temperature (RT), permeabilized with 0.5% Triton X-100 for 5 min at RT and blocked with PBS, 0.1% Triton X-100, 2% Bovine Serum Albumin (BSA). Immunostainings were performed with monoclonal antibodies directed against phospho-histone H2AX (Ser139) (Millipore; clone JBW301) and tubulin (Sigma-Aldrich; catalog number T9026) at a dilution of 1:250 and 1:500, respectively. Goat anti-mouse IgG Alexa-Fluor 594 secondary

Table 2. List of forward (For) and reverse (Rev) primers used for (q)RT-PCR.

Target gene		Primers sequences	Accesion number gene
Cyclin A	For	5'CAAGCTCCAGAATGAAACTC3'	GenBank X72892
	Rev	5'GATGTAGACGAACTCTGCTA3'	
Cyclin B	For	5'GTTCTGTCTCCTGTCCCTAT3'	GenBank X62531
	Rev	5'AGCTCAAGCTGTCTCAGATA3'	
Cyclin D	For	5'GCATTTACACCGACAACTCC3'	GenBank NM_205381
	Rev	5'GCATGTTTACGGATGATCTG3'	
Cyclin E	For	5'TCACCGCTACCAATTCTGGG3'	ENSGALT00000007159
	Rev	5'ACTTCACAAACCTCCATTAG3'	
Cdk1	For	5'GTAGTGACACTGTGGTACAG3'	GenBank NM_205314
	Rev	5'CTGAAGATTCTGAAGAGCTG3'	
Cdk6	For	5'ATGTTGATCAGCTAGGAAAA3'	GenBank NM_001007892
	Rev	5'CGATTTAAGAAGCAAGTCTT3'	
pRb	For	5'GATGTGTTCCATGTATGGCA3'	GenBank NM_204419
	Rev	5'TGAACACTAAGTTGTAGAAG3'	
p53	For	5'CAGCCAAATCGGTCACCTGC3'	Genbank NM_205264
	Rev	5'CCGCACCACTTCGGCCACGT3'	
c-Myc	For	5'CCGAGGACATCTGGAAGAAGTT3'	GenBank J00889.1
	Rev	5'TCGCAGATGAAGCTCTGGTTGA3'	
E2F1	For	5'CGGTGAAGCGGAAGCTGAAC3'	GenBank NM_205219
	Rev	5'GCTCCAGGAAGCGCTTGGTG3'	
ICP4	For	5'TTTCTAGCAAGGAGCGACGC3'	GenBank NC_002229.3
	Rev	5'CTGACTTGCGCTTACGGGAA3'	
UL13	For	5'CTCGGCAAAGCAGTTGTGTTTC3'	Genbank EF523390
	Rev	5'GTCAGACTAAAATCACCAATTAC3'	
US3	For	5'CGCCTGAACTGCTTGCACTTG3'	Genbank EF523390
	Rev	5'TGGATCTCAGCTGAGAACCTG3'	
UL49	For	5'GGAAGACGTTTCGTCTACCAC3'	Genbank EF523390
	Rev	5'CATACGCTGATTAAATGCCACTG3'	
UL37	For	5'GCGCATTGGGTTCGAAAGAAC3'	Genbank EF523390
	Rev	5'CGTTGCGCAACTTATCAGCCG3'	
UL54	For	5'CTCGCGAGTCCGATGACATG3'	Genbank EF523390
	Rev	5'GTTCCTCGTACACGGTGGC3'	
UL48	For	5'CGAAGCATCTCAAATGGGACG3'	Genbank EF523390
	Rev	5'CTGAGCCCGGAGTTCAGGAG3'	
GAPDH	For	5'TGATGATATCAAGAGGGTAGTGAAG3'	GenBank K01458
	Rev	5'TCCTTGGATGCCATGTGGACCAT3'	

antibody (Invitrogen) was used at 1:2000. Cell nuclei were counterstained with Hoechst 33342 dye (Invitrogen). Cells were observed under an Axiovert 200 M inverted epifluorescence microscope equipped with the Apotome imaging system (Zeiss). Images were captured with an Axiocam MRm camera and analyzed by using the Axiovision software (Zeiss). To determine the cellular distribution of eGFP-tagged proteins, a minimum of 100 transfected cells were observed and the results were presented as percentage reflecting the nuclear and/or cytoplasmic distribution of the protein.

Cell Sorting

RecEGFPVP22-infected cells or LMH cells transfected with peGFP vectors were trypsinized 24 h post-transfection and filtered on a 30-μm-pore-size membrane. eGFP positive and negative cells were sorted with a MoFlo (Beckman Coulter, Fort Collins, CO, USA) high-speed cell sorter equipped with a solid-state laser operating at 488 nm and 100 mW. Damaged cells and debris were eliminated on the basis of morphological criteria. eGFP fluorescence was analyzed with a 530/40 nm band-pass filter. The sorting speed was around 15,000 cells/s and cells were collected in appropriate media supplemented with 10% of FBS.

Alkaline Comet Assay

LMH cells transfected with the peGFP, peGFP-UL49 or pUL49-eGFP were harvested 24 h post-transfection and eGFP positive and negative cell were sorted by flow cytometry. After sorting, 2.10^5 cells were used to prepare 3 slides for comet assays, realized as previously described with minor modifications [52]. Electrophoresis was performed at 0.7 volts/cm for 26 min with the Sub-cell GT agarose gel electrophoresis system (Bio-Rad). DNA was then stained with a 20 μg/ml ethidium bromide solution and slides were observed using the Axiovert 200 M inverted epifluorescence microscope (Zeiss). Images were captured with an Axiocam MRm camera (Zeiss) and comets were analyzed with the CometScore software version 1.5 (TriTek). The Tail Extend Moment (TEM) was calculated on the basis of the comet tail length and the relative proportion of DNA contained in the tail. Experiments were carried out 3 times and for each experiment, a minimum of 50 comets was analyzed on each of the 3 slides. Results are presented as the mean (±SD) of the TEM calculated for each condition or as a distribution of the comets with respect to their respective TEM value.

High Salt Extraction of Histones

Salt extraction of histones from chromatin was performed as previously described [53]. Briefly, 1.10^7 cells were resuspended in 1 ml extraction buffer (340 mM Sucrose, 10 mM Hepes pH 7.9, 10 mM KCl, 1.5 mM $MgCl_2$, 10% glycerol containing 0.2% Igepal (Sigma-Aldrich) and 1X protease inhibitors (Complete Mini EDTA free, Roche). After incubation on ice for 10 min, the soluble fraction was separated from the nuclei by centrifugation at 6,500×g for 5 min. Nuclei were resuspended in 1 ml no-salt lysis buffer (3 mM EDTA, 0.2 mM). After incubation at 4°C for 30 min, the chromatin was pelleted by centrifugation at 6,500×g for 5 min, and incubated in 500 μl of high-salt solubilization buffer (50 mM Tris-HCl pH 8.0, 2.5 M NaCl and 0,05% NP40) for 30 min at 4°C. Nuclear debris were pelleted by centrifugation at 16,000×g for 10 min and the supernatant containing the histones fraction was collected. The proteins included in this fraction were separated in a 10% SDS-PAGE gel and revealed with colloidal coomassie blue staining (Sigma-Aldrich). Detection of VP22 was accomplished by immunoblotting using the monoclonal anti-VP22 antibody (L13a, [4]) diluted 1:1000 and an anti-mouse IgG conjugated to horseradish peroxidase (HRP) (Sigma-Aldrich). Specific protein signals were detected with the Pierce ECL2 Western Blotting Substrate (Thermo Scientific) and the Fusion-FX7 imaging system (Vilber Lourmat). Quantification was carried out using the Bio-profil 1D++ software (ChemiSmart 5000).

Statistical Analysis

All graphs and statistics were performed using the GraphPad Prism software version 5.02 (San Diego, USA). Data are presented as means and standard deviations (±SD). Significant differences were determined using Student's t-test P values <0.05 were considered statistically significant.

Results

MDV Infection Delays Cell Cycle Progression in S-phase

In order to analyze the influence of MDV infection on cell cycle progression, chicken embryonic skin cells (CESC) were infected with the parental BAC20 virus. At days 1, 4, and 6 post-infection (pi), mock- and virus-infected cells were fixed in ethanol, DNA was stained with propidium iodide and DNA content was analyzed by flow cytometry (Fig. 1A). While no significant difference in the cell cycle progression was observed in the early steps of infection (1 dpi), at day 4 pi the cell population in S-phase in BAC20-infected cells was about 3-fold higher than in mock-infected cells. At day 6 pi, the proportion of cells in S-phase as well as in the G2-phase remained 3-fold higher in BAC20-infected cells, suggesting that MDV infection activates cell cycle progression of CESC that normally exhibit a low proliferating rate (3 to 4% of cells in S-phase) and that MDV may delay the cell cycle in S-phase. In order to ascertain viral replication, mRNA expression of the early ICP4 viral gene was followed by qRT-PCR (Fig. 1B lower panel). To confirm the activation of the cell cycle progression assessed by DNA content analysis, and to define the molecular mechanisms of this process upon infection, we examined the expression of key factors involved in cell cycle regulation including cyclins and cyclin dependent kinases (cdk). CESC were mock-infected or infected with BAC20, and qRT-PCR analyzes were performed on total mRNAs extracted at 1, 4, and 6 days pi (Fig. 1B). At 4 dpi, BAC20-infected cells showed an increase of the mRNA expression of cyclin D (of about 6 fold), cdk6 (2.3-fold), pRb (3-fold), E2F1 (3.5-fold) and c-myc (3.1-fold) compared to mock-infected control cells. A slight up-regulation of cyclin A (1.7-fold), cyclin B (1.5-fold) and cdk1 (1.6-fold) mRNA expression was also detected at 1 dpi in infected cells compared to non-infected cells, while the level of cyclin E mRNA expression was comparable to that in mock-infected cells. These observations are in good agreement with the DNA content analyses showing an activation of the proliferative capacities of infected CESC, since cell cycle progression markers, especially cellular factors involved in the progression into G1 and S-phases (cyclin D, cyclin A, cdk6, pRb, c-Myc, and E2F1) were up-regulated during MDV infection. Interestingly, analysis of the mRNA expression pattern of p53, a protein crucially involved in cell cycle checkpoints and DNA damage pathways, revealed a strong up-regulation (of about 5.9 fold) of its expression at 4 dpi. Of note, we also observed a down-regulation of the mRNA expression of cdk6, pRb, E2F1 and c-myc at 6 dpi that reflect the non-progression of the cell cycle in G1/S phase.

To specifically determine the regulation of the cell cycle in infected cells and to preclude problems associated to asynchronous infection and moderate infectivity titers, CESC infected with the recEGFPVP22 virus, an MDV recombinant virus expressing an eGFP-VP22 fusion protein, were sorted by flow cytometry [18].

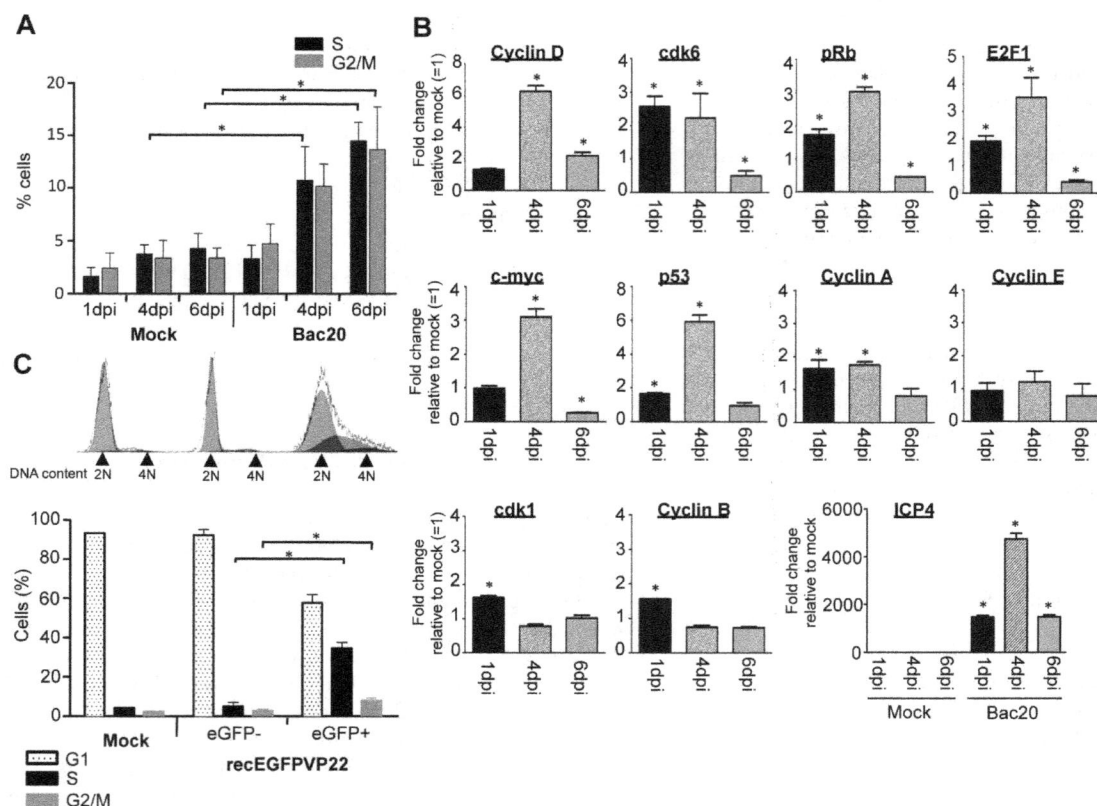

Figure 1. Cell cycle regulation during MDV infection. (A) Cell cycle analysis of CESC infected with the BAC20 virus. Mock-infected or infected CESC were harvested 1, 4, and 6 days pi and the cellular DNA content was analyzed by flow cytometry after staining with propidium iodide. The percentage of cells in S and G2/M phases is represented as bars. *$p<0.05$. (B) mRNA expression of cell cycle regulators during MDV infection. RNAs were extracted from mock-infected and BAC20-infected cells at indicated time points. The mRNA expression levels of cellular key cell cycle regulators were detected by qRT-PCR using specific primers given in Table 2. To assess lytic viral replication, the expression of the ICP4 gene encoded by MDV was analyzed in parallel by qRT-PCR. Expression of GAPDH mRNA was used for the normalization of mRNA expression levels of all target genes and the relative changes in gene expression were determined by the $2(-\Delta\Delta CT)$ method. Results are presented as histograms showing the fold-change of expression of the target gene in infected cells relative to its expression in mock-infected cells (Mock = 1). *$p<0.05$. (C) Cell cycle regulation in MDV infected cells. CESC were mock-infected or infected with the recEGFPVP22 virus. Six days post-infection, cells were trypsinized, fixed, and the DNA was stained with propidium iodide. DNA content was analyzed by flow cytometry in late infected cells (eGFP+) and non-infected cells (eGFP−). Single parameter histograms are shown, and the results corresponding to the percentage of cells in G1, S, and G2/M phases of 3 independent experiments is presented as mean (±SD). *$p<0.05$.

Infected cells monolayers were harvested at 6 days pi, and the DNA content in non-infected (eGFP-negative) and infected cells (eGFP-positive) was analyzed by flow cytometry. Using this approach, we could observe that 34.4% of the eGFP-positive cells were delayed in S-phase (Fig. 1C), while in the eGFP-negative population the percentage of cells in S-phase was equivalent to that in mock-infected cells (up to 5%). In addition, a slight increase of cells in G2-phase was detected in infected cells (8% compared to 2.8% in eGFP-negative cells or mock-infected cells). All together, these data clearly demonstrate that lytic MDV infection drives primary avian cells into an active proliferating state. Furthermore, the significant increase of MDV-infected cells accumulating in S-phase is also indicative of a virus-mediated delay in S-phase progression.

VP22 is a Major Cell Cycle Regulator

To identify viral factors involved in the regulation of the cell cycle during MDV infection, we tested the impact of the overexpression of six different viral proteins in CESC (low rate proliferating primary cells) and LMH cells (a cell line with high proliferative rate). Putative candidates were selected either on the

basis of their biological activities that might influence host cell-encoded cell cycle regulators and/or on the basis of their essential role in the MDV life cycle. Because of the central role of cellular kinases in cell cycle progression, we were interested to test the two kinases encoded by MDV, pUL13 and pUS3. The ICP27 protein, encoded by the UL54 gene, was also included in the study as a multifunctional viral regulatory protein that has previously been shown to contribute to cell cycle modulation during HSV-1 infection [54,55]. Three tegument proteins were also tested: the UL48-encoded viral trans-activator VP16, as well as pUL37 and VP22, both of which were shown to be essential for MDV growth (J-F Vautherot, unpublished data; [5]). Eukaryotic expression vectors harboring the viral candidate genes UL37, UL48, UL49, and UL54 (encoding pUL37, VP16, VP22 and ICP27, respectively) were transiently transfected into LMH or CESC cells. At 48 h post-transfection, the cell cycle status was analyzed as outlined earlier and the expression of each of the transfected MDV genes was verified by RT-PCR from total RNA extractions. No significant differences in the proportion of cells in each cell cycle phase was observed (Fig. 2A, left panel) for transfected CESC, suggesting that none of the overexpressed proteins was able to impact the cell cycle in quiescent CESC. In LMH cells, we also did not observe any cell cycle regulation in response to the expression of UL13, US3, UL37, UL54 and UL48, despite an effective expression of their respective mRNA (Fig. 2A lower panel). However, VP22 (pUL49) overexpression had a substantial effect on the cell cycle in the LMH cell line (Fig. 2A, right panel), as shown by the strong accumulation of cells in S-phase compared to control cells transfected with the empty vector pcDNA (35% versus 18% of cells in S-phase). Next, we tried to confirm our finding that VP22-expression alone results in an increase of cells in S-phase by transfecting LMH cells with plasmids encoding the VP22 protein fused to a eGFP-tag at its N- or C-terminus. Using an N-terminal eGFP-tagged VP22 protein (peGFP-UL49), we could confirm our finding that VP22 modulates the cell cycle, since more than 90% of LMH cells expressing VP22 (eGFP-positive cells) were blocked in S-phase (Fig. 2B). However, cells transfected with the plasmid encoding VP22 tagged at its C-terminus did not show any difference in cell cycle regulation compared to empty vector (peGFP)-transfected cells, which indicates that the location of the eGFP-tag at the carboxy-terminal extremity of the VP22 protein abrogates its activity on the cell cycle. Of note, the dramatic intra S-phase arrest observed with the N-terminal eGFP-tagged VP22 protein could be reproduced after overexpression of VP22 in two other avian cell lines: the chicken fibroblast cell line DF1 and the quail myoblast cell line QM7 (data not shown).

To verify whether the S-phase promoting activity of the MDV-encoded UL49 is conserved in other alphaherpesvirus orthologues, we tested the ability of VP22 encoded by HSV-1 and VZV to regulate the cell cycle. The HSV-1 and VZV-UL49 genes were cloned in-frame with eGFP and transiently overexpressed in the LMH cell line. At 48 hours post-transfection, the flow cytometry-based cell cycle analysis targeting transfected cells (eGFP-positive population) showed a significant S-phase arrest upon expression of all VP22 orthologues tested (Fig. 2C). VP22 orthologues derived from MDV and VZV proofed to be equally efficient, as approximately 80% of the cells expressing these VP22 were blocked in S-phase. Although HSV-1-VP22 substantially blocked the cell cycle progression in S-phase (61.8% of the transfected cells), it appeared slightly less efficient than other VP22 orthologues (especially MDV-VP22) in this process.

We thus identified a novel function for MDV-VP22 as a potent cell cycle modulator, with a strong S-phase promoting activity. We also revealed that an unmodified C-terminal extremity of VP22 is required for this process. Moreover this biological feature seems to be conserved among the human alphaherpesvirus, even though the two VP22 orthologues tested does not exhibit equal activity.

Subcellular Localization of the VP22 Protein Encoded by MDV

We took advantage of the differential cell cycle modulating activities of the C- or N-terminally eGFP-tagged VP22 fusion proteins to decipher which VP22 properties are crucial to mediate S-phase arrest. One hypothesis for different activity patterns could rest on differential subcellular distributions of the two proteins. To test this hypothesis, the two constructs peGFP-UL49 and pUL49-eGFP were transfected in LMH cells and the respective locations of the proteins were analyzed based on the eGFP signal by fluorescence microscopy at 48 h post transfection. In order to visualize more accurately the distribution of the two proteins, nuclei were stained with Hoechst 33342 and the cytoskeleton was stained with an anti-α-tubulin. Upon overexpression in LMH, the control eGFP protein (peGFP) was distributed all over the cells; the two VP22 proteins tagged at the N- or C-terminus did not show

any significant difference in their cellular localization (Fig. 3A), with respectively 74,4% or 72,6% of eGFP-positive cells presenting an exclusive nuclear distribution and 17,2% or 25% showing a combined nuclear/cytoplasmic staining (Fig. 3B). Thus, the location of the eGFP tag, at the amino- or carboxy-terminus of VP22, does not seem to affect VP22 cellular distribution in LMH cells.

Another interesting feature of VP22 is its ability to bind to chromatin, especially to histones as it has previously been shown for the VP22 encoded by BoHV-1 [20,55]. By performing a high-salt histones extraction protocol from cells transfected with either pcDNA-UL49 or pcDNA3.1 (empty vector), we found VP22 to be included in the histones fraction, as it is demonstrated by a 27 kDa band in the colloidal coomassie blue SDS-PAGE gel and by the VP22-specific antibody (L13a)-probed Western blot (shown in Fig. 3C-left panel). This result indicates that MDV-VP22 shares the ability of the VP22 encoded by BoHV-1 to interact with histones. In order to investigate the impact of the position of the eGFP tag on the ability of VP22 to associate with chromatin, we carried out a similar experiment using LMH cells transfected with peGFP, peGFP-UL49 or pUL49-eGFP. We observed that the VP22 tagged at its amino terminus could be co-extracted with histones and visualized as a specific 55 kDa band in coomassie blue-stained SDS-PAGE gel (Fig. 3C-right panel). However, the protein tagged at its carboxy terminus appeared to be significantly less retained in the histones fraction (Fig. 3C). These observations were confirmed by immunoblotting experiments using the anti-VP22 L13a antibody that show the presence of VP22 in the histone extracts prepared from cells expressing peGFP-UL49 and at a far lesser extent (about 4.5 fold) from the pUL49-eGFP transfected cells (Fig. 3C lower panel). All together, these data indicate that VP22 is predominantly targeted to the nucleus of LMH transfected cells independently of the eGFP tag location. However, the fusion of eGFP at the C-terminus of the VP22 protein affects its capacity to associate with chromatin.

Accumulation of DNA Damages in VP22 Overexpressing Cells

Arrest or delay in S phase can arise either from the occurrence of DNA damages, especially double strand breaks (DSB), or replication fork stalling [57,58]. Since VP22 is able to drastically arrest the cell cycle in S phase and moreover seems to be associated to chromatin, we tested whether the overexpression of VP22 in LMH can induce DNA damages. LMH cells were transfected with pcDNA-UL49 or pcDNA3.1 (as a negative control) and DNA damages were analyzed by alkaline comet assay at 24 h post-transfection. This method, based on a single-cell gel electrophoresis, allows the detection of DNA breaks that are visualized as fragmented DNA exhibiting the shape of a comet's tail. We could observe an increased number of comets in the population of cells transfected with VP22 compared to the cells transfected with the empty vector pcDNA3.1 (Fig. 4A). To estimate the extent of DNA damages, a more precise analysis with the Comet Score software was performed and the tail extent moment (TEM) was calculated. This parameter is calculated on the basis of the tail length, reflecting the severity of the damages and the amount of DNA in the tail relative to the head, which is an indicator of DNA break frequencies. The calculation of the TEM could show that cells expressing VP22 presented a significant higher TEM (12.62 ± 0.62) than the cells transfected with pcDNA3.1 (4.11 ± 0.25), indicating that the expression of VP22 seems to be associated with the occurrence of DNA damages in LMH cells (Fig. 4A lower panel). However, it should be stressed that this result reflects the DNA damage analysis on the whole

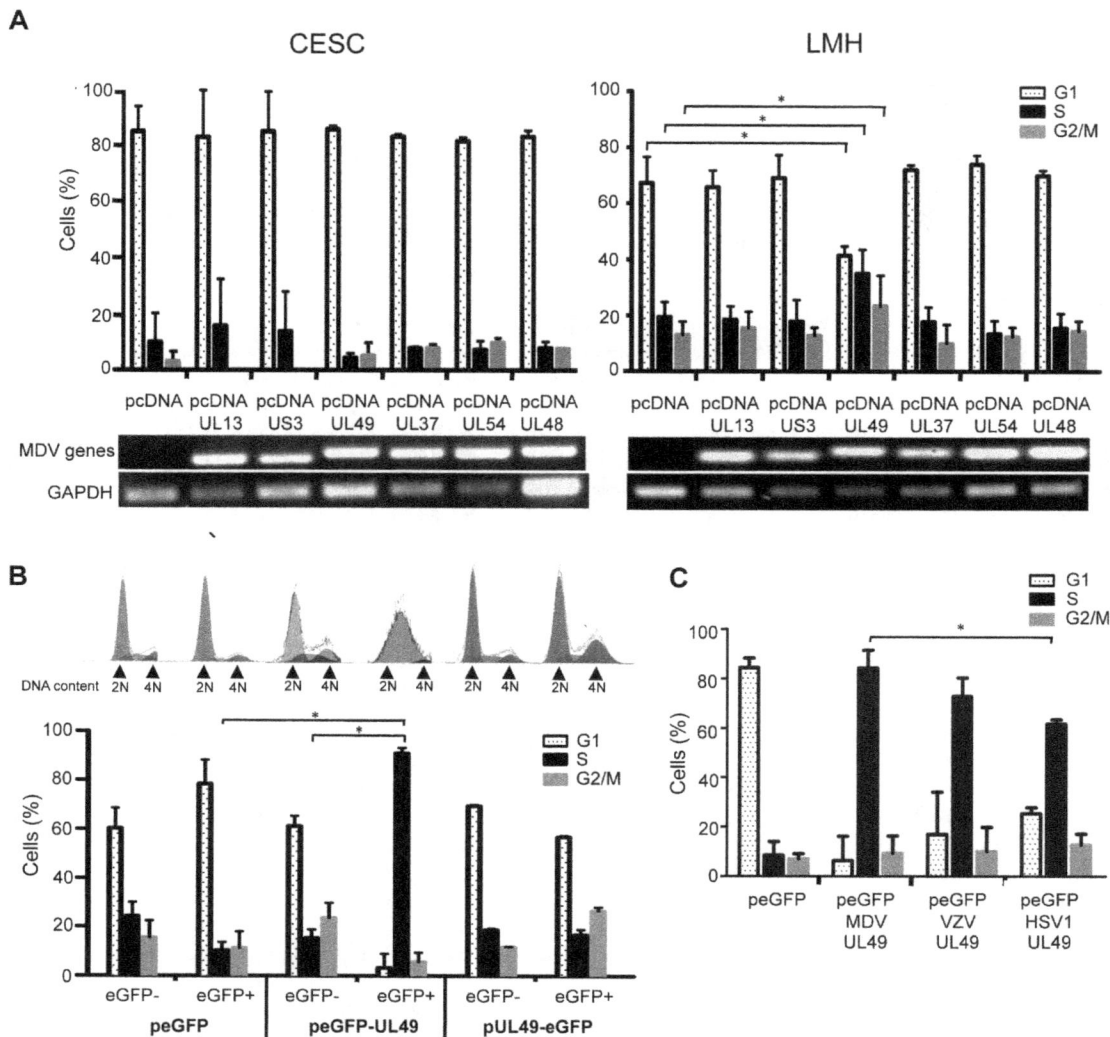

Figure 2. Identification of VP22 as a potent cell cycle modulator in proliferating cells. (A) Impact of the overexpression of different viral proteins on the cell cycle modulation. Primary (CESC) and high proliferating (LMH) cells were transfected with either an empty vector (pcDNA3) or with vectors expressing MDV-UL13, US3, UL49, UL37, UL54, or UL48 under CMV promoter control. Cell cycle analysis was performed at 48 h post-transfection by flow cytometry on the whole cell population (*$p<0.05$). The mRNA expression of the different MDV genes transfected was assayed by RT-PCR from total RNA extracted. GAPDH expression was analyzed for each sample as a control. The RT-PCR products were resolved by agarose (2%) gel electrophoresis. (B) Cell cycle regulation in VP22 expressing cells. LMH cells were transfected with plasmids harboring the UL49 gene in fusion with eGFP either at its N-terminal extremity (peGFP-UL49) or at its C-terminus (pUL49-eGFP) and with the empty vector peGFP. At 48 h post-transfection, cell cycle was analyzed specifically in UL49-expressing cells (eGFP+) and in non-transfected cells (eGFP−). Single parameter cytometry histograms are shown and the percentage of cells in the G1, S, and G2/M phases of the cell cycle is reported as bars. *$p<0.05$. (C) Cell cycle analysis in LMH cells expressing human VP22 orthologues. LMH cells were transfected with the empty vector peGFP and plasmids harboring the UL49 gene encoded by MDV, VZV and HSV-1 cloned in frame with eGFP at the N-terminus (peGFP-UL49). Cell cycle was analyzed at 48 h post-transfection exclusively in UL49-expressing cells (eGFP+) and the percentage of cells in the G1, S, and G2/M phases of the cell cycle is represented as bars. *$p<0.05$.

population of pcDNA-UL49 transfected and non-transfected cells. Consequently to corroborate these findings and to determine whether the VP22 protein tagged at the C- or N-terminus was also able to induce DNA damage in LMH cells, we transfected the peGFP (empty vector), peGFP-UL49 or pUL49-eGFP plasmids in LMH cells and examined the onset of DNA damages by alkaline comet assay at 24 h post-transfection specifically in the eGFP positive cells sorted by flow cytometry. As positive control, LMH cells were treated with etoposide and as negative control, non-treated and non-transfected LMH cells were analyzed. We could readily observe comets from cells treated with etoposide and most of the cells overexpressing eGFP-UL49 and, to a lesser extent,

from cells expressing the UL49-eGFP protein, whereas cells expressing peGFP produced almost no comets or comets with a shorter tail similar to the non-transfected cells (Fig. 4B). Calculation of TEM revealed that the mean tail moments of cells expressing eGFP-UL49 (27,73±2,11) or UL49-eGFP (11,99±1,52) is significantly higher than for cells transfected with peGFP (4,739±0,54), indicating that the expression of both tagged-VP22 proteins increases DNA damage in cells (Fig. 4B left panel). However, the damages were significantly more pronounced in cells expressing the protein tagged at its amino-terminal extremity than in cells expressing the C-terminally tagged version of VP22. Of note cells treated with etoposide showed a TEM of

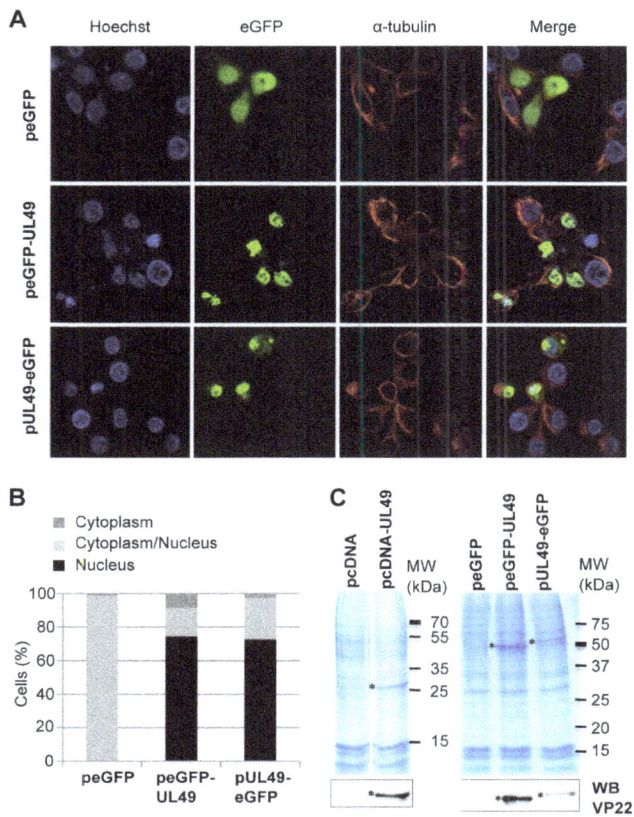

Figure 3. VP22 is predominantly located in the nucleus and associated to histones. (A–B) Subcellular localization of VP22 in LMH cells. The peGFP-UL49 and pUL49-eGFP plasmids as well as the peGFP empty vector were transfected in LMH. At 48 h post-transfection, cells were fixed with PFA 4% and subjected to immunofluorescence using anti-α-tubulin and AlexaFluor594-conjugated goat anti-rabbit IgG secondary antibody in order to demarcate the cytoplasm (red). Nuclei were counterstained with Hoechst 33342 (blue) and the eGFP proteins (green) were visualized directly by fluorescence microscopy. A representative example of the results obtained is shown (A). The nuclear/cytoplasm distribution of the eGFP proteins was estimated on an average of 100 cells and results are represented as stacked bars (B). (C) VP22 associates with histones. At 48 h post-transfection, histones were extracted in high salt conditions from LMH cells transfected with pcDNA, pcDNA3-UL49, peGFP, peGFP-UL49 or pUL49-eGFP. Extracts were separated by SDS-PAGE. Proteins were either directly stained in the gel with colloidal coomassie blue or transferred onto nitrocellulose membrane to perform a western blot analysis with an anti-MDV VP22 antibody (L13a). (*) indicates the presence of the VP22 proteins. The unmodified VP22 shows a molecular weight of 27 kDa and the N- or C-terminus tagged VP22 have a molecular weight of 35 kDa.

77.47±0.55 thereby affirming the drastic induction of DNA damages by this DNA topoisomerase II inhibitor. In addition, we were interested to analyze the frequency distribution of tail moments (i.e. the percentage of cells presenting a defined TEM), which is representative for the number of cells encompassing damages (Fig. 4C). About 63% of cells transfected with peGFP had a tail moment inferior at 5, indicating that the majority of the cells contain non-damaged DNA or DNA with very limited damages. However, this cellular population decreased when VP22 was expressed both with the eGFP tag at the N-terminus or C-terminus (10% and 29%, respectively), and we could observe a marked increase of the proportion of cells presenting TEM values above 5 (89,6% and 71%, respectively). In particular, the expression of

eGFP-UL49 tends to increase the frequency of cells with highly damaged DNA, more than 50% of the cells having a TEM>20 and 13,7% presenting TEM>50. In comparison, 19,7% of cells expressing UL49-eGFP showed a TEM>20 and only 1,9% a TEM>50. These observations indicate that the expression of VP22 in cells leads to an increased incidence DNA damaged cells and that damages are more severe when the VP22 is fused to eGFP at its N-terminal extremity. It should be however stressed that although the expression of VP22 leads to the occurrence of significant DNA damage, those damages are relatively less heavy than the ones induced by drugs such as etoposide that are responsible of potent damages (more than 55% of the comet having a TEM>50).

In order to specify the nature of the DNA damages generated in cells expressing VP22, we monitored by immunofluorescence staining the expression and localization of γ-H2AX in LMH cells transfected with peGFP-UL49, pUL49-eGFP or with the empty vector peGFF. Because histone H2AX is rapidly phosphorylated (γ-H2AX) after generation of DNA double strand breaks (DSB), γ-H2AX is a preferential marker used to reveal these damages [59]. As positive control, the expression of γ-H2AX was also examined in cells exposed to etoposide [60]. We observed an overall increase of the staining intensity of the γ-H2AX DSB-marker in cells treated with etoposide and specifically in cells expressing eGFP-UL49 compared to non-transfected cells, peGFP transfected cells or UL49-eGFP expressing cells (Fig. 5A). Moreover, with higher magnification we could visualize that γ-H2AX formed discrete foci in the nucleus of eGFP-UL49 transfected cells as was also observed in etoposide-treated cells (Fig. 5B). This typical punctuated staining of γ-H2AX reflects its recruitment to sites of DNA damage and thus indicates that cells expressing VP22 tagged at its amino-terminus undergo multiple DSB.

Discussion

In the present report, we show for the first time that MDV lytic infection leads to a dysregulation of the cell cycle progression of the host cell. MDV infection not only promotes the proliferation of primary embryonic skin cells, but also leads to an accumulation of infected cells in S-phase. This modulation of the cell cycle is accompanied by a significant up-regulation of cellular genes involved in G0/G1 transition (cyclin D, cdk6) and in G1 to S-phase progression (pRb, E2F1, c-myc, cyclins A). A substantial mRNA up-regulation of the cell cycle regulator p53 was also observed early after infection. Cell cycle modulation is a mechanism that is frequently exploited by viruses in order to facilitate viral replication. In contrast to small DNA viruses, a cellular S-phase environment is not mandatory for herpesviruses encoding their own DNA polymerase and accessory factors required for optimal viral replication [29]. Consequently, for most of alphaherpesviruses it has been demonstrated that they prevent the S-phase entry and rather activate the G1/S checkpoint. MDV is an alphaherpesvirus that shares a number of biological features with gammaherpesviruses, notably the viral lymphotropism and the ability to induce tumors. With respect to cell cycle modulation, our data suggest that MDV has adapted a similar strategy than EBV and KSHV, both of which were shown to promote cell cycle progression, especially into S-phase [33,61]. It is conceivable that the MDV-mediated cell cycle modulation might also play a role in the multi-factorial events eventually leading to transformation and tumorigenesis. The S-phase is in fact the most vulnerable period of the cell cycle and a defect or inactivation of the key components of the intra S-phase checkpoint may predispose cells to oncogenic transformation (for review [57]).

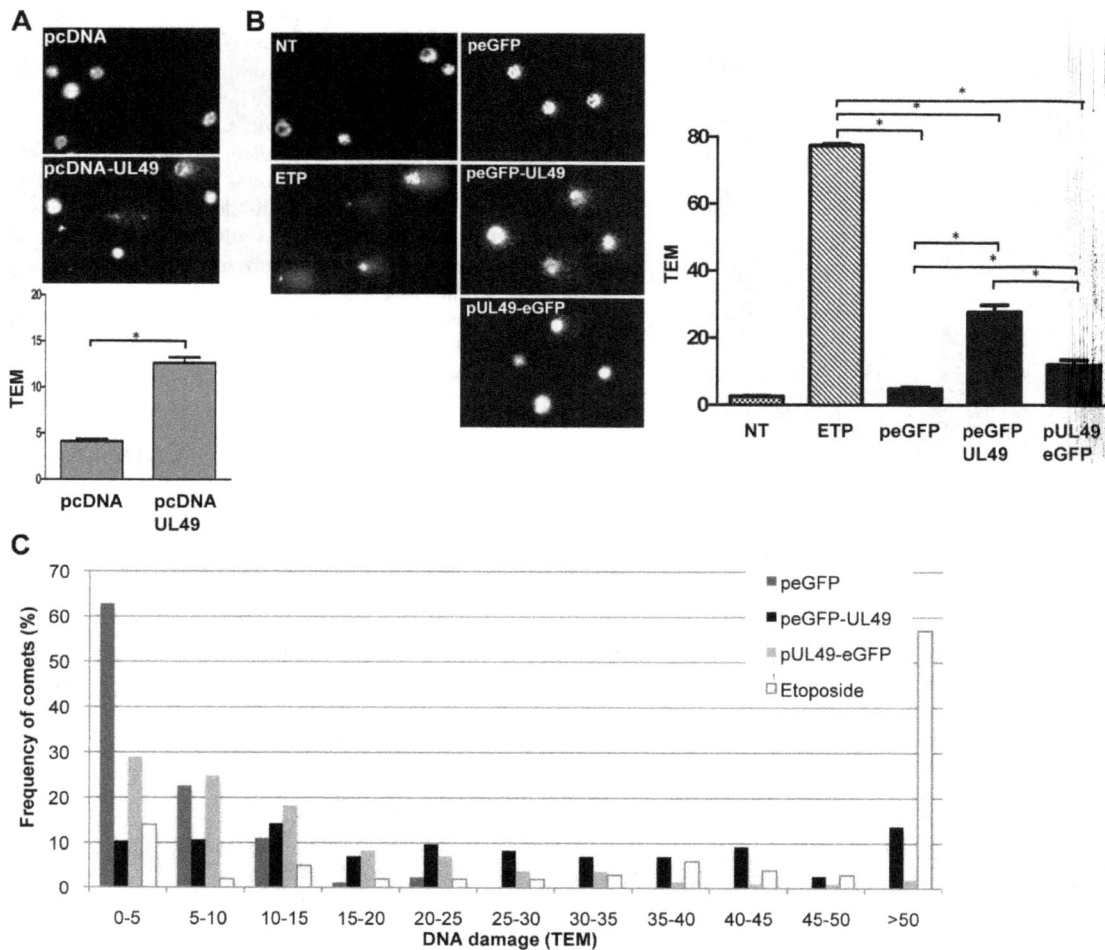

Figure 4. Expression of VP22 leads to cellular DNA damage. Detection of DNA damage in cells overexpressing VP22 by comet assay. (A–B) LMH cells were transfected with pcDNA, pcDNA3-UL49, peGFP, peGFP-UL49 or pUL49-eGFP as indicated. Non-treated (NT) LMH cells and cells treated with 1.5 µM etoposide (ETP) for 24 h were used as negative and positive control, respectively. After 24 h, the whole population of NT, ETP-treated cells, pcDNA and pcDNA3-UL49 transfected cells was directly subjected to comet assay. In the case of eGFP transfected cells, the EGFP positive cells were sorted by flow cytometry prior to comet assay analysis (B). Representative images of comets are shown as photographs. Quantitative and qualitative analyses of the comets are represented as histograms ($\pm$ SD) on the basis of the calculation of the means of the tail extent moments (TEM) parameter measured with the CometScore software. (C) Frequency distribution of the comets with respect to their value of TEM. $*p < 0.05$.

The most unexpected discovery of our study is the identification of the MDV-VP22 protein as a potent trigger of cell cycle arrest in S-phase, as evidenced by the observation that its overexpression in proliferating LMH cells lead to the enrichment of up to 90% of transfected cells in S-phase. VP22 is a major component of the viral tegument of the *Alphaherpesvirinae*. While VP22 orthologs exhibit functional homology, their significance for alphaherpesviruses life cycle varies according to the virus species. This is well illustrated by previous studies showing that VP22 is dispensable for *in vitro* replication of PRV, HSV-1, and BoHV1, whereas it is essential for MDV and VZV replication [5,7,8,10,11,12]. However, the biological properties of VP22 that determine its key role in the life cycle of MDV remain still unknown. One hypothesis is based on the crucial function of VP22 in cell-to-cell spread [5,18]. We can also not exclude that a rapid distribution of VP22 after viral entry might prepare an optimal environment for viral replication by inducing an S-phase arrest.

Several viral proteins encoded by herpesviruses have been shown to have an impact on the cell cycle. Among the ones encoded by the *Alphaherpesvirinae*, the ICP0 protein is probably the best studied. This multifunctional protein required for efficient HSV-1 lytic replication and reactivation from latency, has been identified as a major cell cycle modulator that is able to act either on the G1/S or at the G2/M checkpoints [62]. However, the observation that ICP0 deficient mutant viruses are still capable to elicit proliferation arrest indicates that other viral factors also impact the cell cycle [62]. Notably, the immediate early protein ICP27 was shown to be essential for the G1/S cell cycle arrest triggered by HSV-1, with ICP4, ICP0, and the virion host shutoff protein acting as contributors [55]. Hence, MDV and HSV-1 appear to employ differential cell cycle modulation mechanisms as MDV does not encode a functional ICP0 protein and we could not detect any effect of ICP27-overexpression on cell proliferation. It is an interesting speculation that MDV may have evolved a distinct mechanism for cell cycle modulation that crucially involves VP22 in order to compensate for the absence of ICP0 activities. It should be noted that the overexpression of the VP22 proteins encoded by HSV-1 or VZV also resulted in a dramatic arrest of the cell cycle in S-phase in transfected LMH cells. This finding suggests that VP22 might also contribute to the modulation of the cell cycle in

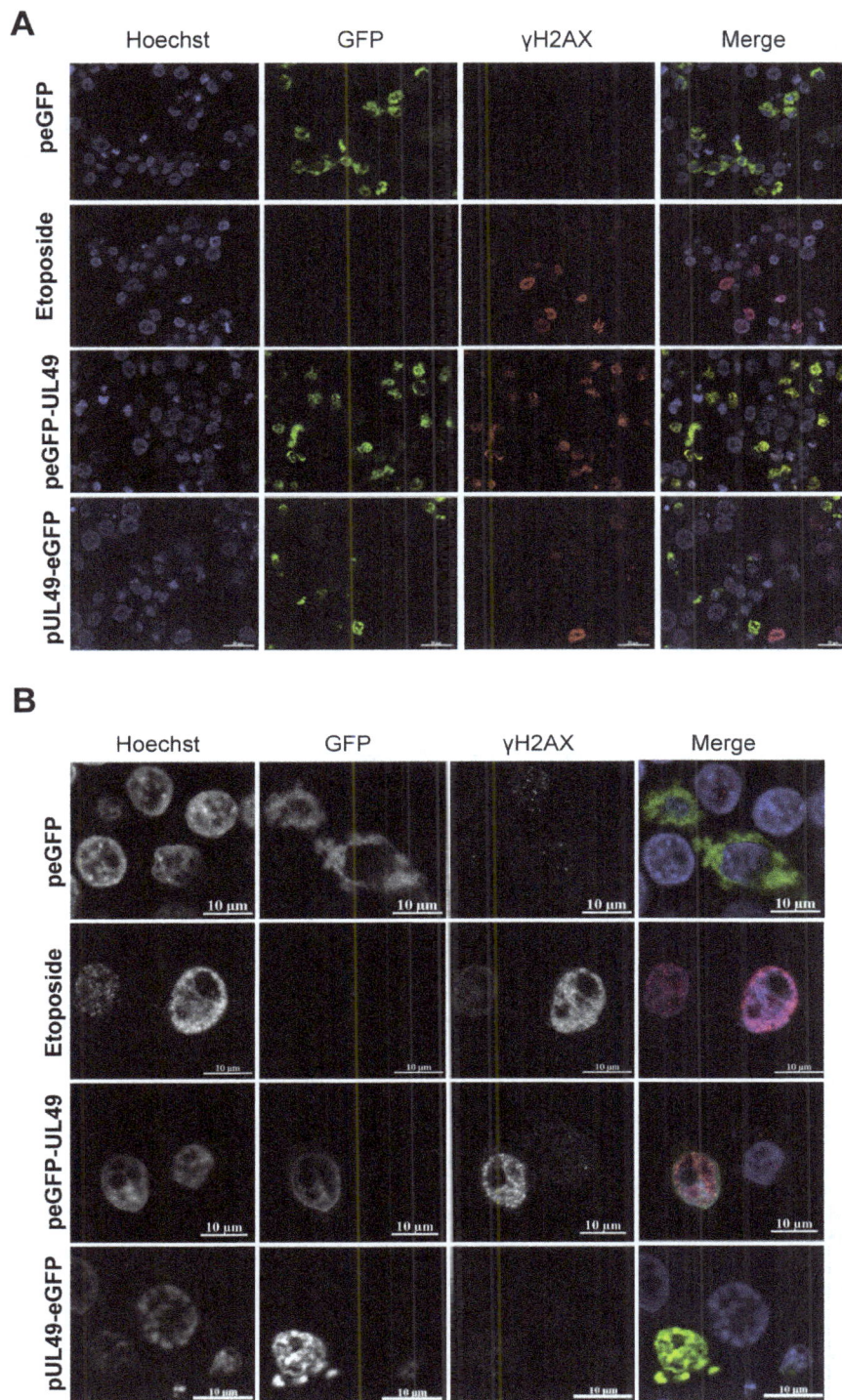

Figure 5. Accumulation of DNA double strand breaks in VP22 expressing cells. At 24 h post-transfection, LMH cells expressing VP22 eGFP-tagged proteins (at N or C-terminus) or the empty vector peGFP were subjected to immunofluorescence using mouse anti-γH2AX and AlexaFluor594-conjugated goat anti-rabbit IgG antibodies (red). LMH cells exposed 24 h to etoposide were stained with the anti-γH2AX antibody and served as a positive control. Nuclei were stained with Hoechst 33342 (blue) and eGFP positive cells expressing VP22 were directly visualized by fluorescent microscopy (green) at low (A) and high magnifications (B).

the context of infections with human herpesviruses. While screening for viral factors that are involved in the MDV-associated cell cycle regulation, we also tested whether the activity of the two MDV-encoded serine-threonine kinases pUS3 and pUL13 could

have a cell cycle regulatory effect. Indeed, it is well known that cell cycle progression is submitted to a tight regulation mediated by kinases and phosphatases. Overexpression of UL13 and/or US3 in low proliferating cells (CESC) or high proliferating cells (LMH) had no effect on the cell cycle, thus excluding a direct involvement of these kinases in the cell cycle modulation. However, pUS3 and pUL13 are able to phosphorylate various cellular and viral proteins, including the VP22 proteins encoded by HSV-1 and -2, as well as BoHV-1 [63,64,65]. So far, the phosphorylation status of MDV-VP22 during infection has not been investigated, and we cannot exclude post-translational modifications of MDV-VP22 by UL13 and/or US3, as previously shown for other alphaherpesviruses.

Intra-S checkpoints activation mainly reflects DNA breaks or stalled replication fork formation [58]. In order to identify the molecular mechanisms underlying the VP22-driven S-phase arrest, we focused on the impact of VP22 expression on the generation of DNA damage in the host cell genome. Following overexpression of VP22 in proliferating cells, we could indeed show by comet assay that the presence of VP22 coincided with the occurrence of massive DNA damage. Moreover, VP22-expressing cells showed an increased staining of the phosphorylated form of H2AX, suggesting that the DNA lesions observed are double strand breaks [59]. Interestingly, the VP22-mediated generation of DNA damages seems to be tightly associated to the cell cycle modulation property of VP22. This was evidenced by our comparative experimental approach using two versions of the VP22 protein tagged either at its N- or C-terminus. The data from this experiment show that virtually all cells expressing eGFP-VP22 (N-terminal eGFP-tag) are arrested in S-phase and present severe DNA damage, whereas cells expressing VP22-eGFP (C-terminal eGFP-tag) are not affected in their cell cycle progression and show significantly less DNA lesions. Of note, all our attempts to generate a LMH stable cell line overexpressing the MDV-VP22 protein failed due to a high level of cellular mortality. These observations raised the question of the potential toxicity of VP22, which might find an explanation in the induction of double strand breaks in cells overexpressing VP22.

The mechanisms by which VP22 induces S-phase arrest and DNA breaks still remain to be elucidated. However, among the characteristics of VP22, we can speculate that its capacity to interact with chromatin and histones might participate to those processes. Interactions of VP22 protein with nucleosomes were previously demonstrated for the BoHV-1-encoded VP22, which physically interacts with nucleosome-associated histones and thereby causes an impaired acetylation of histone H4 [21,56]. In addition, for MDV-VP22, the regions allowing interaction with heterochromatin were previously defined [17]. In the present study, we confirmed that MDV-VP22 is found predominately in the nucleus of cells following overexpression in LMH cells. We also found that an N-terminally eGFP-tagged MDV-VP22 can be extracted from chromatin preparations together with histones. However, for a C-terminally eGFP-tagged MDV-VP22, the efficiency of recovery from histones extracts was far less, suggesting that an unmodified C-terminal extremity of VP22 is necessary for the association of VP22 with chromatin. Together, these observations suggest that the abilities of VP22 to arrest the cell cycle in S-phase and to effect DNA damage are linked to its direct or indirect interaction with histones and/or chromatin. According

to this model, it is conceivable that the interaction of VP22 with chromatin or histones may disturb the unwinding of DNA in a similar fashion than cellular helicases or topoisomerases by preventing access to the DNA replication machinery. Alternatively, it can be speculated that an association of VP22 with DNA/histones may cause physical tension of the DNA double helix eventually leading to DNA breaks.

Activation of DNA damage response (DDR) pathways is well documented for a number of viruses, especially tumorigenic viruses and plays a central role in viral replication [46,47]. While our data only provide clear evidence for a role of VP22 as a powerful inducer of DNA damage in a non-infectious context, it can be assumed that DDR activation might play a role in MDV replication and/or MD pathogenesis. Due to the critical role of VP22 for MDV-replication [5], it is so far impossible to evidence the function of VP22 as a major cell cycle regulatory factor during MDV infection by using a VP22-deleted virus. However, Jarosinsky et al. have demonstrated that a recombinant MDV harboring a VP22 protein tagged with eGFP at the C-terminus (a construct that is identical to the VP22-eGFP used in the present study) showed a drastic decrease in its ability to induce MD in infected chickens, with only 10% of the chicken developing tumors [19]. In addition, we have recently observed that a recombinant virus with the VP22 protein tagged at the N-terminus is also attenuated, but in a lower extent, with 33 to 66% of the infected chickens developing MD lymphoma [6]. Although the impairment of pathogenicity of the recombinant MDV studied by Jarosinski et al. could in parts be explained by a lower viral replication efficiency in vivo, in view of our data, it can also be speculated that the defect in tumor development observed by the authors might be due to the loss of the ability of the C-terminally tagged VP22 protein to induce S-phase arrest and DNA damage.

In conclusion, our findings provide new insights into herpesvirus-cell host interactions by demonstrating that the oncogenic alphaherpesvirus MDV affects the cell cycle progression in infected cells. Moreover, we could assign a novel role to the VP22 tegument protein as a potent cell cycle modulator, property that seems to be associated to its ability to induce DNA damages in cells. Current efforts are under way to elucidate the detailed mechanisms of VP22-induced DNA damage response, and its role during viral infection, especially with respect to a possible involvement of DDR in MDV replication and/or the establishment of MDV-latency and subsequent lymphoma formation.

Acknowledgments

We thank S. Trapp (INRA, Nouzilly, France) and J. Vignard for their constructive comments and corrections on the manuscript.

Author Contributions

Conceived and designed the experiments: LTF. Performed the experiments: LTF DB DCV YLV SR EBR. Analyzed the data: LTF JFV CD GM. Contributed reagents/materials/analysis tools: JFV CD EBR GM. Wrote the paper: LTF.

References

1. Calnek BW, Adldinger HK, Kahn DE (1970) Feather follicle epithelium: a source of enveloped and infectious cell-free herpesvirus from Marek's disease. Avian Dis 14: 219–233.

2. Carrozza JH, Fredrickson TN, Prince RP, Luginbuhl RE (1973) Role of desquamated epithelial cells in transmission of Marek's disease. Avian Dis 17: 767–781.

3. Nazerian K, Burmester BR (1968) Electron microscopy of a herpes virus associated with the agent of Marek's disease in cell culture. Cancer Res 28: 2454–2462.

4. Dorange F, El Mehdaoui S, Pichon C, Coursaget P, Vautherot JF (2000) Marek's disease virus (MDV) homologues of herpes simplex virus type 1 UL49 (VP22) and UL48 (VP16) genes: high-level expression and characterization of MDV-1 VP22 and VP16. J Gen Virol 81: 2219–2230.

5. Dorange F, Tischer BK, Vautherot JF, Osterrieder N (2002) Characterization of Marek's disease virus serotype 1 (MDV-1) deletion mutants that lack UL46 to UL49 genes: MDV-1 UL49, encoding VP22, is indispensable for virus growth. J Virol 76: 1959–1970.

6. Remy S, Blondeau C, Le Vern Y, Lemesle M, Vautherot JF, et al. (2013) Fluorescent tagging of VP22 in N-terminus reveals that VP22 favors Marek's disease virus (MDV) virulence in chickens and allows morphogenesis study in MD tumor cells. Vet Res 44: 125.

7. Che X, Reichelt M, Sommer MH, Rajamani J, Zerboni L, et al. (2008) Functions of the ORF9-to-ORF12 gene cluster in varicella-zoster virus replication and in the pathogenesis of skin infection. J Virol 82: 5825–5834.

8. del Rio T, Werner HC, Enquist LW (2002) The pseudorabies virus VP22 homologue (UL49) is dispensable for virus growth in vitro and has no effect on virulence and neuronal spread in rodents. J Virol 76: 774–782.

9. Duffy C, Lavail JH, Tauscher AN, Wills EG, Elaho JA, et al. (2006) Characterization of a UL49-null mutant: VP22 of herpes simplex virus type 1 facilitates viral spread in cultured cells and the mouse cornea. J Virol 80: 8664–8675.

10. Elliott G, Hafezi W, Whiteley A, Bernard E (2005) Deletion of the herpes simplex virus VP22-encoding gene (UL49) alters the expression, localization, and virion incorporation of ICP0. J Virol 79: 9735–9745.

11. Fuchs W, Granzow H, Klupp BG, Kopp M, Mettenleiter TC (2002) The UL48 tegument protein of pseudorabies virus is critical for intracytoplasmic assembly of infectious virions. J Virol 76: 6729–6742.

12. Liang X, Chow B, Li Y, Raggo C, Yoo D, et al. (1995) Characterization of bovine herpesvirus 1 UL49 homolog gene and product: bovine herpesvirus 1 UL49 homolog is dispensable for virus growth. J Virol 69: 3863–3867.

13. Maringer K, Stylianou J, Elliott G (2012) A network of protein interactions around the herpes simplex virus tegument protein VP22. J Virol 86: 12971–12982.

14. van Leeuwen H, Okuwaki M, Hong R, Chakravarti D, Nagata K, et al. (2003) Herpes simplex virus type 1 tegument protein VP22 interacts with TAF-I proteins and inhibits nucleosome assembly but not regulation of histone acetylation by INHAT. J Gen Virol 84: 2501–2510.

15. Elliott G, O'Hare P (1998) Herpes simplex virus type 1 tegument protein VP22 induces the stabilization and hyperacetylation of microtubules. J Virol 72: 6448–6455.

16. Blondeau C, Marc D, Courvoisier K, Vautherot JF, Denesvre C (2008) Functional homologies between avian and human alphaherpesvirus VP22 proteins in cell-to-cell spreading as revealed by a new cis-complementation assay. J Virol 82: 9278–9282.

17. O'Donnell LA, Clemmer JA, Czymmek K, Schmidt CJ (2002) Marek's disease virus VP22: subcellular localization and characterization of carboxyl terminal deletion Mutations. Virology 292: 235–240.

18. Denesvre C, Blondeau C, Lemesle M, Le Vern Y, Vautherot D, et al. (2007) Morphogenesis of a highly replicative EGFPVP22 recombinant Marek's disease virus in cell culture. J Virol 81: 12348–12359.

19. Jarosinski KW, Arndt S, Kaufer BB, Osterrieder N (2011) Fluorescently tagged pUL47 of Marek's disease virus reveals differential tissue expression of the tegument protein in vivo. J Virol.

20. Ingvarsdottir K, Blaho JA (2010) Association of the herpes simplex virus major tegument structural protein VP22 with chromatin. Biochim Biophys Acta 1799: 200–206.

21. Zhu J, Qiu Z, Wiese C, Ishii Y, Friedrichsen J, et al. (2005) Nuclear and mitochondrial localization signals overlap within bovine herpesvirus 1 tegument protein VP22. J Biol Chem 280: 16038–16044.

22. Elliott G, O'Hare P (2000) Cytoplasm-to-nucleus translocation of a herpesvirus tegument protein during cell division. J Virol 74: 2131–2141.

23. Berthet C, Raj K, Saudan P, Beard P (2005) How adeno-associated virus Rep78 protein arrests cells completely in S phase. Proc Natl Acad Sci U S A 102: 13634–13639.

24. Dickmanns A, Zeitvogel A, Simmersbach F, Weber R, Arthur AK, et al. (1994) The kinetics of simian virus 40-induced progression of quiescent cells into S phase depend on four independent functions of large T antigen. J Virol 68: 5496–5508.

25. Gilbert DM, Cohen SN (1987) Bovine papilloma virus plasmids replicate randomly in mouse fibroblasts throughout S phase of the cell cycle. Cell 50: 59–68.

26. Helt AM, Harris E (2005) S-phase-dependent enhancement of dengue virus 2 replication in mosquito cells, but not in human cells. J Virol 79: 13218–13230.

27. Liang MH, Geisbert T, Yao Y, Hinrichs SH, Giam CZ (2002) Human T-lymphotropic virus type 1 oncoprotein tax promotes S-phase entry but blocks mitosis. J Virol 76: 4022–4033.

28. Tang QH, Zhang YM, Fan L, Tong G, He L, et al. (2011) Classic swine fever virus NS2 protein leads to the induction of cell cycle arrest at S-phase and endoplasmic reticulum stress. Virol J 7: 4.

29. Flemington EK (2001) Herpesvirus lytic replication and the cell cycle: arresting new developments. J Virol 75: 4475–4481.

30. Nascimento R, Costa H, Parkhouse RM (2012) Virus manipulation of cell cycle. Protoplasma 249: 519–528.

31. Kudoh A, Daikoku T, Sugaya Y, Isomura H, Fujita M, et al. (2004) Inhibition of S-phase cyclin-dependent kinase activity blocks expression of Epstein-Barr virus immediate-early and early proteins, preventing viral lytic replication. J Virol 78: 104–115.

32. Moffat JF, McMichael MA, Leisenfelder SA, Taylor SL (2004) Viral and cellular kinases are potential antiviral targets and have a central role in varicella zoster virus pathogenesis. Biochim Biophys Acta 1697: 225–231.

33. Kudoh A, Fujita M, Zhang L, Shirata N, Daikoku T, et al. (2005) Epstein-Barr virus lytic replication elicits ATM checkpoint signal transduction while providing an S-phase-like cellular environment. J Biol Chem 280: 8156–8163.

34. Dheekollu J, Lieberman PM (2011) The Replisome Pausing Factor Timeless is Required for Episomal Maintenance of Latent Epstein-Barr Virus. J Virol.

35. Ma W, Galvin TA, Ma H, Ma Y, Muller J, et al. (2011) Optimization of chemical induction conditions for human herpesvirus 8 (HHV-8) reactivation with 12-O-tetradecanoyl-phorbol-13-acetate (TPA) from latently-infected BC-3 cells. Biologicals.

36. Teissier S, Pang CL, Thierry F (2010) The E2F5 repressor is an activator of E6/E7 transcription and of the S-phase entry in HPV18-associated cells. Oncogene 29: 5061–5070.

37. Lilley CE, Schwartz RA, Weitzman MD (2007) Using or abusing: viruses and the cellular DNA damage response. Trends Microbiol 15: 119–126.

38. Turnell AS, Grand RJ (2012) DNA viruses and the cellular DNA-damage response. J Gen Virol 93: 2076–2097.

39. Gruhne B, Sompallae R, Masucci MG (2009) Three Epstein-Barr virus latency proteins independently promote genomic instability by inducing DNA damage, inhibiting DNA repair and inactivating cell cycle checkpoints. Oncogene 28: 3997–4008.

40. Jha HC, Upadhyay SK, M AJP, Lu J, Cai Q, et al. (2013) H2AX phosphorylation is important for LANA-mediated Kaposi's sarcoma-associated herpesvirus episome persistence. J Virol 87: 5255–5269.

41. Luo MH, Rosenke K, Czornak K, Fortunato EA (2007) Human cytomegalovirus disrupts both ataxia telangiectasia mutated protein (ATM)- and ATM-Rad3-related kinase-mediated DNA damage responses during lytic infection. J Virol 81: 1934–1950.

42. Mohni KN, Dee AR, Smith S, Schumacher AJ, Weller SK (2012) Efficient herpes simplex virus 1 replication requires cellular ATR pathway proteins. J Virol 87: 531–542.

43. Xiao Y, Chen J, Liao Q, Wu Y, Peng C, et al. (2013) Lytic infection of Kaposi's sarcoma-associated herpesvirus induces DNA double-strand breaks and impairs non-homologous end joining. J Gen Virol 94: 1370–1375.

44. Lilley CE, Chaurushiya MS, Boutell C, Everett RD, Weitzman MD (2011) The intrinsic antiviral defense to incoming HSV-1 genomes includes specific DNA repair proteins and is counteracted by the viral protein ICP0. PLoS Pathog 7: e1002084.

45. Tarakanova VL, Stanitsa E, Leonardo SM, Bigley TM, Gauld SB (2010) Conserved gammaherpesvirus kinase and histone variant H2AX facilitate gammaherpesvirus latency in vivo. Virology 405: 50–61.

46. Weitzman MD, Lilley CE, Chaurushiya MS (2010) Genomes in conflict: maintaining genome integrity during virus infection. Annu Rev Microbiol 64: 61–81.

47. Nikitin PA, Luftig MA (2012) The DNA damage response in viral-induced cellular transformation. Br J Cancer 106: 429–435.

48. Nikitin PA, Luftig MA (2011) At a crossroads: human DNA tumor viruses and the host DNA damage response. Future Virol 6: 813–830.

49. Schumacher D, Tischer BK, Fuchs W, Osterrieder N (2000) Reconstitution of Marek's disease virus serotype 1 (MDV-1) from DNA cloned as a bacterial artificial chromosome and characterization of a glycoprotein B-negative MDV-1 mutant. J Virol 74: 11088–11098.

50. Elliott GD, Meredith DM (1992) The herpes simplex virus type 1 tegument protein VP22 is encoded by gene UL49. J Gen Virol 73 (Pt 3): 723–726.

51. Bontems S, Di Valentin E, Baudoux L, Rentier B, Sadzot-Delvaux C, et al. (2002) Phosphorylation of varicella-zoster virus IE63 protein by casein kinases influences its cellular localization and gene regulation activity. J Biol Chem 277: 21050–21060.

52. Lebailly P, Devaux A, Pottier D, De Meo M, Andre V, et al. (2003) Urine mutagenicity and lymphocyte DNA damage in fruit growers occupationally exposed to the fungicide captan. Occup Environ Med 60: 910–917.

53. Shechter D, Dormann HL, Allis CD, Hake SB (2007) Extraction, purification and analysis of histones. Nat Protoc 2: 1445–1457.

54. Sandri-Goldin RM (2011) The many roles of the highly interactive HSV protein ICP27, a key regulator of infection. Future Microbiol 6: 1261–1277.

55. Song B, Yeh KC, Liu J, Knipe DM (2001) Herpes simplex virus gene products required for viral inhibition of expression of G1-phase functions. Virology 290: 320–328.

56. Ren X, Harms JS, Splitter GA (2001) Bovine herpesvirus 1 tegument protein VP22 interacts with histones, and the carboxyl terminus of VP22 is required for nuclear localization. J Virol 75: 8251–8253.

57. Bartek J, Lukas C, Lukas J (2004) Checking on DNA damage in S phase. Nat Rev Mol Cell Biol 5: 792–804.

58. Grallert B, Boye E (2008) The multiple facets of the intra-S checkpoint. Cell Cycle 7: 2315–2320.

59. Rogakou EP, Pilch DR, Orr AH, Ivanova VS, Bonner WM (1998) DNA double-stranded breaks induce histone H2AX phosphorylation on serine 139. J Biol Chem 273: 5858–5868.

60. Schonn I, Hennesen J, Dartsch DC (2010) Cellular responses to etoposide: cell death despite cell cycle arrest and repair of DNA damage. Apoptosis 15: 162–172.

61. Fujimuro M, Wu FY, ApRhys C, Kajumbula H, Young DB, et al. (2003) A novel viral mechanism for dysregulation of beta-catenin in Kaposi's sarcoma-associated herpesvirus latency. Nat Med 9: 300–306.

62. Lomonte P, Everett RD (1999) Herpes simplex virus type 1 immediate-early protein Vmw110 inhibits progression of cells through mitosis and from G(1) into S phase of the cell cycle. J Virol 73: 9456–9467.

63. Coulter LJ, Moss HW, Lang J, McGeoch DJ (1993) A mutant of herpes simplex virus type 1 in which the UL13 protein kinase gene is disrupted. J Gen Virol 74 (Pt 3): 387–395.

64. Geiss BJ, Tavis JE, Metzger LM, Leib DA, Morrison LA (2001) Temporal regulation of herpes simplex virus type 2 VP22 expression and phosphorylation. J Virol 75: 10721–10729.

65. Labiuk SL, Lobanov V, Lawman Z, Snider M, Babiuk LA, et al. (2010) Bovine herpesvirus-1 US3 protein kinase: critical residues and involvement in the phosphorylation of VP22. J Gen Virol 91: 1117–1126.

7

A Novel Gain-Of-Function Mutation of the Proneural *IRX1* and *IRX2* Genes Disrupts Axis Elongation in the Araucana Rumpless Chicken

Nowlan H. Freese[1], Brianna A. Lam[1], Meg Staton[2], Allison Scott[1], Susan C. Chapman[1]*

1 Department of Biological Sciences, Clemson University, Clemson, South Carolina, United States of America, 2 Department of Entomology and Plant Pathology, University of Tennessee, Knoxville, Tennessee, United States of America

Abstract

Axis elongation of the vertebrate embryo involves the generation of cell lineages from posterior progenitor populations. We investigated the molecular mechanism governing axis elongation in vertebrates using the Araucana rumpless chicken. Araucana embryos exhibit a defect in axis elongation, failing to form the terminal somites and concomitant free caudal vertebrae, pygostyle, and associated tissues of the tail. Through whole genome sequencing of six Araucana we have identified a critical 130 kb region, containing two candidate causative SNPs. Both SNPs are proximal to the *IRX1* and *IRX2* genes, which are required for neural specification. We show that *IRX1* and *IRX2* are both misexpressed within the bipotential chordoneural hinge progenitor population of Araucana embryos. Expression analysis of *BRA* and *TBX6*, required for specification of mesoderm, shows that both are downregulated, whereas *SOX2*, required for neural patterning, is expressed in ectopic epithelial tissue. Finally, we show downregulation of genes required for the protection and maintenance of the tailbud progenitor population from the effects of retinoic acid. Our results support a model where the disruption in balance of mesoderm and neural fate results in early depletion of the progenitor population as excess neural tissue forms at the expense of mesoderm, leading to too few mesoderm cells to form the terminal somites. Together this cascade of events leads to axis truncation.

Editor: Moises Mallo, Instituto Gulbenkian de Ciência, Portugal

Funding: Research reported in this publication was supported by National Institute on Deafness (http://www.nidcd.nih.gov/) and other Communication Disorders of the National Institutes of Health under award number R01DC009236 to SCC. Technical Contribution No. 6247 of the Clemson University Experiment Station. This material is based upon work supported by the NIFA/USDA, under project number SC-1700374 to SCC. The funders had no role in study design, data collection and analysis, decision to publish, or preparation of the manuscript.

Competing Interests: The authors have declared that no competing interests exist.

* Email: schapm2@clemson.edu

Introduction

During secondary body formation the regressing primitive streak and Hensen's node are transformed into a bulblike structure, the tailbud, a morphologically uniform mass of mesenchyme [1,2,3]. The tailbud mesenchyme is located adjacent to the posterior end of the neural tube and notochord, an area known as the chordoneural hinge (CNH) [4,5]. The CNH together with the dorso-posterior and ventral tailbud populations gives rise to all the derivatives of the tail [6,7]. These include the precursor to the secondary neural tube (the medullary cord), somite progenitors of the presomitic mesoderm, and the posterior extension of the notochord [3,4,6]. The CNH acts as a bipotential population of long-term axial progenitors, contributing cells to both the somitic mesoderm and medullary cord neuroectoderm [6,7,8]. The bipotential nature of the CNH requires fine control of expression of neural genes such as *SOX2* and mesodermal genes such as *TBX6* in order to maintain the balance of neural and mesodermal cell fates [9,10,11].

Throughout secondary body formation the remaining somites form from the paraxial mesoderm (PM) of the tail [12]. Somites are transitory paired epithelial spheres that differentiate to generate the axial skeleton, including the vertebrae, cartilage, and most of the skeletal musculature and dermis [13]. The number of somites formed is species-specific and highly variable; for example whereas chicken (*Gallus gallus*) has between 51–53 somites, mouse (*Mus musculus*) has approximately 65 somites and the corn snake (*Pantherophis guttatus*) over 300 somites [14,15]. Somitogenesis is a critical process during axis elongation, and interference brought about by teratogenic factors or the existence of congenital mutations can lead to axis truncation. Many model organisms have been used to study axis elongation including chicken, mouse, and zebrafish [7]. Notably in mice, spontaneous mutants such as the *vestigial tail* mouse, hypomorphic for *Wnt3a*, and the *Brachyury (T)* mouse have been studied as models of axis truncation [16,17,18,19]. Null or reduced expression of either *Wnt3a* or *T* leads to a failure to maintain the CNH progenitor population and failure of mesoderm specification. Additionally, gene knockouts of *FGF8* and *CYP26A1* also result in axis truncation, with the latter required to protect the progenitor population from the apoptotic effects of retinoic acid until extension is complete [20,21]. Current models of axis length termination include the elimination of the tailbud progenitor population through programmed cell death and diminution of the presomitic mesoderm (PSM) [22,23,24,25]. However, many gaps in knowledge still exist, requiring a better understanding of the genes, signals and regulators of secondary body formation and its subsequent termination.

The Araucana chicken breed has been maintained as show birds for their rumpless (Rp) and ear tuft morphology [26,27,28,29,30]. Rumplessness is an inherited autosomal dominant disorder, which we had previously mapped to a 740 kb region on chromosome 2 [28,31]. The Araucana model offers an opportunity to further elucidate the morphogenetic and molecular mechanisms required for normal tail development, as well as the cessation of axis elongation, in an accessible model organism. Here, we investigate the mechanism of rumplessness and the identity of the causative mutation.

In the current study we show that misexpression of the *IRX1* and *IRX2* proneural genes, located within our candidate region, precedes a cascade of altered downstream gene expression. This results in a morphogenetic chain reaction including: changes in bipotential progenitor cell fate, premature depletion of progenitors, early termination of somitogenesis, and early apoptosis of the progenitor remnant and posterior axis malformation. Furthermore, we identify two candidate causative mutations, within a narrowed 130 kb region of chromosome 2 through bioinformatics analysis of whole genome resequencing of six Araucana birds. Together, our results provide a greater understanding of the mechanism of secondary body formation, cell fate determination, axial elongation, determination of posterior somite numbers and control of overall tail length.

Materials and Methods

Animals

Clemson University IACUC approved the study, protocol number 2011-041. Fertilized chicken eggs were obtained from SkyBlueEgg (Arkansas, U.S.A.) and the Clemson University Poultry Farm. Eggs were incubated at 38.5°C in a humidified chamber to the desired stage. Embryos were staged according to Hamburger and Hamilton (HH) [32]. Skeletal material was the gift of the Araucana Club of America.

Bone and cartilage staining

Bone and cartilage staining was carried out on E18 Araucana^{Rp} and tailed controls using Alcian blue (Polysciences) and Alizarin red S (Acros Organics) according to standard procedures [33]. Briefly, Embryos were fixed 3×24 hours in 95% EtOH, 100% EtOH, 2×24 h in 100% Acetone. Cartilage staining (20 mg Alcian Blue in 100 ml of 40% acetic acid glacial/EtOH) was performed from a few hours to overnight depending on sample size. Embryos were rinsed in EtOH for 15 min followed by EtOH for 24 hrs. They were then placed in saturated borax solution 2×24 hours (Na$_2$B$_4$O$_7$10H$_2$O in H$_2$O). Trypsin solution (0.45 g purified trypsin in 400 mL of 30% borax dissolved in distilled water) at 30°C was used to clear tissue until flesh became translucent and soft (between 1–4 days, depending on size of sample). Alizarin Red S solution (0.5% KOH and 0.1% Alizarin Red S) was used to stain bones (12–24 hours). Samples were then washed in distilled water, followed by a wash in 0.5% KOH solution for 15 min. Excess Alizarin Red S stain was removed using 0.5% KOH solution for 2×24 hours at room temperature under a light source. Samples then went through series of glycerol 0.5% KOH washes (20% glycerol/0.5% KOH, 50/50 and 75/25

mix). Samples were stored in 100% glycerol with 100 mg thymol crystals.

Somite number counts

Araucana^{Rp} and controls were incubated to between HH16-25. Embryos were harvested and somite counts performed using a Nikon stereoscopic microscope (control n = 73, Araucana^{Rp} n = 83). At later stages, between HH22-25, *DACT2* ISH labeling was used to aid counts of the posterior somites. The number of somites in tailed controls was compared against the expected number of somites as described in the normal stage series, and found to match [32].

Statistical Analysis

Assuming a normal distribution of the data, a two-tailed *t*-test was carried out to test for differences in the average values of samples from experiments for somite counts, proliferation, and TUNEL. Analysis was carried out using Statistical Analysis Software (SAS).

Immunohistochemistry

Embryos were fixed in 4% paraformaldehyde (PFA/PBS) for 48 hours before being cryoembedded in 15% sucrose/7.5% gelatin/PBS and sectioned on a Leica cryotome at 25 μm. Immunostaining was carried out using our standard protocol [34]. Briefly, sections were blocked in PBS with 0.1% TritonX-100 and 0.2% bovine serum albumin (BSA). Then incubated overnight at 4°C with the primary antibodies anti-E-cadherin (cat 610182, BD Bioscience) and anti-laminin (cat L9393, Sigma). Following washing in PBS sections were incubated at secondary antibodies 1:200 Alexa Fluor 488 goat anti-mouse IgG and Alexa Fluor 594 goat anti-rabbit IgG (Invitrogen). Following washing in PBS and mounting with SlowFade (Life Technologies), fluorescent images were captured using a Nikon Ti Eclipse confocal microscope.

In situ hybridization

Whole-mount in situ hybridization (ISH) was performed according to our standard procedures using probes against *BRA* (control = 18, Araucana^{Rp} n = 14), *CYP26A1* (control = 16, Araucana^{Rp} n = 8), *DACT2* (control = 19, Araucana^{Rp} n = 11), *FGF8* (control = 19, Araucana^{Rp} n = 11), *IRX1* (control = 26, Araucana^{Rp} n = 12), *IRX2* (control = 25, Araucana^{Rp} n = 17), *IRX4* (control = 16, Araucana^{Rp} n = 9), *MESO1* (control = 17, Araucana^{Rp} n = 9), *RALDH2* (control = 8, Araucana^{Rp} n = 8), *SOX2* (control = 6, Araucana^{Rp} n = 9), *TBX6* (control = 29, Araucana^{Rp} n = 19), and *WNT3A* (control = 18, Araucana^{Rp} n = 13) [34]. Probes have all been previously described as follows: *DACT2*, *MESO1* and *TBX6* [24], *FGF8* [35], *IRX1*, *IRX2* and *IRX4* [36], *RALDH2* [37] and *WNT3A* [38]. *IRX1*, *IRX2*, and *IRX4* probes were the generous gift of Dr. Cheryll Tickle. Embryos were cryoembedded in 15% sucrose, 7.5% gelatin/PBS and sectioned on a Leica cryotome. Whole mount embryos and sections were imaged on a Nikon Smz1500 stereomicroscope and Nikon Eclipse 80i compound microscope, respectively using a Qimaging Micropublisher 5.0 camera.

EdU and TUNEL labeling

For proliferation analysis, Click-iT EdU 488 Imaging Kit (Invitrogen) was used to carry out labeling of cells as previously described [39]. Briefly, embryos were pulsed with EdU for 60 minutes before harvesting and fixation in 4% PFA overnight. Embryos were then cryoembedded in 15%sucrose/7.5%gelatin/PBS and sectioned on a Leica cryotome at 25 μm. Alternating sections were processed for EdU detection with a counterstain of TO-PRO-3 Iodide, or apoptosis detection using the TUNEL method (In Situ Cell Death Detection Kit, TMR red, Roche) with a Hoechst counterstain, and imaged on a Nikon Ti Eclipse confocal microscope. For each sample, a single matching EdU and TUNEL labeled mediolateral image was selected. Image analysis of EdU, TUNEL, and TO-PRO-3 labeling was carried out using ImageJ (control n = 21, Araucana^{Rp} n = 17). For individual cell counts of EdU, TO-PRO-3, and TUNEL stained cells, a region of interest was manually selected consisting of the tailbud mesenchyme based on location and morphology, excluding the surrounding ectoderm.

Whole genome sequencing and bioinformatics

DNA samples for six Araucana were acquired from our previous study [31]. Each of the six samples was sequenced on six lanes with an Illumina HiSeq 2000 sequencer. Average genomic coverage was 27.63x and average number of bases sequenced was 29.009 Giga base pairs (Table S1). Sequence reads were trimmed using Trimmomatic and aligned to the corresponding region previously identified to be associated with the rumpless phenotype on chromosome 2 of the ICGSC Gallus_gallus-4.0/galGal4 build using Bowtie2 applications [40,41]. The mpileup function of SamTools was used to call variants [42]. The view option of bcftools was used to call the genotype at each variant for each individual bird using the bcftools defaults. Variants that were found to be homozygous in all three homozygous rumpless birds, found to be heterozygous in all two heterozygous birds, and were not found in the homozygous tailed bird were considered fixed in the population and targets for genomic variation that may result in the rumpless phenotype. Identified variants were compared against known variants in the Beijing Genomics Institute (BGI) database, and variants that were previously identified not to be involved in rumplessness were removed [43].

Results

Araucana^{Rp} lack the caudal-most vertebrae

The current North American breed standard of the Araucana chicken requires that they lack the caudal vertebrate and other tail structures, appearing rumpless (Fig. 1A). This distinctive morphology is revealed by comparison of adult tailed control and rumpless Araucana (Araucana^{Rp}) skeletons, where Araucana^{Rp} lack the free caudal vertebrae and pygostyle of the tail (Fig. 1B–C). By performing cartilage and bone staining (Alcian Blue and Alizarin Red S) from embryonic day E5 to E18, we determined that failure to form the vertebrae, rather than reabsorption occurs. At E18, 5 free caudal vertebrae and 6 fused vertebrae of the pygostyle are observed in controls (Fig. 1D), but are missing in Araucana^{Rp} (Fig. 1E). From these data we conclude that the rumpless phenotype observed in the Araucana^{Rp} adult chicken arises during early posterior development due to the lack of free caudal vertebrae and the pygostyle, matching previous observations of rumpless chickens [28].

Araucana^{Rp} embryos display truncated tail morphology at the tail organizer stage

To examine potential mechanisms resulting in axis truncation, we first determined the earliest stage at which Araucana^{Rp} embryos displayed morphological differences by examining the gross morphology of the tail region. Formation of the tailbud occurs at HH14-15 and is identical between controls and Araucana^{Rp} [1,2,3]. At HH16, the Araucana^{Rp} embryo tail appeared less elongated and more pointed than rounded

compared to controls (Fig. 2A–B). Moreover, a lesser angle of curvature between the Ventral Ectodermal Ridge (VER) and extraembryonic ectoderm was apparent (arrow, Fig. 2A–B). This became more pronounced at HH18, with reduced distance between the most recently formed somite pair (asterisk) and the end of the tail in Araucana^{Rp} (Fig. 2C–D). The tip of the tail appeared pointed and the mesenchymal cells underlying the ectoderm cap had dense apoptotic morphology. By HH20 elongation of the tail has ceased in Araucana^{Rp}, with the most recently formed somite located more posteriorly in the tail than controls (asterisk, Fig. 2E–F).

Araucana^{Rp} embryos downregulate TBX6 and BRA at the tail organizer stage

Having identified the critical stage at which morphology was affected we next analyzed TBX6 and BRA expression. Both T-box transcription factors are important for proper axial elongation, with mutations leading to changes in either gene's expression causing axis truncation [10,11,17,18,19,44,45,46]. TBX6 is a marker of the presomitic mesoderm (PSM) and undifferentiated mesenchyme in the tailbud (asterisk, Fig. 2G), acting to indirectly repress the neural transcription factor SOX2, in order to specify mesoderm (Fig. 2G,I,K) [9,11]. Up to HH15 TBX6 expression was as expected in control and Araucana^{Rp} embryos. At HH15 (26 pairs of somites), TBX6 expression was downregulated in the undifferentiated mesenchyme of the Araucana^{Rp} tailbud (asterisk, Fig. 2H, inset). The tailbud comprises all mesenchyme posterior to the neural tube and notochord, consisting of the chordoneural hinge population (CNH) and tailbud mesenchyme (TBM) that lies directly underneath the ectoderm capping the tail [3,6]. Expression of TBX6 was maintained in the PSM in HH15-HH17 embryos, but not the undifferentiated mesenchyme of the TBM (Fig. 2H,J). However, by HH20 TBX6 expression within the PSM of Araucana^{Rp} embryos was no longer visible (Fig. 2K,L). This suggests that following downregulation of TBX6 within the tailbud, Araucana^{Rp} fail to specify new PSM.

BRA is a marker of the TBM and notochord, and is required for mesoderm specification in the tail (Fig. 2M,O,Q) [17,18,44]. Heterozygous T mutant mice display malformed sacral vertebrae and shortening of the tail due to a failure of axis elongation and somite formation during embryogenesis [18,19,44]. Loss of BRA expression in Araucana^{Rp} followed downregulation of TBX6 within the tailbud mesenchyme at HH16, whereas notochord expression of BRA remained unaffected (Fig. 2N,P,R). These results show that the defect leading to Araucana^{Rp} rumplessness arises early in tailbud development, and involves the downregulation of transcription factors required for specification of the PSM. Observation of the downregulation of TBX6 and BRA within the TBM, which contributes to the PSM, suggests a failure of new PSM specification in Araucana^{Rp} embryo tails as early as HH15 (26 somites) [6].

Somitogenesis ends prematurely in Araucana^{Rp} embryos

In tailed chicken embryos axial elongation and somitogenesis continue until HH24-25, when 51–53 somite pairs have formed [24]. Due to the downregulation of TBX6 and BRA and their requirement to specify PSM, we predicted Araucana^{Rp} would fail to form the correct number of somite pairs [11,44]. To that end we analyzed MESO1 expression. MESO1 is the homolog of mouse Mesp2, a transcription factor expressed in the anterior presomitic mesoderm that plays a role in somite segment border formation, and is required for the formation of the next pair of epithelial somites (Fig. 3A,C) [47,48]. TBX6 binds to regulatory elements of Mesp2 and is required for its expression [49]. MESO1

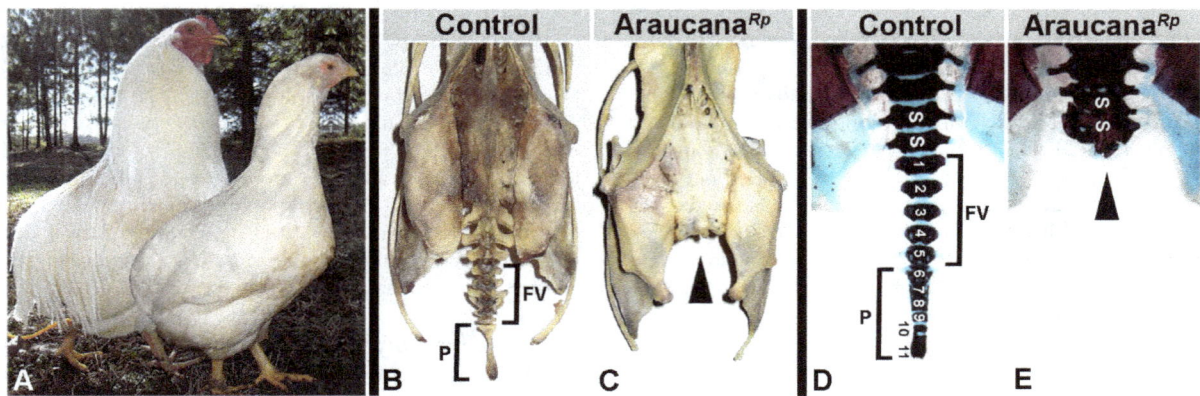

Figure 1. Skeletal analysis shows Araucana^{Rp} lack caudal vertebrae. Adult Araucana^{Rp} male and female birds shown in a composite image (A) (courtesy of Fritz Ludwig). Note the characteristic rounded rump, lacking tail structures. Skeletons of control (B) and Araucana^{Rp} (C) birds (courtesy of the ACA). The free vertebrae and pygostyle are missing in the Araucana^{Rp} skeleton (arrow). E18 embryos stained with Alcian Blue in control (D) and Araucana^{Rp} (E). Araucana^{Rp} embryos lack the free vertebrae and pygostyle. Arrowheads indicate lateral processes. Vertebral elements are numbered from the first free vertebrae (1–5). The more posterior vertebral elements (6–11) fuse to form the mature pygostyle after hatching. FV-free caudal vertebrae, P-pygostyle, S-sacral vertebrae.

expression is normally downregulated at HH24-25, marking the end of somitogenesis [24]. We found that *MESO1* expression matched controls (Fig. 3A–B) until HH19, at which point *MESO1* expression was lost in Araucana^{Rp} (Fig. 3C–D). *MESO1* expression was lost as *TBX6* was downregulated within the remaining PSM in Araucana^{Rp} (Fig. 2L and 3D). This result suggests that Araucana^{Rp} somite formation is arrested as early as HH19.

Next we examined expression of the homolog of zebrafish *dpr2*, *DACT2*, a regulator of *WNT* and *TGFβ* signaling that is expressed in the anterior primitive streak, neural crest cells, and most recently formed somites after HH17 [24,50,51,52]. *DACT2* expression in Araucana^{Rp} labeled a similar number of recently formed somites when compared to controls at HH19 (Fig. 3E–F).

However, the distance from the most posterior somite to the tip of the tail was shortened, suggesting less PSM in Araucana^{Rp} (Fig. 3F). At HH20, *DACT2* labeled fewer somites in Araucana^{Rp} than controls (Fig. 3G,H). This suggests that somite formation is reduced in HH20 Araucana^{Rp} as fewer *DACT2* labeled somites equates to a lack of further somite formation. In controls, *DACT2* continued to label recently formed somites through the end of somitogenesis at HH24-25 [24,52]. The downregulation of *MESO1* and *DACT2* expression 22.5 hours earlier than in controls indicates that the most posterior somites, which give rise to the free caudal vertebrae and pygostyle, fail to form.

To confirm the lack of somite formation we performed somite counts for Araucana^{Rp} and controls from HH16-25 (Fig. 3I).

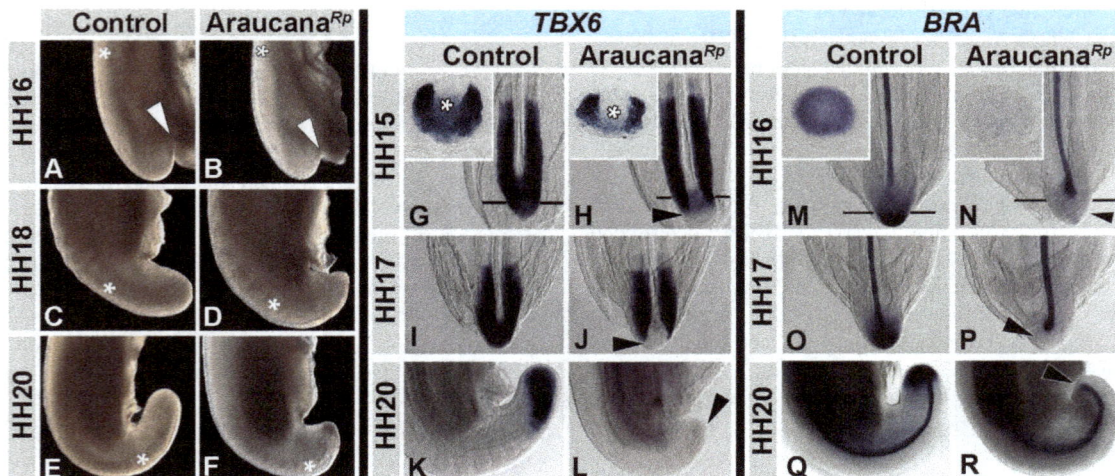

Figure 2. Araucana^{Rp} embryo tailbud is truncated and downregulates *TBX6 and BRA*. Whole mount tails in lateral view, with posterior at the bottom, dorsal to the left (A–F). At HH16, the angle of tail curvature is narrower in controls (A) compared to Araucana^{Rp} embryos (B-arrowhead). By HH18, the reduced length and pointed shape of the Araucana^{Rp} tail is dramatic (D) compared to controls (C). The control tail has curved ventrally by HH20 (E), whereas the Araucana^{Rp} tail has failed to extend (F). Asterisks denote level of posterior-most somite pair. Somite formation in Araucana^{Rp} is near the tip of the tail (F-asterisk). Expression patterns of *TBX6* (G–L) and *BRA* (M–R) during tailbud development. ISH expression of *TBX6* in control (G,I,K) and Araucana^{Rp} (H,J,L) at HH15, 17 and 20. Inset in G and H are transverse sections of respective embryos at level of tailbud, asterisk denotes undifferentiated mesenchyme. Note downregulation of *TBX6* expression in Araucana^{Rp} (arrowheads) compared to controls. ISH expression of *BRA* in control (M,O,Q) and Araucana^{Rp} (N,P,R) at HH16, 17, and 20. Inset in M and N are transverse sections of respective embryos at level of tailbud. Note loss of *BRA* expression in tailbud mesenchyme in Araucana^{Rp} (arrowheads) versus controls.

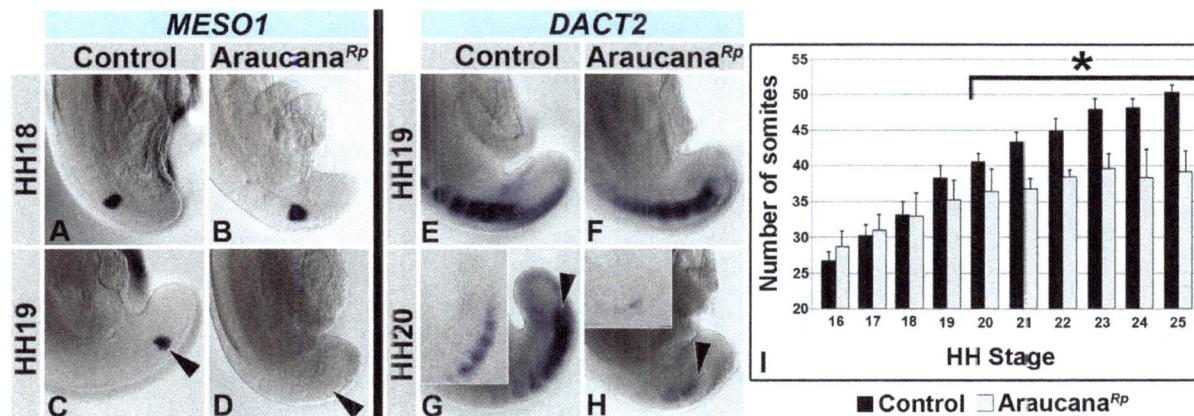

Figure 3. Araucana^{Rp} form fewer somites. *MESO1* expression in control (A,C) and Araucana^{Rp} (B,D) embryos at HH18-19. Note downregulation of *MESO1* in HH19 Araucana^{Rp} (D-arrowhead) compared to controls (C-arrowhead). *DACT2* expression in control (E,G) and Araucana^{Rp} (F,H) from HH19-20. Inset in G and H are sagittal sections through the paraxial mesoderm from respective embryo. Note decrease in somites labeled by *DACT2* at HH20 in Araucana^{Rp} (H-arrowhead) compared to control (H-arrowhead). Anterior up in whole mount and section insets. (I) Graph showing number of somites compared to embryonic stage (HH stages 16–25). Araucana^{Rp} have significantly fewer somites beginning at HH20 (T-test, asterisk marks $p < 0.01$) than controls. Control-Black, Araucana^{Rp}-Grey.

Beginning at HH20, the number of somites in Araucana^{Rp} embryos was significantly different than in controls ($p < 0.01$, Fig. 3I). At HH20 Araucana^{Rp} embryos averaged 4 fewer somites than controls, and by the end of somitogenesis at HH25 Araucana^{Rp} embryos had on average 11 fewer somites than controls. These data are consistent with our bone and cartilage labeling data in which embryos lacked the 5 free caudal vertebrae and the 6 vertebrae that will later fuse to form the mature pygostyle.

In summary, these data indicate that a mechanism capable of ending somitogenesis prematurely is triggered within Araucana^{Rp} embryos. The main question that arises from these observations is what is the functional mechanism leading to axial truncation? Several possibilities present themselves, including: changes in proliferation and apoptosis in the CNH population, changes in the balance of cell fates within the bi-potential progenitor population, and changing molecular signaling within the tissues [9,22,23,24,25]. While none of these is mutually exclusive it is logical to assume that there is a single molecular event that triggers a cascade of downstream changes that eventually results in axial truncation. We next examined each of these possibilities to determine the sequence of events taking place in affected Araucana^{Rp} embryos.

IRX1 and IRX2 are misexpressed in the Araucana^{Rp} embryo tailbud

We previously identified a critical region associated with rumplessness in Araucana, containing two genes, *iroquois 1* (*IRX1*) and *iroquois 2* (*IRX2*) [31]. Iroquois-class homeodomain proteins play multiple roles during pattern formation, with one of their primary roles being the initial specification of the vertebrate neuroectoderm [53,54]. Importantly, *IRX1* and *IRX2* expression is normally excluded from the tailbud. To investigate expression of *IRX1* and *IRX2* in Araucana^{Rp} embryos we performed in situ hybridization from HH10-HH20.

Expression of *IRX1* in control embryos is restricted to the neural tube of the elongating posterior axis (Fig. 4A,C,E). Analysis of *IRX1* expression up to HH14 revealed identical expression between controls and Araucana^{Rp} (Fig. 4A–B). At HH15, *IRX1* expression in the neural tube remains unaffected in Araucana^{Rp}

(Fig. 4C–D,M), whereas *IRX1* is misexpressed medially at the level of the CNH progenitor population (Fig. 4C–D,N–P). *IRX1* misexpression occurred one somite pair earlier than downregulation of *TBX6* expression (Fig. 2H,4D). *IRX1* misexpression in Araucana^{Rp} tailbud continued at HH16 (Fig. 4E–F) and HH17 (not shown), but by HH18 normal expression was restored (data not shown). These results illustrate a critical time window of *IRX1* misexpression at the onset of tailbud formation and the tail organizer stage.

We then analyzed the expression of *IRX2*, which is also expressed in the neural tube, but restricted more anteriorly than *IRX1* (Fig. 4G,I,K). Analysis of *IRX2* at HH14 revealed no difference in expression between controls and Araucana^{Rp} (Fig. 4G–H). However, *IRX2* was misexpressed in Araucana^{Rp} tails beginning at HH15 (Fig. 4I–J). The subpopulation of cells appeared to overlap that of *IRX1* to a limited degree, with a more ventro-lateral subpopulation of mesenchyme labeled (Fig. 4Q). *IRX2* misexpression continues at HH16, but is no longer observed from HH17 (Fig. 4K–L and data not shown).

Although *IRX4* does not fall within the defined critical region, it is within the same genomic cluster as *IRX1* and *IRX2* [31,55]. *IRX4* expression is restricted primarily to the heart at these stages, suggesting that the regulatory elements controlling its expression are not as closely shared as for *IRX1* and *IRX2* [55,56,57,58]. ISH analysis from HH14-16 revealed identical expression between control and Araucana^{Rp} embryos (data not shown).

In summary, both *IRX1* and *IRX2* are misexpressed within the Araucana^{Rp} tailbud at the level of the CNH beginning at HH15. This aberrant expression raises two important questions. First, what is the causative mutation resulting in the gain-of-function of two genes simultaneously? Second, what is the effect of misexpression of two proneural genes within the mesenchyme progenitor population, which commits cells to both mesodermal and neuroectodermal lineages?

Sequencing of the critical region reveals candidate mutations

We previously identified a 740 kb critical region associated with the rumplessness containing both *IRX1* and *IRX2* [31]. In order to identify candidate causative mutations within this region we

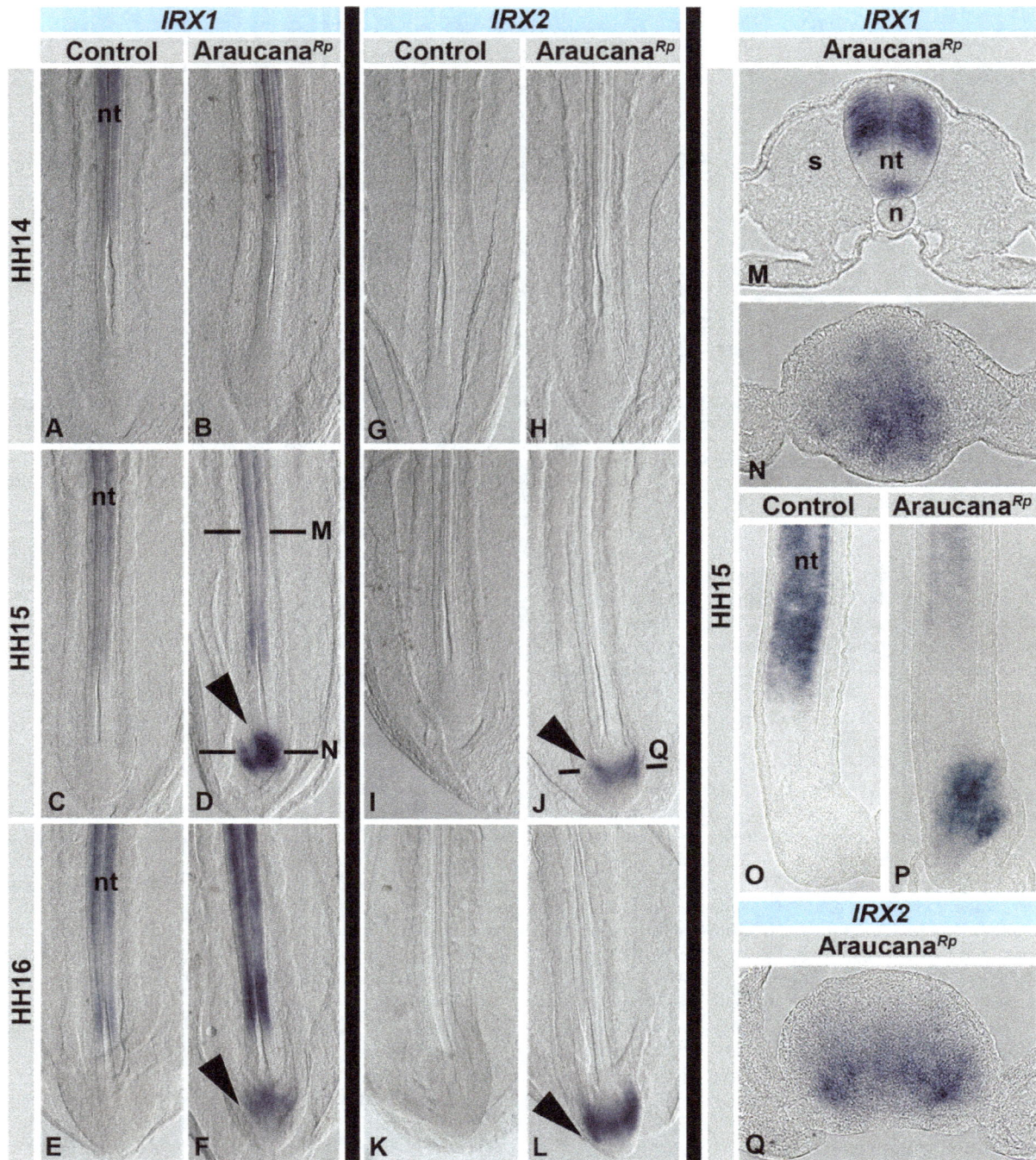

Figure 4. *IRX1* and *IRX2* are misexpressed in Araucana*Rp* tailbud. *IRX1* expression pattern (A–F,M–P). (A,C,E) Control embryos express *IRX1* in the neural tube. *IRX1* expression in controls matches Araucana*Rp* at HH14 (A,B), but is misexpressed in Araucana*Rp* tailbud at HH15 (D-arrowhead). Normal expression at level of somites is within neural tube (transverse section, M). Misexpression in Araucana*Rp* can be seen at the level of the chordoneural hinge (transverse section-N and sagittal section-P) compared to expression in HH15 control (sagittal section-O). Misexpression is maintained through HH16 in Araucana*Rp* (F-arrow) as compared to control (E). *IRX2* expression pattern (G–L,Q). (G,I,K) Control embryos do not express *IRX2* in the tailbud. No difference in expression between controls and Araucana*Rp* seen at HH14 (G,H). *IRX2* is misexpressed in HH15 (J-arrowhead) and HH16 (L-arrowhead) Araucana*Rp*. Transverse section of *IRX2* (Q) shows expression similar to *IRX1* at level of chordoneural hinge. A–L - anterior to top. M,N,Q - dorsal to top. O,P - dorsal to left, anterior to top. Abbreviations: nt-neural tube, s-somite, n-notochord.

performed whole genome sequencing on DNA from six Araucana birds. After aligning reads to chicken chromosome 2 (GGA2), small variants (insertions, deletions, and SNPs) were called using the mpileup function of SamTools [42]. Variants were separated into three haplotype groups: three homozygous rumpless, two heterozygous rumpless, and one homozygous tailed Araucana. A total of 2092 unique small variants were identified within the candidate region when compared to the chicken Galgal4 reference sequence. We reduced the list of variants by excluding those that did not hold true to type; keeping only those that occurred in both alleles in the homozygote, a single allele in the heterozygotes, and not present in the tailed Araucana. A total of 316 small variants matched this pattern within the critical region. A further 18 variants that lined up with previously reported variants in tailed birds were also removed [43]. Of the remaining 298 small variants, we identified 274 SNPs and 24 insertion/deletions unique to rumpless Araucana (Fig. 5 and Table S2). None of the identified 298 small variants fell within the sequenced exons or introns of *IRX1* or *IRX2*, suggesting that the causative mutation is within the surrounding regulatory region.

During analysis of the small variants, a promising 55 base pair deletion on chromosome 2 between base pairs 86,830,975–86,831,030 was found to segregate with rumplessness. PCR analysis of the deletion was carried out on 131 Araucana^{Rp}, 30 tailed Araucana, and 38 non-Araucana tailed birds. In 198 cases, the deletion segregated with the phenotype correctly. However, in one case the genotype of a tailed Araucana male (RB1) did not segregate correctly. Six SNPs throughout the previously identified critical region and three newly identified SNPs from the whole genome sequencing were analyzed to identify the haplotype of RB1 (Fig. 5). RB1 was found to share only part of the rumpless haplotype, delimiting a new critical interval of 125,146 base pairs (86,476,559–86,601,705). Importantly, this interval contains neither *IRX1* nor *IRX2*, indicating that the causative mutation falls outside of either gene's coding sequence. On analysis of the 298 small variants unique to rumpless Araucana, only two SNPs fall within the new 125 kb critical interval. PCR analysis found both SNPs segregate by phenotype in 6-rumpless Araucana, 6-tailed Araucana, and 5-mixed breed tailed birds. Comparison of the surrounding 300 bp regions of both SNPs revealed that only the region for the SNP at bp location 86,594,449 was conserved in turkey (melGal1), zebra finch (taeGut2), and the medium ground finch (geoFor1). However, further functional testing is required to determine if either of the two identified SNPs within the 125 kb critical region is the heritable cause of rumplessness, however, this is beyond the scope of the current study.

In summary, these results identify two causative candidate SNPs by narrowing down the size of the critical region to 125 kb, as well as supporting the idea that the candidate mutation lies in the flanking regulatory region of *IRX1* and *IRX2* [55].

Expansion of neural tissue within Araucana^{Rp} embryo tailbud

Iroquois genes are involved in specifying and patterning neural domains [53,54]. In *Xenopus*, over-expression of *iroquois* causes the neural plate to expand, and promotes the onset of neural differentiation [59,60]. Importantly, cells of the tailbud have a bipotential fate, becoming mesoderm (paraxial mesoderm) or neuroectoderm (neural tube) [6,8,9,61]. Considering the misexpression of the proneural genes *IRX1* and *IRX2* and the premature downregulation of PSM marker *TBX6*, we predicted that the remaining unspecified progenitor cells would be pushed towards a neural fate. As *TBX6* plays an indirect role in repressing the enhancer of the neural marker *SOX2* we tested the expression of *SOX2* in Araucana^{Rp} embryos [9,62]. We found that beginning at HH18, Araucana^{Rp} embryos displayed ectopic *SOX2* expression, labeling an expanded domain of neural tissue (Fig. 6A–D). In addition, the Araucana^{Rp} neural tube displayed expanded epithelial structures with multiple open and irregular lumens (Fig. 6E–F). *WNT3A* is expressed in the dorsal midline of the neural tube. At HH22, expression of *WNT3A* in control embryos was limited to the dorsal neural tube, whereas in Araucana^{Rp} embryos *WNT3A* expression was dramatically altered within the expanded neuroepithelium of the tailbud region (Fig. 6G–H).

Thus, our results demonstrate aberrant neural differentiation in the Araucana^{Rp} tail as demonstrated by ectopic neural tissue labeled with both *SOX2* and *WNT3A*. This ectopic expression suggests an increase in the number of cells specified to a neural fate. This opens up the question of whether the shift in fate is at the expense of the mesoderm, and therefore the cause of the reduction in PSM in Araucana^{Rp}.

Gene expression in the retinoic acid pathway is altered in Araucana^{Rp} embryos

As the tail elongates and tail mesenchyme cells are displaced anteriorly they are exposed to an increasing concentration of retinoic acid (RA) from the somites [13,25]. This exposure to RA induces differentiation, as well as plays a role in defining somite segment boundaries at early stages [63,64]. However, ectopic exposure of the tailbud to RA leads to downregulation of *WNT3A* and *FGF8*, neural expansion, and a decrease in the PSM [24,25,65]. Therefore, we evaluated the expression of *RALDH2*,

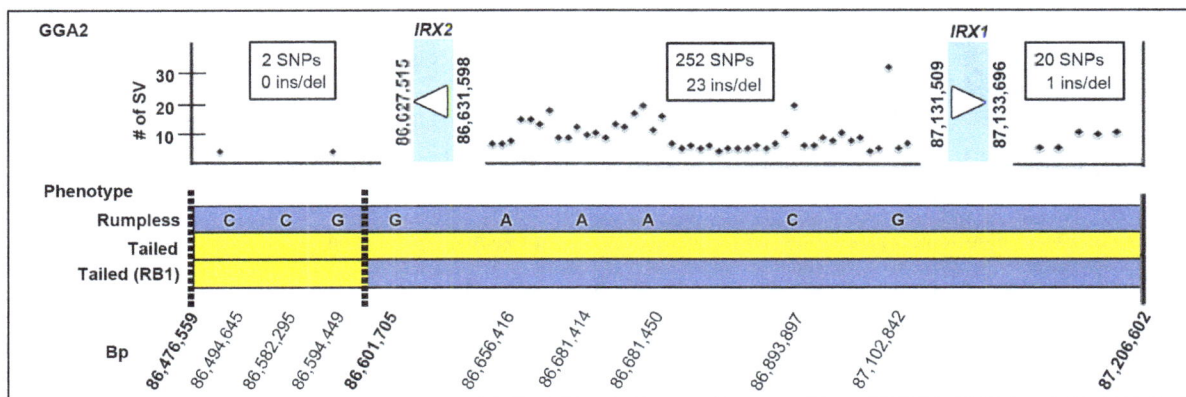

Figure 5. Location of small variants associated with rumplessness in the 740 kb critical region. Top: positional map of the 740 kb critical region located on *Gallus gallus* chromosome 2 (GGA2). Location of *IRX1* and *IRX2* is indicated, with numbers of small variants (SV) (SNPs, insertions and deletions) and their general location and frequency indicated. No small variants were found within *IRX1* or *IRX2* coding sequence. Bottom: relative position of SNPs used to build new critical interval within the established Araucana^{Rp} haplotype block. Rumpless Araucana^{Rp} haplotype in blue, with associated SNPs, haplotype from tailed Araucana in yellow. New critical interval defined by RB1 tailed Araucana highlighted by dashed lines.

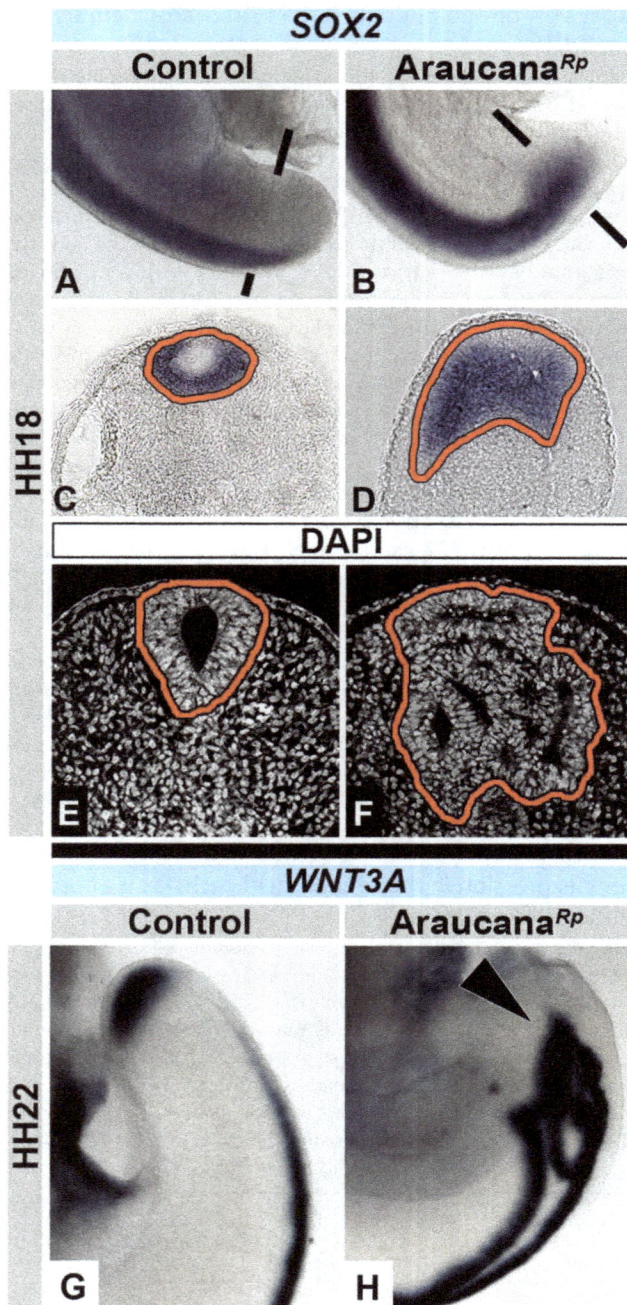

Figure 6. Ectopic neural tissue in Araucana^{Rp} embryo tailbud. Wholemount images and respective transverse sections showing ISH expression of *SOX2* in control (A,C) and Araucana^{Rp} (B,D) embryos at HH18. Note the presence of ectopic neural tissue and lumens in Araucana^{Rp} (highlighted in red) (D) compared to single neural tube and lumen in control (C). DAPI stained transverse sections of control (E) and Araucana^{Rp} (F) embryo tailbud, highlighting (in red) differences in number of neural tube lumens. Wholemount images of *WNT3A* expression in control (G) and Araucana^{Rp} (H). Arrowhead denotes ectopic expression of *WNT3A* in ectopic neural tubes of Araucana (H). Anterior to top in wholemount images. Dorsal to top in transverse sections. Sections taken at approximate level of black bars.

which encodes a dehydrogenase involved in endogenous production of RA [66]. Expression of *RALDH2* in the anterior PSM and the newly formed somites of Araucana^{Rp} embryos appeared

normal compared to the pattern displayed in control embryos at HH18 (Fig. 7A–B). Therefore, there does not appear to be a posterior shift in the production of RA in the tail. However, because of the reduced distance between the rostral PSM and tip of the tail in Araucana^{Rp} embryos, RA is likely able to influence cells closer to the tip of the tail at earlier stages than normal.

A cytochrome p450 enzyme, *CYP26A1*, is involved in the degradation of RA in the posterior tail, providing a protective effect against RA in the progenitor population during elongation [20,66,67]. In *Cyp26a1$^{-/-}$* mouse embryos the progenitor population is unprotected against RA and the result is the generation of embryos suffering severe caudal truncation [20]. Therefore, we investigated whether a change in RA degradation could be affecting Araucana^{Rp} embryos, and found that as early as HH16, *CYP26A1* is downregulated in the Araucana^{Rp} progenitor mesenchyme (Fig. 7C–D). Thus, although no differences in *RALDH2* expression were observed, lack of *CYP26A1* in Araucana^{Rp} could lead to increased levels of RA within the progenitor mesenchyme. However, if levels of RA in the tail were higher in Araucana^{Rp}, we would expect to see a downregulation of *WNT3A* and *FGF8* within the tailbud [16,24,25].

WNT3A is expressed within the tailbud where it is necessary to specify mesoderm from the bipotential population, and is involved in the proliferation of mesoderm within the tailbud [16,67,68,69,70]. WNT3A also regulates expression of *FGF8* within the tailbud, where FGF8 acts to maintain expression of *CYP26A1* as well as inhibit expression of *RALDH2*, thereby creating an opposing gradient to RA [64,71,72,73]. This can be disrupted through ectopic exposure to RA in the tailbud, which leads to the downregulation of both *WNT3A* and *FGF8*, and cessation of further axial elongation [24,25,65,74,75].

We found that until HH16, *WNT3A* expression within the tailbud was indistinguishable between control and Araucana^{Rp} embryos, but at HH17 (Fig. 7E,F) *WNT3A* expression became significantly downregulated in Araucana^{Rp} embryos and was lost by HH18 (Fig. 7G,H). We next examined the expression of *FGF8* within the tailbud, where it is expressed throughout tail elongation (Fig. 7I,K). In Araucana^{Rp}, expression was downregulated as early as HH17 (Fig. 7J), and by HH18 *FGF8* transcripts were undetectable (Fig. 7L).

These data suggest that although there is no change in expression of *RALDH2*, the downregulation of the RA degrading enzyme *CYP26A1* would allow for the exposure of the tailbud to RA. Exposure of the tailbud to RA following the downregulation of *CYP26A1* would explain the subsequent downregulation of *WNT3A* and *FGF8*.

Araucana^{Rp} tails have reduced proliferation and increased apoptosis

The premature downregulation of mesodermal markers *TBX6* and *BRA*, the loss of *WNT3A* which plays a role in proliferation of the progenitor population, and the expanded domain of expression of the neural marker *SOX2* suggests that the progenitor population has been critically jeopardized by forced differentiation towards a neural cell fate. Analysis of total cell numbers by TO-PRO-3 Iodide labeling of the TBM posterior to the hindgut revealed Araucana^{Rp} have consistently fewer total cells than controls, and this difference was statistically significant by HH19-20 (p<0.05, Fig. 8I). To examine potential drivers of the change in overall cell numbers we compared the levels of proliferation and apoptosis in the tailbud.

Quantification by image analysis showed that proliferation was similar between control and Araucana^{Rp} TBM up to HH15 (Fig. 8A–B,J). After this a dramatic increase in proliferating cells

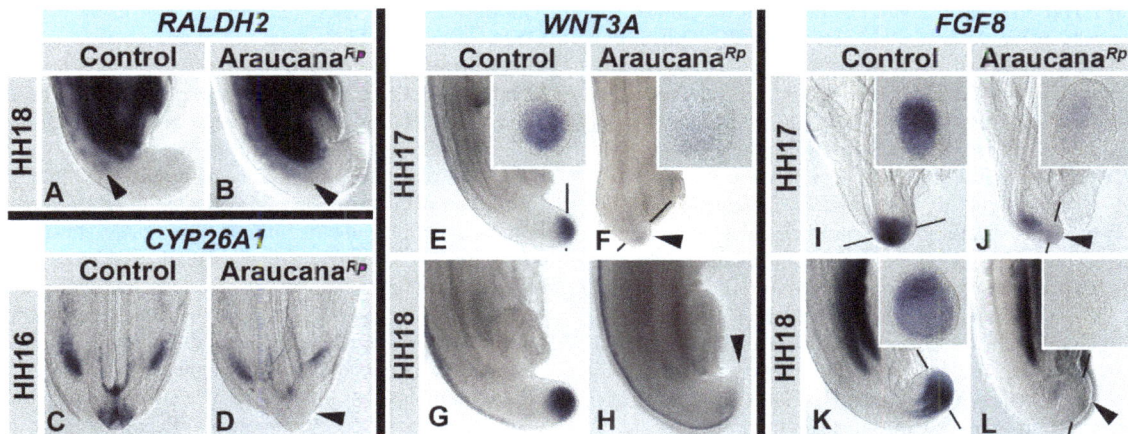

Figure 7. ISH expression pattern of *RALDH2, CYP26A1, WNT3A* **and** *FGF8* **during tail development.** *RALDH2* expression in control (A) and Araucana[Rp] (B) embryos at HH18. Note in Araucana[Rp] the truncated tail at HH18 coupled with the close expression of *RALDH2* in the formed somites. Posterior-most expression of *RALDH2* is marked with arrowhead. *CYP26A1* expression in control (C) and Araucana[Rp] (D) embryos at HH16. Note the lack of expression in Araucana[Rp] tailbud (arrowhead). *WNT3A* expression in control (E,G) and Araucana[Rp] (F, H) embryos from HH17-18. Note lack of expression in tailbud mesenchyme in Araucana[Rp] (arrowheads). *FGF8* expression in control (I,K) and Araucana[Rp] (J,L) embryos from HH stages 17–18. Note the downregulation of expression in tailbud beginning at HH17 in Araucana[Rp] (arrowheads). Insets are transverse sections at level of tailbud. Anterior is top in whole mount images. Dorsal is top in transverse section insets. Sections were taken at approximate level of black bars.

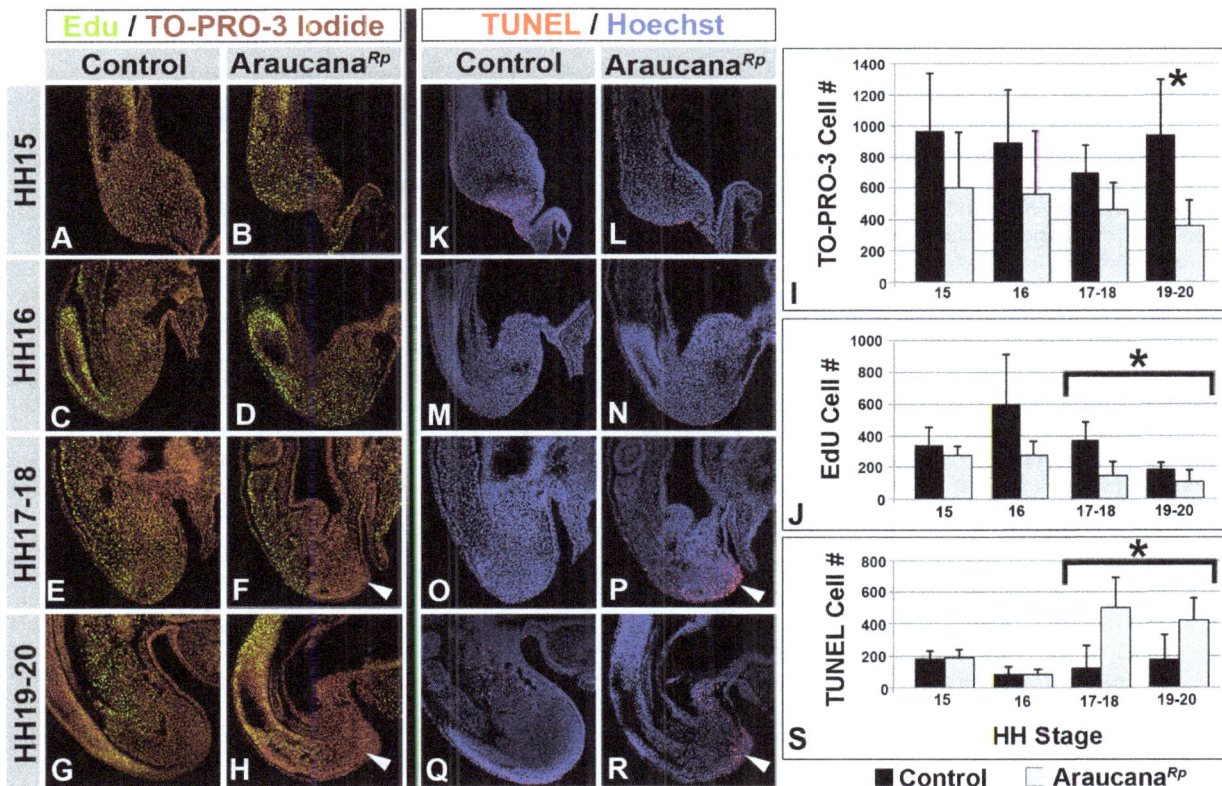

Figure 8. Role of proliferation and apoptosis in Araucana[Rp] tailbud development. EdU labeled proliferating cells in tailbud sagittal sections at HH15-20 in controls. (A,C,E,G) and Araucana[Rp] (B,D,F,H) embryo tailbud. Green is EdU labeled proliferating cells, red is TO-PRO-3 Iodide labeled nuclei. Arrows denote decreased regions of proliferation in Araucana[Rp] compared to controls. (I) Quantification of all TBM cells for control and Araucana[Rp] labeled with TO-PRO-3 Iodide. (J) Quantification of proliferating TBM cells for control and Araucana[Rp] labeled with EdU. TUNEL labeled apoptotic cells in tailbud sagittal sections from HH stages 15–20 of controls. (K,M,O,Q) and Araucana[Rp] (L,N,P,R) embryo tailbud. Red is TUNEL labeled apoptotic cells, blue is Hoechst labeled nuclei. Arrows denote increased areas of apoptosis in Araucana[Rp] compared to controls. (S) Quantification of apoptotic TUNEL labeled cells in control and Araucana[Rp] TBM. Araucana[Rp] have significantly more TUNEL positive cells beginning at HH17 than controls. T-test, asterisk marks $p < 0.05$. Control-Black, Araucana[Rp]-Grey. Same embryo (two different sections) was used for both TUNEL and EdU labeling. Anterior is towards the top, dorsal is towards the left in all sections.

was noted at HH16 in controls, with less than half the number of proliferating cells in the Araucana^{Rp} tailbud (Fig. 8C–D,J). From HH17 to HH20 the reduced levels of proliferation are statistically significant (p<0.05, Fig. 8E–H,J). One possible explanation for the reduced rate of proliferation is that the fate change away from mesoderm to neural described previously will reduce the number of progenitor cells as neural cells move out of the cell cycle [76].

Exposure of the tailbud to RA leads to increased levels of apoptosis [25,65]. Historical studies in rumpless chickens identified degenerating cells within the tailbud and morphologically the tailbud appears to have apoptotic cells in the tip of the tail (Fig. 2D,F) [28]. Therefore, we examined the levels of apoptosis within the tailbud region using TUNEL labeling of sagittal sections. Apoptosis was similar up to HH16 between control and Araucana^{Rp} embryos (Fig. 8 K–N,S). At HH17-18, however, there was a dramatic rise in apoptosis within the tailbud of Araucana^{Rp} embryos compared to controls (Fig. 8O–P,S). The increased Apoptosis in Araucana^{Rp} occurred within the posterior TBM and surface ectoderm, matching the same area in Araucana^{Rp} that had decreased proliferation (Fig. 8F,P). By HH19-20, apoptosis had increased in the ventral ectodermal ridge of controls, whereas Araucana^{Rp} apoptosis was still elevated within the remaining posterior TBM, surface ectoderm, and VER (Fig. 8Q–S). These apoptotic Araucana^{Rp} cells appear in the same position as cells that would normally be expressing TBX6 and WNT3A (Fig. 8P,R). Interestingly, the pattern of apoptosis seen in Araucana^{Rp} matches that observed by Tenin and coworkers in chicken embryo tailbud at HH26, the end of somitogenesis [24].

Discussion

The Araucana rumpless phenotype results from a failure to form the most posterior axial somites. Two candidate mutations proximal to the genes IRX1 and IRX2 have been identified as associated with the rumpless phenotype. Araucana^{Rp} exhibit a gain-of-function of both IRX1 and IRX2 genes within the tailbud during the tail organizer stage. This gain-of-function precedes observed changes in mesoderm and neural cell specification, maintenance and proliferation of the progenitor population, and regulation of the RA pathway.

Our results indicate that the Araucana^{Rp} causative mutation is one of two (or both) SNPs within the proximal regulatory region of IRX1 and IRX2. Tena and colleagues found that the 3D architecture of IRX1 and IRX2 bring the promoters into close proximity allowing them to share enhancers [55]. The dual misexpression of IRX1 and IRX2 supports this co-regulation model in which the Araucana^{Rp} phenotype results from a gain-of-function mutation in the regulatory region of the IRX1 and IRX2 genes. Until further functional testing is done the nature of the causative mutation is speculative, but a mutation could act by removing regionally specific repression/silencing of IRX1 and IRX2 within the tailbud, or by creating a new CNH specific enhancer. Importantly, the misexpression of IRX1 and IRX2 only occurs post gastrulation at the beginning of the tailbud stages, indicating that the upstream trigger may hold a key to understanding the change between primary and secondary body formation.

The misexpression of IRX1 and IRX2 within the tailbud precedes all observed genetic and morphological changes. The iroquois genes encode homeodomain-containing transcription factors within the TALE (three amino acid loop extension) family and are involved in proneural fate and patterning through transcriptional repression of neural antagonists [53,60,77,78,79]. Given the role of iroquois genes in specifying neural identity, we hypothesized that IRX gene misexpression at the level of the CNH, which is bipotential in chickens, and in the tailbud mesenchyme, a population that has been shown to be bipotential in other vertebrate models including mice and zebrafish, would disrupt the delicate balance between progenitor cell maintenance and mesoderm/neural specification, potentially through the repression of neural antagonists such as TBX6 [6,8,9,10,25,70]. Indeed, immediately upon misexpression of IRX1 and IRX2 there is a cascade of gene disruption; immediate downregulation of TBX6 quickly followed by downregulation of BRA, indicating loss of mesoderm identity and an arrest in mesoderm specification [9,10,11,17,19], loss of CYP26A1 that is required to degrade RA (protecting progenitor cells from differentiation and apoptosis) [20,66,67], upregulation of the neural marker SOX2 [9], and concomitant loss of WNT3A and FGF8 that are required for maintenance and proliferation of the progenitor population (Fig. 9) [16,21,64,68,69,72]. Therefore, we propose a model where proneural gene misexpression overrides the balance of factors within the bipotential progenitor population, steering presumptive mesoderm cells toward the neural lineage.

Morphological changes in the shape of the Araucana^{Rp} embryo tail are evident as early as HH16 and coincide with the downregulation of the mesoderm markers TBX6 and BRA (T in mice), the loss of which lead to axial truncation [11,17]. Tbx6 knockout in mice leads to a loss of mesoderm specification, and the upregulation of the neural marker Sox2 within ectopic neural tubes [10,11]. Tbx6 indirectly regulates Sox2 through repression of its N1 enhancer, making it both necessary to repress neural fate as well as push cells towards a mesoderm fate [9]. The downregulation of TBX6 expression in Araucana^{Rp} is the first observed change in gene expression following the misexpression of IRX1 and IRX2. TBX6 expression is initially lost within the tailbud, not affecting more anterior, previously specified TBX6-positive cells. At the anterior boundary of the PSM the cyclic formation of somites continued unaffected, consuming the specified PSM. When these remaining TBX6 positive cells were incorporated into the most recent somite, somitogenesis arrested prematurely as no further PSM has been specified. BRA is required for proper mesoderm cell fate and in its absence embryos fail to form the proper mesoderm structures [18,80,81,82]. T homozygote knockout mice embryos fail to form both somites and the notochord, with heterozygotes forming a truncated axis [18,81]. In Araucana^{Rp} embryos, notochord expression is maintained, but is absent from tailbud cells by HH17. Interestingly, the downregulation of BRA in Araucana^{Rp} resembles zebrafish no-tail mutants (ntl, T ortholog), as well as zebrafish treated with RA, which downregulates ntl expression [70,82]. Importantly, as ntl both activates cyp26a1 and functions within an autoregulatory loop with wnt, the loss of expression of any of the three genes is sufficient to cause axis truncation [82,83]. Similarly, the loss of expression of BRA in Araucana^{Rp} could explain the downregulation of CYP26A1 observed at HH16, as well as the downregulation of WNT3A at HH17. Without TBX6 to repress a neural fate in Araucana^{Rp} embryos, as well as both TBX6 and BRA to promote a presomitic mesoderm fate, cells that would normally form PSM instead form ectopic neural tissue, or remain in an undifferentiated transition state. Neural cells then leave the progenitor population to join the secondary neural tube (medullary cord), reducing the available progenitor pool and starving the paraxial mesodermal of additional cells required for ongoing somite production.

During somitogenesis, the posterior region of the vertebrate embryo including the PSM is exposed to RA expressed from the somites and FGF8 expressed from the tailbud, generating two opposing gradients [64,66,84,85,86]. The tailbud progenitor

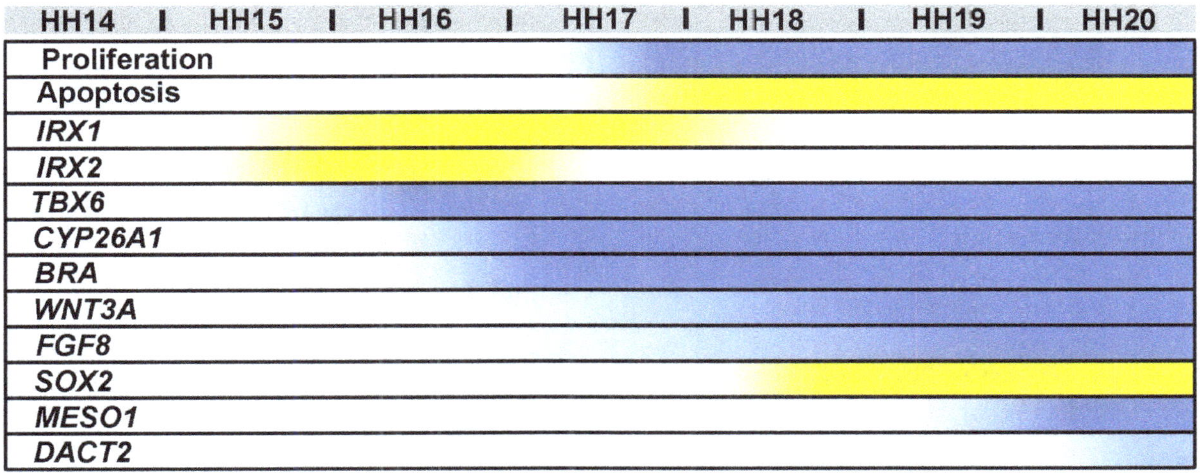

Figure 9. Timeline of changes in proliferation, apoptosis, and gene expression in Araucana[Rp]. Visual timeline summarizing changes in proliferation, apoptosis, and gene expression within Araucana[Rp] embryo tail compared to control embryos. Yellow is increased expression/prevalence, and blue is decreased expression/prevalence. The WNT3A summary represents change in expression observed in posterior tailbud, not neural tube.

population is protected from the effects of RA by the action of CYP26A1, which metabolizes RA [20,66]. One of the first events following IRX1/IRX2 misexpression is the loss of CYP26A1 within the tailbud, though whether the downregulation of CYP26A1 is a direct consequence of the ectopic IRX1/IRX2 expression is still unknown. In the tailbud of Araucana[Rp], without the protection of CYP26A1 we would predict that the levels of RA would be increased, although we have not directly shown this. However, data from chicken and mice studies show that ectopic RA leads to the downregulation of WNT3A and FGF8, axis truncation, and ectopic neural tissue, which is precisely what we observed in Araucana[Rp] embryos [24,25,64,65,74].

Based on our results changes in proliferation and apoptosis occur following the initial morphological changes of Araucana[Rp] embryos at HH16. As cells are constantly moving from the TBM to form the PSM and then somites, a constant supply of new cells is required. Controls display a spike in proliferation at HH16, followed by a declining trend in proliferation. The number of TBM cells as labeled by TO-PRO-3 Iodide in controls appears relatively stable through these stages, indicating that proliferation may not be the only contributor to the cells of the TBM. Cells that migrate through the ventral ectodermal ridge into the TBM are most likely helping to maintain the TBM population in controls [22]. Araucana[Rp] do not display the spike in proliferation seen in controls at HH16, with the number of proliferating cells continuously decreasing. A significant decrease in proliferation within the TBM occurs in Araucana[Rp] at HH17-18, which coincides spatiotemporally with the downregulation of FGF8 and WNT3A. We are unable to determine from these results whether the downregulation of FGF8 and/or WNT3A directly causes the decrease in proliferation. However, studies in mice have shown that Wnt3a plays a role in the generation and proliferation of cells within the tailbud, and loss of Wnt3a signaling leads to axis truncation and formation of ectopic neural tissue [16,68,69]. This suggests that WNT3A is required for proper patterning of presumptive presomitic mesoderm cells and without WNT3A there is an expansion of neural tissue, as observed by the ectopic SOX2-positive neural tubes in Araucana[Rp] embryos [10,69]. Furthermore, as cells within the TBM of Araucana[Rp] are pushed towards a neural fate, they would exit the cell cycle ceasing

proliferation [76]. As Araucana[Rp] display ectopic neural tissue by HH18, cell cycle exiting due to neural differentiation could also explain the differences in proliferation observed. With the decrease in proliferation within the TBM relative to controls, Araucana[Rp] have fewer overall TBM cells, and are unable to produce additional PSM.

Apoptosis occurs within the tailbud of control embryos and by the end of somitogenesis (HH26) is localized to the posterior-most TBM and surface ectoderm [23,24,87]. Tenin and coworkers proposed that the localized apoptosis in the posterior TBM contributes to the removal of the remaining progenitor cells [24]. The increase and localization of apoptosis in Araucana[Rp] beginning at HH17-18 mirrors that seen in controls at HH26. In both controls and Araucana[Rp] this coincides with the downregulation of CYP26A1, FGF8, and WNT3A expression. Furthermore, exposure to ectopic RA leads to downregulation of WNT3A and apoptosis in the tailbud [24,65,74]. It is therefore likely that in Araucana[Rp] embryos the apoptotic event that is triggered at HH17-18 results from exposure to increased levels of RA within the tailbud. Thus, changes in both apoptosis and proliferation do not appear to be the initial cause of changes in the morphology of the Araucana[Rp] phenotype, but rather occur following the initial changes in gene expression within the TBM. As proliferation decreases, and apoptosis increases, the diminution of the TBM ensues, leading to a diminished progenitor population in Araucana[Rp] embryos by HH19-20. As cells from this progenitor pool populate both the neural tube and PSM, Araucana[Rp] are unable to contribute enough cells to the PSM to continue axis elongation, as there is both a decrease in the number of progenitor cells, and an increase in cells forming neural tissue.

In conclusion, we have provided evidence that a novel gain of function is responsible for the Araucana rumpless phenotype. Hence, the Araucana rumpless mutation (Rp) is separate from the spontaneous mutations that cause the lack of a tail in chickens [88]. This study has highlighted the fine control required to maintain axis elongation and has added additional evidence that avian tailbud cells are bipotential, continuing to make germ layer decisions between neural and mesoderm post gastrulation similar to mice and zebrafish [8,70]. Considering that mutants such as T and Wnt3a mice are a loss of function, it was surprising that a similar phenotype occurred in Araucana as a gain of function. These results provide insight into a novel developmental mechanism controlling the termination of axis elongation and therefore total axis length. Future studies are required to better understand how the Araucana gain of function mutation drives expression to unbalance the bipotential fate, and whether the same mechanism can control somite numbers in other organisms.

Supporting Information

Table S1   Number of reads, bases, coverage, SNPs, and INDELS for each Araucana following WGS.

Table S2   Complete list of 298 unique small variants found from WGS.

Acknowledgments

We thank Ann Charles of Sky Blue Egg for supplying Araucana eggs, Fritz and Joyce Ludwig and Jocelyn Clarke for Araucana skeletons, as well as the Araucana Club of America and Clemson University Poultry Farm for providing samples and assistance. We would like to thank the Clemson University Light Imaging Facility for technical advice concerning microscopy. We would like to thank Cheryll Tickle, J. Kim Dale, and Olivier Pourquié for reagents. We thank Miguel Maroto and Leigh Anne Clark for discussions related to the project.

Author Contributions

Conceived and designed the experiments: NHF SCC. Performed the experiments: NHF AS SCC. Analyzed the data: NHF BAL MS AS SCC. Contributed reagents/materials/analysis tools: NHF BAL MS AS SCC. Contributed to the writing of the manuscript: NF BAL MS AS SCC.

References

1. Schoenwolf GC (1979) Histological and ultrastructural observations of tail bud formation in the chick embryo. The Anatomical record 193: 131–148.
2. Schoenwolf GC (1981) Morphogenetic processes involved in the remodeling of the tail region of the chick embryo. Anatomy and Embryology 162: 183–197.
3. Catala M, Teillet M-A, Le Douarin NM (1995) Organization and development of the tail bud analyzed with the quail-chick chimaera system. Mechanisms of Development 51: 51–65.
4. Cambray N, Wilson V (2002) Axial progenitors with extensive potency are localised to the mouse chordoneural hinge. Development 129: 4855–4866.
5. Cambray N, Wilson V (2007) Two distinct sources for a population of maturing axial progenitors. Development 134: 2829–2840.
6. McGrew MJ, Sherman A, Lillico SG, Ellard FM, Radcliffe PA, et al. (2008) Localised axial progenitor cell populations in the avian tail bud are not committed to a posterior Hox identity. Development 135: 2289–2299.
7. Wilson V, Olivera-Martinez I, Storey KG (2009) Stem cells, signals and vertebrate body axis extension. Development 136: 1591–1604.
8. Tzouanacou E, Wegener A, Wymeersch FJ, Wilson V, Nicolas J-F (2009) Redefining the Progression of Lineage Segregations during Mammalian Embryogenesis by Clonal Analysis. Developmental Cell 17: 365–376.
9. Takemoto T, Uchikawa M, Yoshida M, Bell DM, Lovell-Badge R, et al. (2011) Tbx6-dependent Sox2 regulation determines neural or mesodermal fate in axial stem cells. Nature 470: 394–398.
10. Nowotschin S, Ferrer-Vaquer A, Concepcion D, Papaioannou VE, Hadjanto-nakis AK (2012) Interaction of Wnt3a, Msgn1 and Tbx6 in neural versus paraxial mesoderm lineage commitment and paraxial mesoderm differentiation in the mouse embryo. Dev Biol 367: 1–14.
11. Chapman DL, Papaioannou VE (1998) Three neural tubes in mouse embryos with mutations in the T-box gene Tbx6. Nature 391: 695–697.
12. Brand-Saberi B, Christ B (2000) Evolution and development of distinct cell lineages derived from somites. Curr Top Dev Biol 48: 1–42.
13. Dequeant M-L, Pourquie O (2008) Segmental patterning of the vertebrate embryonic axis. Nat Rev Genet 9: 370–382.
14. Gomez C, Ozbudak EM, Wunderlich J, Baumann D, Lewis J, et al. (2008) Control of segment number in vertebrate embryos. Nature 454: 335–339.
15. Richardson MK, Allen SP, Wright GM, Raynaud A, Hanken J (1998) Somite number and vertebrate evolution. Development 125: 151–160.
16. Greco TL, Takada S, Newhouse MM, McMahon JA, McMahon AP, et al. (1996) Analysis of the vestigial tail mutation demonstrates that Wnt-3a gene dosage regulates mouse axial development. Genes & Development 10: 313–324.
17. Yamaguchi TP, Takada S, Yoshikawa Y, Wu N, McMahon AP (1999) T (Brachyury) is a direct target of Wnt3a during paraxial mesoderm specification. Genes & Development 13: 3185–3190.
18. Chesley P (1935) Development of the short-tailed mutant in the house mouse. Journal of Experimental Zoology 70: 429–459.
19. Herrmann BG, Labeit S, Poustka A, King TR, Lehrach H (1990) Cloning of the T gene required in mesoderm formation in the mouse. Nature 343: 617–622.
20. Abu-Abed S, Dollé P, Metzger D, Beckett B, Chambon P, et al. (2001) The retinoic acid-metabolizing enzyme, CYP26A1, is essential for normal hindbrain patterning, vertebral identity, and development of posterior structures. Genes & Development 15: 226–240.
21. Ciruna B, Rossant J (2001) FGF Signaling Regulates Mesoderm Cell Fate Specification and Morphogenetic Movement at the Primitive Streak. Developmental Cell 1: 37–49.
22. Ohta S, Suzuki K, Tachibana K, Tanaka H, Yamada G (2007) Cessation of gastrulation is mediated by suppression of epithelial-mesenchymal transition at the ventral ectodermal ridge. Development 134: 4315–4324.
23. Mills CL, Bellairs R (1989) Mitosis and cell death in the tail of the chick embryo. Anatomy and Embryology 180: 301–308.
24. Tenin G, Wright D, Ferjentsik Z, Bone R, McGrew M, et al. (2010) The chick somitogenesis oscillator is arrested before all paraxial mesoderm is segmented into somites. BMC Developmental Biology 10: 24.
25. Olivera-Martinez I, Harada H, Halley PA, Storey KG (2012) Loss of FGF-Dependent Mesoderm Identity and Rise of Endogenous Retinoid Signalling Determine Cessation of Body Axis Elongation. PLoS Biol 10: e1001415.
26. Dunn LC, Landauer W (1934) The genetics of the rumpless fowl with evidence of a case of changing dominance. Journal of Genetics 29: 217–243.
27. Dunn LC, Landauer W (1936) Further data on genetic modification of rumplessness in the fowl. Journal of Genetics 33: 401–405.
28. Zwilling E (1942) The development of dominant rumplessness in chick embryos. Genetics 27: 641–656.
29. Landauer W (1945) Recessive rumplessness of fowl with kyphoscioliosis and supernumerary ribs. Genetics 30: 403–428.
30. Somes RG Jr, Pabilonia MS (1981) Ear tuftedness: a lethal condition in the Araucana fowl. Journal of Heredity 72: 121–124.
31. Noorai RE, Freese NH, Wright LM, Chapman SC, Clark LA (2012) Genome-wide association mapping and identification of candidate genes for the rumpless and ear-tufted traits of the Araucana chicken. PLoS ONE 7: e40974.
32. Hamburger V, Hamilton LH (1951) A series of normal stages in the development of the chick embryo. Journal of Morphology 88: 49–92.
33. Yamazaki Y, Yuguchi M, Kubota S, Isokawa K (2011) Whole-mount bone and cartilage staining of chick embryos with minimal decalcification. Biotechnic & Histochemistry 86: 351–358.
34. Wood JL, Hughes AJ, Mercer KJ, Chapman SC (2010) Analysis of chick (Gallus gallus) middle ear columella formation. BMC Dev Biol 10: 16.
35. Chapman SC, Schubert FR, Schoenwolf GC, Lumsden A (2002) Analysis of Spatial and Temporal Gene Expression Patterns in Blastula and Gastrula Stage Chick Embryos. Developmental Biology 245: 187–199.
36. McDonald LA, Gerrelli D, Fok Y, Hurst LD, Tickle C (2010) Comparison of Iroquois gene expression in limbs/fins of vertebrate embryos. J Anat 216: 683–691.
37. Quinlan R, Gale E, Maden M, Graham A (2002) Deficits in the posterior pharyngeal endoderm in the absence of retinoids. Dev Dyn 225: 54–60.
38. Chapman SC, Brown R, Lees L, Schoenwolf GC, Lumsden A (2004) Expression analysis of chick Wnt and frizzled genes and selected inhibitors in early chick patterning. Dev Dyn 229: 668–676.
39. Warren M, Puskarczyk K, Chapman SC (2009) Chick embryo proliferation studies using EdU labeling. Developmental Dynamics 238: 944–949.
40. Lohse M, Bolger AM, Nagel A, Fernie AR, Lunn JE, et al. (2012) RobiNA: a user-friendly, integrated software solution for RNA-Seq-based transcriptomics. Nucleic Acids Res 40: W622–627.
41. Langmead B, Salzberg SL (2012) Fast gapped-read alignment with Bowtie 2. Nat Methods 9: 357–359.
42. Li H, Handsaker B, Wysoker A, Fennell T, Ruan J, et al. (2009) The Sequence Alignment/Map format and SAMtools. Bioinformatics 25: 2078–2079.
43. Consortium ICPM (2004) A genetic variation map for chicken with 2.8 million single-nucleotide polymorphisms. Nature 432: 717–722.
44. Herrmann BG (1991) Expression pattern of the Brachyury gene in whole-mount TWis/TWis mutant embryos. Development 113: 913–917.
45. Chapman DL, Agulnik I, Hancock S, Silver LM, Papaioannou VE (1996) Tbx6,a Mouse T-Box Gene Implicated in Paraxial Mesoderm Formation at Gastrulation. Developmental Biology 180: 534–542.
46. White PH, Farkas DR, McFadden EE, Chapman DL (2003) Defective somite patterning in mouse embryos with reduced levels of Tbx6. Development 130: 1681–1690.
47. Buchberger A, Seidl K, Klein C, Eberhardt H, Arnold HH (1998) cMeso-1, a novel bHLH transcription factor, is involved in somite formation in chicken embryos. Dev Biol 199: 201–215.

48. Buchberger A, Bonneick S, Arnold H (2000) Expression of the novel basic-helix-loop-helix transcription factor cMespo in presomitic mesoderm of chicken embryos. Mech Dev 97: 223–226.

49. Yasuhiko Y, Haraguchi S, Kitajima S, Takahashi Y, Kanno J, et al. (2006) Tbx6-mediated Notch signaling controls somite-specific Mesp2 expression. Proceedings of the National Academy of Sciences of the United States of America 103: 3651–3656.

50. Brott BK, Sokol SY (2005) Frodo proteins: modulators of Wnt signaling in vertebrate development. Differentiation 73: 323–329.

51. Waxman JS, Hocking AM, Stoick CL, Moon RT (2004) Zebrafish Dapper1 and Dapper2 play distinct roles in Wnt-mediated developmental processes. Development 131: 5909–5921.

52. Alvares LE, Winterbottom FL, Jorge EC, Rodrigues Sobreira D, Xavier-Neto J, et al. (2009) Chicken dapper genes are versatile markers for mesodermal tissues, embryonic muscle stem cells, neural crest cells, and neurogenic placodes. Dev Dyn 238: 1166–1178.

53. Cavodeassi F, Modolell J, Gomez-Skarmeta JL (2001) The Iroquois family of genes: from body building to neural patterning. Development 128: 2847–2855.

54. Gomez-Skarmeta JL, Modolell J (2002) Iroquois genes: genomic organization and function in vertebrate neural development. Curr Opin Genet Dev 12: 403–408.

55. Tena JJ, Alonso ME, de la Calle-Mustienes E, Splinter E, de Laat W, et al. (2011) An evolutionarily conserved three-dimensional structure in the vertebrate Irx clusters facilitates enhancer sharing and coregulation. Nat Commun 2: 310.

56. Bao ZZ, Bruneau BG, Seidman JG, Seidman CE, Cepko CL (1999) Regulation of chamber-specific gene expression in the developing heart by Irx4. Science 283: 1161–1164.

57. López-Sánchez C, Bártulos O, Martínez-Campos E, Gañán C, Valenciano AI, et al. (2010) Tyrosine hydroxylase is expressed during early heart development and is required for cardiac chamber formation. Cardiovascular Research 88: 111–120.

58. Diaz-Hernandez ME, Bustamante M, Galvan-Hernandez CI, Chimal-Monroy J (2013) Irx1 and Irx2 are coordinately expressed and regulated by retinoic acid, TGFbeta and FGF signaling during chick hindlimb development. PLoS ONE 8: e58549.

59. Bellefroid EJ, Kobbe A, Gruss P, Pieler T, Gurdon JB, et al. (1998) Xiro3 encodes a Xenopus homolog of the Drosophila Iroquois genes and functions in neural specification. EMBO J 17: 191–203.

60. Gomez-Skarmeta JL, Glavic A, de la Calle-Mustienes E, Modolell J, Mayor R (1998) Xiro, a Xenopus homolog of the Drosophila iroquois complex genes, controls development at the neural plate. EMBO J 17: 181–190.

61. Kondoh H, Takemoto T (2012) Axial stem cells deriving both posterior neural and mesodermal tissues during gastrulation. Curr Opin Genet Dev 22: 374–380.

62. Rex M, Orme A, Uwanogho D, Tointon K, Wigmore PM, et al. (1997) Dynamic expression of chicken Sox2 and Sox3 genes in ectoderm induced to form neural tissue. Dev Dyn 209: 323–332.

63. Goldbeter A, Gonze D, Pourquié O (2007) Sharp developmental thresholds defined through bistability by antagonistic gradients of retinoic acid and FGF signaling. Developmental Dynamics 236: 1495–1508.

64. Diez del Corral R, Olivera-Martinez I, Goriely A, Gale E, Maden M, et al. (2003) Opposing FGF and retinoid pathways control ventral neural pattern, neuronal differentiation, and segmentation during body axis extension. Neuron 40: 65–79.

65. Shum ASW, Poon LLM, Tang WWT, Koide T, Chan BWH, et al. (1999) Retinoic acid induces down-regulation of Wnt-3a, apoptosis and diversion of tail bud cells to a neural fate in the mouse embryo. Mechanisms of Development 84: 17–30.

66. Swindell EC, Thaller C, Sockanathan S, Petkovich M, Jessell TM, et al. (1999) Complementary Domains of Retinoic Acid Production and Degradation in the Early Chick Embryo. Developmental Biology 216: 282–296.

67. Sakai Y, Meno C, Fujii H, Nishino J, Shiratori H, et al. (2001) The retinoic acid-inactivating enzyme CYP26 is essential for establishing an uneven distribution of retinoic acid along the anterio-posterior axis within the mouse embryo. Genes & Development 15: 213–225.

68. Takada S, Stark KL, Shea MJ, Vassileva G, McMahon JA, et al. (1994) Wnt-3a regulates somite and tailbud formation in the mouse embryo. Genes Dev 8: 174–189.

69. Yoshikawa Y, Fujimori T, McMahon AP, Takada S (1997) Evidence that absence of Wnt-3a signaling promotes neuralization instead of paraxial mesoderm development in the mouse. Dev Biol 183: 234–242.

70. Martin BL, Kimelman D (2012) Canonical Wnt Signaling Dynamically Controls Multiple Stem Cell Fate Decisions during Vertebrate Body Formation. Developmental Cell 22: 223–232.

71. Aulehla A, Wehrle C, Brand-Saberi B, Kemler R, Gossler A, et al. (2003) Wnt3a Plays a Major Role in the Segmentation Clock Controlling Somitogenesis. Developmental Cell 4: 395–406.

72. Olivera-Martinez I, Storey KG (2007) Wnt signals provide a timing mechanism for the FGF-retinoid differentiation switch during vertebrate body axis extension. Development 134: 2125–2135.

73. Wahl MB, Deng C, Lewandoski M, Fourquié O (2007) FGF signaling acts upstream of the NOTCH and WNT signaling pathways to control segmentation clock oscillations in mouse somitogenesis. Development 134: 4033–4041.

74. Iulianella A, Beckett B, Petkovich M, Lohnes D (1999) A Molecular Basis for Retinoic Acid-Induced Axial Truncation. Developmental Biology 205: 33–48.

75. Zhao X, Duester G (2009) Effect of retinoic acid signaling on Wnt/[beta]-catenin and FGF signaling during body axis extension. Gene Expression Patterns 9: 430–435.

76. Shimokita E, Takahashi Y (2011) Secondary neurulation: Fate-mapping and gene manipulation of the neural tube in tail bud. Development, Growth & Differentiation 53: 401–410.

77. Gomez-Skarmeta JL, Diez del Corral R, de la Calle-Mustienes E, Ferre-Marco D, Modolell J (1996) Araucan and caupolican, two members of the novel iroquois complex, encode homeoproteins that control proneural and vein-forming genes. Cell 85: 95–105.

78. Rodriguez-Seguel E, Alarcon P, Gomez-Skarmeta JL (2009) The Xenopus Irx genes are essential for neural patterning and define the border between prethalamus and thalamus through mutual antagonism with the anterior repressors Fezf and Arx. Dev Biol 329: 258–268.

79. Gomez-Skarmeta J, de La Calle-Mustienes E, Modolell J (2001) The Wnt-activated Xiro1 gene encodes a repressor that is essential for neural development and downregulates Bmp4. Development 128: 551–560.

80. Showell C, Binder O, Conlon FL (2004) T-box genes in early embryogenesis. Dev Dyn 229: 201–218.

81. Gluecksohn-Schoenheimer S (1944) The Development of Normal and Homozygous Brachy (T/T) Mouse Embryos in the Extraembryonic Coelom of the Chick. Proc Natl Acad Sci U S A 30: 134–140.

82. Martin BL, Kimelman D (2010) Brachyury establishes the embryonic mesodermal progenitor niche. Genes Dev 24: 2778–2783.

83. Martin BL, Kimelman D (2008) Regulation of canonical Wnt signaling by Brachyury is essential for posterior mesoderm formation. Developmental Cell 15: 121–133.

84. Dubrulle J, McGrew MJ, Pourquié O (2001) FGF Signaling Controls Somite Boundary Position and Regulates Segmentation Clock Control of Spatiotemporal Hox Gene Activation. Cell 106: 219–232.

85. Delfini MC, Dubrulle J, Malapert P, Chal J, Pourquie O (2005) Control of the segmentation process by graded MAPK/ERK activation in the chick embryo. Proc Natl Acad Sci U S A 102: 11343–11348.

86. Naiche LA, Holder N, Lewandoski M (2011) FGF4 and FGF8 comprise the wavefront activity that controls somitogenesis. Proc Natl Acad Sci U S A 108: 4018–4023.

87. Miller SA, Briglin A (1996) Apoptosis removes chick embryo tail gut and remnant of the primitive streak. Developmental Dynamics 206: 212–218.

88. Dunn LC (1925) The inheritance of rumplessness in the domestic fowl. Journal of Heredity 16: 127–134.

8

Characterization of a Novel Hemolytic Activity of Human IgG Fractions Arising from Diversity in Protein and Oligosaccharide Components

Shaoying Min[1][9], Fang Yan[1,2][9], Yueling Zhang[1]*, Xiangqun Ye[3], Mingqi Zhong[1], Jinsong Cao[1], Haiying Zou[3], Jiehui Chen[1]

1 Department of Biology and Guangdong Provincial Key Laboratory of Marine Biotechnology, Shantou University, Shantou, China, 2 Mariculture Institute of Shandong Province, Qingdao, China, 3 Medical College, Shantou University, Shantou, China

Abstract

Human IgG is a well-established multifunctional antigen specific immunoglobulin molecule of the adaptive immune system. However, an antigen nonspecific immunological function of human IgG has never been reported. In this study, human IgG was isolated using ammonium sulfate fractional precipitation and diethylaminoethanol (DEAE) cellulose 52 ion exchange chromatography, from which h-IgG and hs-IgG fractions were purified on the basis of their differential binding to rabbit anti-shrimp hemocyanin antibody (h) and rabbit anti-shrimp hemocyanin's small subunit antibody (hs), respectively. We found that h-IgG had a higher hemolytic activity than hs-IgG against erythrocytes from humans, rabbits, mice and chickens, whereas the control IgG showed negligible activity. h-IgG could interact directly with erythrocyte membranes, and this interaction was suppressed by high molecular weight osmoprotectants, showing that it may follow a colloid-osmotic mechanism. In comparative proteomics and glycomics studies, h-IgG and hs-IgG yielded 20 and 5 significantly altered protein spots, respectively, on a 2-D gel. The mean carbohydrate content of h-IgG and hs-IgG was approximately 3.6- and 2-fold higher than that of IgG, respectively, and the α-D-mannose/α-D-glucose content was in the order of h-IgG>hs-IgG>IgG. In this study, a novel antigen nonspecific immune property of human IgG was investigated, and the diversity in the protein constituents and glycosylation levels may have functional signficance.

Editor: Érika Martins Braga, Universidade Federal de Minas Gerais, Brazil

Funding: This work was sponsored by National Natural Science Foundation of China (No. 31072237), New Century Excellent Talents Program of Ministry of Education of China (No. NCET-11-0922), Natural Science Foundation of Guangdong Province (No. 10251503101000002), Guangdong Province College Talent Introduction Project Special Funds (2011), The Key Innovation Project of Science and Technology of Guangdong Province College (No. 2012CXZD0025) and Academic Innovation Team Construction Project of Shantou University (2010). The funders had no role in study design, data collection and analysis, decision to publish, or preparation of the manuscript.

Competing Interests: The authors have declared that no competing interests exist.

* E-mail: zhangyl@stu.edu.cn

[9] These authors contributed equally to this work.

Introduction

Immunoglobulins consist of four polypeptides, namely, two heavy and two light chains, each of which contains a variable region at the amino terminus and a constant region at the carboxyl terminus [1,2]. Five human immunoglobulin classes, namely IgM, IgG, IgA, IgE and IgD, have been identified in blood. Among them, IgG is the most abundant, constituting over 75% of the immunoglobulins in serum. IgG plays a primary role amongst immunoglobulins in immune defense. Moreover, human IgG is multifunctional, with an involvement in neutralization of toxins and viruses, agglutination and precipitation, complement binding and activation, binding to macrophage Fc-receptors and antigen-dependent cellular cytotoxicity [3–5].

It is generally accepted that human IgG only possesses antigen specific immunological functions. Interestingly, recent evidence indicates that human IgG and a large variety of antigen nonspecific immunological molecules such as rabbit C-reactive protein (CRP) [6], *Lymnea stagnalis* molluscan defence molecule [7], *Hyalophora cecropia* hemolin [8], *Aplysia californica* aplysia cell

adhesion molecule (apCAM) [9], *Drosophila* Down syndrome cell adhesion molecule (Dscam) [10], *Branchiostoma floridae* variable region-containing chitin-binding proteins (VCBPs) [11] and *Biomphalaria glabrata* fibrinogen-related proteins (FREPs) [12] may share a common evolutionary origin. In particular, our previous research found that human IgG not only contained several conserved regions with shrimp hemocyanin, but could also cross-react with rabbit anti-shrimp hemocyanin antibodies [13,14]. In addition, hemocyanin in mollusks and arthropods has been documented to possess several immune functions, such as phenoloxidase [15], antiviral [16], antibacterial [17], antitumoral [18], hemolytic [19] and hemagglutinative [14] activities. Therefore, it is of great interest to further investigate whether human IgG also possesses some antigen nonspecific immune properties.

In the present study, two human IgG fractions were purified using affinity chromatography, and these molecules showed hemolytic activity, which was likely associated with the diversity in amino acid sequence and glycosylation of the IgG fractions. The results will contribute to the investigation of the evolutionary relationships between the adaptive immune molecules in vertebrates and the innate immune molecules in invertebrates.

Materials and Methods

Purification and identification of human IgG fractions

Human blood was harvested from 50 healthy individuals in the First Affiliated Hospital of Shantou University, Shantou, Guangdong Province, China. The study protocol was approved by the Institutional Animal Care and Use Committee of Shantou University. Written informed consent was also obtained from the healthy individuals before the start of the study. Serum was collected by centrifugation at 3,000 g for 20 min, pooled and stored at $-20°C$ until needed. Human IgG fractions were isolated using ammonium sulfate fractional precipitation, diethylaminoethanol (DEAE) cellulose 52 ion exchange chromatography and affinity chromatography as described previously, with modifications [14]. Briefly, IgG was precipitated with ammonium sulfate at room temperature, and fractions between 33–50% saturation were collected by centrifugation at 12,000 g for 10 min at 4°C. The pellet was redissolved in a minimum volume of 0.01 M phosphate-buffered saline (PBS; pH 7.4) and subjected to extensive dialysis against the buffer overnight. The dialyzed fraction was purified using DEAE cellulose 52 ion exchange chromatography. The first peak was collected and used as the control IgG, which was then further purified by affinity chromatography. Two affinity chromatography columns were used, one containing rabbit anti-shrimp hemocyanin antibody (h), and the other containing the rabbit anti-shrimp hemocyanin's small subunit antibody (hs). DEAE-purified IgG (400 µl) was loaded onto each affinity column. After incubation for 3 h at room temperature, the columns were washed with 0.01 M PBS (pH 7.4) and eluted with 0.1 M glycine-HCl buffer (pH 2.4). Both unbound IgG fractions (named as unh-IgG and unhs-IgG) from the wash fractions and bound IgG fractions (named as h-IgG and hs-IgG) from the eluted fractions were collected and subjected to further analysis. Protein concentration was determined using the Bradford assay [20]. Unbound and bound IgG fractions were characterized using sodium dodecyl sulfate polyacrylamide gel electrophoresis (SDS-PAGE) under reducing conditions (3% stacking gel, 10% separating gel). For immunoblotting assays, proteins were transferred onto a polyvinylidene fluoride (PVDF) membrane with a semi-dry transfer apparatus according to the manufacturer's instructions. The membrane was blocked for 1 h with 5% skimmed milk in Tris-buffered saline (TBS; 20 mM Tris, 0.15 M NaCl, pH 7.4) at room temperature, then incubated with rabbit anti-human IgG antisera (1:1,000 dilution) and goat anti-rabbit IgG-horseradish peroxidase (HRP; 1:2,000 dilution) antibodies at room temperature for 1 h. Finally, the membrane was washed and developed with substrate 3,3'-diaminobenzidine (DAB) until optimum color was developed.

Preparation of erythrocyte suspensions

Mature mice (CD1 strain), White New Zealand rabbits and White Leghorn chickens were housed in a temperature- and light-controlled environment with free access to regular food and water. Blood was harvested from the healthy humans, mice, rabbits and chickens. Of these, human blood was acquired from the same healthy individuals as described in the above section on the purification and identification of human IgG fractions. All animal procedures were approved by the Institutional Animal Care and Use Committee of Shantou University. The collected blood was centrifuged at 2,000 g for 10 min to obtain erythrocytes. Cells were washed three times with 0.01 M PBS (pH 7.4) and centrifuged at 500 g for 5 min. Erythrocytes were then diluted with 0.01 M PBS (pH 6.0) containing 0.15 M NaCl and 10 mM $CaCl_2$ to obtain a 0.5% (v/v) suspension.

Determination of hemolytic activity

Hemolytic activity was determined as previously described [21]. In brief, each IgG fraction (0.9 ml, 0.01 mg/ml) was mixed with 0.5% (v/v) erythrocyte suspension (0.3 ml). After incubation at 37°C for 1 h, unbroken cells and cell debris were removed by centrifugation at 3,500 g for 10 min, and hemolysis was determined by measuring the absorbance at 540 nm in supernatants. Next, 0.5% (v/v) erythrocyte suspensions were treated with double distilled water (ddH_2O) or 0.01 M PBS-Ca^{2+} (pH 6.0) and used as 100% and 0% hemolysis controls, respectively. All samples were prepared in triplicate. The percentage of hemolysis was calculated as $(A - A_0)/(A_{100\%} - A_0) \times 100\%$. To further investigate the process of IgG-dependent hemolysis, kinetic analysis of chicken erythrocyte hemolysis was performed. A 0.5% (v/v) chicken erythrocyte suspension (10 µl) was mixed with 5 µg/ml h-IgG (10 µl) on a slide. After incubation at 37°C for 15, 30, 45 and 60 min, digital photomicrographs were taken with an Olympus BH-2 microscope (Olympus Company, Tokyo, Japan).

Interaction of IgG fractions with erythrocyte membranes during hemolysis

To further confirm whether the human IgG fractions are involved in hemolysis, SDS-PAGE and immunoblotting were performed as described by Promdonkoy and Ellar with modification [22]. Briefly, A 0.5% (v/v) erythrocyte suspension (0.3 ml) was incubated with 0.1 mg/ml h-IgG (0.9 ml) at 37°C for 1 h, and erythrocyte membranes were collected by centrifugation at 3,500 g for 10 min and washed twice with 0.01 M PBS (pH 7.4) buffer. Erythrocyte membrane pellets were solubilized in 2 X protein loading buffer at room temperature before SDS-PAGE (3% stacking gel, 10% separating gel). A suspension of erythrocytes incubated with ddH_2O was used as a control. The subsequent immunoblotting analysis was carried out as described above for the identification of human IgG fractions. Following the first SDS-PAGE, the band binding to anti-human IgG specifically around 121 kDa was further purified as Kang and Tong described [23]. In brief, the first SDS-PAGE gel was stained by Coomassie Brilliant Blue fast staining reagent (45% ethanol, 10% acetic acid, 1 mg/ml Coomassie Brilliant Blue R-250) for 20 min and then destained by 250 mM KCl. The band was excised and transferred into a dialysis bag. After electrophoresis in the SDS-PAGE electrode buffer overnight at 4°C, the eluate in the bag was concentrated by precooling acetone at $-20°C$ for 30 min, then solubilized in protein loading buffer and heated for 5 min. And the following procedures including the second SDS-PAGE and immunoblotting were same as above described.

Osmotic protection assay

The osmotic protection assay was performed as described [24] with modifications. Chicken erythrocytes (0.5%, v/v) were suspended in 0.01 M PBS (pH 6.0) containing 0.15 M NaCl, 0.01 M $CaCl_2$ and one of the following osmoprotectants: 0.015 M polyethylene glycol (PEG) 4000, 6000 and 8000. Hemolysis was

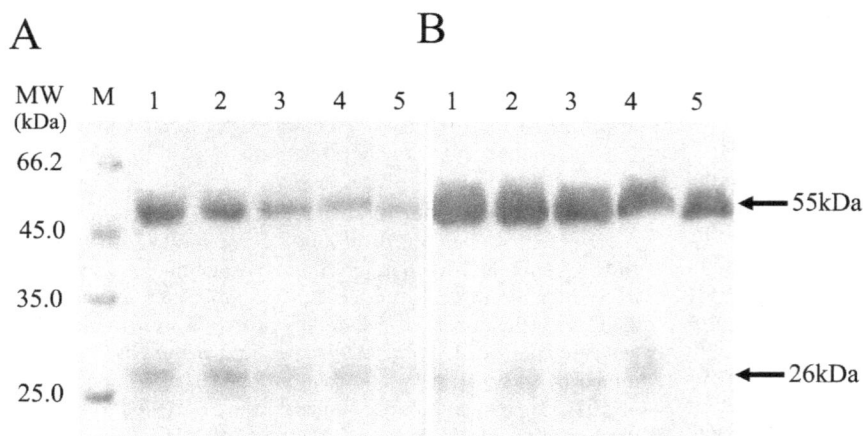

Figure 1. Detection of human IgG fractions purified by ion exchange and affinity chromatography. (A) SDS-PAGE and (B) immunoblotting analysis of different human IgG fractions. Rabbit anti-human IgG antisera (1:1,000) and goat anti-rabbit IgG-HRP (1:2,000) were used as the primary and secondary antibodies, respectively. M, molecular mass markers; 1, IgG; 2, unh-IgG; 3, unhs-IgG; 4, h-IgG; 5, hs-IgG.

initiated by the addition of 0.9 ml h-IgG (0.01 mg/ml) to a 0.3 ml erythrocyte suspension with or without an osmoprotectant. The hemolysis was assessed as described above.

Two-dimensional gel electrophoresis (2-DE)

We performed 2-DE of the IgG fractions as described previously [25]. A total of 20 μg of IgG, h-IgG or hs-IgG in rehydration buffer (7 M urea, 2 M thiourea, 4% CHAPS, 0.2% DTT and 3.4 μl of immobilized pH gradient [IPG] buffer, pH 3–10) was used to rehydrate the IPG strip (7 cm, pH 7–10, Bio-Rad, USA) for 16 h. Isoelectric focusing (IEF) was performed at a constant temperature of 20°C using a continuous increase in voltage (up to 4000 V) for 65,000 Vh. Prior to the second dimension, the IPG

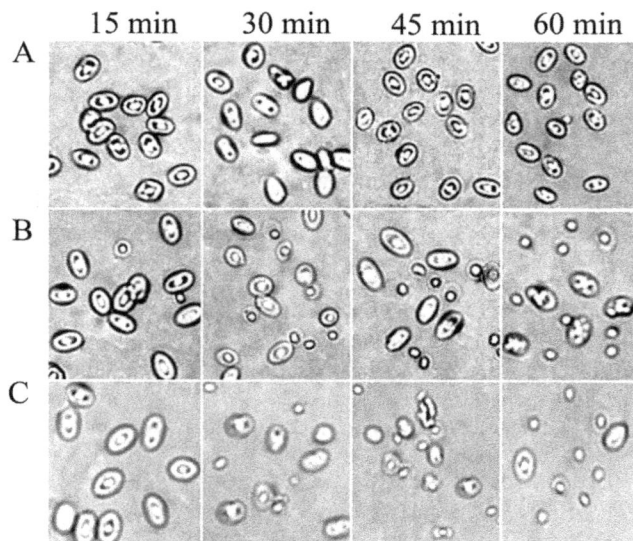

Figure 2. Kinetics of hemolysis of chicken erythrocytes by h-IgG. h-IgG was incubated with erythrocytes for 15, 30, 45 and 60 min at 37°C. The 0.5% (v/v) chicken erythrocyte suspensions were treated with (A) 0.01 M PBS, 10 mM CaCl₂ (pH 6.0; negative control), (B) 5 μg/ml h-IgG and (C) ddH₂O (positive control). The original digital images were magnified 172-fold.

strip was incubated for 15 min in equilibration buffer (20% [w/v] glycerol, 2% SDS, 2% DTT, 0.375 M Tris-HCl, pH 8.8) then further equilibrated for 15 min in equilibration buffer containing 2.5% iodoacetamide instead of 2% DTT. The strip was placed onto a 10% SDS-PAGE gel. Molecular weight markers were loaded onto a filter paper and placed next to the IPG strip. Low-melting point agarose was used to cover the IPG strip and filter paper. Proteins were separated using the same conditions described above for SDS-PAGE analysis of human IgG fractions.

Imaging analysis

The 2-DE gel images were analyzed with PDQuest software version 8.0 (Bio-Rad, CA). Comparative analysis of protein spots was performed by matching corresponding spots across different gels. Each of the matched protein spots was rechecked manually. The intensity of individual spots was normalized against the total intensity of all spots present in each gel, and subjected to statistical analysis to compare the normalized intensity of individual h-IgG or hs-IgG spots to those of control IgG. Significantly altered proteins were chosen using the following criteria: (1) P values<0.05, (2) means of both groups in the unpaired Student's t-test were increased or decreased >2.0-fold and (3) the change was consistent in all replicates for each group of IgG fractions.

Sequence alignments of human IgG and hemolysins

For the prediction of potential specific amino acid sequences related to the hemolytic properties in human IgG, the amino acid sequences of human IgG heavy chain (AAA02914) and three hemolysins of *Serratia marcescens*, *Edwardsiella tarda* and *Xenorhabdus bovienii* from the NCBI database were aligned using sequence alignment programs (Clustal X and BioEdit).

Determination of the carbohydrate content of human IgG fractions

The carbohydrate content was determined by the colorimetric method [26]. Glucose (0.01 mg/ml) was serially diluted with distilled water to a final volume of 2.0 ml, and 1.0 ml phenol (6% m/v) and 5.0 ml sulfuric acid were added immediately. After standing at room temperature for 25 min, the absorbance was

Table 1. Hemolysis of inhomogeneous erythrocytes by different human IgG fractions.

IgG Fractions	Hemolytic activity (%) to inhomogeneous erythrocytes						
	Human (A)	Human (B)	Human (O)	Human (AB)	Rabbit	Mouse	Chicken
IgG	0.00±0.01	1.46±0.03	5.77±0.05	0.60±0.01	0.00±0.00	1.12±0.01	0.00±0.00
unh-IgG	0.00±0.00	2.34±0.04	9.92±0.09	1.81±0.02	0.36±0.01	0.00±0.00	0.00±0.01
unhs-IgG	0.00±0.00	1.46±0.03	8.33±0.07	0.00±0.00	0.85±0.01	0.00±0.00	0.00±0.01
h-IgG	78.53±0.06**	70.04±0.08**	89.24±0.07**	75.90±0.07**	83.55±0.07**	93.89±0.06**	100.00±0.08**
hs-IgG	30.99±0.01**	34.99±0.06**	77.51±0.14**	62.68±0.05**	75.37±0.06**	5.29±0.01	57.97±0.10**

These data represent the mean ± SD of at least three separate experiments. We incubated 0.9 ml of 0.01 mg/ml IgG fractions with 0.3 ml of 0.5% (v/v) erythrocyte suspension at 37°C for 1 h.
**indicates a significant difference as compared to the control IgG ($P<0.01$).

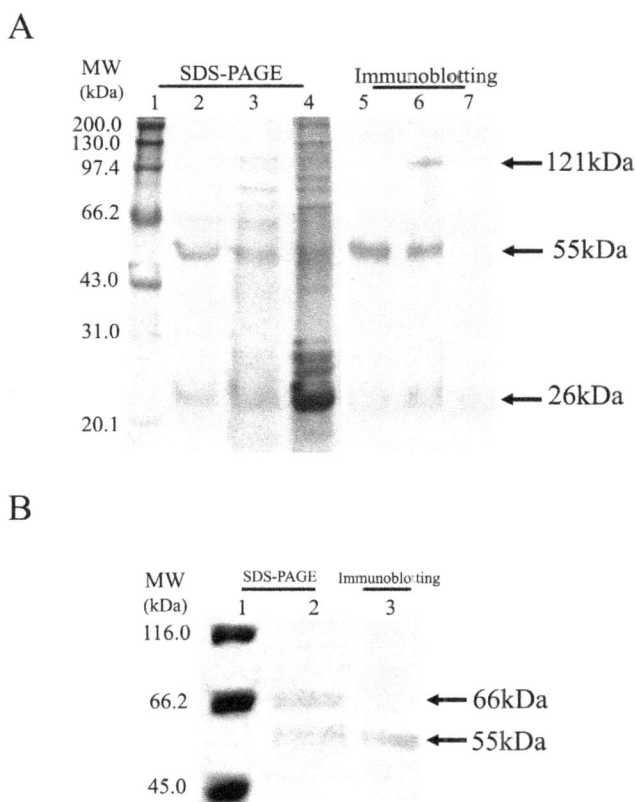

Figure 3. SDS-PAGE and immunoblotting of IgG fractions involved in hemolysis. Immunoblotting was performed using rabbit anti-human IgG antisera (1:1,000) and goat anti-rabbit IgG-HRP (1:2,000) as the primary and secondary antibodies, respectively. (A) analysis of proteins solubilized from the chicken erythrocyte membranes treated with h-IgG. 1, molecular mass markers; 2, h-IgG; 3 and 4, proteins from chicken erythrocyte membranes treated with h-IgG and double distilled water, respectively; 5–7, immunoblotting analysis of lanes 2–4. (B) analysis of the protein around 121 kDa following elution with electrophoresis. 1, molecular mass markers; 2, purified protein around 121 kDa; 3, immunoblotting analysis of lane 2.

measured at 490 nm, and the average of duplicate samples was used to plot a standard curve.

The α-D-mannose/α-D-glucose content was further measured using dot-blotting for lectin method [27]. Nitrocellulose (NC) membranes were cut into the desired size, soaked in TBS (20 mM Tris, 150 mM NaCl, pH 7.4) for 30 min and dried on a filter paper. Different concentrations of human IgG fractions were prepared (0.3 and 0.15 mg/ml), and 2-μl aliquots of each concentration were spotted onto the NC membrane. After drying, the NC membrane was blocked at room temperature for 2 h with 2% polyvinylpyrrolidone (PVP) 360,000 in TBS, washed three times with TBS for 10 min each time and reacted with a 1:10 dilution of biotinylated concanavalin A (ConA) and with a 1:1,000 dilution of avidin-peroxidase for 2 h at room temperature. The membrane was washed with TBS for 10 min a further three times, then reacted with DAB until optimum color was developed. The map of the precipitation dots was scanned and analyzed by the analytic system of GDS8000PC.

Results

Isolation of two human IgG fractions

Two IgG fractions, h-IgG and hs-IgG, were isolated from healthy human sera. The purity and apparent molecular weight of the proteins were analyzed by SDS-PAGE. Two bands at molecular weights of approximately 26 kDa and 55 kDa were observed for the two IgG fractions (Fig. 1A). The identity of these two bands was verified with rabbit anti-human IgG antibody (Fig. 1B). No other nonspecific bands was detected. As a control, the other three IgG species, namely, IgG, unh-IgG and unhs-IgG, were included in the analysis.

Both IgG fractions possessed hemolytic activity

Interestingly, the hemolysis assay showed that the two human IgG fractions (h-IgG and hs-IgG) could lyse human, rabbit, mouse and chicken erythrocytes within 1 h, and their hemolytic activities ranged from 5.29% to 100%, while hemolysis (except for type O human erythrocytes) by IgG, unh-IgG and unhs-IgG was almost negligible (Table 1). To determine the kinetics of the lysis reaction, chicken erythrocytes were incubated with h-IgG (5 μg/ml) for 15, 30, 45 and 60 min. After a 30-min exposure, some erythrocytes were lysed, and the extent of hemolysis increased with time until no cells were intact after the full 1-h incubation (Fig. 2). This confirmed that the h-IgG and hs-IgG fractions possessed hemolytic activity.

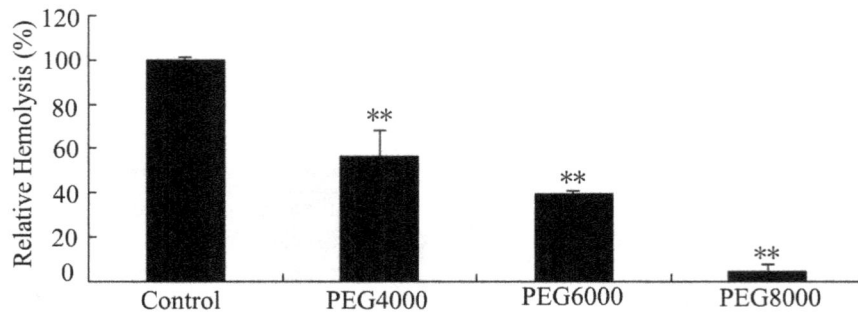

Figure 4. Osmotic protection against hemolysis by various osmoprotectants. 0.9 ml h-IgG (0.01 mg/ml) was incubated with 0.3 ml of 0.5% (v/v) chicken erythrocytes at 37°C for 1 h in the presence of various osmoprotectants. h-IgG incubated with chicken erythrocytes without osmoprotectant was used as a 100% hemolysis control. Data are expressed as means ± SD (n≥3). ** indicates a significant difference as compared to the control ($P<0.01$).

The IgG fractions might bind to erythrocyte membranes directly via a colloid osmotic pressure mechanism

To investigate whether the IgG fractions could bind to erythrocyte membranes directly, proteins solubilized from the hemolyzed chicken erythrocyte membranes were examined by SDS-PAGE and immunoblotting analysis. After immunobloting with anti-human IgG antibodies, 26-kDa and 55-kDa bands corresponding to the light and heavy chain of h-IgG, respectively, were observed in the erythrocyte membranes treated with h-IgG, but not with ddH$_2$O (Fig. 3A). Importantly, a positive band about 121 kDa was also found (Fig. 3A). To further confirm the protein, it was cut out from the first SDS-PAGE gel and subjected to the second SDS-PAGE and immunoblotting analysis following elution with electrophoresis. Fig. 3B showed that the 121 kDa protein was separated into two bands of 66 and 55 kDa, and the 55 kDa band also could react with anti-human IgG antibody specifically, suggesting that the 121 kDa protein might be a compound of h-IgG heavy subunit and an erythrocyte membrane protein. Thus, it led to be deduced that the IgG fractions might bind to erythrocyte membrane directly.

Further, the osmotic protection assay was performed to investigate the possible mechanism of the IgG fractions-induced hemolysis. When chicken erythrocytes were incubated with h-IgG in the presence of PEG 4000, 6000 and 8000, hemolysis was inhibited by 43.75±11.62%, 60.41±1.27% and 95.44±3.16%, respectively (Fig. 4). Nevertheless hemolysis was not completely inhibited by the osmoprotectants. The result suggested that erythrocytes were likely ruptured by colloid osmotic pressure shock after the IgG fractions had formed ion-permeable pores in the membranes.

IgG protein polymorphism might be responsible for hemolytic activity

To investigate the molecular basis for the hemolytic activity, comparative proteomics strategy was carried out. Following 2-DE, approximately 30 spots on each gel were distinguished by PDQuest software, mainly around 26 and 55 kDa. Twenty protein spots on the h-IgG gel were significantly altered compared with the IgG gel. Of these, spots 1–3 were present only on the h-IgG gel, spots 4–14 were upregulated, and spots 15–20 were downregulated (Figs. 5A and 5B). Similarly, a total of five significantly altered protein spots were observed in the hs-IgG gel.

Of these, spot 10 was upregulated while spots 18, 20, 21 and 22 were downregulated (Figs. 5A and 5C). Regretfully, due to the highly diversity, some specific amino acid sequences related to the hemolytic properties in the specific IgG fractions have not been obtained by mass spectrometry (data not shown). Accordingly, an amino acid sequence alignment between human IgG and hemolysins was further produced. Figure 6 showed that homologies between human IgG heavy chain (AAA02914) and hemolysins from *Serratia marcescens*, *Edwardsiella tarda* and *Xenorhabdus bovienii* were about 28%, 24% and 20%, respectively. These findings imply that the protein polymorphism might be one of the molecular bases and responsible for hemolytic activity of IgG fractions, meanwhile, some potential specific sequences involved in the antigen nonspecific immunological function was also likely existed.

The glycosylation diversity of the IgG fractions might also facilitate hemolytic property

Based on the above analysis, carbohydrate content analysis was further investigated. As shown in Table 2, the mean values for h-IgG and hs-IgG were approximately 3.6- and 2-fold higher than IgG, respectively. The α-D-mannose/α-D-glucose content of the IgG fractions was also different (Fig. 7A). All of the human IgG fractions could react specifically with ConA, and by measuring the intensity of the blotting, the α-D-mannose/α-D-glucose content was found to be in the order h-IgG>hs-IgG>IgG (Fig. 7B), which was in agreement with the carbohydrate content results. These evidences suggest that the diversity of IgG fraction in glycosylation level, like polymorphism in protein level, might also facilitate hemolytic property.

Discussion

It was generally accepted that no close evolutionary relationship existed between the antigen nonspecific and antigen specific immunological molecules [28–29]. However, recent reports showed a significant association [30–32]. Antigen nonspecific proteins possess molecular polymorphism that exhibit primary adaptive immune functions [10–12]. At the same time, many antigen specific immunological molecules such as immunoglobulin share homologous domains or epitopes with antigen nonspecific immune factors [6–14]. This implies that human immunoglobulin

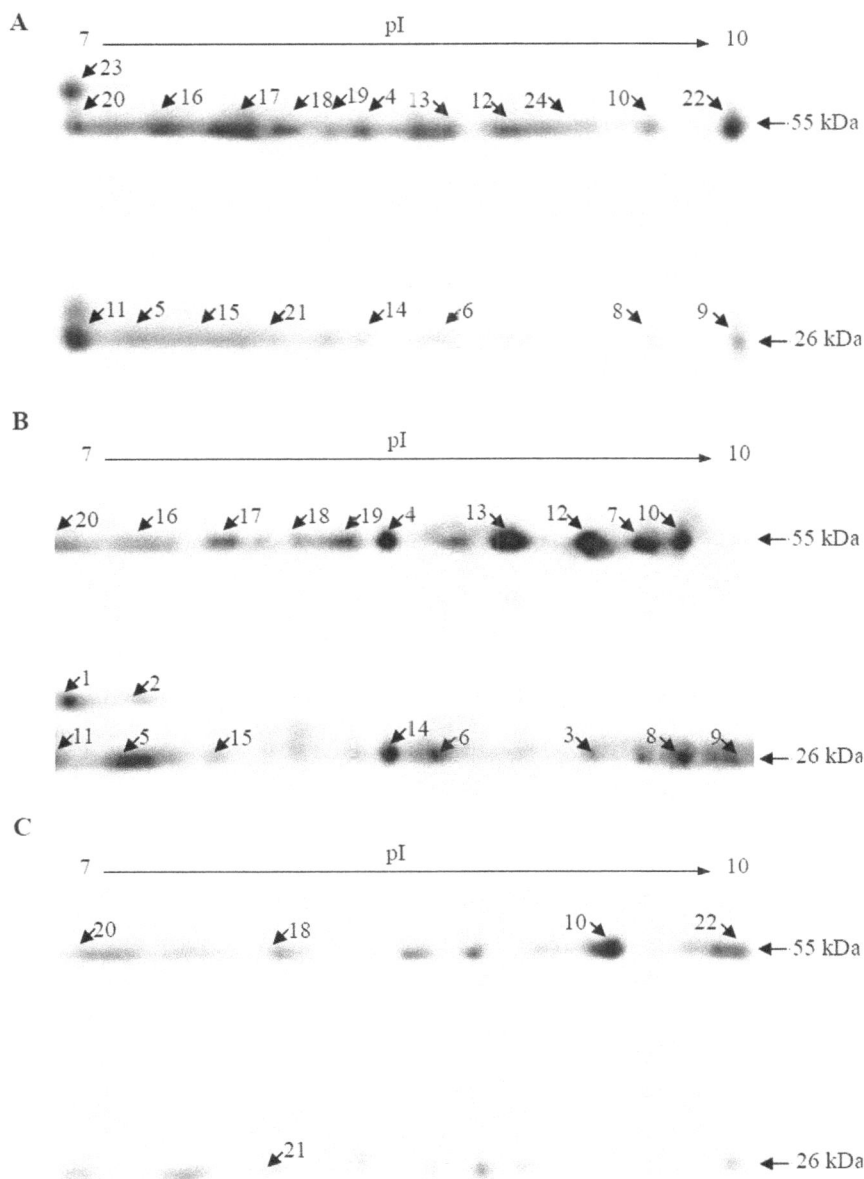

Figure 5. 2-DE analysis of different human IgG fractions. The 2-DE gel images were analyzed with PDQuest software version 8.0 (Bio-Rad, CA). A, IgG; B, h-IgG; C, hs-IgG.

may also share similar functions with antigen nonspecific immunological molecules.

In this study, we separated two human IgG fractions, h-IgG and hs-IgG, by affinity chromatography (Fig. 1) and characterized their hemolytic activity. h-IgG and hs-IgG showed obvious hemolytic activity against a variety of erythrocytes, while the control IgG, unhs-IgG and unh-IgG fractions did not (Table 1 and Fig. 2). This is somewhat in agreement with our previous findings that the agglutinative activity of *Litopenaeus vannamei* a-hemocyanin (purified by affinity chromatography) was increased 37-fold compared with that of *L. vannamei* s-hemocyanin (purified by size-exclusion chromatography) [14]. Furthermore, we found that the IgG

fractions could bind to erythrocyte membranes directly (Fig. 3A). It is worth emphasizing that an approximately 121 kDa protein reacted with rabbit anti-human IgG antibody specifically was also found (Fig. 3A), then it was speculated as a compound of human IgG heavy chain and erythrocyte membrane protein based on the findings from the second SDS-PAGE and immunoblotting analysis (Fig. 3B). This result resemble those found for human IgG autoantibodies or Portuguese Man-of-War toxin binding to four or one erythrocyte membrane proteins in the hemolytic actions [33–34]. Besides, osmotic protection analysis indicated that the hemolysis mediated by h-IgG could be inhibited by PEG with different molecular weights in a degree (Fig. 4). These cumulative

```
human IgG     1  MDWTWRFLFVVAAATG--------------VQSQMQVVQSGAEVKKPGSSVTVSCKASGGT
smHemolysin   1  AGGVLDLLIAPKIDSRGEVIVQDFKQSNGKVTSAAINAISGLNRVARDGTVQAS-----GS
etHemolysin   1  AERVLDLVAPRIESNG----------AVRAEAINAISGNNRLTRDLTQLEA-----GS
xbHemolysin   1  HTDVLNLLIAPKIDSHG----------REITAKIININTGNNQISADGKILAS-----GS

human IgG    48  FSNYAISWVRQAPGQGLEWMGGIIPLFGTPTYSQNFQGRVTITADKSTSAHMELISLRSE
smHemolysin  56  MQAGRINIINTAQGSGVKLAG----VDDA----QNYRGGIYWNDRSS--AHLTATDIRGE
etHemolysin  44  MQAGRIRIINTAEGSGVKLAG----ITER----QNYRGGIDWHASQQ--NHLTATDLRGD
xbHemolysin  45  MQAGRIRLLNTAEGSGVWLQG-----LNQS----QGSFSGLYAQQKQI--NRVMATDIVGD

human IgG   108  DTAVYYCATDRYRQANFDRARVGMFDPWGQGTLVTVSSASTKGPSVFPLAPSSKSTSGGT
smHemolysin 106  DIT--------------MFYSWQYDVTREREQLQQASSTVAAGSAKLISTQE-
etHemolysin  94  DVT--------------MFYSWQ-DETRRNSQQQAQTGRIEARHNVNLSATRG-
xbHemolysin  95  NVT--------------WKLSWQYDVTEESEKYHHEGNTIKPHENVHLTASEG-

human IgG   168  AALGCLVKDYFPEPVTVSWNSGALTSGVHTFPAVLQSSGLYSLSSVVTVPSSSLGTQTYI
smHemolysin 145  ---RGYQRNHTSSLRTGRW--GGIGGG----ASELTSGGTLRLN--VTITGS--GTAFNI
etHemolysin 133  ---TGYKRNEGASLNSGRW--GGIGGG----ASEVESGGTLTIS--VTISGS--GGAFNI
xbHemolysin 134  ---KGYKKNHTVGLQTGNWN--GGIAGG----ISDIVANGHLLS--VTITGS--GTIFNI

human IgG   228  CNVNHKPSNTKVDKKVEPKSCDKTHTCPPCPAPELLGGPSVFLFPPKPKDTLMISRTPEV
smHemolysin 192  TSSSHKADNSYQSSTASELKSDTNLTLVSHKDADVIGS--WNGYAKEAGD--KLKAEKDV
etHemolysin 180  TTASQQRKES-EQLRASQLTSDVNLRVQSAGDIQVQGS--ASAYAKKQGD--ALTAGEDV
xbHemolysin 182  TKNSNQE-HNTQQTTSNSQLVSDTDLKLLSNKDINIIGS--WQYYAHEAED--ITLNASENV

human IgG   288  TCVVVDVSHEDPEVKFNWYVDGVEVHNAKTKPREEQYNSTYRVVSVLTVLHQDWLNGKEY
smHemolysin 248  TFSGSKLVADKGDAS--YYTGGIDKLG----QSSKAITSGSDVKGNLTINARDKL---NI
etHemolysin 236  QFTGSRLQANAGDAQ--YYTAGIDKAG----KQTRARGSSSEVKGNLIINAQDTL---EV
xbHemolysin 239  TVTGSKLVTTQGDAK--FLTGGMDRLG----NQTTAVWSKTQVTGNLELNATGEL---EI

human IgG   348  KCKVSNKALPAPIEKTISKAKGQPREPQVYTLPPSRDELTKNQVSLTCLVKGFYPSDIAV
smHemolysin 299  GANVDYSAVTRPVERAVGKA---TADSHRSEAAANRQDEQSPDTRGSAGRVYTTTGSDL
etHemolysin 287  GANVDYSAVTRPLAGAAQKV---QADNHRFEAASDSSSSQSSAVQGGATLRVYTTTGKDI
xbHemolysin 290  SGSADYSATTRPAEKTAKAA---TAGSHTSEAAYNRKDS-SSDTTGNARLRVYTSTGSDI

human IgG   408  EWESNGQPENNYKTTPPVLDSDGSFFLYSKLTVDKSRWQQGNVFSCSVMHEALHNHYTQK
smHemolysin 356  TVDAKGEGGTQRSNSSASQAVTGSIDA--NMNVKKDAIYQGTALNGGRGKTAVN--LDQA
etHemolysin 344  GVTASGQGENRREEQHSQAQSGSIQA--DIRLGQANYQGTALDGGQGKTQIH-HIIAQA
xbHemolysin 347  SVNGKGDGSYQATSNSSANAVTSGVQA--DIQVTRDARYQGTSMNAGSGSTNIN--FDQA

human IgG   468  SLSLSPGK
smHemolysin 412  SDKQSESR
etHemolysin 401  TDS-STRH
xbHemolysin 403  TNRNEQNR
```

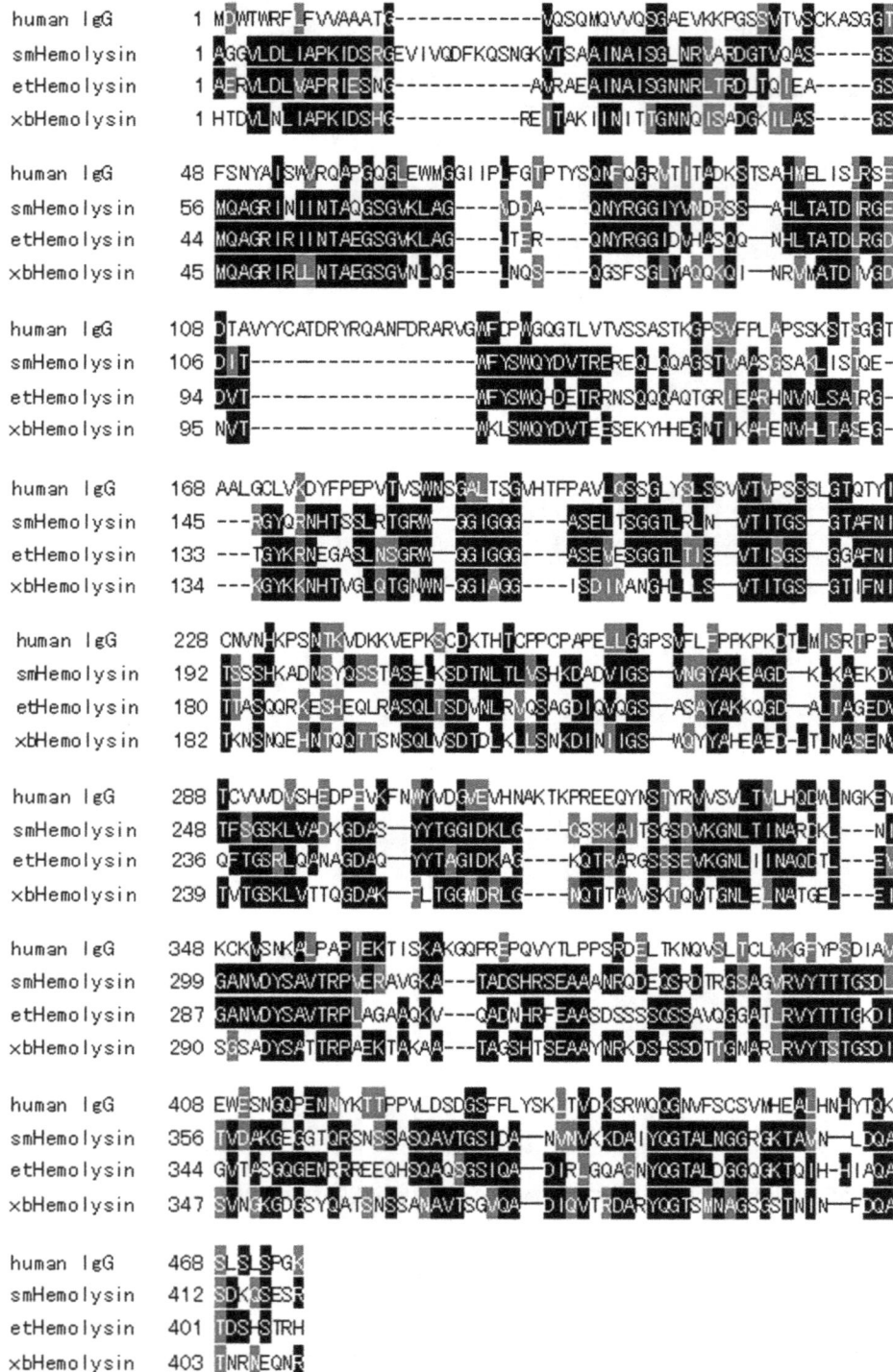

Figure 6 Multiple amino acid sequence alignments between human IgG and hemolysins. Black and gray show 100% and above 50% homology, respectively. GenBank accession numbers are: human IgG (AAA02914), hemolysin from *Serratia marcescens* (AAA50323), hemolysin from *Edwardsiella tarda* (ZP_06713740) and hemolysin from *Xenorhabdus bovienii* (YP_003466207).

evidence implied that the IgG fractions might bind to erythrocyte membranes directly via a colloid osmotic pressure mechanism. This is similar to the results obtained for shrimp hemocyanin [19], sea anemone Equinatoxin II [35] and *Staphylococcus aureus* a-hemolysin [36] during hemolysis. However, direct evidence for the formation of ion-permeable pores is still missing, further investigations are required in the future.

To date, many hemolytic mechanisms have been found in human pathologic conditions [37–39]. For example, IgG and complement could mediate autoimmune hemolytic anemia [39].

A

B

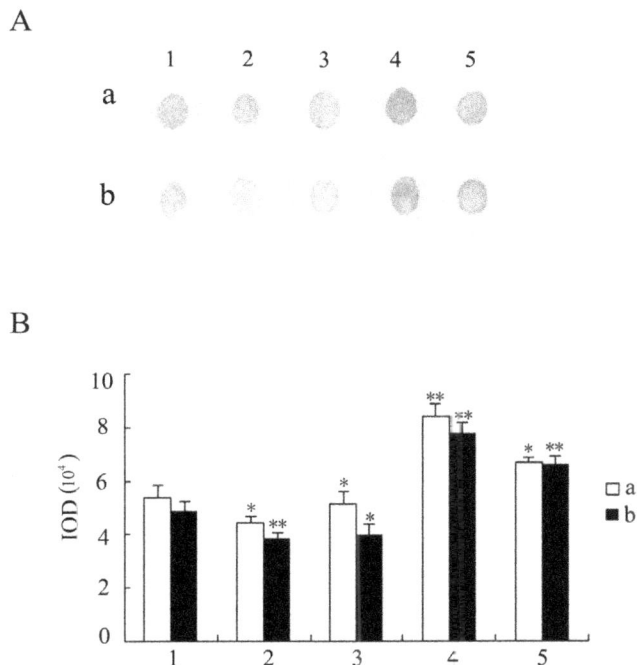

Figure 7. Analysis of the α-D-mannose/α-D-glucose content of different human IgG fractions by dot-blotting for lectin. Biotinylated ConA (1:10) and avidin-peroxidase (1:1,000) were used as the primary and secondary reagent, respectively. (A) dot-blotting for lectin. (B) bar graph showing the changes in spot intensity of human IgG fractions. Significant differences between the four different IgG fractions and the control IgG at each concentration were indicated with one ($P<0.05$) or two ($P<0.01$) asterisks. a and b, the concentration of the human IgG fractions was 0.3 and 0.15 mg/ml, respectively. 1, IgG; 2, unh-IgG; 3, unhs-IgG; 4, h-IgG; 5, hs-IgG.

However, the mechanism through which the erythrocyte meets its destruction in normal conditions remains controversial [40–42]. The present findings that two IgG fractions possessed nonspecific hemolytic properties against erythrocytes might provide an alternative approach to induce hemolysis in normal conditions in human. Although the exact physiological function of the hemolytic activity of the IgG fractions in vivo is yet to be determined, we believe this study provided strong evidences for the first time to associate nonspecific immune properties with human IgG, and offered an opportunity to learn more about the characterization of human IgG.

To investigate the origins of the h-IgG and hs-IgG hemolytic properties, a proteomic approach was employed. Significantly altered protein spots were found for h-IgG (20 spots) and hs-IgG (5 spots) compared to the control IgG (Fig. 5). The similar polymorphism of the IgG fractions to the nonspecific immune molecules of Toll-like receptor 4 (TLR4) [43], Pacifastacus leniusculus Dscam [44] and Ciona VCBPs [45], suggests that different isoforms are capable of recognizing and differentiating different pathogens.

Regretfully, due to the highly diversity of IgG fractions, we failed to obtain some specific amino acid sequences related to the antigen nonspecific immune properties by proteomics analysis. Thereby, we decided to compare the amino acid sequence between human IgG and hemolysins using bioinformatics analysis. Interestingly, the results indicated that human IgG presented a significant degree of homology to three hemolysins from Serratia marcescens, Edwardsiella terda and Xenorhabdus bovienii (Fig. 6). Collectively, these provides evidence that the protein polymorphism of human IgG fractions is likely to be responsible for the hemolytic property. At the same time, there might also exist some characteristic sequences involved in the antigen nonspecific immunological function in the IgG fractions, but further investigations are needed to identify the specific sequences.

To further investigate whether the hemolytic activity of IgG fractions was also related to its glycosylated modification, a glycomic strategy was performed. Interestingly, we found that both the carbohydrate and α-D-mannose/α-D-glucose content of h-IgG and hs-IgG were significantly higher ($P<0.05$) than that of control IgG (Table 2; Fig. 7), suggesting that higher glycosylation levels in IgG fractions might also contribute to the higher hemolytic activity. It led to be deduced that diversity in glycosylation level, as well as polymorphism in protein level, might together constitute the molecular bases of hemolytic properties of the human IgG fractions. Notably, although the h-IgG carbohydrate content was about 3.6-fold higher than that of IgG, no significant difference was observed between the unh-IgG and the control IgG (Table 2). This implied that only a subpopulation of the IgG fractions is likely to be involved in the hemolysis, the exact proportion is needed to further investigated.

Recent reports reveal that specific glycoforms of human IgG can be produced in different immune cells, and this is relevant to many physiological and pathological processes. In particular, recombinant IgG proteins produced in different host cells display different patterns of oligosaccharides [46]. Additionally, agalacto-sylated glycoforms of aggregated IgG may induce rheumatoid arthritis [47]. High mannosylation of surface IgM creates a functional bridge between human follicular lymphoma and microenvironmental lectins [48]. In this study, we demonstrated diversity in the carbohydrate and α-D-mannose/α-D-glucose content among IgG fractions. Furthermore, we also identified two protein spots (spot 1 and spot 2) that were present only in the h-IgG (Fig. 5) and had molecular weights that were higher than the light chains of IgG, indicating diversity in glycosylation. Unfortunately, although we tried to identify the specific glycans in h-IgG or hs-IgG for the relation to the hemolytic activity, no good data was obtained by mass spectrometry because of a limited carbohydrate content isolated from the specific IgG fractions. Thus, in the future, it will be interesting to investigate whether the oligosaccharides of human IgG fractions recognize pathogens nonspecifically and directly, and which oligosaccharides are involved in the hemolytic properties.

In summary, we show for the first time that human IgG fractions possess hemolytic activity, and identify polymorphisms in

Table 2. Carbohydrate content of different human IgG fractions.

IgG Fractions	IgG	unh-IgG	unhs-IgG	h-IgG	hs-IgG
Carbohydrate content (μg/mg)	22.97±2.61	21.79±1.17	18.44±2.03	82.92±13.14**	47.58±3.32**

These data represent the mean ± SD of at least three separate experiments.
**indicates a significant difference as compared to the control IgG ($P<0.01$).

both the protein and oligosaccharide components that may be responsible for the antigen nonspecific immune properties. However, the identification of specific amino acid sequences or glycans of the IgG fractions involved in the function is left untested, further studies are required in the future. This work will assist future investigations into the evolutionary relationships between antigen nonspecific and antigen specific immunological molecules.

Acknowledgments

We thank Dr. Chiju Wei, and one of the highly qualified native English speaking editors at ELIXIGEN (www. elixigen.com) for their editorial assistance.

Author Contributions

Conceived and designed the experiments: YLZ SYM FY. Performed the experiments: SYM FY. Analyzed the data: SYM FY YLZ MQZ JSC HYZ. Contributed reagents/materials/analysis tools: JHC XQY. Wrote the paper: YLZ FY SYM.

References

1. Fleischman JB, Porter RR, Press EM (1963) The arrangement of the peptide chains in γ-globulin. Biochem J 88: 220–228.
2. Titani K, Whitley E, Avogardo L, Putnam FW (1965) Immunoglobulin structure: partial amino acid sequence of a Bence-Jones protein. Science 149: 1090–1092.
3. Mimura Y. Church S, Ghirlando R, Ashton PR, Dong S, et al. (2000) The influence of glycosylation on the thermal stability and effector function expression of human IgG1-Fc: properties of a series of truncated glycoforms. Mol Immunol 37: 697–706.
4. Gala FA, Morrison SL (2002) The role of constant region carbohydrate in the assembly and secretion of human IgD and IgA1. J Biol Chem 277: 29005–29011.
5. Mimura Y, Sondermann P, Ghirlando R, Lund J, Young SP, et al. (2001) Role of oligosaccharide residues of IgG1-Fc in FcγRIIb binding. J Biol Chem 276: 45539–45547.
6. Osmand AP, Gewurz H, Friedenson B (1977) Partial amino-acid sequences of human and rabbit C-reactive proteins: homology with immunoglobulins and histocompatibility antigens. Proc Nat Acad Sci USA 74: 1214–1218.
7. Hoek RM, Smit AB, Frings H, Vink JM, de Jong-Brink M, et al. (1996) A new Ig-superfamily member, molluscan defence molecule (MDM) from *Lymnaea stagnalis*, is down-regulated during parasitosis. Eur J Immunol 26: 939–944.
8. Lanz-Mendoza H, Bettencourt R, Fabbri M, Faye I (1996) Regulation of the insect immune response: the effect of hemolin on cellular immune mechanisms. Cell Immunol 169: 47–54.
9. Schachner M (1997) Neural recognition molecules and synaptic plasticity. Curr Opin Cell Biol 9: 627–634.
10. Schmucker D, Clemens JC, Shu H, Worby CA, Xiao J, et al. (2000) Drosophila Dscam is an axon guidance receptor exhibiting extraordinary molecular diversity. Cell 101: 671–684.
11. Cannon JP, Haire RN, Litman GW (2002) Identification of diversified genes that contain immunoglobulin-like variable regions in a protochordate. Nat Immunol 3: 1200–1207.
12. Zhang SM, Adema CM, Kepler TB, Loker ES (2004) Diversification of Ig superfamily genes in an invertebrate. Science 305: 251–254.
13. Zhang YL, Wang SY, Peng XX (2004) Identification of a type of human IgG-like protein in shrimp *Penaeus vannamei* by mass spectrometry. J Exp Mar Biol Ecol 301: 39–54.
14. Zhang YL, Wang SY, Xu AL, Chen J, Lin BK, et al. (2006) Affinity proteomic approach for identification of an IgA-like protein in *Litopenaeus vannamei* and study on its agglutination characterization. J Proteome Res 5: 815–821.
15. Nagai T, Osaki T, Kawabata S (2001) Functional conversion of hemocyanin to phenoloxidase by horseshoe crab antimicrobial peptides. J Biol Chem 276: 27166–27170.
16. Zhang X, Huang C, Qin Q (2004) Antiviral properties of hemocyanin isolated from shrimp *Penaeus monodon*. Antiviral Res 61: 93–99.
17. Jiang N, Tan NS, Ho B, Ding JL (2007) Respiratory protein–generated reactive oxygen species as an antimicrobial strategy. Nat Immunol 8: 1114–1122.
18. Becker MI, Aliaga A, Ferreira J, de Ioannes AE, de Ioannes P, et al. (2009) Immunodominant role of CCHA subunit of *Concholepas* hemocyanin is associated with unique biochemical properties. Int Immunopharmacol 9: 330–339.
19. Zhang YL,Yan F, Hu Z, Zhao XL, Min SY, et al. (2009) Hemocyanin from shrimp *Litopenaeus vannamei* shows hemolytic activity. Fish Shellfish Immunol 27: 330–335.
20. Bradford MM (1976) A rapid and sensitive method for the quantitation of microgram quantities of protein utilizing the principle of protein-dye binding. Anal Biochem 72: 248–254.
21. Hatakeyama T, Nagatomo H, Yamasaki N (1995) Interaction of the hemolytic lectin CEL-III from the marine invertebrate *Cucumaria echinata* with the erythrocyte membrane. J Biol Chem 270: 3560–3564.
22. Promdonkoy B, Ellar DJ (2003) Investigation of the pore-forming mechanism of a cytolytic δ-endotoxin from *Bacillus thuringiensis*. Biochem J 374: 255–259.
23. Kang B, Tong Z (2000) A fast staining-destaining method for SDS-PAGE which favours the recovery of protein. Prog Biochem Biophys 27: 210–211.
24. Aranda FJ, Teruel JA, Ortiz A (2005) Further aspects on the hemolytic activity of the antibiotic lipopeptide iturin A. Biochim Biophys Acta 1713: 51–56.
25. Qiao J, Du ZH, Zhang YL, Du H, Guo LL, et al. (2011) Proteomic identification of the related immune-enhancing proteins in shrimp *Litopenaeus vannamei* stimulated with vitamin C and Chinese herbs. Fish Shellfish Immunol 31: 736–745.
26. Dubois M, Gilles KA, Hamilton JK, Rebers PA, Smith F (1951) A colorimetric method for the determination of sugars. Nature 168: 167.
27. Figueroa-Soto CG, Calderón de la Barca AM, Vázquez-Moreno L, Higuera - Ciapara I, Yepiz-Plascencia G (1997) Purification of hemocyanin from white shrimp (*Penaeus vannamei Boone*) by immobilized metal affinity chromatography. Comp Biochem Physiol 117B: 203–208.
28. Schluter SF, Bernstein RM, Bernstein H, Marchalonis JJ (1999) 'Big Bang' emergence of the combinatorial immune system. Dev Comp Immunol 23: 107–111.
29. Hughes AL, Yeager M (1997) Molecular evolution of the vertebrate immune system. BioEssays 19: 777–786.
30. Kasahara M, Suzuki T, Pasquier LD (2004) On the origins of the adaptive immune system: novel insights from invertebrates and cold-blooded vertebrates. Trends Immunol 25: 105–111.
31. Kurtz J, Franz K (2003) Evidence for memory in invertebrate immunity. Nature 425: 37–38.
32. Thomas P (2009) Immune system: "Big Bang" in question. Science 325 (5939): 393.
33. Leddy JP, Falany JL, Kissel GE, Passador ST, Rosenfeld SI (1993) Erythrocyte membrane proteins reactive with human (warm-reacting) anti-red cell autoantibodies. J Clin Invest 91: 1672–1680.
34. Lin D, Hessinger DA (1979) Possible involvement of red cell membrane proteins in the hemolytic action of Portuguese Man-of-War toxin. Biochem Biophys Res Commun 91(3): 761–769.
35. Hong Q, Gutiérrez-Aguirre I, Barlič A, Malovrh P, Kristan K, et al. (2002) Two-step membrane binding by Equinatoxin II, a pore-forming toxin from the sea anemone, involves an exposed aromatic cluster and a flexible helix. J Biol Chem 277: 41916–41924.
36. Gouaux JE, Braha O, Hobaugh MR, Song L, Cheley S, et al. (1994) Subunit stoichiometry of staphylococcal α-hemolysin in crystals and on membranes: a heptameric transmembrane pore. Proc Natl Acad Sci USA 91: 12828–12831.
37. Skals M, Jorgensen NR, Leipziger J, Praetorius HA (2009) α-Hemolysin from *Escherichia coli* uses endogenous amplification through P2X receptor activation to induce hemolysis. Proc Natl Acad Sci USA 106: 4030–4035.
38. Hou SZ, Su ZR, Chen SX, Ye MR, Huang S (2011) Role of the interaction between puerarin and the erythrocyte membrane in puerarin-induced hemolysis. Chem Biol Interact 192(3): 184–192.
39. Danchaivijitr P, Yared J, Rapoport AP (2011) Successful treatment of IgG and complement-mediated autoimmune hemolytic anemia with bortezomib and low-dose cyclophosphamide. Am J Hematol 86 (3): 331–332.
40. Stefanini M, Xefteris E, Moschides E, Blumenthal W (1961) The role of a hemolytic system in human tissues in some hemolytic states. Am J Med Sci 242 (3): 303–314.
41. Cappellini MD, Fiorelli G (2008) Glucose-6-phosphate dehydrogenase deficiency. Lancet 371: 64–74.
42. Laser H (1951) The hemolytic substance present in animal tissues. Science 113: 609.
43. Gu S, Wang T, Chen X (2008) Quantitative proteomic analysis of LPS-induced differential immune response associated with TLR4 polymorphisms by multiplex amino acid coded mass tagging. Proteomics 8: 3061–3070.
44. Watthanasurorot A, Jiravanichpaisal P, Liu HP, Söderhäll I, Söderhäll K (2011) Bacteria-induced Dscam isoforms of the crustacean, *Pacifastacus leniusculus*. PLoS pathog 7: e1002062.
45. Dishaw LJ, Giacomelli S, Melillo D, Zucchetti I, Haire RN, et al. (2011) A role for variable region-containing chitin-binding proteins (VCBPs) in host gut-bacteria interactions. Proc Natl Acad Sci USA 108: 16747–1652.
46. Raju TS, Briggs JB, Borge SM, Jones AJS (2000) Species-specific variation in glycosylation of IgG: evidence for the species-specific sialylation and branch-specific galactosylation and importance for engineering recombinant glycoprotein therapeutics. Glycobiology 10: 477–486.
47. Rudd PM, Elliott T, Cresswell P, Wilson IA, Dwek RA (2001) Glycosylation and the immune system. Science 291: 2370–2376.
48. Coelho V, Krysov S, Ghaemmaghami AM, Emara M, Potter KN, et al. (2010) Glycosylation of surface Ig creates a functional bridge between human follicular lymphoma and microenvironmental lectins. Proc Natl Acad Sci USA 107: 18587–18592.

9

Tissue Distribution, Gender- and Genotype-Dependent Expression of Autophagy-Related Genes in Avian Species

Alissa Piekarski[9], Stephanie Khaldi[9], Elizabeth Greene, Kentu Lassiter, James G. Mason, Nicholas Anthony, Walter Bottje[9], Sami Dridi*

Center of Excellence for Poultry Science, University of Arkansas, Fayetteville, Arkansas, United States of America

Abstract

As a result of the genetic selection of broiler (meat-type breeders) chickens for enhanced growth rate and lower feed conversion ratio, it has become necessary to restrict feed intake. When broilers are fed *ad libitum,* they would become obese and suffer from several health-related problems. A vital adaptation to starvation is autophagy, a self-eating mechanism for recycling cellular constituents. The autophagy pathway has witnessed dramatic growth in the last few years and extensively studied in yeast and mammals however, there is a paucity of information in avian (non-mammalian) species. Here we characterized several genes involved in autophagosome initiation and elongation in Red Jungle fowl (*Gallus gallus*) and Japanese quail (*coturnix coturnix Japonica*). Both complexes are ubiquitously expressed in chicken and quail tissues (liver, leg and breast muscle, brain, gizzard, intestine, heart, lung, kidney, adipose tissue, ovary and testis). Alignment analysis showed high similarity (50.7 to 91.5%) between chicken autophagy-related genes and their mammalian orthologs. Phylogenetic analysis demonstrated that the evolutionary relationship between autophagy genes is consistent with the consensus view of vertebrate evolution. Interestingly, the expression of autophagy-related genes is tissue- and gender-dependent. Furthermore, using two experimental male quail lines divergently selected over 40 generations for low (resistant, R) or high (sensitive, S) stress response, we found that the expression of most studied genes are higher in R compared to S line. Together our results indicate that the autophagy pathway is a key molecular signature exhibited gender specific differences and likely plays an important role in response to stress in avian species.

Editor: Graham R. Wallace, University of Birmingham United Kingdom

Funding: This study was supported by Arkansas state start-up funding (to SD) and Arkansas Bioscience 339 Institute (ABI) equipment grant (to SD) and FY2012-15 (to WB). The funders had no role in study design, data collection and analysis, decision to publish, or preparation of the manuscript.

Competing Interests: The authors have declared that no competing interests exist.

* Email: dridi@uark.edu

[9] These authors contributed equally to this work.

Introduction

Autophagy or cellular self-digestion, a lysosomal degradation pathway that is conserved from yeast to human, plays a key role in recycling cellular constituents, including damaged organelles [1]. Based on their mechanisms and functions, there are various types of autophagy, including micro- and macro-autophagy, as well as chaperone-mediated autophagy [2,3]. The first two types have the capacity to engulf large structures through both selective (specific organelles such as mitochondria or endoplasmic reticulum referred to as mitophagy or reticulophagy, respectively [4,5]) and non-selective mechanisms (bulk cytoplasm), whereas chaperone-mediated autophagy degrades only soluble proteins [6]. Micro-autophagy refers to the sequestration of cytosolic components directly by lysosomes through invaginations in their limiting membrane. However macro-autophagy refers to the sequestration within an autophagosome, a unique double-membrane cytosolic vesicle. Autophagosomes fuse with late endosomes and lysosomes, promoting the delivery of organelles, aggregated proteins and cytoplasm to the luminal acidic degradative milieu that enables their breakdown into constituent molecular building blocks that can be recycled by the cell [7].

A number of protein complexes (more than 30) and signaling pathways that regulate autophagy have been identified in yeast and many of these have mammalian orthologs (for review see [8]). These proteins can be grouped according to their functions at key stages of the autophagy pathway. The beclin-Vps34 complex is involved in the initiation of autophagosome formation. Beclin-1 enhances Vps34 activity [9] and binds to several partners that induce autophagy including ambra-1 [10], UVRAG [11], and bif-1 [12]. The second complex implicated in the initiation step of autophagosome formation is the ULK1/Atg1-Atg13-FIP200 complex [13]. Indeed, Atg13 binds ULK1 and when they are dephosphorylated they activate ULK1 that phosphorylates FIP200 to induce autophagosome formation [14–16]. For the autophagosome elongation, two ubiquitin-like systems are involved. The E1 ubiquitin activating enzyme-like, Atg7, activates Atg12 that is transferred to Atg10. Atg12 binds to Atg5 and then form a conjugate with Atg16L1 resulting in an 800-kDa complex [17] that is essential for the elongation of the pre-autophagosomal membrane. The second ubiquitin-like complex involves the protein microtubule-associated protein 1 light chain 3 (LC3/Atg8). LC3 is cleaved by Atg4B to form the cytosolic isoform LC3-I [18]. LC3-I is conjugated to phosphatidylethanolamine in a reaction involving Atg7 and Atg3 to form LC3-II which in turn targeted to elongating autophagosome membrane [19]. For the maturation and fusion stage, autophagosome moves, via dynein motor proteins [20,21] towards the microtubule organizing center where the lysosomes are enriched. Autophagosome fuses with lysosome in a reaction involving several proteins including ESCRT [22], SNAREs [23,24], Rab7 [25,26], and the class C Vps proteins [27]. Recently, it has been reported that beclin-1 functions in the maturation of autophagosome through interaction with Rubicon [28].

Autophagy is essential for maintaining cellular homeostasis and autophagy malfunction is associated with diverse diseases such as neurodegeneration [29], cancer [30], immunity [31] and metabolic syndrome [32]. The amount of research focused on the autophagy pathway has witnessed dramatic growth in the last few years and the bulk of data are mainly originated from yeast and mammals. There is, however, a paucity of information on avian (non-mammalian) species. Therefore, the present study aimed firstly to characterize autophagy-related genes and their tissue distribution in chicken and quail, and secondly to determine their regulation by gender and genotype.

Materials and Methods

Ethic Statement

The present study was conducted in accordance with the recommendations in the guide for the care and use of laboratory animals of the National Institutes of Health and the protocol was approved by the University of Arkansas Animal Care and Use Committee under protocols 13039 and 10025.

Animals

Chickens. Red Jungle fowl male and female chickens (*Gallus gallus*) (body weight average 1049±52 and 1654±67 g for female and male, respectively) were reared in floor pen under environmentally controlled facilities and under standard poultry rearing conditions (22±3°C for temperature and 50±5% relative humidity). Chickens were supplied with food (12.6 MJ·kg−1, 22% protein) and water available *ad libitum*.

Japanese quails. In order to assess whether the expression of autophagy-related genes is regulated by genotype, two lines of male Japanese quails (*coturnix coturnix Japonica*) were used. These two lines were established by long-term divergent selection for circulating corticosterone response to restraint stress, after which the low stress line (resistant, R) had 66% low plasma corticosterone levels compared to their high stress (sensitive, S) counterpart [33]. Quails of each genetic line were reared separately in floor pen under environmentally controlled facilities and were allowed *ad libitum* access to water and food (12.6 MJ·kg−1, 22% protein).

Animals were killed by cervical dislocation and tissues (liver, leg and breast muscle, brain, gizzard, intestine, heart, lung, kidney, adipose tissue, ovary and testis) were removed, immediately snap frozen in liquid nitrogen, and stored at −80°C until use.

RNA isolation, reverse transcription and quantitative real-time PCR

Total RNA was extracted from chicken and quail tissues by Trizol reagent (catalog #15596018, Life Technologies) according to manufacturer's recommendations, DNAse treated and reverse transcribed (catalog #95048-100, Quanta Biosciences). RNA integrity and quality was assessed using 1% agarose gel electrophoresis and RNA concentrations and purity were determined for each sample by Take 3 micro volume plate using

Synergy HT multi-mode microplate reader (BioTek, Winooski, VT). The RT products (cDNAs) were amplified by real-time quantitative PCR (Applied Biosystems 7500 Real-Time PCR system) with Power SYBR green Master Mix (catalog #4312074, Life Technologies). Oligonucleotide primers used for avian autophagy-related genes are summarized in Table 1. The qPCR cycling conditions were 50°C for 2 min, 95°C for 10 min followed by 40 cycles of a two-step amplification program (95°C for 15 s and 58°C for 1 min). At the end of the amplification, melting curve analysis was applied using the dissociation protocol from the Sequence Detection system to exclude contamination with unspecific PCR products. The PCR products were also confirmed by agarose gel and showed only one specific band of the predicted size. For negative controls, no RT products were used as templates in the qPCR and verified by the absence of gel-detected bands. Relative expressions of target genes were determined by the $2^{-\Delta\Delta Ct}$ method [34].

Multiple alignement and molecular evolution

Sequence alignments and percentage of amino acid conservation were assessed with the Clustal W and MUSCLE multiple alignment algorithms [35,36] using chicken (non-mammalian) and mammalian autophagy-related gene (beclin1, Atg3, Atg5, Atg9a, Atg10, Atg12, Atg14, Atg13, Atg7, Atg4b, Atg4a, Atg16L1, UVRAG, Ambra1) sequences from database (see Table 2 for GenBank accession numbers). The phylogenetic tree based on these nucleotide sequence alignments was constructed using the neighbor-joining method of the MEGA6 program [37].

Statistical analyses

Data were analysed by two-factor ANOVA with tissue and gender (for chicken) and tissue and genotype (for quail) as classification variables. If ANOVA revealed significant effects, the means were compared by Tukey multiple range test using the Graph Pad Prism version 6.00 for Windows, Graph Pad Software, La Jolla California USA. Differences were considered significant at $P<0.05$.

Results

Tissue distribution of autophagy-related genes in chickens and quails

Since the role of autophagy-related genes is still unknown in avian species, we classified them in the following sections based on their roles in yeast and mammals. Gene complexes involved in autophagosome initiation (beclin1, Ambra1, UVRAG, Atg9a, Atg13 and Atg14) and elongation (Atg3, Atg4A, Atg4B, Atg5, Atg7, Atg10, Atg12 and Atg16L1) were ubiquitously expressed in chicken and quail. Only one band of the predicted size for each gene was observed in liver, leg and breast muscle, brain, gizzard, intestine, heart, lung, kidney, adipose tissue, ovary and testis (Fig. 1a, b and Fig. 2a, b). The sequences of the fragments were identical (100%) to these previously described in GenBank (for accession number, see Table 1 and 2).

Expression of autophagosome initiation complex in different tissues of female and male chickens

The autophagosome initiation complex was expressed in all tissues examined in female and male Red Jungle fowl chickens. In female chickens, beclin 1 was highly expressed in the heart, brain and leg muscle, followed by the ovary (Fig. 3a). Ambra 1 mRNA abundance was higher in the ovary followed by kidney, lung, heart, brain and the liver (Fig. 3b). UVRAG and Atg13 genes

Table 1. Oligonucleotide PCR primers.

Gene	Accession number[a]	Primer sequence (5′ → 3′)	Orientation	Product size (bp)
Beclin 1	NM_001006332	TGCATGCCCTTGCTAACAAA	Forward	61
		CCATACGGTACAAGACGGTATCTTT	Reverse	
Atg3	NM_001278070	GAACGTCATCAACACGGTGAA	Forward	65
		TGAGGACGGGAGTGAGGTACTC	Reverse	
Atg5	NM_001006409	TCACCCCTGAAGATGGAGAGA	Forward	66
		TTTCCAGCATTGGCTCAATTC	Reverse	
Atg9A	NM_001034821	AGTATGCCTCCACTGAGATGAGTCT	Forward	65
		GGCATGCTGCTTGTGCAA	Reverse	
Atg10	XM_424902	CATCTCACCAGATCTCAAGAAGGA	Forward	62
		CGACATGCGTAAGCAACGTT	Reverse	
Atg12	XM_003643073	GCACCCGCACCATCCA	Forward	61
		GAGGCCATCAGCTTCAGGAA	Reverse	
Atg14	XM_426476	GCGCTGCGAGGGTGTTAAT	Forward	61
		TTCTGTTACAAAAGCGTTCCTTGA	Reverse	
Atg13	XM_003641387	GGTCCCCGAGCCAAATA	Forward	55
		ATGAGGTGCGGGAGCTGTAG	Reverse	
Atg7	NM_001030592	ACTGGCAATGCGTGTTTCAG	Forward	57
		CGATGAACCCAAAAGGTCAGA	Reverse	
Atg4B	NM_213573	CCCCGATGAAAGCTTCCA	Forward	56
		GCTCAGCGATGCTCATTCTG	Reverse	
Atg4A	NM_001271986	CACAGCAGTGCACATTTGCA	Forward	62
		CAGAGTCCTGCTGCGTTCCT	Reverse	
Atg16L1	XM_003641751	TGCATCCAGCCAAACCTTTC	Forward	57
		CGACGCTGGTGGCTTGTC	Reverse	
UVRAG	NM_001030839	GGGCTCATGGTCAGATGTGA	Forward	57
		CTTTGGAACGGGAATTGCA	Reverse	
Ambra 1	XM_001233288	GGGATGTTGTGCCTTTGCA	Forward	67
		CCTGGTGTGGGAAGAGAGAAGA	Reverse	
18S	AF173612	TCCCCTCCCGTTACTTGGAT	Forward	60
		GCGCTCGTCGGCATGTA	Reverse	

[a]Accession number refer to Genbank (NCBI).

were highly expressed in the brain and the liver (Fig. 3c and d). Atg9a mRNA levels were greater in the brain followed by breast muscle, ovary and liver (Fig. 3e). The highest amount of Atg14 mRNA was found in the ovary followed by the liver, brain, kidney, heart and the breast muscle (Fig. 3f). In males, however, the highest levels of beclin1 mRNA were observed in brain and testis, followed by leg and breast muscle and liver (Fig 3a). The greatest expression of Ambra 1 and Atg13 genes was found in kidney and testis (Fig. 3b). UVRAG gene was highly expressed in liver and testis followed by the brain (Fig. 3c). Atg9a mRNA abundance was high in the liver, brain, leg and breast muscle followed by the testis (Fig. 3e). Atg14 gene expression was high in kidney followed by testis, intestine, brain, liver and breast muscle (Fig. 3f).

Interestingly, when tissues from the two genders were plotted together, female chickens exhibited greater hepatic abundance of beclin 1 (3.57 fold, $P<0.05$), UVRAG (1.5 fold, $P<0.01$), Atg13 (6.25 fold, $P<0.01$), and Atg9a mRNA (1.35 fold, $P<0.05$) compared to males (Fig. 3a, c, d and e). Females also exhibited significant higher expression of the following genes: beclin1 in leg muscle (4.25 fold), brain (2 fold) and heart (6.24 fold, Fig. 3a),

Ambra 1in lung and heart (3.67 and 4.72 fold respectively, Fig. 3b), UVRAG in leg muscle, brain, and heart (4.4, 4.75, and 8.75 fold respectively, Fig. 3c), Atg13 in brain (3.24 fold, Fig. 3d), Atg9a in breast muscle, brain and ovary (2.71, 5.42, and 1.87 fold respectively, Fig. 3e) and Atg14 in ovary (1.94 fold, Fig. 3f) compared to males. However, in male chickens, Ambra 1 and Atg14 gene expression was higher in kidney (1.62 and 2.76 fold respectively, $F<0.01$, Fig. 3d and f) and Atg13 mRNA levels were higher in testis and kidney (4.4 and 14.6 fold respectively, $P<0.01$, Fig. 3d) compared to female.

Expression of autophagosome elongation complex in different tissues of female and male chickens

As the initiation complex, the autophagosome elongation complex is ubiquitously expressed in both male and female chickens. In females, the highest amount of Atg3 mRNA was found in the ovary followed by brain and kidney. Atg4A mRNA was abundant in the liver, heart, brain, leg and breast muscle. Atg5 gene expression was higher in the liver, brain, gizzard, heart and breast muscle (Fig. 4a, b and d). Atg4B gene was highly

Table 2. Multiple alignment of the amino acid sequences of chicken autophagy-related genes with their mammalian orthologs.

GENE	SPECIES					
	Human	**Mouse**	**Rat**	**Horse**	**Pig**	**Bovine**
Chicken Beclin 1	75.97	73.87	74.03	75.26	72.28	75.36
(NM_001006332)	(NM_003766)	(NM_019584)	(NM_001034117)	(XM_005597370)	(XM_005668792)	(NM_001033627)
Chicken Atg3	81.84	81.6	79.7	75.57	80.33	91.53
(NM_001278070)	(NM_001278712)	(NM_026402)	(NM_134394)	(XM_005601995)	(XM_003132682)	(NM_001075364)
Chicken Atg5	80.11	75.55	72.09	68.96	68.84	79.53
(NM_001006409)	(NM_004849)	(NM_053069)	(NM_001014250)	(XM_005596852)	(NM_001037152)	(NM_001034579)
Chicken Atg9A	71.97	71.37	66.47	74.7	80.01	73.46
(NM_001034821)	(BC_065534)	(NM_001288612)	(NM_001014218)	(XM_001493040)	(NM_001190275)	(NM_001034706)
Chicken Atg10	59.76	64.64	65.6	62.35	69.42	59.12
(XM_424902)	(NM_001131028)	(NM_025770)	(NM_001109505)	(XM_005599592)	(NM_001190281)	(NM_001083531)
Chicken Atg12	65.87	52.69	54.48	75.89	61.67	66.99
(XM_003643073)	(NM_004707)	(NM_026217)	(NM_001038495)	(XM_003362836)	(NM_001190282)	(NM_001076982)
Chicken Atg14	65.13	64.74	65.33	74.53	65.34	64.05
(XM_426476)	(NM_014924)	(NM_172599)	(NM_001107258)	(XM_001914860)	(XM_001924990)	(NM_001192099)
Chicken Atg13	65.84	67.14	75.47	69.47	61.84	63.58
(XM_003641387)	(NM_001205119)	(NM_145528)	(NM_001271212)	(NM_001242529)	(XM_003122826)	(NM_C01076812)
Chicken Atg7	75.77	71.52	76.28	74.05	76.15	69.17
(NM_001030592)	(NM_006395)	(NM_001253717)	(NM_001012097)	(XM_005600372)	(NM_001190285)	(NM_001142967)
Chicken Atg4B	79.19	78.26	50.76	78	75.8	76.06
(NM_213573)	(NM_013325)	(NM_174874)	(NM_001025711)	(XM_005610806)	(NM_001190283)	(NM_001001170)
Chicken Atg4A	69.32	66.82	65.76	73.15	72.51	78.03
(NM_001271986)	(NM_052936)	(NM_174875)	(NM_001126298)	(XM_005614404)	(XM_005657911)	(NM_001001171)
Chicken Atg16L1	67.77	68.31	70.67	65.24	78.4	67.59
(XM_003641751)	(NM_030803)	(NM_001205391)	(NM_001108809)	(XM_005610723)	(NM_001190272)	(NM_001191389)
Chicken UVRAG	77.23	76.16	73.96	76.6	-	75.73
(NM_001030839)	(AB.12958)	(NM_178635)	(NM_001107536)	(XM_001917231)		(NM_001193026)
Chicken AMBRA1	70.69	73.38	73.4	77.09	74.5	72.66
(XM_001233288)	(NM_001267782)	(NM_172669)	(NM_001134341)	(XM_005598075)	(XM_003122844)	(NM_001034522)

Genbank accession number is indicated for each gene and each species between brackets.

expressed in brain, followed by ovary and liver (Fig. 4c). Atg7 and Atg12 mRNA were abundant in liver followed by ovary, brain, lung, kidney and leg muscle for Atg7 and ovary and brain for Atg12 (Fig. 4e and g). Atg16L1 mRNA levels were high in heart followed by breast muscle, brain, liver, ovary and leg muscle (Fig. 4h). In males, however, the highest amount of Atg3, Atg4B, Atg7, Atg10, Atg12 and Atg16L1 mRNA was found in testis followed by kidney, brain and liver for Atg3, intestine, brain, liver and kidney for Atg4B, brain, liver and kidney for Atg7, kidney, liver, brain, intestine and heart for Atg10, liver, brain and kidney for Atg12, and kidney, breast muscle, intestine, brain and liver for Atg16L1 (Fig. 4a, c, e, f, g and h). Atg4A and Atg5 mRNA levels were high in kidney and intestine followed by testis, liver, brain and breast muscle for Atg4A and testis, liver and brain for Atg5 (Fig. 4b and d). Importantly, when we profile the autophagosome elongation complex for each tissue within the two genders, only a few genes showed gender- and tissue-dependent pattern. Female chickens displayed significant high expression of Atg4A in leg muscle, heart and ovary (4.18, 2.85 and 1.68 fold, respectively, Fig. 4b), Atg4B in the brain (3.41 fold, Fig. 4c), Atg7 in liver, brain and lung (5.88, 2.59 and 14 fold, respectively, Fig. 4e), and

Atg16L1 in liver, leg muscle, brain and heart (1.69, 5, 1.85 and 5.24 fold, respectively, Fig. 4h) compared to males. However, male chickens exhibited significant higher levels of Atg4A mRNA in kidney (3.12 fold, Fig. 4b), Atg4B mRNA in intestine (13.11 fold, Fig. 4c), Atg5 mRNA in kidney and intestine (3.43 and 4.61 fold, respectively, Fig. 4d), Atg7 mRNA in testis (2.17 fold, Fig. 4e), Atg10 in testis and kidney (4.61 and 4.65 fold, respectively, Fig. 4f), Atg12 in testis (9.25 fold, Fig. 4g), and Atg16L1 in testis and kidney (1.52 and 5.31 fold, Fig. 4h). Atg3 gene expression did not differ between male and female in every studied tissue (Fig. 4a).

Expression of autophagosome initiation complex in different tissues of S and R quail lines

The autophagosome initiation complex was expressed in all tissues examined in quail. Beclin 1 mRNA levels were abundant in testis, heart and leg muscle of R line and in adipose tissue and testis of S line (Fig. 5a). The highest amount of Ambra1 was found in lung, heart and leg muscle of R line and in intestine, lung, heart, kidney, and breast muscle of S line (Fig. 5b). The UVRAG expression was high in lung and adipose tissue of R quail and in

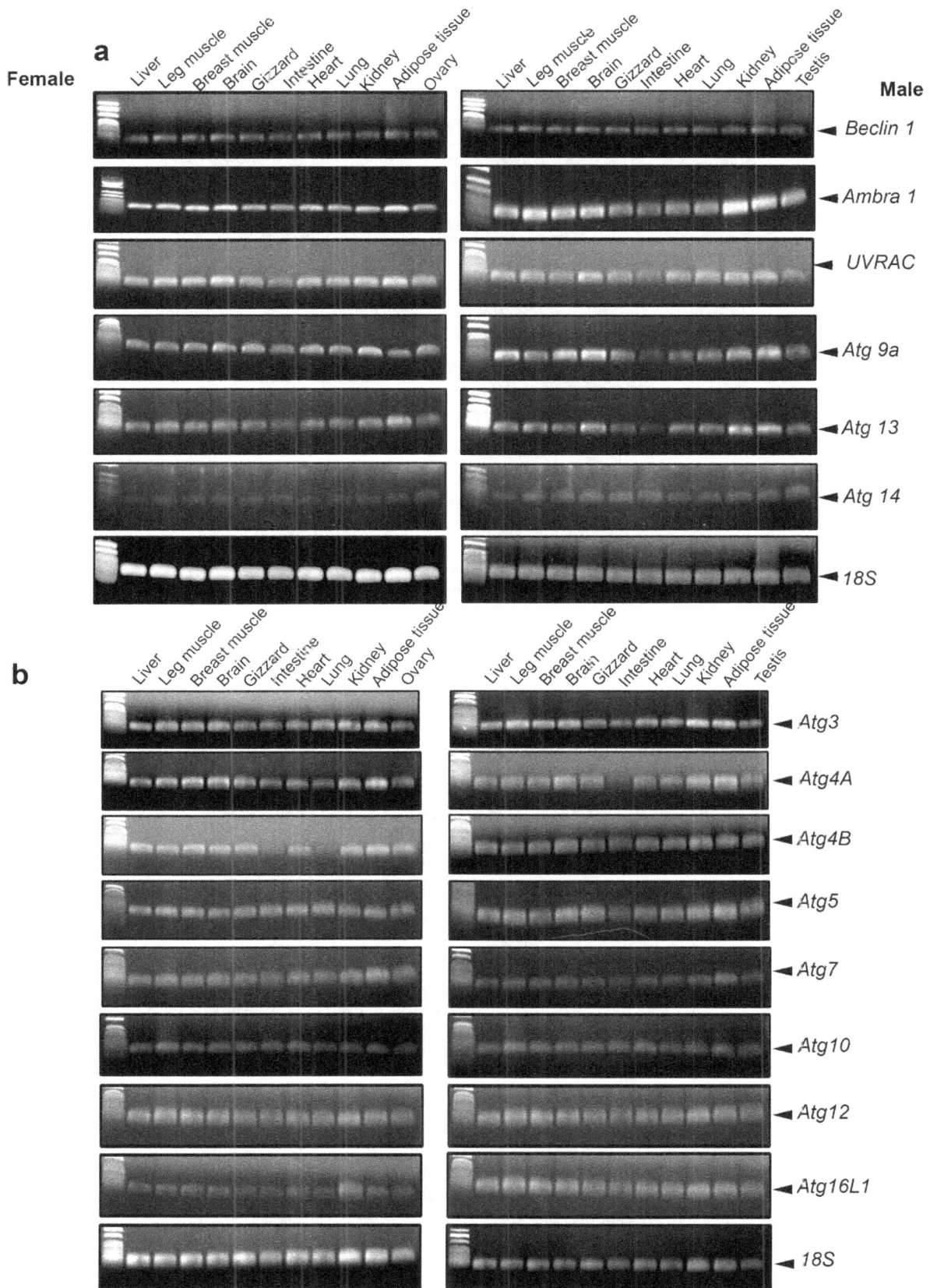

Figure 1. Characterization of autophagosome initiation- (a) and elongation-related genes (b) in various tissues of male and female Red Jungle Fowl (*Gallus gallus*) by RT-qPCR as described in materials and methods. Signals were visualized by agarose gel electrophoresis.

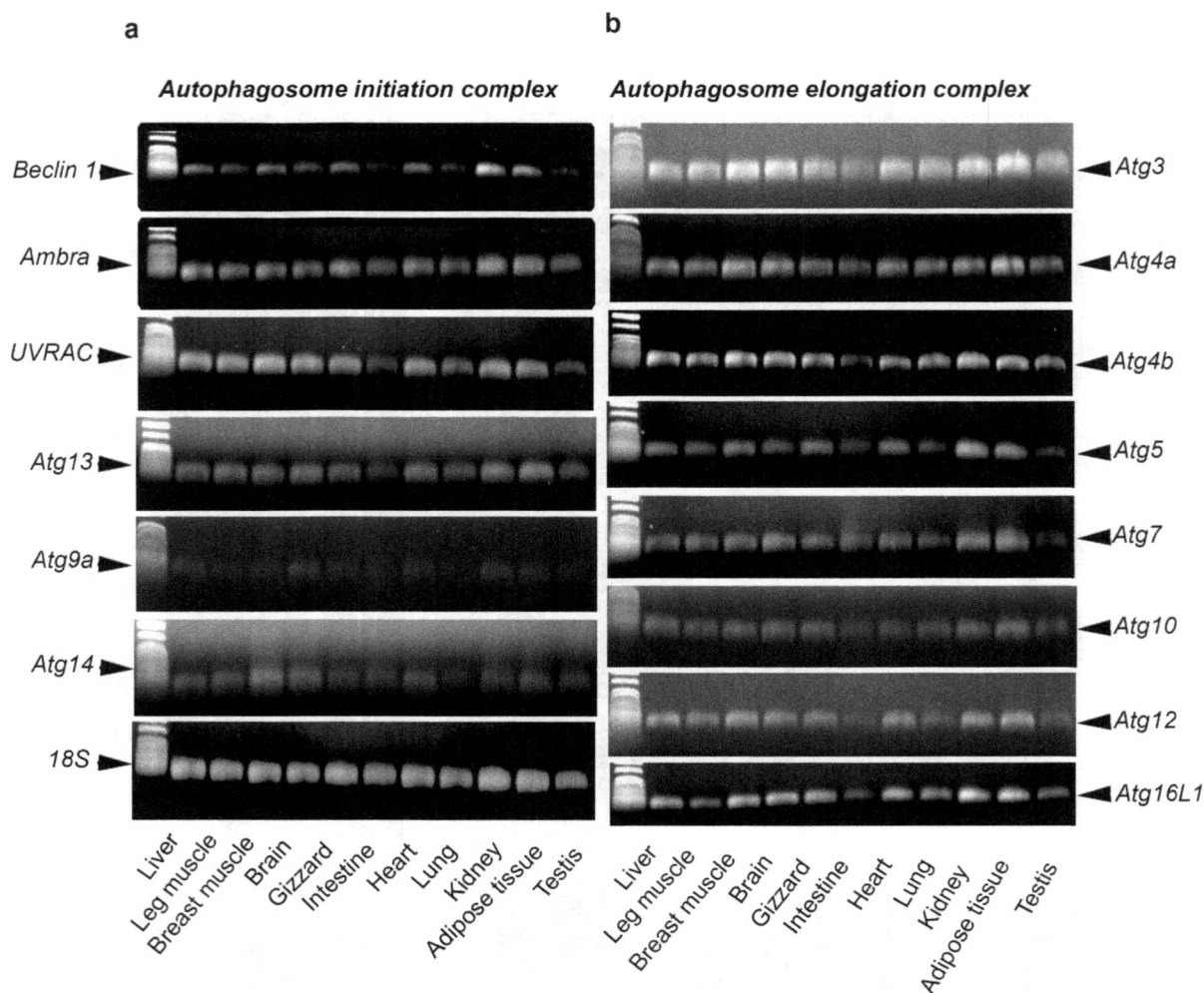

Figure 2. Characterization of autophagosome initiation- (a) and elongation-related genes (b) in various tissues of stress-sensitive (S) and stress-resistant (R) male Japonica quail (*Coturnix coturnix Japonica*) using RT-qPCR. Signals were visualized by agarose gel electrophoresis.

lung followed by testis, lung, heart and intestine in S line (Fig. 5c). The highest amount of Atg13 was found in leg muscle and kidney of R line and in intestine followed by kidney and testis in S line (Fig. 5d). Atg9a gene was highly expressed in leg and breast muscle followed by intestine in R line and in intestine and adipose tissue in S line (Fig. 5e). Atg14 gene expression was found to be high in lung, adipose tissue and intestine in both lines (Fig. 5f). When tissues from the two lines were plotted together, R line exhibited significant higher mRNA abundance of beclin 1 in leg muscle, heart, and testis (2.76, 2.72 and 1.46 fold, respectively), Ambra 1 in lung and adipose tissue (1.46 and 2.37 fold, respectively), UVRAG in gizzard and adipose tissue (17.3 and 7.1 fold, respectively), Atg13 in liver, leg muscle, brain, heart, lung and kidney (2.56, 46.5, 11.9, 3.4, 11.4 and 3.2 fold, respectively), Atg9a in lung, leg and breast muscle (3.86, 8.8 and 8.3 fold, respectively), and Atg14 in in lung (1.77 fold) compared to S line (Fig. 5a, b, c, d, e and f). However, S line exhibited significant higher levels of beclin 1 in adipose tissue (17.8 fold), Ambra 1 in breast muscle and intestine (9 and 12.5 fold, respectively), UVRAG in intestine and heart (17 and 1.7 fold, respectively), Atg13 in breast muscle, intestine and adipose tissue (7.5, 4.8 and 4.6 fold, respectively), Atg9a in intestine and adipose tissue (3.4 and 4.2 fold, respectively), and

Atg14 in adipose tissue (2.6 fold) compared to R line (Fig. 5a, b, c, d, e and f).

Expression of autophagosome elongation complex in different tissues of R and S quail lines

Atg3 gene was highly expressed in adipose tissue of both lines followed by testis, heart, leg muscle, gizzard and liver in R line and by testis, lung and heart in S line (Fig. 6a). The highest amount of Atg4a mRNA was found in leg muscle and adipose tissue of R line and in intestine of S line (Fig. 6b). Atg4b mRNA levels, however, was high in adipose tissue, leg muscle, brain and lung of R line and in intestine and brain of S line (Fig. 6c). Atg5 was highly expressed in intestine, heart, adipose tissue and lung of R line and its expression remain unchanged between the examined tissues of S line (Fig. 6d). Atg7 mRNA abundance was found to be high in adipose tissue, leg muscle, brain and lung of R line and in brain, adipose tissue and intestine of S line (Fig. 6e). Atg10 was highly expressed in leg muscle in R line and in brain of S line (Fig. 6f). The highest amount of Atg12 mRNA was found in lung of R line but did not differ between tissues in S line (Fig. 6g). Atg16L1 gene expression was high in adipose tissue followed by leg muscle and lung in R line and in leg and breast muscle of S line (Fig. 6h).

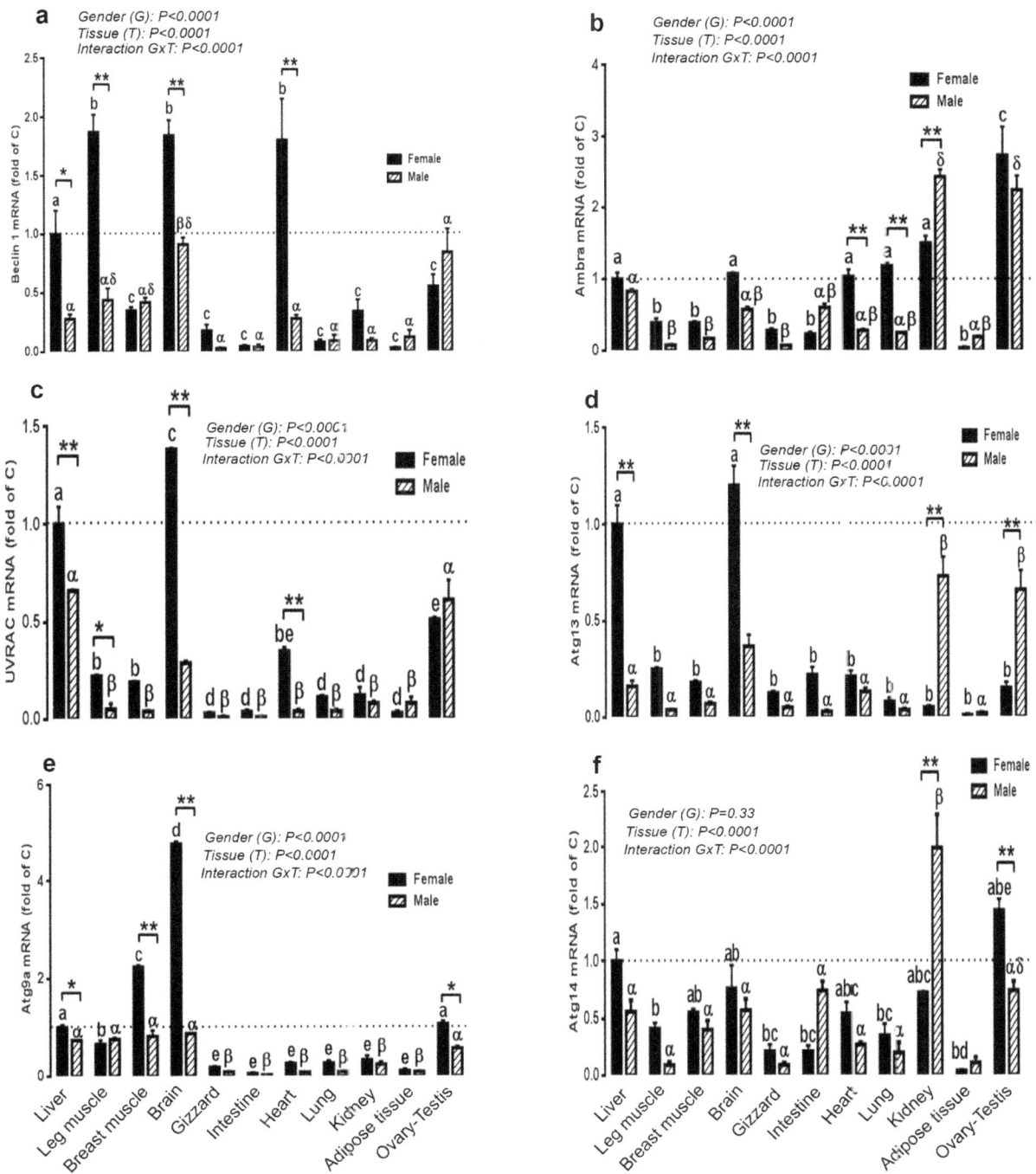

Figure 3. Comparison of relative expression of autophagosome initiation-related genes in various tissues of male and female Red Jungle Fowl. Total RNA from each tissue was DNAse-treated, reverse transcribed, and subjected to real-time quantitative PCR as described in material and methods. Samples were run in duplicate, and the average threshold cycle (Ct) values were determined for the target and housekeeping genes. Relative quantity of autophagy genes was determined by the $2^{-\Delta\Delta Ct}$ method [58]. Data are presented as mean $\pm$ SEM (n = 6 for each gender and each tissue). * Sex-matched differences among tissues (*$P<0.05$ and **$P<0.01$). Different letters indicate tissue-matched differences among gender (a–e, difference between tissues within female and α-δ indicate differences between male tissues).

Interestingly, when the two genotypes are plotted together, R line displayed significant high levels of Atg3 in liver, leg muscle, gizzard, heart, and testis (6.6, 7.4, 10, 2.7, and 3.3 fold, respectively), Atg4a in leg muscle and adipose tissue (9.4 and 5.6 fold, respectively), Atg4b in leg muscle, lung and adipose tissue (13.9, 7, and 8.7 fold, respectively), Atg5 in intestine, heart and

adipose tissue (33, 8, and 5 fold, respectively), Atg7 and Atg10 in leg muscle (9.7 and 12.3 fold, respectively), Atg12 in lung (77 fold), and Atg16L1 in liver, leg muscle, brain, lung, and adipose tissue (6.25, 3.89, 5.2, 25, and 39 fold, respectively) compared to S line (Fig. 6a–h). However S line exhibited higher mRNA levels of Atg3

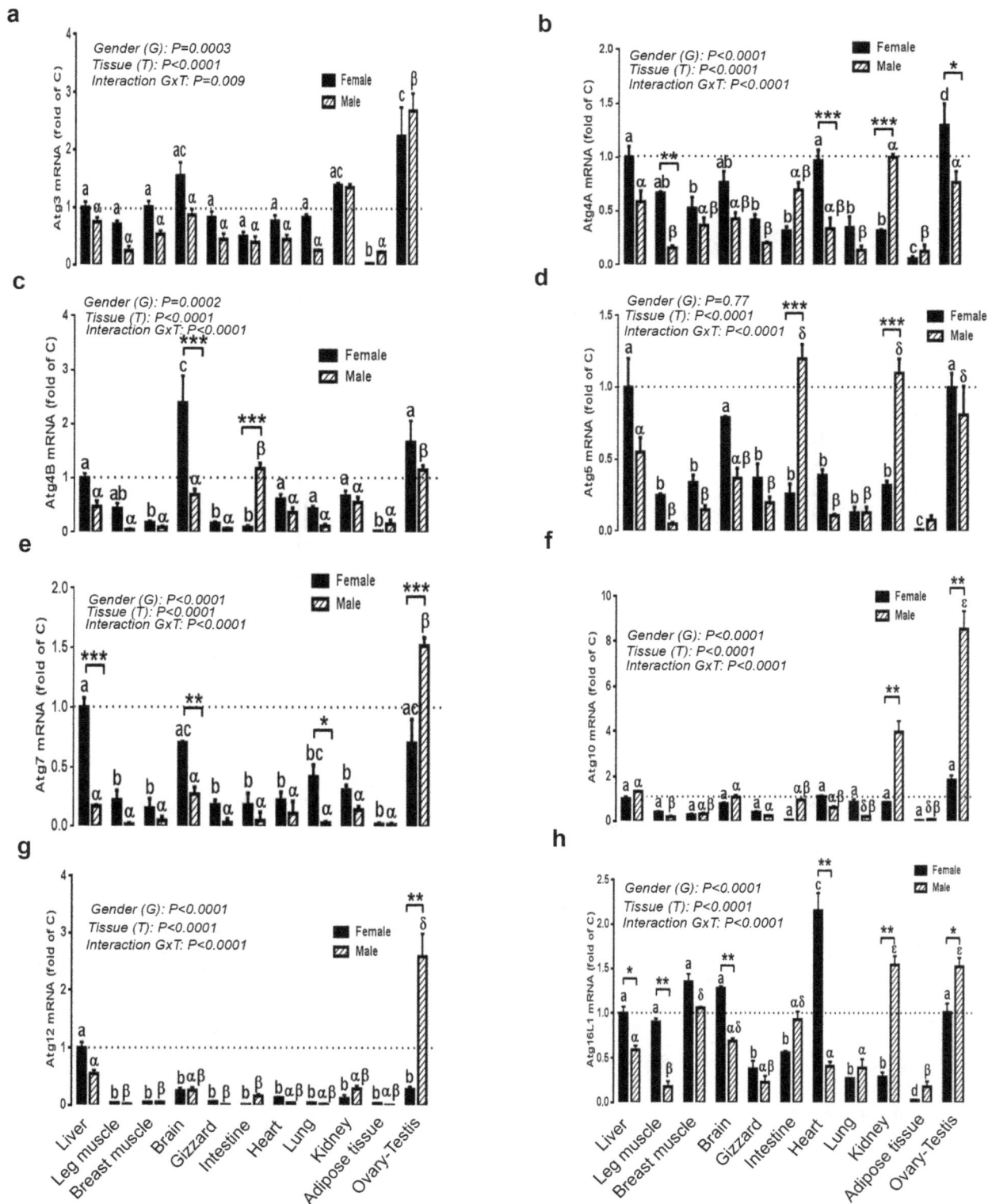

Figure 4. Comparison of relative expression of autophagosome elongation-related genes in various tissues of male and female Red Jungle Fowl. Total RNA from each tissue was DNAse-treated, reverse transcribed, and subjected to real-time quantitative PCR as described in material and methods. Sample were run in duplicate, and the average threshold cycle (Ct) values were determined for the target and houskeeping genes. Relative quantity of autophagy genes was determined by the $2^{-\Delta\Delta Ct}$ method [58]. Data are presented as mean ± SEM (n = 6 for each gender and each tissue). * Sex-matched differences among tissues (*$P<0.05$ and **$P<0.01$). Different letters indicate tissue-matched differences among gender (a–e, difference between tissues within female and α–δ indicate differences between male tissues).

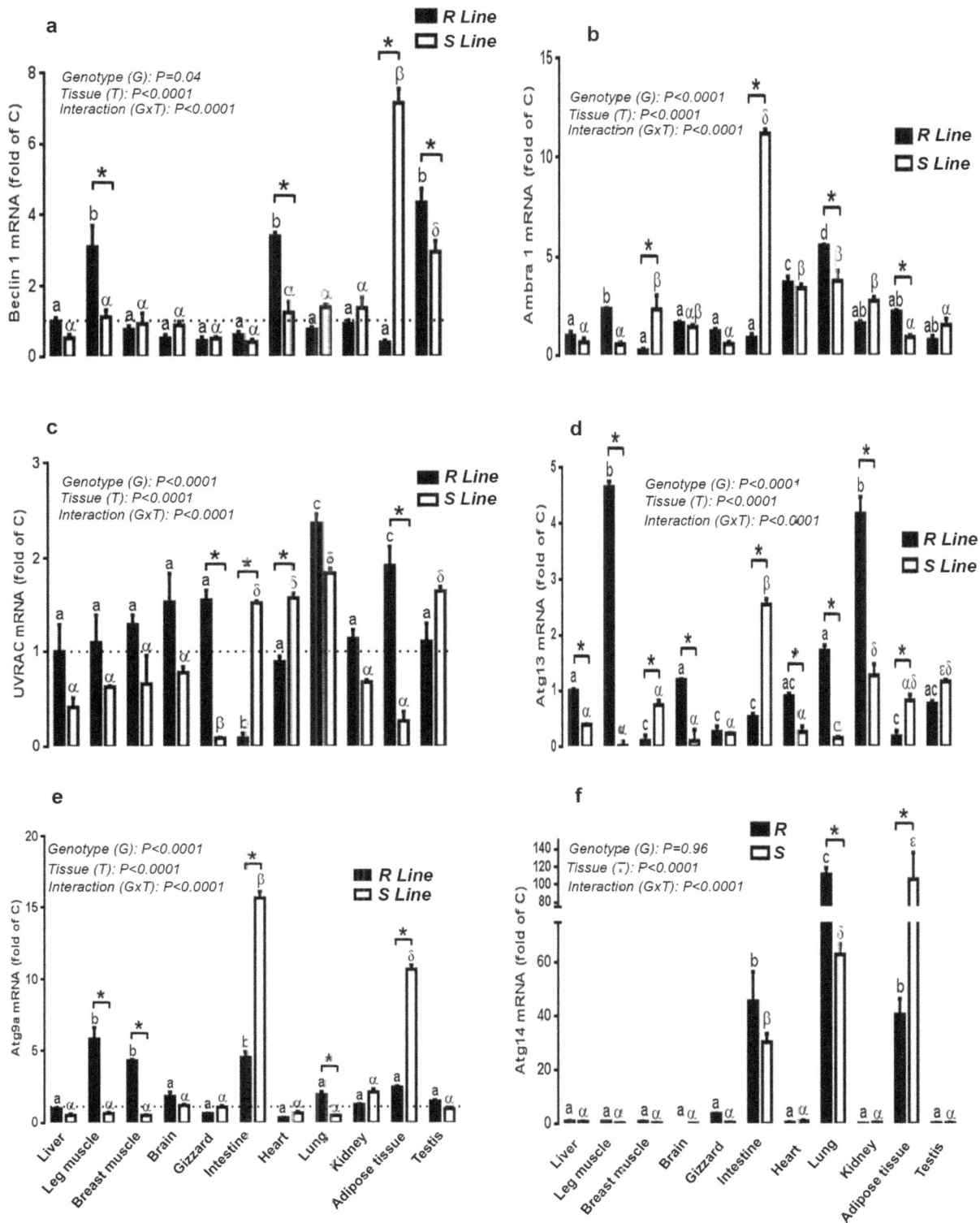

Figure 5. Relative expression of autophagosome initiation-related genes in various tissues of R ans S male Japonica quail lines.
Total RNA from each tissue was DNAse-treated, reverse transcribed, and subjected to real-time quantitative PCR. Sample were run in duplicate, and the average threshold cycle (Ct) values were determined for the target and houskeeping genes. Relative quantity of autophagy genes was determined by the $2^{-\Delta\Delta Ct}$ method [58]. Data are presented as mean ± SEM (n = 6 for each ine and each tissue). * Line-matched differences among tissues (*P<0.05). Different letters indicate tissue-matched differences among Lines (a–c, d fference between tissues within R line and α-ε indicate differences between tissues within S line).

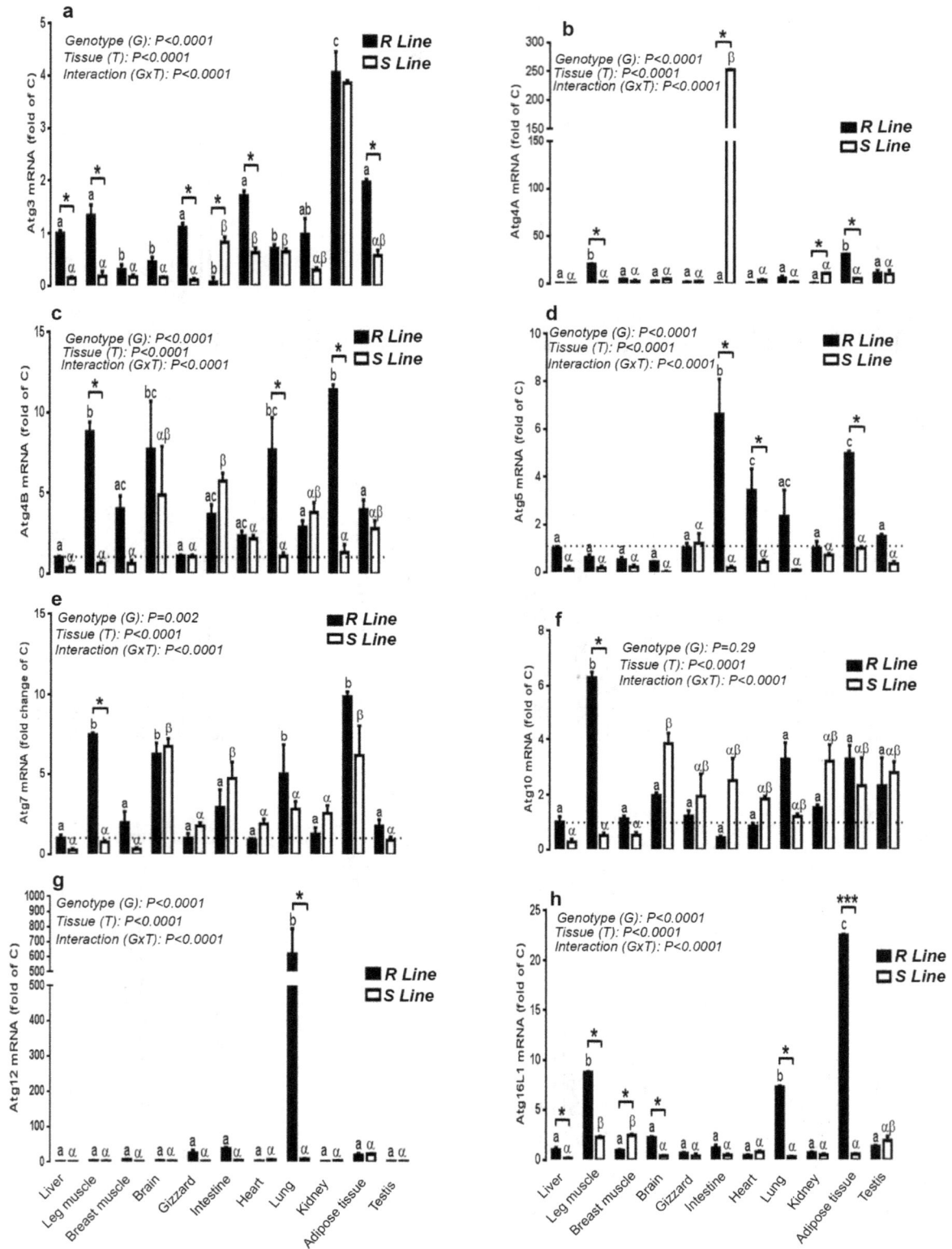

Figure 6. Comparison of relative expression of autophagosome elongation-related genes in various tissues of R and S male Japonica quail lines. Total RNA from each tissue was DNAse-treated, reverse transcribed, and subjected to real-time quantitative PCR. Sample were run in duplicate, and the average threshold cycle (Ct) values were determined for the target and houskeeping genes. Relative quantity of autophagy genes was determined by the $2^{-\Delta\Delta Ct}$ method [58]. Data are presented as mean ± SEM (n = 6 for each line and each tissue). * Genotype-matched differences among tissues (*$P<0.05$ and ***$P<0.001$). Different letters indicate tissue-matched differences among genotype (a, b, difference between tissues within R line and α-β indicate differences between tissues within S line). 28.

in intestine (4.8 fold), Atg4a in kidney (8.4 fold), and Atg16L1 in breast muscle (2.6 fold) (Fig. 6a–h).

Alignment and phylogenetic tree analysis of chicken autophagy-related genes with other sources

Comparison of the nucleotide sequences of autophagosome initiation and elongation-related genes between chickens and other species showed low to high similarity (52.6%–91.5%) (Table 2). Phylogenetic analysis indicates that chicken Atg4b, Atg7, Atg9, Atg10, Atg14, Atg16L1, and Ambra1 are more closely related to the mouse orthologs however Atg3 and Atg4a are closely related to the pig orthologs, beclin1 is closely related to the horse ortholog, UVRAG is closely related to the rat ortholog and Atg5 is closely related to the bovine ortholog (Fig. 7).

Discussion

Autophagy is an evolutionary conserved catabolic process regulating the degradation of a cell's own components through the lysosomal machinery [38]. It plays a key homeostatic role in every cell type to preserve the balance between the synthesis, degradation, and subsequent recycling of cellular components [39]. Currently, more than thirty different autophagy-related genes have been identified by genetic screening in yeast, and many of these genes are conserved in plants, flies and mammals [40]. However, data in birds are scarce. Here, we report for the first time the characterization of fourteen avian genes involved in the autophagosome initiation and elongation. All genes had high basal expression levels in every examined tissue from chicken and quail maintained under normal (low stress) physiological conditions. These data indicate that avian cells are also equipped with the autophagy system which may be involved in numerous vital cell

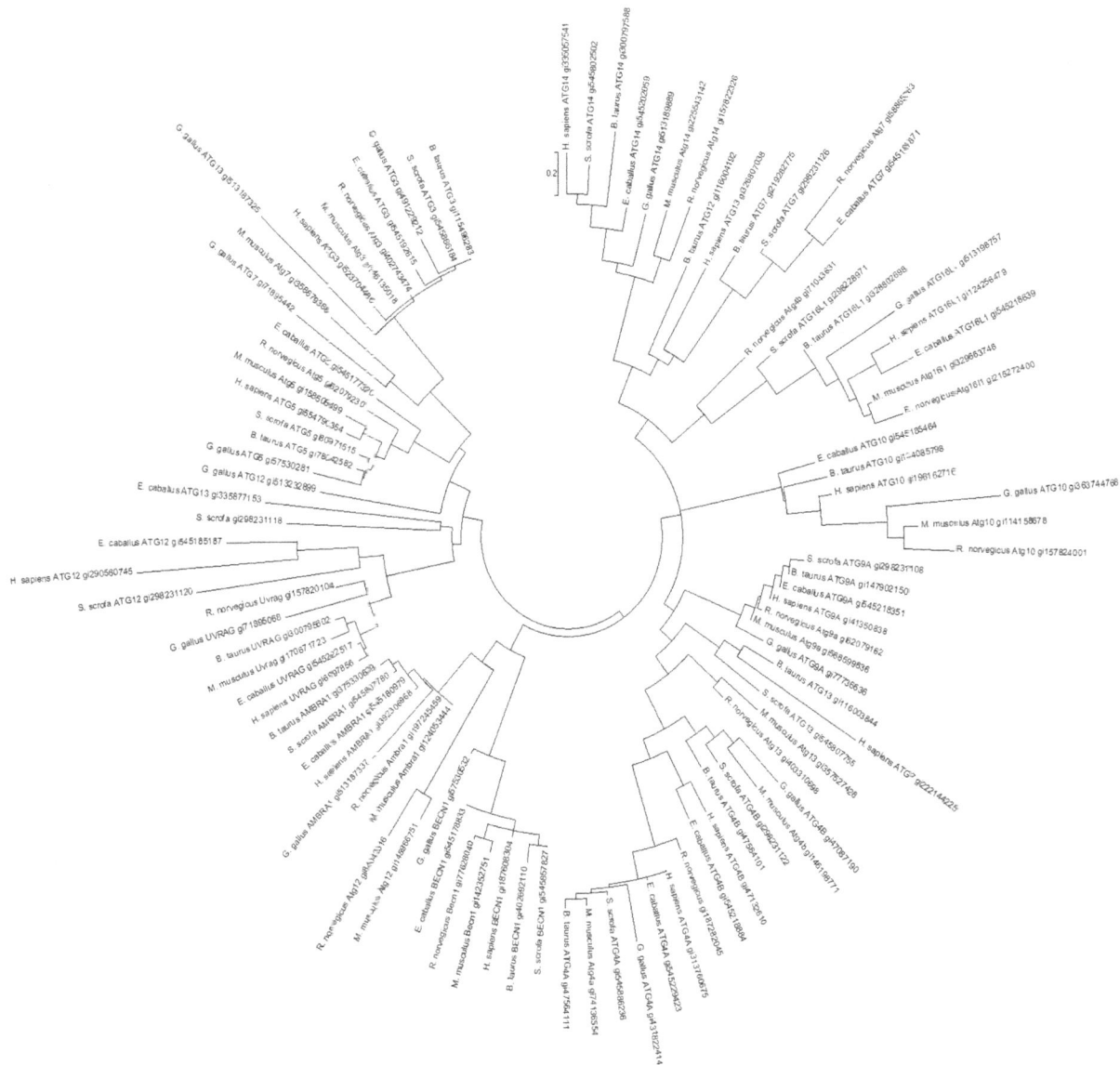

Figure 7. Phylogenetic relationships among chicken autophagy-related genes and their mammalian orthologs were inferred using the neighbor-joining method in MUSCLE alignment and MEGA6. Scale bar indicates the substitution rate per residue. Genbank accession numbers are included in the phenogram and in Table 2.

processes including cellular homeostasis, tissue development and a defense mechanism against aggregated proteins, damaged organelles and infectious agents.

The MUSCLE alignment and the phenogram construction of nucleotide sequences of chicken autophagy-related genes and their mammalian orthologs show high homology and indicate that the evolutionary relationship is consistent with the consensus view of vertebrate evolution. Although the evolutionary conservation of the autophagy pathway, many of the mechanistic breakthroughs in delineating how autophagy is regulated and executed at the molecular level have been made in yeast [41]. Here, our quantitative real-time PCR analysis revealed that the expression of avian autophagy-related genes is tissue specific suggesting a tissue-dependent modulation mechanism of autophagy under physiological conditions and corroborating results from previous studies in rats [42] and mice [43]. This may reflect fundamental differences in the fate of the tissue and/or cells, the role of the autophagy-related genes and their interactions in rapidly dividing versus post-mitotic cells. For instance, inhibition of autophagy using Atg7 small interfering RNA inhibited cell death during starvation in neuronal cells, but increased cell death in fibroblasts [42]. Most likely there is no single autophagy pathway across all tissues, and even within the same tissue, multiple effectors and mediators may exist.

We have also demonstrated that avian autophagy-related gene expression is gender dependent. Although the underlying mechanism(s) for this apparent sexual dimorphic expression is (are) unknown, the results are not surprising because sex-dependent differences in the activation of the autophagic cytoprotection pathway have been reported in mammals [42,44,45]. The gender-associated differences in autophagy-related genes observed in our study could be a result of physiological, morphological, and hormonal differences between both sexes. Indeed, variations in androgen levels have been shown to be an important factor for the development of autophagy [44]. Sobolewska and co-workers showed that 17 beta-estradiol and progesterone exerted stimulatory effects on autophagy in bovine mammary epithelial cells [46]. Furthermore, a sexual dimorphism of estrogen receptor (ERα and ERβ) expression was observed in basal vascular smooth muscle cells (VSMC) and their regulation by oxidative stress was also found to be gender-dependent (alteration in female but not in male VSMC) [47]. Additionally, The downstream cascades mediating the cardio-protective effects of ERβ in tumor necrosis factor receptor-2 (TNFR2) knockout mice has been shown to be gender-dependent with activation and translocation of signal transducer and activator of transcription 3 (STAT3) in female and decrease of c-jun N-terminal kinase (JNK) in male [48]. Because STAT3 activation has been associated with autophagy processes [49], and JNK phosphorylates B-cell lymphoma 2 (Bcl-2) triggering its release from beclin 1 in response to various stimuli [50], the existence of sexual dimorphic autophagy signaling cascades is very likely. As autophagy is tightly linked to starvation and fatty acid metabolism [42], it is possible that other sex-dependent hormones known to be involved in the regulation of energy homeostasis and lipid metabolism such as leptin, insulin and ghrelin, may affect autophagy-related gene expression as previously reported in mammals [51–54].

Interestingly, we also found that the expression of autophagy-related genes is tissue- and genotype-dependent in male Japanese quails. These quail lines were divergently selected for circulating corticosterone response to restraint stress [33]. The high stress or sensitive line had, in general, high plasma corticosterone levels, high mortality, increased bacterial colonization, high fearfulness, low sexual activity and high stress-induced osteoporosis compared to their low stress (resistant) counterpart [33,55–58]. Although the role and the regulation of the autophagy-related genes are still unknown in avian species, the high expression of most studied genes in the R line suggest that the autophagy might be a protective mechanism in response to stress and might be involved in the aforementioned behavior and physiological differences between the two quail lines. For instance, inhibition of autophagy has been shown to aggravate the effect of glucocorticoid on cell viability of chicken primary osteocytes [59] which may explain the high stress-induced corticosterone levels and osteoporosis in S lines. Furthermore, sperm quality has been linked to autophagy (LC3B) processing [60] which may support the heightened reproductive efficiency of male R quails which are characterized by increased testis expression of beclin1 and Atg3 genes. Since autophagy has been, recently, reported to be associated with the development of learning and memory in fear conditioning [61], the low expression of Atg13 and Atg16L1 in the brain of S line might be involved in their high fearfulness. Intriguingly, the expression of autophagosome initiation (Ambra 1, UVRAG, Atg13, Atg9a) and elongation genes (Atg3 and Atg4a), except Atg5, was higher in the intestine of S line compared to the R line. The biological significance of this differential expression is not known at this time and further studies are warranted.

In conclusion, the characterization herein of several genes involved in autophagosome initiation and elongation will open new research avenues to understand the regulation and the roles of autophagy in avian species maintained under physiological and pathophysiological conditions. Further studies are warranted to identify and characterize genes involved in autophagosome maturation in birds. The present study also provides proof of principle evidence supporting gender- and genotype-dependent differences in autophagy in avian species and better insight into the underlying mechanisms may ultimately help to develop new management tools for poultry production improvement. The quail lines may be a useful model to study stress-related disorder in human and develop therapeutic strategies.

Author Contributions

Conceived and designed the experiments: SD. Performed the experiments: AP SK EG KL JM. Analyzed the data: AP SK EG. Contributed reagents/materials/analysis tools: NA WB SD. Contributed to the writing of the manuscript: SD.

References

1. Levine B, Kroemer G (2008) Autophagy in the pathogenesis of disease. Cell 132: 27–42.
2. Klionsky DJ (2005) The molecular machinery of autophagy: unanswered questions. J Cell Sci 118: 7–18.
3. Massey AC, Zhang C, Cuervo AM (2006) Chaperone-mediated autophagy in aging and disease. Curr Top Dev Biol 73: 205–235.
4. Hasson SA, Kane LA, Yamano K, Huang CH, Sliter DA, et al. (2013) High-content genome-wide RNAi screens identify regulators of parkin upstream of mitophagy. Nature 504: 291–295.
5. Tasdemir E, Maiuri MC, Tajeddine N, Vitale I, Criollo A, et al. (2007) Cell cycle-dependent induction of autophagy, mitophagy and reticulophagy. Cell Cycle 6: 2263–2267.
6. Mizushima N, Levine B, Cuervo AM, Klionsky DJ (2008) Autophagy fights disease through cellular self-digestion. Nature 451: 1069–1075.
7. Hamasaki M, Furuta N, Matsuda A, Nezu A, Yamamoto A, et al. (2013) Autophagosomes form at ER-mitochondria contact sites. Nature 495: 389–393.
8. Yorimitsu T, Klionsky DJ (2005) Autophagy: molecular machinery for self-eating. Cell Death Differ 12 Suppl 2: 1542–1552.

9. Furuya N, Yu J, Byfield M, Pattingre S, Levine B (2005) The evolutionarily conserved domain of Beclin 1 is required for Vps34 binding, autophagy and tumor suppressor function. Autophagy 1: 46–52.

10. Fimia GM, Stoykova A, Romagnoli A, Giunta L, Di Bartolomeo S, et al. (2007) Ambra1 regulates autophagy and development of the nervous system. Nature 447: 1121–1125.

11. Liang C, Feng P, Ku B, Dotan I, Canaani D, et al. (2006) Autophagic and tumour suppressor activity of a novel Beclin1-binding protein UVRAG. Nat Cell Biol 8: 688–699.

12. Takahashi Y, Coppola D, Matsushita N, Cualing HD, Sun M, et al. (2007) Bif-1 interacts with Beclin 1 through UVRAG and regulates autophagy and tumorigenesis. Nat Cell Biol 9: 1142–1151.

13. Hara T, Takamura A, Kishi C, Iemura S Natsume T, et al. (2008) FIP200, a ULK-interacting protein, is required for autophagosome formation in mammalian cells. J Cell Biol 181: 497–510.

14. Chang YY, Neufeld TP (2009) An Atg1/Atg13 complex with multiple roles in TOR-mediated autophagy regulation. Mo Biol Cell 20 2004–2014.

15. Ganley IG, Lam du H, Wang J, Ding X, Chen S, et al. (2009) ULK1.ATG13.FIP200 complex mediates mTOR signaling and is essential for autophagy. J Biol Chem 284: 12297–12305.

16. Jung CH, Jun CB, Ro SH, Kim YM, Otto NM, et al. (2009) ULK-Atg13-FIP200 complexes mediate mTOR signaling to the autophagy machinery. Mol Biol Cell 20: 1992–2003.

17. Mizushima N, Kuma A, Kobayashi Y, Yamamoto A, Matsubae M, et al. (2003) Mouse Apg16L, a novel WD-repeat protein, targets to the autophagic isolation membrane with the Apg12-Apg5 conjugate. J Cell Sci 116: 1679–1688.

18. Hemelaar J, Lelyveld VS, Kessler BM, Ploegh HL (2003) A single protease, Apg4B, is specific for the autophagy-related ubiquitin-like proteins GATE-16, MAP1-LC3, GABARAP, and Apg8L. J Biol Chem 278 51841–51850.

19. Kabeya Y, Mizushima N, Ueno T, Yamamoto A, Kirisako T, et al. (2000) LC3, a mammalian homologue of yeast Apg8p, is localized in autophagosome membranes after processing. EMBO J 19: 5720–5728.

20. Kimura S, Noda T, Yoshimori T (2008) Dynein-dependent movement of autophagosomes mediates efficient encounters with lysosomes. Cell Struct Funct 33: 109–122.

21. Ravikumar B, Acevedo-Arozena A, Imarisio S, Berger Z, Vacher C, et al. (2005) Dynein mutations impair autophagic clearance of aggregate-prone proteins. Nat Genet 37: 771–776.

22. Lee JA, Beigneux A, Ahmad ST, Young SG, Gao FB (2007) ESCRT-III dysfunction causes autophagosome accumulation and neurodegeneration. Curr Biol 17: 1561–1567.

23. Itakura E, Kishi-Itakura C, Mizushima N (2012) The hairpin-type tail-anchored SNARE syntaxin 17 targets to autophagosomes for fusion with endosomes/lysosomes. Cell 151: 1256–1269.

24. Nair U, Jotwani A, Geng J, Gammoh N, Richerson D et al. (2011) SNARE proteins are required for macroautophagy. Cell 146: 290–302.

25. Gutierrez MG, Munafo DB, Beron W, Colombo MI (2004) Rab7 is required for the normal progression of the autophagic pathway in mammalian cells. J Cell Sci 117: 2687–2697.

26. Jager S, Bucci C, Tanida I, Ueno T, Kominami E, et al. (2004) Role for Rab7 in maturation of late autophagic vacuoles. J Cell Sci 117: 4837–4848.

27. Liang C, Lee JS, Inn KS, Gack MU, Li Q, et al. (2008) Beclin1-binding UVRAG targets the class C Vps complex to coordinate autophagosome maturation and endocytic trafficking. Nat Cell Biol 10: 776–787.

28. Matsunaga K, Saitoh T, Tabata K, Omori H, Satoh T, et al. (2009) Two Beclin 1-binding proteins, Atg14L and Rubicon, reciprocally regulate autophagy at different stages. Nat Cell Biol 11: 385–396.

29. Hara T, Nakamura K, Matsui M, Yamamoto A, Nakahara Y, et al. (2006) Suppression of basal autophagy in neural cells causes neurodegenerative disease in mice. Nature 441: 885–889.

30. Rosenfeldt MT, O'Prey J, Morton JP, Nixon C, MacKay G, et al. (2013) p53 status determines the role of autophagy in pancreatic tumour development. Nature 504: 296–300.

31. Levine B, Mizushima N, Virgin HW (2011) Autophagy in immunity and inflammation. Nature 469: 323–335.

32. Singh R, Kaushik S, Wang Y, Xiang Y, Novak I, et al. (2009) Autophagy regulates lipid metabolism. Nature 458: 1131–1135.

33. Satterlee DG, Johnson WA (1988) Selection of Japanese quail for contrasting blood corticosterone response to immobilization. Poult Sci 67: 25–32.

34. Schmittgen TD, Livak KJ (2008) Analyzing real-time PCR data by the comparative C(T) method. Nat Protoc 3: 1101–1108.

35. Thompson JD, Higgins DG, Gibson TJ (1994) CLUSTAL W: improving the sensitivity of progressive multiple sequence alignment through sequence weighting, position-specific gap penalties and weight matrix choice. Nucleic Acids Res 22: 4673–4680.

36. Edgar RC (2004) MUSCLE: multiple sequence alignment with high accuracy and high throughput. Nucleic Acids Res 32: 1792–1797.

37. Tamura K, Stecher G, Peterson D, Filipski A, Kumar S (2013) MEGA6: Molecular Evolutionary Genetics Analysis version 6.0. Mol Biol Evol 30: 2725–2729.

38. Kroemer G, Levine B (2008) Autophagic cell death: the story of a misnomer. Nat Rev Mol Cell Biol 9: 1004–1010.

39. Ohsumi Y (2001) Molecular dissection of autophagy: two ubiquitin-like systems. Nat Rev Mol Cell Biol 2: 211–216.

40. Nakatogawa H, Suzuki K, Kamada Y, Ohsumi Y (2009) Dynamics and diversity in autophagy mechanisms: lessons from yeast. Nat Rev Mol Cell Biol 10: 458–467.

41. Amaravadi RK, Yu D, Lum JJ, Bui T, Christophorou MA, et al. (2007) Autophagy inhibition enhances therapy-induced apoptosis in a Myc-induced model of lymphoma. J Clin Invest 117: 326–336.

42. Du L, Hickey RW, Bayir H, Watkins SC, Tyurin VA, et al. (2009) Starving neurons show sex difference in autophagy. J Biol Chem 284: 2383–2396.

43. Komatsu M, Waguri S, Koike M, Sou YS, Ueno T, et al. (2007) Homeostatic levels of p62 control cytoplasmic inclusion body formation in autophagy-deficient mice. Cell 131: 1149–1163.

44. Coto-Montes A, Tomas-Zapico C, Martinez-Fraga J, Vega-Naredo I, Sierra V, et al. (2009) Sexual autophagic differences in the androgen-dependent flank organ of Syrian hamsters. J Androl 30: 113–121.

45. Vega-Naredo I, Caballero B, Sierra V, Huidobro-Fernandez C, de Gonzalo-Calvo D, et al. (2009) Sexual dimorphism of autophagy in Syrian hamster Harderian gland culminates in a holocrine secretion in female glands. Autophagy 5: 1004–1017.

46. Sobolewska A, Gajewska M, Zarzynska J, Gajkowska B, Motyl T (2009) IGF-I, EGF, and sex steroids regulate autophagy in bovine mammary epithelial cells via the mTOR pathway. Eur J Cell Biol 88: 117–130.

47. Straface E, Vona R, Gambardella L, Ascione B, Marino M, et al. (2009) Cell sex determines anoikis resistance in vascular smooth muscle cells. FEBS Lett 583: 3448–3454.

48. Wang M, Crisostomo PR, Markel TA, Wang Y, Meldrum DR (2008) Mechanisms of sex differences in TNFR2-mediated cardioprotection. Circulation 118: S38–45.

49. Yoon S, Woo SU, Kang JH, Kim K, Kwon MH, et al. (2010) STAT3 transcriptional factor activated by reactive oxygen species induces IL6 in starvation-induced autophagy of cancer cells. Autophagy 6: 1125–1138.

50. Mehrpour M, Esclatine A, Beau I, Codogno P (2010) Overview of macroautophagy regulation in mammalian cells. Cell Res 20: 748–762.

51. Slupecka M, Wolinski J, Gajewska M, Pierzynowski SG (2014) Enteral leptin administration affects intestinal autophagy in suckling piglets. Domest Anim Endocrinol 46: 12–19.

52. Malik SA, Marino G, BenYounes A, Shen S, Harper F, et al. (2011) Neuroendocrine regulation of autophagy by leptin. Cell Cycle 10: 2917–2923.

53. Rodriguez A, Gomez-Ambrosi J, Catalan V, Rotellar F, Valenti V, et al. (2012) The ghrelin O-acyltransferase-ghrelin system reduces TNF-alpha-induced apoptosis and autophagy in human visceral adipocytes. Diabetologia 55: 3038–3050.

54. Yansong W, Wei W, Dongguo L, Mi L, Peipei W, et al. (2014) IGF-1 Alleviates NMDA-Induced Excitotoxicity in Cultured Hippocampal Neurons Against Autophagy via the NR2B/PI3K-AKT-mTOR Pathway. J Cell Physiol.

55. Huff GR, Huff WE, Wesley IV, Anthony NB, Satterlee DG (2013) Response of restraint stress-selected lines of Japanese quail to heat stress and Escherichia coli challenge. Poult Sci 92: 603–611.

56. Satterlee DG, Roberts ED (1990) The influence of stress treatment on femur cortical bone porosity and medullary bone status in Japanese quail selected for high and low blood corticosterone response to stress. Comp Biochem Physiol A Comp Physiol 95: 401–405.

57. Satterlee DG, Marin RH (2006) Stressor-induced changes in open-field behavior of Japanese quail selected for contrasting adrenocortical responsiveness to immobilization. Poult Sci 85: 404–409.

58. Davis KA, Schmidt JB, Doescher RM, Satterlee DG (2008) Fear responses of offspring from divergent quail stress response line hens treated with corticosterone during egg formation. Poult Sci 87: 1303–1313.

59. Xia X, Kar R, Gluhak-Heinrich J, Yao W, Lane NE, et al. (2010) Glucocorticoid-induced autophagy in osteocytes. J Bone Miner Res 25: 2479–2488.

60. Bolanos JM, Moran AM, da Silva CM, Davila MP, Munoz PM, et al. (2014) During cooled storage the extender influences processed autophagy marker light chain 3 (LC3B) of stallion spermatozoa. Anim Reprod Sci 145: 40–46.

61. Yang DS, Stavrides P, Mohan PS, Kaushik S, Kumar A, et al. (2011) Reversal of autophagy dysfunction in the TgCRND8 mouse model of Alzheimer's disease ameliorates amyloid pathologies and memory deficits. Brain 134: 258–277.

10

Vertebrate Ssu72 Regulates and Coordinates 3′-End Formation of RNAs Transcribed by RNA Polymerase II

Shotaro Wani[1,◊], Masamichi Yuda[2,◊], Yosuke Fujiwara[1], Masaya Yamamoto[1], Fumio Harada[2], Yoshiaki Ohkuma[1], Yutaka Hirose[1]*

1 Laboratory of Gene Regulation, Graduate School of Medicine and Pharmaceutical Sciences, University of Toyama, Sugitani, Toyama, Japan, 2 Department of Molecular and Cellular Biology, Cancer Research Institute, Kanazawa University, Kakuma-machi, Kanazawa, Japan

Abstract

In eukaryotes, the carboxy-terminal domain (CTD) of the largest subunit of RNA polymerase II (Pol II) is composed of tandem repeats of the heptapeptide YSPTSPS, which is subjected to reversible phosphorylation at Ser2, Ser5, and Ser7 during the transcription cycle. Dynamic changes in CTD phosphorylation patterns, established by the activities of multiple kinases and phosphatases, are responsible for stage-specific recruitment of various factors involved in RNA processing, histone modification, and transcription elongation/termination. Yeast Ssu72, a CTD phosphatase specific for Ser5 and Ser7, functions in 3′-end processing of pre-mRNAs and in transcription termination of small non-coding RNAs such as snoRNAs and snRNAs. Vertebrate Ssu72 exhibits Ser5- and Ser7-specific CTD phosphatase activity *in vitro*, but its roles in gene expression and CTD dephosphorylation *in vivo* remain to be elucidated. To investigate the functions of vertebrate Ssu72 in gene expression, we established chicken DT40 B-cell lines in which Ssu72 expression was conditionally inactivated. Ssu72 depletion in DT40 cells caused defects in 3′-end formation of U2 and U4 snRNAs and *GAPDH* mRNA. Surprisingly, however, Ssu72 inactivation increased the efficiency of 3′-end formation of non-polyadenylated replication-dependent histone mRNA. Chromatin immunoprecipitation analyses revealed that Ssu72 depletion caused a significant increase in both Ser5 and Ser7 phosphorylation of the Pol II CTD on all genes in which 3′-end formation was affected. These results suggest that vertebrate Ssu72 plays positive roles in 3′-end formation of snRNAs and polyadenylated mRNAs, but negative roles in 3′-end formation of histone mRNAs, through dephosphorylation of both Ser5 and Ser7 of the CTD.

Editor: Akio Kanai, Keio University, Japan

Funding: This work was supported in part by a Grant-in-Aid for Scientific Research on Priority Areas from the Ministry of Education, Science, Sports, and Culture of Japan and a Grant-in-Aid for Solution-Oriented Research for Science and Technology from the Japan Science and Technology Agency. The funders had no role in study design, data collection and analysis, decision to publish, or preparation of the manuscript.

Competing Interests: The authors have declared that no competing interests exist.

* Email: yh620@pha.u-toyama.ac.jp

◊ These authors contributed equally to this work.

Introduction

In eukaryotes, all protein-coding genes and many non-coding RNA genes are transcribed by RNA polymerase II (Pol II), which consists of 12 subunits. The largest subunit of Pol II possesses the catalytic activity and also contains a unique C-terminal domain (CTD) composed of multiple repeats of the evolutionarily conserved heptapeptide sequence Tyr1-Ser2-Pro3-Thr4-Ser5-Pro6-Ser7 (YSPTSPS) [1]. The repeat number varies between species, ranging from 26 in yeast to 52 in vertebrates [1]. The CTD, which is essential for cell viability, is subjected to reversible phosphorylation during the transcription cycle, predominantly at Ser2, Ser5, and Ser7 of the repeats [2,3]. Multiple kinases and phosphatases act on the CTD in a transcription stage-specific manner, thereby generating different CTD phosphorylation patterns along transcribed genes [2,3]. Various nuclear factors involved in RNA processing, histone modification, and transcription elongation/termination can bind the CTD in a phosphorylation pattern-specific manner, providing a basis for coordination between transcription and other processes related to gene expression, such as histone modification and RNA processing [4–6].

Prior to transcription initiation, the pre-initiation complex preferentially recruits Pol II enzymes with a hypophosphorylated CTD [7]. Upon initiation, Ser5 is phosphorylated by CDK7, a subunit of the general transcription factor TFIIH [4,6]. Phosphorylated Ser5 (Ser5P) promotes the recruitment of the capping enzyme and histone methyltransferase Set1 to the early transcription complex [4,6]. During the transition from initiation to early elongation, Ser2 is phosphorylated by P-TEFb (CDK9/Cyclin T) [8]. As transcription proceeds from 5′ to 3′ direction, Ser2P levels are gradually increased through the actions of P-TEFb [8] and CDK12/13 [9]; concurrently, Ser5P levels decline [2,5,10]. Ser2P promotes the recruitment of a histone methyltransferase, 3′-end processing factors, and transcription termination factors to the elongating Pol II [2,3,6,10]. TFIIH also phosphorylates Ser7 residues near promoters [11–13]. Ser7P participates in snRNA transcription and 3′-end processing by specifically recruiting Integrator complex and the putative CTD phosphatase RPAP2

[14,15]. Furthermore, a recent study suggested that Thr4P is involved in 3'-end processing of replication-dependent histone mRNAs [16]. Thus, the dynamically phosphorylated CTD temporally couples transcription with other nuclear processes by serving as a scaffold for recruitment of various proteins involved in transcription, chromatin modification, and RNA processing [2,3,10]. Therefore, regulation of CTD phosphorylation patterns during the transcription cycle by CTD kinases and phosphatases is crucial for proper gene expression.

Ssu72 is a well-studied CTD phosphatase in yeast. The *Ssu72* (suppressor of *Sua7* 2) gene was originally identified in budding yeast as an essential gene that genetically and physically interacts with the general transcription factor TFIIB (*Sua7*) and affects the precision of transcription start site selection [17]. Subsequently, Ssu72 was shown to be a subunit of cleavage and polyadenylation factor (CPF) holo-complex that is involved in 3'-end processing of some pre-mRNAs and in transcription termination of small non-coding RNAs such as snoRNAs, snRNAs, and cryptic unstable transcripts (CUTs) [18–22]. Although yeast Ssu72 was initially shown to be a Ser5P-specific CTD phosphatase [23], recent studies have demonstrated that it also exhibits Ser7P phosphatase activity *in vitro*. Consistent with this, loss of Ssu72 *in vivo* results in an increase in the phosphorylation level of both Ser5 and Ser7, both at snoRNA genes and in the 3' regions of mRNA genes [24,25].

The mammalian ortholog of yeast Ssu72 was originally identified as a binding partner of the tumor suppressor RB [26]. Although mammalian Ssu72 is very similar at the sequence level to yeast Ssu72 and can also associate with TFIIB, it is unable to rescue a lethal *ssu72* mutation in yeast, and its suppression does not affect cell proliferation or viability of mammalian cultured cells [26]. Thus, mammalian Ssu72 may share a subset of the functions of the yeast protein but also exert specific functions in mammalian cells. Although recent studies demonstrated that human Ssu72, like its yeast counterpart, exhibits Ser5P and Ser7P-specific CTD phosphatase activity *in vitro* [27,28], its *in vivo* roles in CTD dephosphorylation and gene expression remain to be elucidated.

To investigate the functions of vertebrate Ssu72 at the cellular level, we developed chicken DT40 B-cell lines [29] in which Ssu72 expression is conditionally inactivated. Ssu72 depletion caused defects in 3'-end formation of U2 and U4 snRNAs and *GAPDH* mRNA. Unexpectedly, however, Ssu72 inactivation increased the efficiency of 3'-end formation of non-polyadenylated replication-dependent histone mRNA. Furthermore, Ssu72 depletion caused a significant increase in both Ser5 and Ser7 phosphorylation on all genes in which 3'-end processing was affected. These results suggest that vertebrate Ssu72 plays positive roles in 3'-end formation of snRNAs and polyadenylated mRNAs, but a negative role in 3'-end formation of histone mRNAs, by dephosphorylating both Ser5P and Ser7P of the RNA Pol II CTD.

Results

Establishment of conditional Ssu72-knockout DT40 cell lines

To investigate the Ssu72 functions in vertebrate cells by gene targeting, we utilized the chicken B-cell line DT40, which exhibits exceptionally high homologous recombination rates and has been widely used as a model cell line for genetic studies [29]. Because our initial attempts to establish homozygous knockouts failed repeatedly, we concluded that Ssu72 is essential for cell viability. Therefore, we generated DT40 cell lines in which Ssu72 expression was conditionally inactivated using the Tet-off regulatory system. To prepare the targeting constructs, we PCR-

amplified two genomic DNA fragments, a 2 kb region upstream of the first exon of the Ssu72 gene (ENSGALG00000001489) and a 3.9 kb region downstream of the second exon, and then cloned these fragments into the flanking regions of drug-resistance cassettes (Fig. 1A). These knockout constructs were sequentially introduced into a parental DT40 cell line that stably expresses the Tet repressor fused to the herpes simplex virus VP16 activation domain, as well as a chicken Ssu72 transgene under control of the tetracycline-repressive promoter ($Ssu72^{+/+}$/FLAG-Ssu72). Hereafter, we refer to this parental strain as WT (+/+).

After the second round of gene targeting, we established three independent cell lines whose two wild-type alleles of the *Ssu72* gene were replaced by targeting constructs, but which still expressed the Tet-repressible Ssu72 transgene ($Ssu72^{-/-}$/FLAG-Ssu72). In this report, we refer to these lines as P1 (−/−), P3 (−/−), and P4 (−/−). Southern blot analysis revealed that the first (lane 1) and second (lanes 2–4) wild-type alleles of the Ssu72 gene had been successfully targeted (Fig. 1B). Expression of Ssu72 was completely abolished at the RNA level in the three homozygous mutants (Fig. 1C). Protein expression from the Ssu72 transgene in one of the homozygous mutant cell lines, P3 (−/−), was tightly repressed within 2 days after doxycycline (Dox) administration (Fig. 1D), and the kinetics of Ssu72 repression following Dox addition were similar in the other two mutants, P1 (−/−) and P4 (−/−) (data not shown).

Ssu72 is essential for DT40 cell growth

Ssu72 is highly conserved among eukaryotes and essential for yeast viability [17]. However, in mammals, siRNA-mediated knockdown of Ssu72 expression does not affect cell proliferation or viability [26]. To investigate the functional roles of Ssu72 in proliferation of vertebrate cells, we used the conditional knockout DT40 mutant cells described above to examine the effect of Ssu72 depletion on cell growth and viability. In the absence of Dox, the proliferation rate of the P1 (−/−) cell line was comparable to that of WT (+/+) (Figs. 2A and 2B, open squares). However, in the presence of Dox, the growth rate of this mutant was markedly reduced after 3 days, and the mutant cells died 6 days after Dox administration (Fig. 2B, filled squares). The other two mutants, P3 (−/−) and P4 (−/−) also exhibited similar growth patterns (data not shown). Proliferation of the mutants recovered if Dox was removed from culture media, resulting in restoration of Ssu72 protein expression (data not shown). The Dox concentration used in this study did not affect cell growth of WT (+/+), as shown in Fig. 2A.

We also examined the effects of Ssu72 depletion on cell-cycle distribution by fluorescence-activated cell sorting (FACS) analysis. Dox treatment for 6 days significantly increased the proportion of the sub-G1 population in the conditional mutants (from 3.1% to 38.4%), but not in wild-type cells (from 3.3% to 5.3%) (Fig. 2C), suggesting that Ssu72 depletion may induce apoptosis in DT40 cells. Taken together, these results suggest that Ssu72 is essential for cell proliferation and viability, at least in cultured chicken cells.

Ssu72 depletion causes a modest increase in CTD phosphorylation

Yeast Ssu72 exhibits CTD phosphatase activity toward Ser5P and Ser7P both *in vitro* and *in vivo* [24,25]. Although human and fly Ssu72 also exhibit Ser5P and Ser7P phosphatase activity *in vitro* [27,28,30,31], the involvement of Ssu72 in dephosphorylation of the phosphorylated CTD *in vivo* has not been examined in vertebrate cells. Therefore, we next investigated whether Ssu72 depletion affects the CTD phosphorylation level *in vivo*. Total proteins extracted from P3 (−/−) cells every 24 hours after Dox

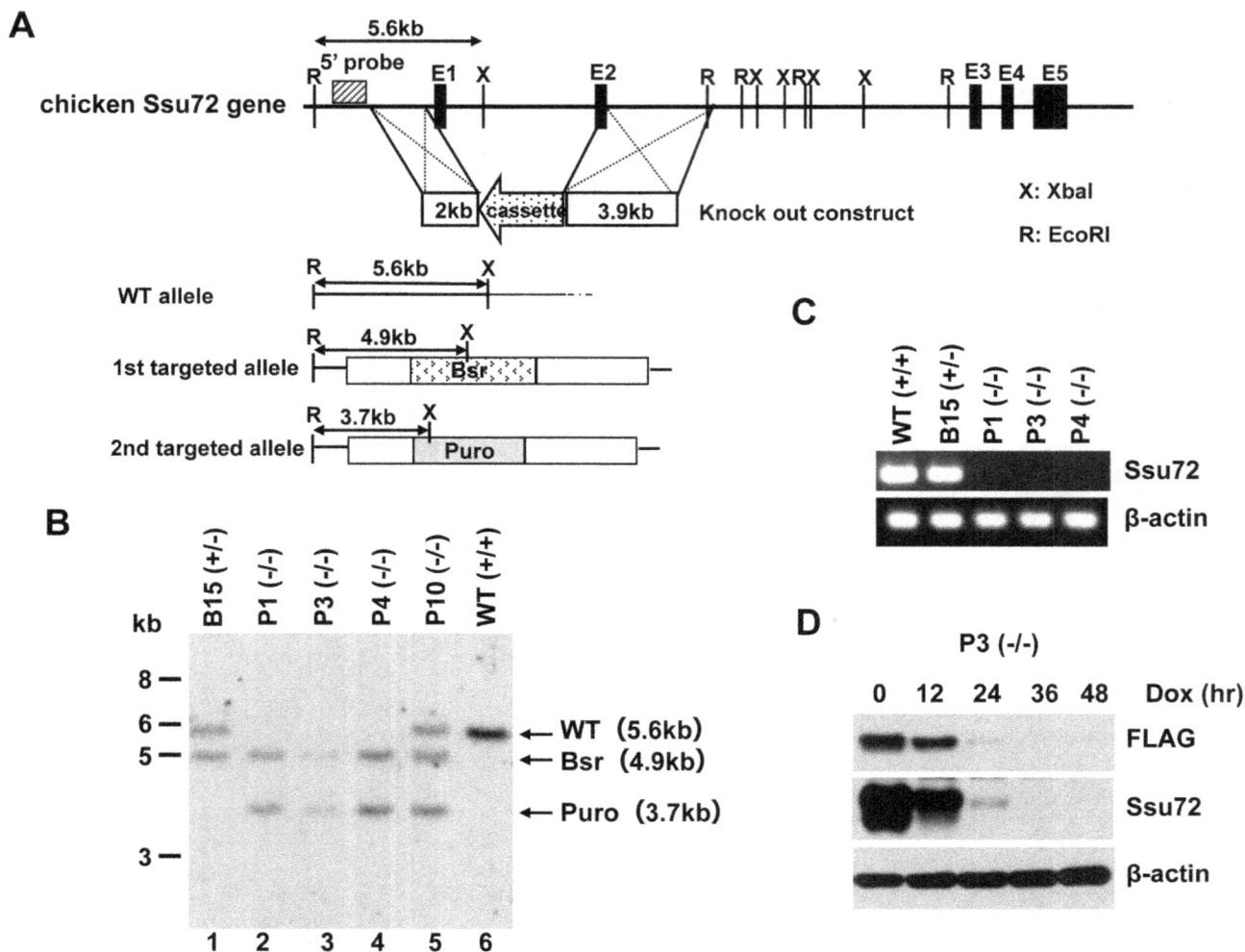

Figure 1. Establishment of conditional Ssu72-knockout DT40 cell lines. (A) Schematic representations of the chicken Ssu72 genomic fragment, knockout constructs, and configuration of the targeted alleles. Exons are shown as black boxes (E1–5), and the location of the 5′ probe used for Southern blotting is shown as a grey box. The double-headed arrows above the genes indicate length (in nucleotides). The *XbaI* and *EcoRI* restriction sites are indicated by vertical lines labeled X and R, respectively. (B) Southern blot analysis of wild-type (WT), heterozygous mutant (B15), homozygous mutant (P1, P2, P3), and unanticipated rearranged mutant (P10) clones. Genomic DNA obtained from each clone was digested with *XbaI* and *EcoRI*, and then hybridized with the 5′ probe shown in panel A. (C) RT-PCR analysis of the wild-type and mutant clones using primer pairs specific for the indicated gene. (D) Immunoblotting analysis of DT40 P3 (−/−) whole-cell extracts treated with Dox for the indicated times, using the indicated antibodies. Western blotting of β-actin was used as to confirm equal protein loading.

addition were subjected to Western blotting using phosphorylation-dependent monoclonal antibodies specific for Ser2P (3E10), Ser5P (3E8), and Ser7P (4E12) [32]. Total levels of the largest subunit of Pol II (Rpb1) were detected using antibodies against the N-terminal region of Rpb1 (N-20) (Figure 3A). Quantification of Western blots by using a luminescent image analyzer revealed that Ssu72 depletion caused a gradual increase of Ser5/7P levels whereas Ser2P levels were relatively constant after Dox addition (Figure 3B). Significantly, Ser7P levels at day 4 increased about 1.7-fold whereas Ser5P levels increased 1.2-fold. These results suggest that Ssu72 may contribute to Ser7P dephosphorylation more significantly than Ser5P *in vivo*. In a previous report, Ssu72 depletion in budding yeast resulted in a significant increase of the Ser5P level on total Pol II *in vivo* [23], indicating that Ssu72 may be a major Ser5P-CTD phosphatase. However, in chicken cells, Ssu72 depletion exerted only a modest effect on the phosphorylation levels of Ser5. These results suggested that in vertebrate cells, although Ssu72 may participate to some extent in

dephosphorylation of Ser5P, this serine residue is primarily dephosphorylated *in vivo* by other CTD phosphatases, such as SCP1.

Ssu72 is required for efficient 3′-end formation of snRNAs and mRNA in DT40 cells

In yeast, Ssu72 participates in 3′-end formation of small non-coding RNAs transcribed by Pol II such as snoRNAs, snRNAs, and CUTs *in vivo* [18–21]. Yeast Ssu72 is also implicated in 3′ processing of specific mRNA both *in vitro* and *in vivo* [18–22]. Although a recent *in vitro* study demonstrated that mammalian Ssu72 is involved in transcription-coupled polyadenylation of model pre-mRNAs [28], it remains unclear whether vertebrate Ssu72 is involved in 3′-end formation of Pol II transcripts *in vivo*. Using the mutant DT40 cell lines, we next investigated whether Ssu72 inactivation affects 3′-end formation of two types of Pol II transcripts, poly(A)-containing mRNA and spliceosomal snRNA, whose 3′ ends are formed by distinct mechanisms (Fig. 4A) [5]. To

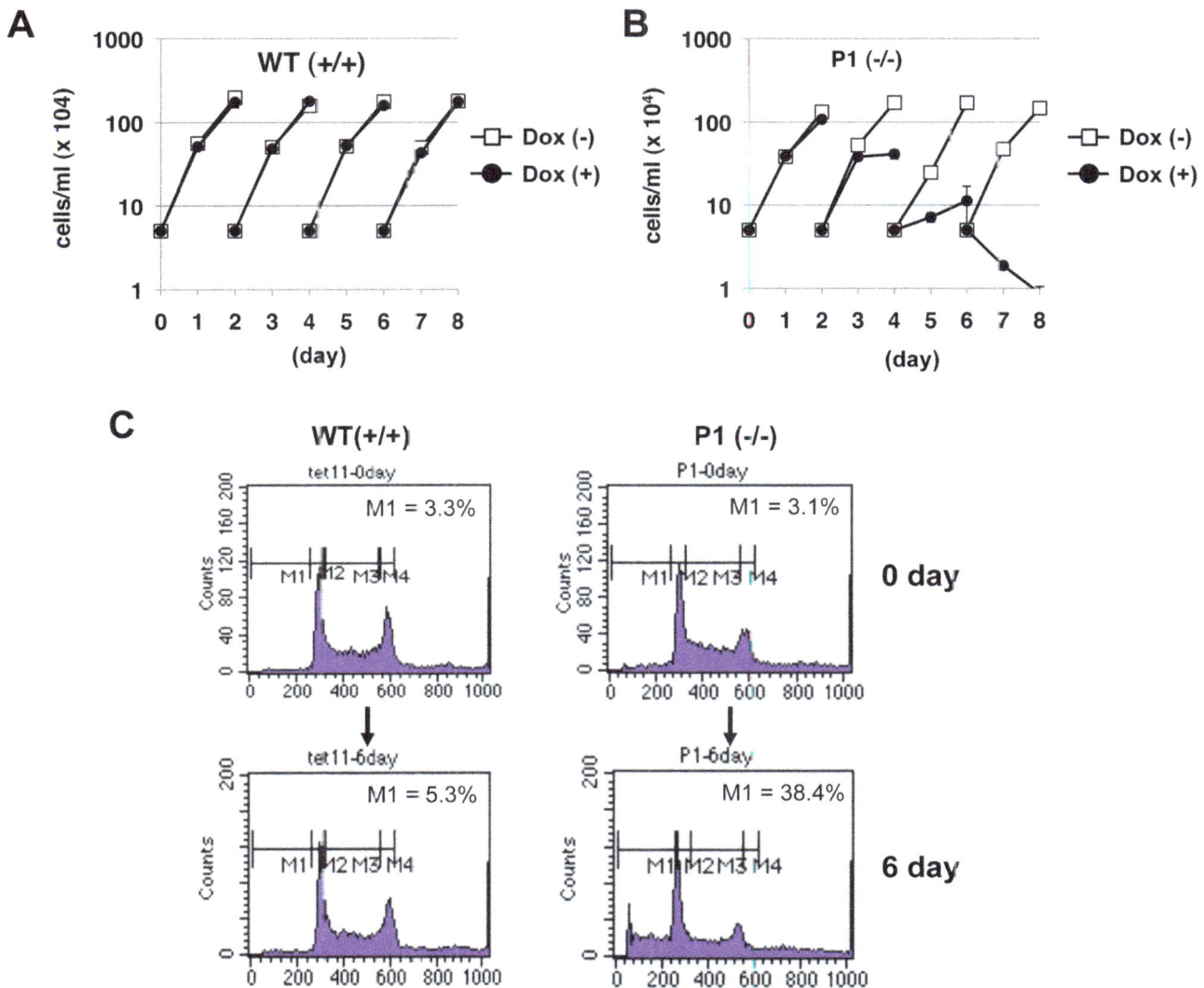

Figure 2. Ssu72 is essential for cell proliferation and viability in DT40 cells. (A, B) Growth curve of DT40 wild-type (WT) and conditional mutant [P1 (−/−)] DT40 cells. WT (A) or P1 (−/−) (B) cells were seeded in triplicate in 12-well plates (5×10³ cells/well) in media with or without doxycycline (Dox), and split every 2 days Concentrations of live cells were determined by Giemsa staining and counted at the indicated time points. Cell density is shown on a logarithmic scale. (C) WT cells (left) or P1 (−/−) cells (right) treated with Dox for 6 days (lower) or untreated (upper) were subjected to FACS analysis after staining with propidium iodide. Gates M1, M2, M3, and M4, delimited by the vertical bars above the FACS traces, indicate the cell-cycle distributions of the sub-G1, G0/G1, S and G2 populations, respectively.

precisely evaluate the efficiency of the 3' processing of Pol II transcripts, we measured the expression levels of both total and unprocessed RNAs by quantitative RT-PCR analyses, and then determined the ratio of unprocessed RNA to total RNAs (i.e., precursor + mature RNAs). To detect unprocessed RNAs, we synthesized first-strand cDNAs by reverse transcription using a primer complementary to sequences downstream of the processing site in each precursor RNA (Fig. 4A, arrows), and then PCR-amplified the cDNAs using primer pairs spanning the processing sites (Fig. 4A, dashed lines). Total RNA levels were detected by amplification of coding regions from cDNAs synthesized using random hexamers.

To investigate the effects of Ssu72 depletion on 3'-end formation of Pol II genes, we first examined *GAPDH* mRNA and U2 and U4 snRNAs (Fig. 4). In metazoans, the 3' end of poly(A)-containing mRNA is formed by endonucleolytic cleavage of pre-mRNA 20–30 nucleotides (nts) downstream of the poly(A)

signal sequence, followed by polyadenylation [5]. By contrast, the primary snRNA transcript is processed by endonucleolytic cleavage just upstream of the 3' box sequence (Fig. 4A) [14]. We isolated RNA from P3 (−/−) cells every 24 hours following Dox administration, and then analyzed both precursor and total RNA levels by quantitative RT-PCR. For all three RNAs tested (*GAPDH* mRNA, U2 snRNA, and U4 snRNA), the ratio of unprocessed RNA to total RNA gradually increased as Ssu72 was depleted (Figs. 4B–4D). The ratios 4 days upon Dox addition were 2- to 4-fold higher than the ratios before Dox addition. Importantly, the levels of total RNAs were not significantly influenced by Dox treatment (Fig. S1). These data suggest that vertebrate Ssu72, like its yeast ortholog, is required for efficient 3'-end formation of at least some snRNAs and poly(A)+ mRNA. This is the first demonstration that vertebrate Ssu72 participates in 3'-end formation of Pol II transcripts *in vivo*.

A

B

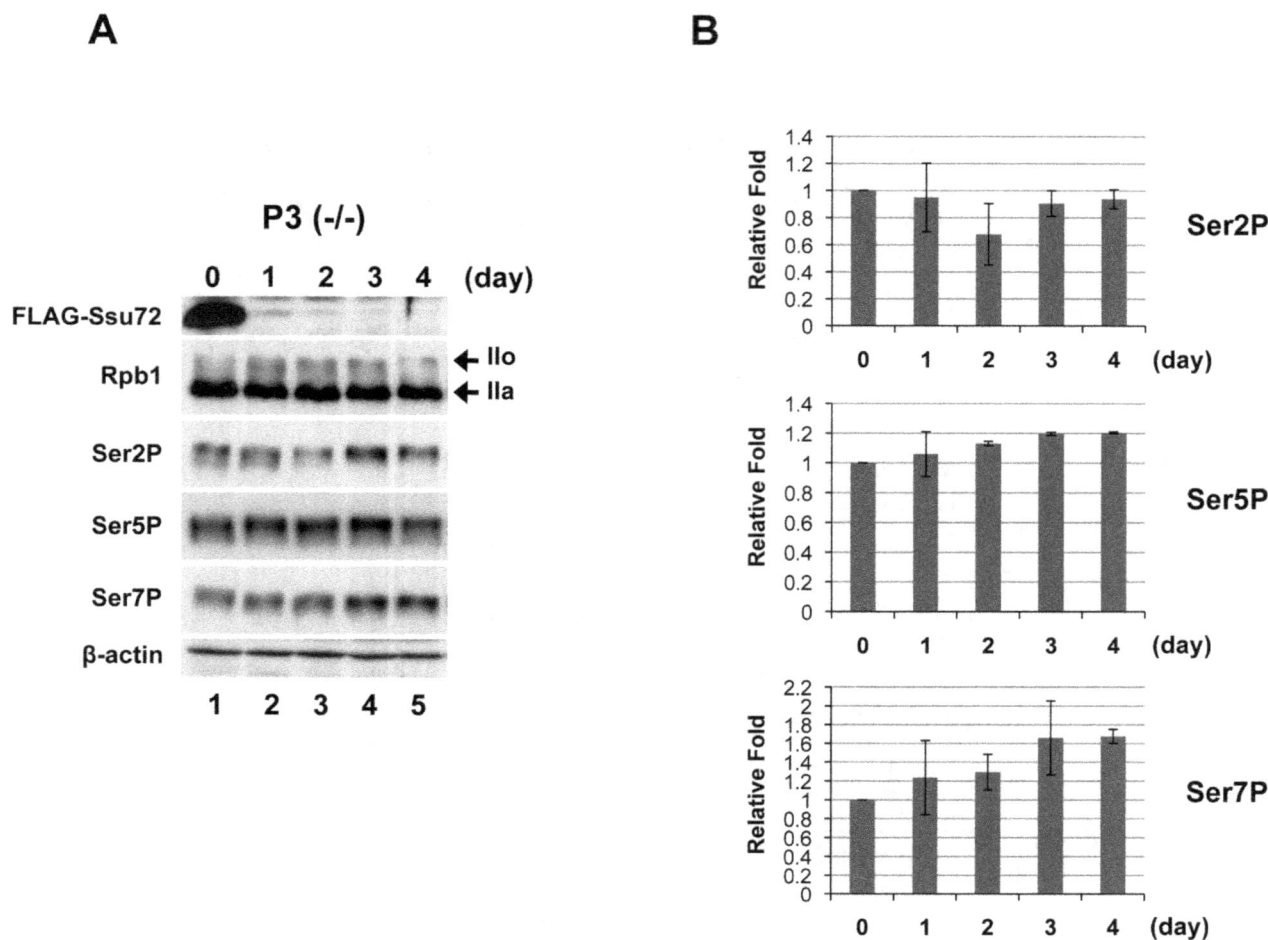

Figure 3. Ssu72 depletion results in a modest increase in CTD phosphorylation. (A) Whole-cell extracts from DT40 P3 (−/−) cells treated with Dox for the indicated number of days were separated by SDS-PAGE, and then analyzed by Immunoblotting with antibodies against FLAG-Ssu72 (anti-FLAG M2), the largest subunit (Rpb1) of Pol II (N20), Ser2P (3E10), Ser5P (3E8), Ser7P (4E12), and β-actin. Western blotting of β-actin was used to confirm equal protein loading. The positions corresponding to the hyper-phosphorylated form of Rpb1 (IIo) and hypo-phosphorylated form of Rpb1 (IIa) are indicated by arrows. (B) Fold change of phosphorylation levels of each serine (Ser2/5/7P) relative to the corresponding levels on day 0. The Western blot signals were quantified by using a luminescent image analyzer LAS-4000 mini (Fujifilm). The each value was normalized by total Pol II signal (Rpb1). Error bars represent standard deviations of two independent experiments.

Ssu72 suppresses 3′-end formation of replication-dependent histone mRNAs in DT40 cells

The results described above suggest that the functions of Ssu72 in 3′-end formation of snRNAs and mRNAs are evolutionarily conserved from yeast to chicken. The 3′ ends of most mRNAs are formed by cleavage and polyadenylation. However, in metazoans but not yeast, the 3′ ends of replication-dependent histone mRNAs are generated by a distinct process [33]. Primary histone mRNAs are cleaved just downstream of a conserved 3′ stem-loop sequence, but no poly(A) tail is added at the 3′ end of the upstream cleavage product (Fig. 5A). Recent studies showed that regulation of CTD phosphorylation participates in 3′-end processing of replication-dependent histone pre-mRNA [16]. Therefore, we examined whether Ssu72 is involved in 3′-end formation of histone mRNAs (Figs. 5B–5E). Using the assay described above, we used quantitative RT-PCR to determine the ratios of precursor mRNAs to total mRNAs of chicken histone H3 and H4 genes following Ssu72 depletion. Surprisingly, in contrast to the cases of snRNAs and polyadenylated mRNAs, the ratios of precursor RNAs to total RNAs for both H3 and H4 mRNAs dramatically

decreased upon depletion of Ssu72 (Figs. 5B and 5C). Notably, the relative precursor levels dropped rapidly, mirroring the reduction in Ssu72 protein level (Fig. 3). Even 1 day after Dox addition, the precursor ratio had decreased to ~20% of the ratio in untreated cells. Importantly, the levels of total RNAs (precursor + mature) were not dramatically influenced by Dox treatment (Fig. S2). Thus, Ssu72 depletion results in a significant increase in the efficiency of 3′-end formation of histone mRNAs.

In vertebrates, some replication-dependent histone genes contain not only a stem-loop sequence but also a typical polyadenylation signal, generally located downstream of the stem-loop sequence. When histone-specific processing is inhibited or regulated under certain cellular conditions, a portion of transcripts from these histone genes are processed by cleavage and polyadenylation at the downstream polyadenylation signal [16,34]. To further investigate the possible involvement of Ssu72 in 3′-end formation of histone mRNAs, we examined the effect of Ssu72 depletion on the level of polyadenylated histone H3 and H4 mRNAs, which can be detected using oligo dT-primed first-strand cDNAs. Although polyadenylated histone mRNAs were detectable in the absence of Dox (Figs. 5D and 5E, day 0), their levels quickly

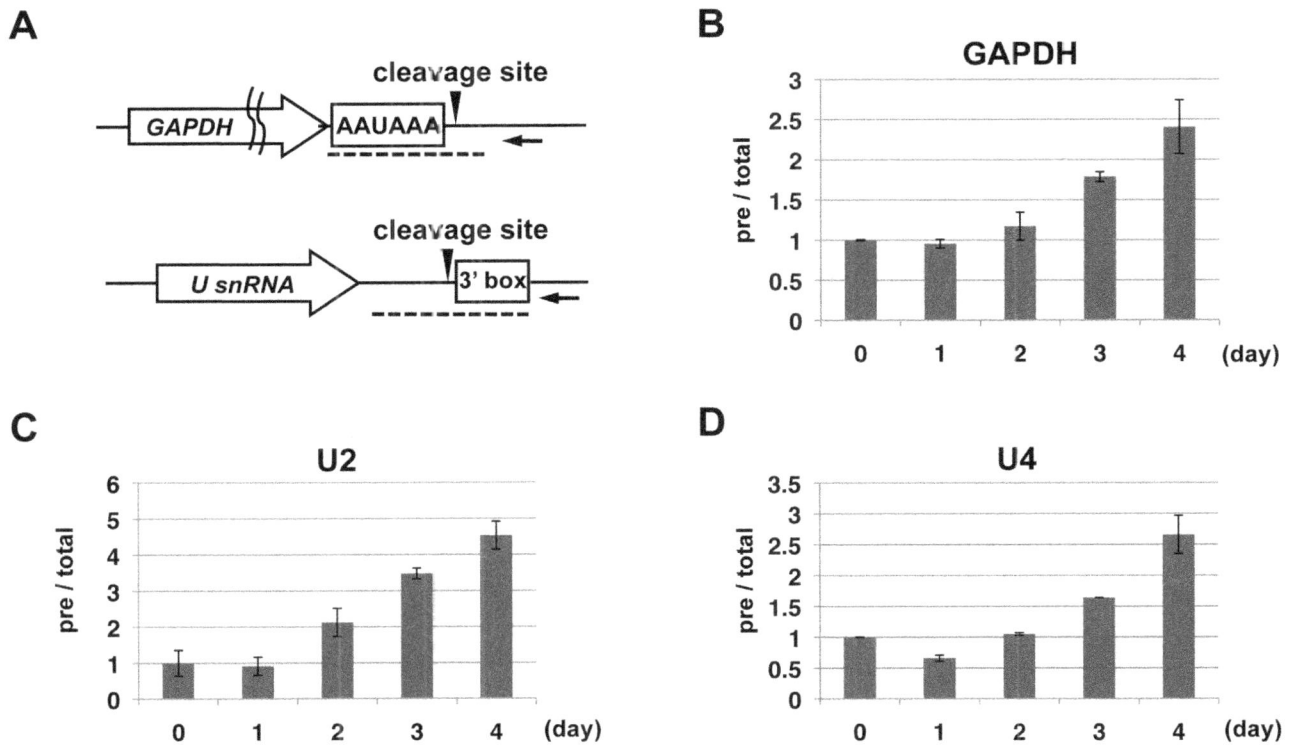

Figure 4. Ssu72 is required for 3'-end formation of *GAPDH* mRNA and U2 and U4 snRNAs. (A) Diagrams of the *GAPDH* (upper) and *U snRNA* genes (lower), shown as open arrows, with the cleavage sites depicted as arrowheads and RNA processing elements indicated in the open boxes. The black arrows represent the gene-specific primers used for reverse transcription of precursor transcripts. Dotted lines indicate the RT-qPCR amplicons used to quantitate the precursor transcripts. (B–D) RT-qPCR analysis of the relative expression levels of precursor (pre) and total transcripts of *GAPDH* (B), U2 snRNA (C) or U4 snRNA (D) in DT40 P3 (−/−) cells treated with Dox for the indicated number of days. Relative expression levels were normalized the corresponding levels on day 0. Error bars represent standard deviations of at least two independent experiments.

decreased after Dox addition (days 1–4), with kinetics similar to those of the relative precursor levels. These results imply that Ssu72 intrinsically suppresses 3'-end formation of stem-loop–dependent histone mRNAs in vertebrates.

Ssu72 depletion causes hyperphosphorylation of the CTD on Pol II-transcribed genes

The results described above suggest that vertebrate Ssu72 is required for efficient 3'-end formation of at least some snRNAs and poly(A)+ mRNAs, but inhibits 3'-end formation of replication-dependent histone mRNAs. Numerous studies have shown that regulation of CTD phosphorylation plays an important role in 3'-end formation of most Pol II transcripts [5]. Although Ssu72 depletion exhibited only a modest effect on the phosphorylation levels of total Pol II in whole-cell extracts of DT40 cells, we hypothesized that the effects of Ssu72 inactivation on 3'-end formation of Pol II-transcribed genes were caused by aberrant CTD phosphorylation on the affected genes. Therefore, we next examined the effect of Ssu72 depletion on the phosphorylation patterns of Pol II on the affected genes. In chromatin immunoprecipitation (ChIP) analyses using well-characterized phosphorylated CTD-specific antibodies (Fig. 6), we compared the distributions of total and phosphorylated Pol II at two distinct regions on the affected genes before and after Dox treatment of P3 (−/−) cells (Figs. 6A–6C). To evaluate the phosphorylation levels of Ser2, Ser5, and Ser7 on the affected genes, we calculated the ratios of the ChIP signals obtained using phosphorylated CTD-specific antibodies to the signals obtained using antibodies against the N-

terminal portion of Rpb1 (total Pol II). We first examined the Pol II distribution in the chicken *GAPDH* gene (Fig. 6A). Four days after Dox addition, when the 3'-end formation of three classes of Pol II transcripts was significantly affected (Figs. 4 and 5), the total Pol II level was elevated by 2-fold near the poly(A) signal sequence (region 1) (Fig. 6A, Pol II). Remarkably, phosphorylation of Ser5 increased about 5-fold in region 1 and region 2 (Fig. 6A, Ser5P); likewise, Ser7P levels increased 5-fold in region 1 and 11-fold in region 2 (Fig. 6A, Ser7P). By contrast, Ser2 phosphorylation decreased slightly in both regions (Fig. 6A, Ser2P).

We next examined the distribution of Pol II on the chicken U2 and U4 snRNA genes (Fig. 6B). Upon Ssu72 depletion, the phosphorylation levels of all serine residues were significantly elevated in 5' regions (region 1). In the region downstream of the 3'-end processing site (region 2), phosphorylation of Ser5 and Ser7 was elevated to a greater extent (4- to 5-fold) than phosphorylation of Ser2 (less than 2-fold) (Fig. 6B, Ser2P, Ser5P, and Ser7P). We then analyzed the distribution of Pol II on the replication-dependent histone H3 and H4 genes in the 5' portion of the coding region (region 1) and 1 kb downstream of the first codon (region 2) (Fig. 6C). Ssu72 depletion caused a significant increase in phosphorylation of all serine residues in the 5' coding regions of both histone genes (Fig. 6C, region 1). Notably, phosphorylation of Ser5 and Ser7 was elevated more than Ser2P levels. On the other hand, in the downstream regions, phosphorylation was markedly increased at Ser5 and Ser7, but not at Ser2 (Fig. 6C, region 2). Furthermore, Ssu72 depletion also led to a significant decrease in Pol II levels in the downstream regions, relative to the

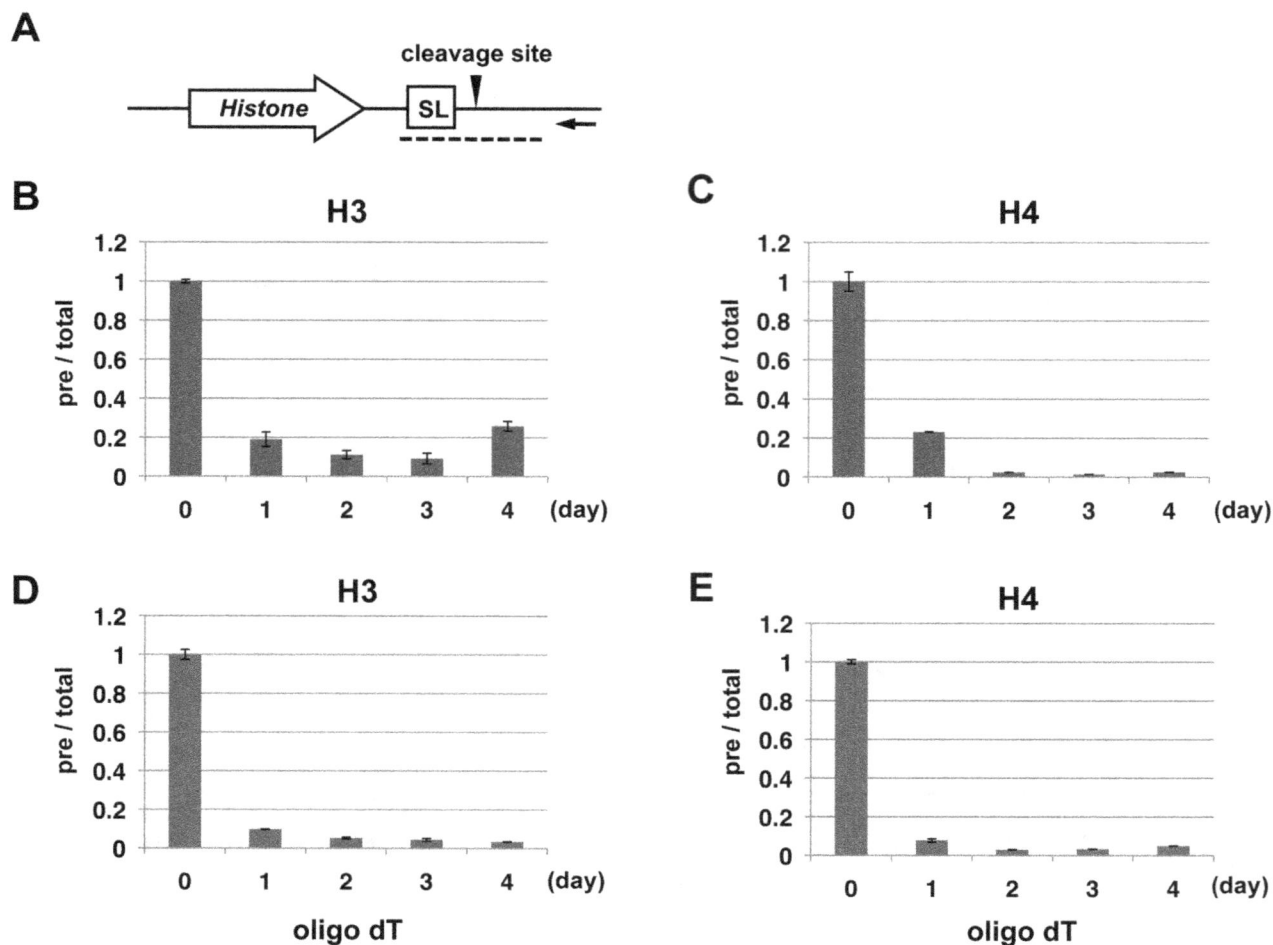

Figure 5. Ssu72 inhibits 3′-end formation of replication-dependent histone mRNAs. (A) Diagram of the histone gene, shown as an open arrow, with the cleavage site depicted an arrowhead and the stem-loop sequence (SL) indicated by the open box. The black arrow represents the gene-specific primer used for reverse transcription of precursor transcripts. Dotted lines indicate the RT-qPCR amplicons used to quantitate the precursor transcripts. (B-E) RT-qPCR analysis of the relative expression levels of precursor (pre) and total transcripts of histone H3 (B, D) or H4 (C, E) in DT40 P3 (−/−) cells treated with Dox for the indicated number of days. Gene-specific primers (B, C) or an oligo-dT primer (D, E) was used for reverse transcription to detect the expression levels of precursor transcripts. Relative expression levels were normalized the corresponding levels on day 0. Error bars represent standard deviations of at least two independent expriments.

5′ regions (Fig. 6C, Pol II, region 2), indicating that Ssu72 inactivation leads to increased efficiency of transcription termination at the 3′ ends of these histone genes.

Recent studies by Xiang *et al.* and our results shown below demonstrated that human Ssu72, like its yeast counterpart, can dephosphorylate Ser5P and Ser7P *in vitro* [27]. Thus, the ChIP results suggest that Ssu72 maintains the appropriate phosphorylation status of the CTD on different classes of Pol II-transcribed genes by directly dephosphorylating both Ser5P and Ser7P during transcription. To our knowledge, this is the first study to demonstrate that vertebrate Ssu72 plays a role in regulating CTD phosphorylation *in vivo*.

Vertebrate Ssu72 can dephosphorylate Ser5P and Ser7P, but not Thr4P, *in vitro*

The elevation of Ser2P both in the 5′ portion of snRNAs and the coding region of histone genes upon Ssu72 inactivation (Figs. 6B and 6C) raised the possibility that Ssu72 can also directly dephosphorylate Ser2P. To investigate this issue, we tested whether purified recombinant Ssu72 could dephosphorylate Ser2P

in an *in vitro* dephosphorylation assay using GST-fused phosphorylated CTD as a substrate (Fig. 7) [35]. The CTD dephosphorylation was monitored by Western blot using phosphorylation-dependent CTD antibodies used in Figure 3. In this experiment, we used recombinant human Ssu72 protein because it is almost identical to chicken Ssu72 (99% identity), and the two proteins have the same specific activity. We also used, as a positive control, recombinant human small CTD phosphatase 1 (SCP1), which belongs to a family of the FCP1-like enzyme and preferentially dephosphorylates Ser5P [36,37]. As expected, SCP1 has strong phosphatase activity towards Ser5P and, to a lesser extent, towards Ser7P, whereas no activity was detected towards Ser2P (Figure 7A, lane 2). Increasing amounts of bacterially-expressed and purified human Ssu72 efficiently dephosphorylated both Ser5P and Ser7P, but did not affect Ser2P; instead, the phosphorylation level of Ser2 increased (Fig. 7A). The apparent increase in Ser2P may have been caused by the increased affinity of our Ser2P-specific antibody (3E10) for CTDs dephosphorylated at both Ser5P and Ser7P, relative to its affinity for CTDs phosphorylated at either residue. This property of the 3E10

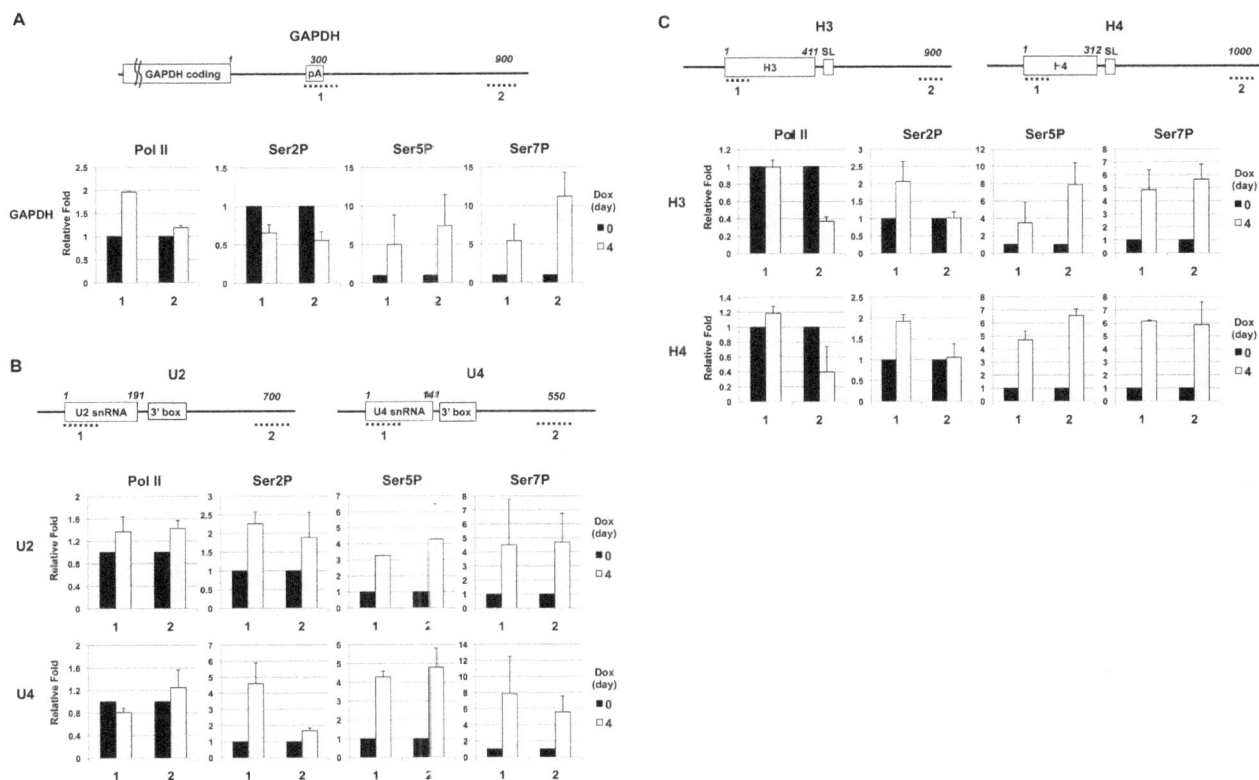

Figure 6. Ssu72 depletion causes hyperphosphorylation of the CTD on Pol II–transcribed genes. (A–C) ChIP analysis of the downstream region of the *GAPDH* gene (A), U2 and U4 snRNA genes (B), and histone H3 and H4 genes (C). (Upper) Diagrams of these genes with the coding regions, polyadenylation signals (pA), 3' box signals, and stem-loop sequences (SL) shown as open boxes. The numbers above these genes represent distance (in nucleotides) from the end of the coding region (for *GAPDH*) or the transcription start site (for snRNA and histone genes). Dotted lines indicate the ChIP amplicons. (Lower) ChIP analysis of region 1 or 2 on the indicated genes, using antibodies against Pol II (N20), Ser2P (3E10), Ser5P (3E8), or Ser7P (4E12), in DT40 P3 (−/−) cells treated with Dox for 4 days (grey bars) or untreated (black bars). The y axe represents the fold change relative to the corresponding levels on day 0. The values for Ser2P, Ser5P or Ser7P signals were normalized by the total Pol II signal. Error bars represent standard deviations of two independent experiments.

antibody may account for some changes of Ser2P ChIP signals obtained in the above experiments (Figure 6).

We next examined the activity and specificity of human Ssu72 for CTD dephosphorylation using the Pol II core complex purified from HeLa cells, which is a more biologically relevant substrate. Wild-type Ssu72, but not a catalytically dead mutant (C12A), efficiently dephosphorylated Ser5P and Ser7P. However, wild-type Ssu72 did not exhibit any phosphatase activity toward Ser2P (Fig. 7B). These results clearly demonstrate that human Ssu72 possesses Ser5P/Ser7P but not Ser2P phosphatase activity *in vitro*. Therefore, the elevation of Ser2P in the 5' regions of snRNA and histone genes may be caused by abnormally high levels of Ser5 and Ser7 phosphorylation, which may indirectly affect the Ser2P levels, possibly by inhibiting a Ser2P phosphatase or activating a Ser2 kinase only in the promoter regions of specific genes.

A recent study using a DT40 cell line bearing a conditional knockout of Rpb1 revealed that Thr4P is required for efficient histone mRNA 3'-end processing [16]. This observation raised the possibility that Ssu72 plays some inhibitory roles in histone mRNA 3'-end processing by directly dephosphorylating Thr4P. Therefore, we investigated whether human Ssu72 has Thr4P phosphatase activity, using the phosphatase assays described above and an antibody specific for Thr4P (6D7) [38]. After incubation with either wild-type Ssu72 or the catalytically inactive mutant, Thr4P levels of purified Pol II were not significantly altered (Fig. 7B,

Thr4P). Thus, human Ssu72 has no phosphatase activity toward Thr4P *in vitro*.

Discussion

In this study, we established Ssu72 conditional-knockout chicken DT40 cell lines and used them to investigate the functions of Ssu72 in gene expression at the cellular level. Our results demonstrated that chicken Ssu72, similar to its yeast ortholog, is essential for cell viability and efficient 3'-end formation of at least some snRNAs, as well as polyadenylated mRNAs. Unexpectedly, however, inactivation of Ssu72 caused a rapid and marked decrease in the levels of unprocessed replication-dependent histone mRNAs, indicating that chicken Ssu72 normally suppresses the stem-loop–type 3'-end formation of histone mRNA. Furthermore, Ssu72 inactivation led to a marked increase in both Ser5 and Ser7 phosphorylation on all genes in which 3'-end formation was affected. This is the first demonstration that Ssu72 functions as a Ser5P/Ser7P phosphatase in vertebrate cells. Together, our findings suggest that vertebrate Ssu72 is involved in various ways in 3'-end formation of Pol II transcripts, and that its function is mediated by dephosphorylation of CTD at both Ser5 and Ser7.

Chicken Ssu72 is essential for cell proliferation

Our results demonstrated that chicken Ssu72 is indispensable for cell proliferation (Fig. 2). Although Ssu72 depletion affected 3'-

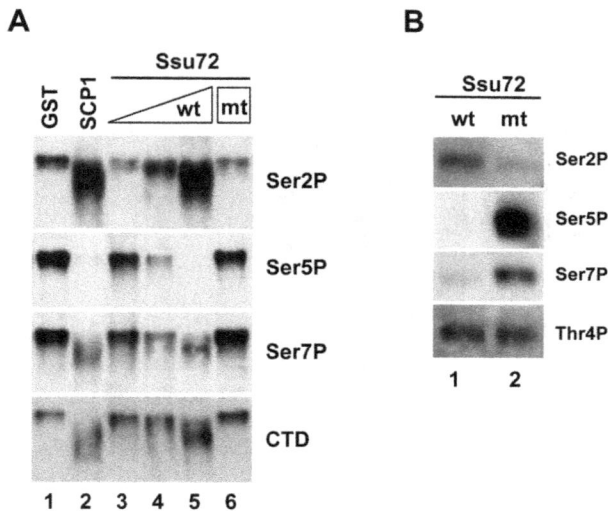

Figure 7. Ssu72 dephosphorylates Ser5P and Ser7P *in vitro*. (A) Fifty ng phosphorylated GST-CTD was incubated with 3.4 µg GST (lane 1), 0.2 µg SCP1 (lane 2), 0.2 µg, 1 µg, or 5 µg wild-type human Ssu72 (lanes 3–5), or 5 µg mutant (C12S) Ssu72 (lane 6), followed by Immunoblotting using the phospho CTD–specific antibodies 3E10 (Ser2P), 3E8 (Ser5P), and 4E12 (Ser7P) and an antibody against the CTD (8WG16). (B) Purified human Pol II was incubated with wild-type (lane 1) or mutant (C12S) Ssu72 (lane 2) and analyzed by Western blotting using phospho CTD–specific antibodies 3E10 (Ser2P), 3E8 (Ser5P), 4E12 (Ser7P), and 6D7 (Thr4P).

end formation of particular snRNAs and mRNAs, it did not significantly influence the steady-state levels of these RNAs (Figs. S1 and S2). At present, we do not know which aspects of Ssu72 functions are required for cell viability. A recent report showed that replacement of the Ser7 residue with phospho-mimetic glutamate in all CTD heptapeptide repeats is lethal, both in budding yeast [24] and in mammals [32], suggesting that persistent elevation of the Ser7P level dramatically decreases cell viability. As shown in this study, Ssu72 depletion results in dramatic elevation of Ser7P levels on several Pol II genes, even though depletion exerts only a modest effect on overall levels of Ser7P (Figs. 3). We assume that Ssu72 depletion significantly suppresses the expression of particular genes, possibly including genes essential for cell proliferation, by sustaining high Ser7P levels on those genes. To identify such specifically affected genes, global gene expression analysis will be required.

Alternatively, apart from its possible function in gene expression, the essential nature of Ssu72 in vertebrates might be explained by the recent findings of Kim et al [39]. Their results indicated that human Ssu72 might regulate the resolution of sister chromatid arm cohesion through direct interactions with the cohesion subunits Rad21 and SA2. Because we did not investigate whether Ssu72 depletion causes abnormalities in sister chromatid cohesion in DT40 cells, our results cannot elucidate whether the lethality of Ssu72-deficient cells can be attributed to such defects.

Chicken Ssu72 stimulates 3′-end formation of snRNAs and polyadenylated mRNAs

In this study, we demonstrated that chicken Ssu72 inactivation resulted in inefficient 3′-end formation of both snRNAs and polyadenylated mRNAs (Fig. 4), concomitant with elevation of Ser5P and Ser7P levels in the 3′ regions of these genes (Fig. 6A and 6B). Our results are in accordance with a study of cultured

human cells by Egloff et al. [15] showing that substituting either the Ser5 or Ser7 residue with phospho-mimetic glutamate (Ser5E or Ser7E) in all CTD heptapeptide repeats led to defects in 3′-end formation of both snRNAs and polyadenylated mRNAs [15]. Why does elevation of either Ser5P or Ser7P levels cause the defects of 3′-end formation of these types of Pol II-transcribed genes? For polyadenylated mRNAs, our ChIP results demonstrated that Ssu72 inactivation also resulted in reduction of Ser2P levels at the 3′-end processing site and downstream region of the *GAPDH* gene (Fig. 6A). This observation is also consistent with the study by Egloff et al., who showed that either the Ser5E or Ser7E mutation resulted in a dramatic decrease in Ser2P levels [15]. These results indicate that aberrant elevation of either Ser5P or Ser7P may somehow decrease the level of Ser2P, which is required for the efficient recruitment of 3′-end processing machinery to the 3′ regions of polyadenylated mRNAs.

In metazoans, 3′-end processing of snRNA requires the Integrator complex, which recognizes the 3′ box sequence just downstream of the processing site and binds CTD phosphorylated at both Ser2 and Ser7 (Fig. 4A) [40,41]. A recent study suggested that the Integrator complex is recruited to elongating Pol II via an interaction with a putative Ser5P-phosphatase, RPAP2, which recognizes the Ser7P mark [14]. Therefore, the elevation in Ser7P resulting from Ssu72 inactivation could potentially promote the recruitment of RPAP2 and the Integrator complex to Pol II–transcribed snRNA genes. However, simultaneous elevation of Ser7P and Ser5P levels caused by Ssu72 depletion might interfere with the interaction between the CTD and Integrator complex, as implied in a previous study [40]. We hypothesize that dephosphorylation of both Ser5P and Ser7P, are carried out either Ssu72 and RPAP2 or by Ssu72 alone, is required for the efficient 3′-end formation of vertebrate snRNAs.

Chicken Ssu72 suppresses 3′-end formation of replication-dependent histone mRNAs

To our surprise, we observed that Ssu72 depletion dramatically decreased levels of the unprocessed form of replication-dependent histone pre-mRNAs in DT40 cells (Fig. 5). We interpret these results as indicating that Ssu72 normally suppresses 3′-end formation of histone mRNA; consequently, Ssu72 inactivation leads to de-repression. Consistent with this view, our ChIP experiments revealed reductions in Pol II occupancies downstream of the processing site of both H3 and H4 genes (Fig. 6C), indicating that the efficiency of processing and/or transcription termination was higher in Ssu72-depleted cells.

Interestingly, a recent genome-wide ChIP-seq study in mammalian cells demonstrated that profiles of Pol II occupancy downstream of 3′-end processing site of protein-coding genes markedly differed between core histone genes and polyadenylated genes [42]. The Pol II occupancy downstream of 3′-end processing site of core histone genes exhibited sharp drop whereas those of polyadenylated genes persisted throughout much longer region, suggesting that difference in 3′-end processing mechanism of these RNAs influence the transcription termination process [42]. We assume that Ssu72 may be a key factor determining which types of 3′-end formation process is selected during transcription by Pol II.

How does Ssu72 specifically suppress histone mRNA 3′-end formation while playing a positive role in 3′-end formation in other types of Pol II–transcribed RNAs? The 3′-ends of most protein-coding mRNAs are formed by a two-step reaction consisting of an endonucleolytic cleavage and a tightly-coupled polyadenylation. However, in metazoan but not yeast, the 3′-ends of replication-dependent histone mRNAs are processed by different mechanism, in which precursor RNAs are endonucleo-lytically cleaved at a site between a conserved stem-loop sequence and a histone gene-specific downstream element recognized by U7 snRNP but are not followed by polyadenylation. Despite distinct complexes participate in each type of processing reaction, histone mRNA 3'-end formation shares many factors with the canonical polyadenylation reaction, including the endonuclease CPSF73 [33]. Another common factor is symplekin, the metazoan ortholog of budding yeast Pta1, which functions as a scaffold protein for both types of 3'-end processing [33]. Both symplekin and Pta1 strongly bind Ssu72, via an interaction with its N-terminal domain (NTD). Multiple authors have suggested that the NTD exerts an inhibitory effect on 3'-end processing of poly(A)-containing mRNAs, both in budding yeast and human, and that Ssu72 promotes 3'-end processing of polyadenylated mRNAs by counteracting the negative effects of symplekin/Pta1 [28,43]. Inversely, we speculate that vertebrate Ssu72 inhibits histone 3'-end processing by counteracting a positive effect of symplekin.

In recent study using the DT40 conditional-knockout system, replacement of Thr4 with Val in all repeats of the CTD caused a defect in efficient 3'-end processing, but not in transcription of replication-dependent histone mRNAs [16]. Moreover, inhibition of CDK9 activity in chicken or human cells compromised the 3'-end processing of histone mRNAs [34] and reduced the level of Thr4P [16], suggesting that Thr4P is involved in 3'-end processing of histone mRNA. Based on our observation (Fig. 7) that Ssu72 did not exhibit any phosphatase activity toward Thr4P in vitro, it is less likely that Ssu72 inhibits histone mRNA processing by directly dephosphorylating Thr4P. Instead, Thr4P may inhibit dephosphorylation of Ser5P and/or Ser7P by Ssu72 and thereby interfere with the suppressive activity of Ssu72 in histone mRNA processing. This possibility is consistent with a recent finding that the in vitro phosphatase activity of Drosophila Ssu72 toward Ser5P is reduced 4-fold by Thr4 phosphorylation within the same repeat [31]. Further studies will be required to decipher the underlying mechanism.

Chicken Ssu72 functions as a Ser5P and Ser7P phosphatase on Pol II–transcribed genes in vivo

Recent in vitro studies by Xiang et al. demonstrated that human Ssu72 exhibits CTD phosphatase activity toward Ser5P and Ser7P [27,28]. Their results showed that whereas Ser7P phosphatase activity of Ssu72 was much lower (~4000-fold) than its Ser5P phosphatase activity when CTD phosphopeptide was used as a substrate, Ssu72 exhibited comparable phosphatase activity toward Ser5P and Ser7P on full-length mammalian CTD. This observation suggests that multiple heptad repeats or non-consensus heptad repeats somehow help Ssu72 to recognize its substrate or catalyze dephosphorylation of Ser7P. Consistent with these findings, our results in this study demonstrated that chicken Ssu72 inactivation led to a comparable increase in both Ser5 and Ser7 phosphorylation on all affected genes (Fig. 6). Furthermore, our ChIP results showed that the fold increase in Ser7P levels after Ssu72 depletion was much higher than the increase in Ser5P levels in the downstream regions of GAPDH (Figs. 6A), indicating that CTD phosphatase activity of Ssu72 toward Ser7P may be specifically activated in the 3' regions of at least some Pol II–transcribed genes. A recent study showed that the N-terminal region of the metazoan mRNA 3'-end processing factor symplekin stimulates, to similar extents, both the Ser5P and Ser7P phosphatase activity of Ssu72 in vitro [27]. Therefore, preferential activation of Ssu72 phosphatase activity towards Ser7P may be caused by association of novel factors other than symplekin, e.g., transcription termination factors. Alternatively, a conformational

change of the CTD, such as proline isomerization, in the 3'-end region may facilitate dephosphorylation of Ser7P by Ssu72. In any case, our study provides the first demonstration that Ssu72 functions as a Ser5P/Ser7P phosphatase in vertebrate cells.

Although Ssu72 depletion in budding yeast resulted in a significant increase in the Ser5P level of total Pol II in vivo [23], both Ser5P and Ser7P levels of total Pol II in DT40 cells were slightly increased by Ssu72 depletion, indicating that other Ser5 and Ser7 phosphatases may exist in vertebrate cells. Indeed, in mammals, the small CTD phosphatase SCP1 has been reported to exhibit Ser5P-specific phosphatase activity in vitro [36,37]. FCP1 (TFIIF-associating CTD phosphatase), considered to be a Ser2P-specific phosphatase [44,45], can also efficiently dephosphorylate Ser5P in vitro [27,46]. Furthermore, SCP1 (Fig. 7A), and FCP1 [27] exert potent CTD phosphatase activity towards Ser7P in vitro. These observations imply that Ser5P and Ser7P levels of total cellular Pol II in vertebrate cells are either primarily regulated by CTD phosphatases other than Ssu72 or redundantly regulated by several CTD phosphatases including Ssu72. Nevertheless, our results suggest that Ssu72 functions as a Ser5/7P phosphatase during the transcription of a subset of Pol II genes. We speculate that after Ssu72 ceases to act or dissociates from Pol II, other CTD phosphatases such as FCP1 or SCP1 remove most of the remaining phosphate groups from Ser2/5/7P before the next round of transcription.

Ssu72-knockout DT40 cells provide a valuable tool for studying gene looping in vertebrate cells

As a component of the budding yeast pre-mRNA 3'-end processing complex, Ssu72 is required for the formation of gene loops [47], which are established by physical interactions between promoters and terminators of transcribed genes [48]. Gene loops have been proposed to function in transcription re-initiation and transcriptional memory [47,49]. Furthermore, Tan-Wong et al. recently showed that Ssu72-mediated gene-loop formation in budding yeast is restricted the production of divergent non-coding transcripts from bidirectional promoters, indicating that gene loops help to maintain the appropriate directionality of transcription [50]. Furthermore, they demonstrated in a human cell line that a polyadenylation-signal mutation of an integrated artificial β-globin gene resulted in suppression of both gene-loop formation and induction of divergent transcription, indicating that gene looping and its role in maintaining transcription directionality are conserved phenomena [50]. Although gene looping has also been detected in HIV-1 provirus [51] and some mammalian endogenous genes such as CD68 [52] and BRCA1 [53], the factors involved in gene-loop formation in vertebrate cells remain unknown. The Ssu72-knockout DT40 cells we established may provide a valuable tool for studying gene looping in vertebrate cells.

Experimental Procedures

Cell culture and transfection

Chicken DT40 cell lines were provided by Dr. K. Yamamoto (Department of Molecular Pathology, Cancer Research Institute, Kanazawa University, Kanazawa, Japan) [54]. DT40 cells were cultured in RPMI 1640 medium (Nissui) supplemented with 10% fetal bovine serum (Cell Culture Bioscience), 1% chicken serum (Gibco), 50 mM 2-mercaptoethanol (Sigma), penicillin, streptomycin, and 2 mM L-glutamine at 39°C in a humidified 5% CO_2 incubator. Transfections were carried out by electroporation using a GENE Pulser II (Bio-Rad) at 25 mF and 550 V. Drug-resistant clones were selected with medium containing 1.5 mg/ml G418,

25 µg/ml blasticidin S, 0.5 µg/ml puromycin, or 2.5 mg/ml hygromycin.

Plasmid constructs

Two genomic DNA fragments, a 2 kb fragment upstream of the first exon of the Ssu72 gene (ENSGALG00000001489) and a 3.9 kb fragment downstream of the second exon, were amplified from DT40 genomic DNA by long-range PCR. These fragments were cloned into the flanking regions of the drug-resistance cassettes (puromycin or blasticidin S) [55], which replaced the genomic region spanning the first and second exons (Fig. 1). The full-length chicken Ssu72 cDNA was amplified by RT-PCR from DT40 total RNA, and then cloned into the doxycycline-inducible vector supplied in the Tet-Off Advanced Inducible Gene Expression System (Clontech). Plasmids expressing GST-fused human SCP1 were prepared essentially as described [35]. For construction of the plasmid expressing His-tagged human Ssu72 in E. coli, the full-length human Ssu72 cDNA was amplified by RT-PCR from HeLa cell total RNA using primer set: hSsu72ATG-KP and hSsu72TGA-SL, and then cloned into a Kpn I/Sal I-digested pCold-II vector (TaKaRa). The plasmid expressing the mutant human Ssu72 (C12S) was made by site-directed mutagenesis using the KOD Plus Mutagenesis Kit (TOYOBO). Sequences of all cDNA inserts derived from PCR amplification were verified by DNA sequencing.

Generation of mutant DT40 cells

Wild-type DT40 cells were transfected with the pTet-off vector (Clontech), and drug-resistant clones were selected by culturing in G418-containing medium. One of the drug-resistant clones constitutively expressing tetR-VP16 protein was selected and used to establish stable clones expressing chicken Ssu72. In the resultant clones, Ssu72 expression is tightly repressed by addition of tetracycline. One of those clones was then used to generate the conditional Ssu72-deficient DT40 cell lines by sequentially introducing two types of linearized targeting constructs. At each step of gene disruption, genomic DNA was isolated from drug-resistant clones and subjected to Southern blotting and genomic PCR to confirm gene replacement by homologous recombination.

Southern blotting

Genomic DNA isolated from drug-resistant clones was digested with EcoRI and XbaI, and the resultant fragments were electrophoresed in 1.0% agarose gels and transferred to nylon membranes. Blots were hybridized with a ^{32}P-labeled 5′ DNA probe (see Fig. 1A) and analyzed by autoradiography.

Growth curves and cell-cycle distribution analysis

Cells were seeded in triplicate in 12-well plates (5×10^3 cells/well) and cultured under normal conditions for 8 days, during which they were diluted at 2-day intervals. Every day, concentrations of live cells were determined by Giemsa staining and counting on a hemocytometer. To examine cell-cycle distribution, $1-5 \times 10^6$ cells were fixed in 70% ethanol, treated with RNase A, and stained with propidium iodide. DNA contents were measured on a FACSCalibur instrument, and cell-cycle profiles were analyzed using the CellQuest software (BD Biosciences).

Immunoblotting and antibodies

Immunoblotting was performed as described previously [35]. The following antibodies were used: phosphorylation-specific anti-CTD [3E8 (pSer2), 3E10 (pSer5), 4E12 (pSer7), and 6D7 (pThr4)] (Ascension); anti-Rpb1 (ARNA-3, PROGEN); anti-β-actin (Sigma);

and anti-DNA polymerase delta and anti-exportin-1 (Transduction). Affinity-purified anti-Ssu72 antibodies were prepared by our laboratory.

RT-PCR

Total RNA was isolated from wild-type and mutant DT40 cells using TRIzol (Invitrogen). First-strand cDNA was synthesized from total RNA using Superscript III reverse transcriptase (Invitrogen) with random hexamer primers. The indicated cDNAs were amplified by PCR using specific primers, and then analyzed by agarose gel electrophoresis followed by staining with SYBR Green I (Invitrogen). For RT-qPCR, total RNA (500 ng) purified from DT40 cells using the Nucleo Spin RNA II kit (TaKaRa) was subjected to reverse transcription (RT) using the PrimeScrip RT Master Mix (TaKaRa). Synthesized first-strand cDNA was quantitated using SYBR Premix Ex Taq II (TaKaRa) on an Mx3000P Real-Time PCR System (Stratagene).

Chromatin immunoprecipitation analysis (ChIP)

Cells (1×10^7) cultured in medium were cross-linked by addition of formaldehyde (final concentration, 1%); after incubation for 10 minutes at room temperature, the reaction was quenched with 125 mM glycine. Cells were washed twice with 10 ml of ice-cold PBS, resuspended in 250 µl of SDS Lysis buffer (50 mM Tris-HCl [pH 8.1], 10 mM EDTA, and 1% SDS) containing protease inhibitor cocktail, and incubated for 10 minutes on ice. Cell lysates were sonicated using a Bioruptor UCD-200 (Cosmobio) to generate DNA fragments of 200–500 bp. Samples were centrifuged for $20,000 \times g$ for 10 minutes, and the resultant supernatant was diluted 10-fold with Dilution buffer (16.7 mM Tris-HCl [pH 8.1], 167 mM NaCl, 1.2 mM EDTA, 1.1% Triton X-100) containing protease inhibitor cocktail; 500 µl of the diluted sample was mixed with antibodies and rotated overnight at 4°C. The following day, 90 µg of Dynabeads Protein G (Invitrogen), pre-coated with salmon sperm DNA, was added to the mixture. The sample was then incubated for 1.5 hours at 4°C, and then washed twice with 1 ml of Low Salt buffer (20 mM Tris-HCl [pH 8.1], 150 mM NaCl, 2 mM EDTA, 0.1% SDS, and 1% Triton X-100), twice with High Salt buffer (20 mM Tris-HCl [pH 8.1], 500 mM NaCl, 2 mM EDTA, 0.1% SDS, and 1% Triton X-100), and once with LiCl buffer (10 mM Tris-HCl [pH 8.1], 250 mM LiCl, 1 mM EDTA, 1% sodium deoxycholate, and 1% Nonidet P-40). After the final wash, the beads were agitated with 200 µl of Elution buffer (100 mM NaHCO$_3$ and 1% SDS) at room temperature for 30 minutes. To reverse cross-links, 8 µl of 5 M NaCl was added, and the samples were incubated at 65°C overnight. The following day, the samples were incubated with 10 µg of RNase A at 37°C for 1 hour, and then with 10 µg of Proteinase K at 45°C for 2 hours. DNA fragments were purified using Wizard SV Gel and the PCR Clean-Up System (Promega), and subjected to RT-PCR using SYBR Premix Ex Taq II (TaKaRa) on an Mx3000P Real-time PCR system.

Protein expression and purification

The cold-induced expression of the His-tagged human Ssu72 in E. coli was performed according to the instruction of pCold-II vector (TaKaRa). The purification of His-tagged recombinant protein was performed with Ni-NTA agarose beads (QIAGEN) according to the instruction manual. All GST fused recombinant proteins were expressed in E. coli and purified as described previously [56]. Purification and separation of the phosphorylated Pol II from HeLa cell nuclear-extract pellets were done as described [56]. Protein concentration was determined by the Bradford method using bovine serum albumin (BSA) as a standard.

In vitro CTD dephosphorylation assay

In vitro phosphorylation of GST-CTD by HeLa cell nuclear extracts (NE) was performed as described [56]. CTD phosphatase assay were performed essentially as previously described [35]. Briefly, 50 ng phosphorylated GST-CTD or 25ng purified Pol II was incubated with 400 ng His-tagged human wild-type or C12A mutant Ssu72 in 10 ml of CTD phosphatase buffer (50 mM Tris-HCl [pH 7.9], 10 mM $MgCl_2$, 5mM DTT, 5% glycerol, 0.025% Tween 20, and 0.1 mM EDTA) at 30°C for 25 min. The reaction was terminated by addition of SDS loading buffer, and the reaction products were analyzed by immunoblotting with phospho-specific CTD antibodies.

Primers lists used in this study

Primers for constructing expression plasmids

cSsu72NheATG ACGAATTCGCTAGCCATG-CCTTCTTCGC-CGCTCCGTG

cSsu72BglTAA CTCTAGATCTTAATAAAAGCAGACAG-TATGAAGAAAAGTCCT

hSsu72ATG-KP TACCGGTACCATGCCGTCGTCCCCGC-TGCGG

hSsu72TGA-SL GACGTCGACTCAGTAGAAGCA-GACGGTGTGCAGAAAG

Primers for RT

GAPDH dw_R2 CAGAACACTTC-CTGGAGTTG.
TFIIE-alpha RT-R1 TCCTAAAGGAGGGCCTCCGATGG
U2 R3 GCTCAGCGCAACCGAACGCAG
U4B R3 CCGGCTCCCATTCATCCGTTGCTCAG
H3 R3 TCGCTGCCAGCACAGCGATAC
H4 R3 CAATGGGTGGTCTCTTCAAG

Primers for real-time PCR (RT, total)

GAPDH ex2-3 F TATCTTCCAGGAGCGTGACC
GAPDH ex2-3 R TCTCCATGGTGGTGAAGACA
TFIIE-alpha 2F TCACAGAATGTACAGCGGAAA
TFIIE-alpha GATCCCGATGAAATCAAGGA
U2 F1 ATCGCTTCTCGGCCTTTTGGC
U2 RT-R1 GTCCTCTCATCGAGGACGTATC
U4 F CGGAGAGAAGGGAGGTCAG
U4 R CCTCGGATTAACCTCATTGG
H3 F1 ATGGCGCGTACGAAGCAGACG
H3 R1 CTTGGTGGCCAGCTGCTTG
H4 F1 ATGTCTGGCAGAGGCAAGGG
H4 R1 TGTTGTCGCGCAGCACCTTG

Primers for real-time PCR (RT, precursor)

GAPDH dw1_F TCACCATCTGAAGTGCCTTG
GAPDH dw1_R CAGGGTGGCTGTAAGCATTT
TFIIE-alpha dw1_F CCCTTCGTTGTCTTCCATGT
TFIIE-alpha dw1_R TGGGTTGTTTCCTTGAATCG
U2 F2 CCGGGGAGGAATGTGGCGTG
U2 R2 GCGGGACACAGCGCCACCCTC
U4 2F TTGGTGTCCAACAGCAGAAA
U4 2R CCATTCATCCGTTGCTCAG
H3 F2 CTGAGGCTGCTGTGCTTTCAC

H3 R2 GCGATACAGCTCTCCCCAGTG
H4 F2 TCTCTGACGTTAGGCTTGGC
H4 R2 CAGCTCTTTTCCAGGCTAAG

Primers for real-time PCR (ChIP)

GAPDH dw1_F TCACCATCTGAAGTGCCTTG
GAPDH dw1_R CAGGGTGGCTGTAAGCATTT
GAPDH dw2_F AGCTCTGACTGCAGGAGTGG
GAPDH dw2_R CTGGCAGAGGGCTGATACTT
U2 F1 ATCGCTTCTCGGCCTTTTGGC
U2 RT-R1 GTCCTCTCATCGAGGACGTATC
U2 4F CCCCGTCTCCAGCCTGTCCA
U2 4R GGGACCCTGAGCCTGCAGGT
U4 F AGCTTTGCGCAGTGGCAGTATC
U4 R TCGTCATGGCGGGGTATTGG
U4 dw-2F ATCGGCACTGGCTTCACTT
U4 dw-2R CGCGACAAGGAGTTAAGGAG
H3 F1 ATGGCGCGTACGAAGCAGACG
H3 R1 CTTGGTGGCCAGCTGCTTG
H3 dw1_F GGCGCTGATAAGAGAAGTGC
H3 dw1_R TCTGTTCTGAGAGCCACGTC
H4 F1 ATGTCTGGCAGAGGCAAGCG
H4 R1 TGTTGTCGCGCAGCACCTTG
H4 dw1_F GCAAAACCTCCAGTCAACAAA
H4 dw1_R GGACAGCATAGTGGGGAAGA

Supporting Information

Figure S1   The levels of total snRNAs and mRNA are not significantly changed by Ssu72 depletion. (A) The expression levels of chicken U2 snRNA, U4 snRNA, and GAPDH in DT40 P3 (−/−) cells treated with Dox for the indicated days were measured by RT-qPCR analysis. The relative expression levels were normalized to those of 0 day. Error bars indicate standard deviation. (B) The expression levels of chicken U4, U13, and GAPDH in DT40 P3 (−/−) cells treated with Dox for the indicated days were measured by Northern blot analysis probed with the ^{32}P-labeled specific cDNA fragments.

Figure S2   The levels of total histone mRNAs are not dramatically changed by Ssu72 depletion. The expression levels of chicken H4 and H3 histone mRNAs in DT40 P3 (−/−) cells treated with Dox for the indicated days were measured by RT-qPCR analysis. The relative expression levels were normalized to those of 0 day. Error bars indicate standard deviation.

Acknowledgments

We thank Dr. Kazuyuki Kuroki and Dr. Yukiharu Kido (Kanazawa Univ.) for technical advice. We also thank Dr. Hiroshi Arakawa and Dr. Jean-Marie Buerstedde for providing plasmids and Dr. Ken-ichi Yamamoto for providing DT40 cell line.

Author Contributions

Conceived and designed the experiments: YH. Performed the experiments: SW M. Yuda M. Yamamoto YH YF. Analyzed the data: SW M. Yuda YH. Contributed reagents/materials/analysis tools: FH. Contributed to the writing of the manuscript: SW YO YH.

References

1. Corden JL (1990) Tails of RNA polymerase II. Trends Biochem Sci 15: 383–387.

2. Eick D, Geyer M (2013) The RNA polymerase II carboxy-terminal domain (CTD) code. Chem Rev 113: 8456–8490.

3. Egloff S, Dienstbier M, Murphy S (2012) Updating the RNA polymerase CTD code: adding gene-specific layers. Trends Genet 28: 333–341.

4. Hirose Y, Ohkuma Y (2007) Phosphorylation of the C-terminal domain of RNA polymerase II plays central roles in the integrated events of eucaryotic gene expression. J Biochem 141: 601–608.

5. Hsin JP, Manley JL (2012) The RNA polymerase II CTD coordinates transcription and RNA processing. Genes Dev 26: 2119–2137.

6. Buratowski S (2009) Progression through the RNA polymerase II CTD cycle. Mol Cell 36: 541–546.

7. Lu H, Flores O, Weinmann R, Reinberg D (1991) The nonphosphorylated form of RNA polymerase II preferentially associates with the preinitiation complex. Proc Natl Acad Sci U S A 88: 10004–10008.

8. Zhou Q, Li T, Price DH (2012) RNA polymerase II elongation control. Annu Rev Biochem 81: 119–143.

9. Bartkowiak B, Liu P, Phatnani HP, Fuda NJ, Cooper JJ, et al. (2010) CDK12 is a transcription elongation-associated CTD kinase, the metazoan ortholog of yeast Ctk1. Genes Dev 24: 2303–2316.

10. Jeronimo C, Bataille AR, Robert F (2013) The writers, readers, and functions of the RNA polymerase II C-terminal domain code. Chem Rev 113: 8491–8522.

11. Kim M, Suh H, Cho EJ, Buratowski S (2009) Phosphorylation of the yeast Rpb1 C-terminal domain at serines 2, 5, and 7. J Biol Chem 284: 26421–26426.

12. Glover-Cutter K, Larochelle S, Erickson B, Zhang C, Shokat K, et al. (2009) TFIIH-associated Cdk7 kinase functions in phosphorylation of C-terminal domain Ser7 residues, promoter-proximal pausing, and termination by RNA polymerase II. Mol Cell Biol 29: 5455–5464.

13. Akhtar MS, Heidemann M, Tietjen JR, Zhang DW, Chapman RD, et al. (2009) TFIIH kinase places bivalent marks on the carboxy-terminal domain of RNA polymerase II. Mol Cell 34: 387–393.

14. Egloff S, Zaborowska J, Laitem C, Kiss T, Murphy S (2012) Ser7 phosphorylation of the CTD recruits the RPAP2 Ser5 phosphatase to snRNA genes. Mol Cell 45: 111–122.

15. Egloff S, O'Reilly D, Chapman RD, Taylor A, Tanzhaus K, et al. (2007) Serine-7 of the RNA polymerase II CTD is specifically required for snRNA gene expression. Science 318: 1777–1779.

16. Hsin JP, Sheth A, Manley JL (2011) RNAP II CTD phosphorylated on threonine-4 is required for histone mRNA 3'-end processing. Science 334: 683–686.

17. Sun ZW, Hampsey M (1996) Synthetic enhancement of a TFIIB defect by a mutation in SSU72, an essential yeast gene encoding a novel protein that affects transcription start site selection in vivo. Mol Cell Biol 16: 1557–1566.

18. Dichtl B, Blank D, Ohnacker M, Friedlein A, Roeder D, et al. (2002) A role for SSU72 in balancing RNA polymerase II transcription elongation and termination. Mol Cell 10: 1139–1150.

19. Ganem C, Devaux F, Torchet C, Jacq C, Quevillon-Cheruel S, et al. (2003) Ssu72 is a phosphatase essential for transcription termination of snoRNAs and specific mRNAs in yeast. Embo J 22: 1588–1598.

20. Steinmetz EJ, Brow DA (2003) Ssu72 protein mediates both poly(A)-coupled and poly(A)-independent termination of RNA polymerase II transcription. Mol Cell Biol 23: 6339–6349.

21. Nedea E, He X, Kim M, Pootoolal J, Zhong G, et al. (2003) Organization and function of APT, a subcomplex of the yeast cleavage and polyadenylation factor involved in the formation of mRNA and small nucleolar RNA 3'-ends. J Biol Chem 278: 33000–33010.

22. He X, Khan AU, Cheng H, Pappas DL, Jr., Hampsey M, et al. (2003) Functional interactions between the transcription and mRNA 3'-end processing machineries mediated by Ssu72 and Sub1. Genes Dev 17: 1030–1042.

23. Krishnamurthy S, He X, Reyes-Reyes M, Moore C, Hampsey M (2004) Ssu72 Is an RNA polymerase II CTD phosphatase. Mol Cell 14: 387–394.

24. Zhang DW, Mosley AL, Ramisetty SR, Rodriguez-Molina JB, Washburn MP, et al. (2012) Ssu72 phosphatase-dependent erasure of phospho-Ser7 marks on the RNA polymerase II C-terminal domain is essential for viability and transcription termination. J Biol Chem 287: 8541–8551.

25. Bataille AR, Jeronimo C, Jacques PE, Laramee L, Fortin ME, et al. (2012) A universal RNA polymerase II CTD cycle is orchestrated by complex interplays between kinase, phosphatase, and isomerase enzymes along genes. Mol Cell 45: 158–170.

26. St-Pierre B, Liu X, Kha LC, Zhu X, Ryan O, et al. (2005) Conserved and specific functions of mammalian ssu72. Nucleic Acids Res 33: 464–477.

27. Xiang K, Manley JL, Tong L (2012) An unexpected binding mode for a Pol II CTD peptide phosphorylated at Ser7 in the active site of the CTD phosphatase Ssu72. Genes Dev 26: 2265–2270.

28. Xiang K, Nagaike T, Xiang S, Kilic T, Beh MM, et al. (2010) Crystal structure of the human symplekin-Ssu72-CTD phosphopeptide complex. Nature 467: 729–733.

29. Caldwell RB, Fiedler P, Schoetz U, Buerstedde JM (2007) Gene function analysis using the chicken B-cell line DT40. Methods Mol Biol 408: 193–210.

30. Werner-Allen JW, Lee CJ, Liu P, Nicely NI, Wang S, et al. (2011) cis-Proline-mediated Ser(P)5 dephosphorylation by the RNA polymerase II C-terminal domain phosphatase Ssu72. J Biol Chem 286: 5717–5726.

31. Luo Y, Yogesha SD, Cannon JR, Yan W, Ellington AD, et al. (2013) novel modifications on C-terminal domain of RNA polymerase II can fine-tune the phosphatase activity of Ssu72. ACS Chem Biol 8: 2042–2052.

32. Chapman RD, Heidemann M, Albert TK, Mailhammer R, Flatley A, et al. (2007) Transcribing RNA polymerase II is phosphorylated at CTD residue serine-7. Science 318: 1780–1782.

33. Marzluff WF, Wagner EJ, Duronio RJ (2008) Metabolism and regulation of canonical histone mRNAs: life without a poly(A) tail. Nat Rev Genet 9: 843–854.

34. Pirngruber J, Shchebet A, Schreiber L, Shema E, Minsky N, et al. (2009) CDK9 directs H2B monoubiquitination and controls replication-dependent histone mRNA 3'-end processing. EMBO Rep 10: 894–900.

35. Hirose Y, Iwamoto Y, Sakuraba K, Yunokuchi I, Harada F, et al. (2008) Human phosphorylated CTD-interacting protein, PCIF1, negatively modulates gene expression by RNA polymerase II. Biochem Biophys Res Commun 369: 449–455.

36. Zhang Y, Kim Y, Genoud N, Gao J, Kelly JW, et al. (2006) Determinants for dephosphorylation of the RNA polymerase II C-terminal domain by Scp1. Mol Cell 24: 759–770.

37. Yeo M, Lin PS, Dahmus ME, Gill GN (2003) A novel RNA polymerase II C-terminal domain phosphatase that preferentially dephosphorylates serine 5. J Biol Chem 278: 26078–26085.

38. Hintermair C, Heidemann M, Koch F, Descostes N, Gut M, et al. (2012) Threonine-4 of mammalian RNA polymerase II CTD is targeted by Polo-like kinase 3 and required for transcriptional elongation. Embo j 31: 2784–2797.

39. Kim HS, Baek KH, Ha GH, Lee JC, Kim YN, et al. (2010) The hsSsu72 phosphatase is a cohesin-binding protein that regulates the resolution of sister chromatid arm cohesion. Embo j 29: 3544–3557.

40. Egloff S, Szczepaniak SA, Dienstbier M, Taylor A, Knight S, et al. (2010) The integrator complex recognizes a new double mark on the RNA polymerase II carboxyl-terminal domain. J Biol Chem 285: 20564–20569.

41. Baillat D, Hakimi MA, Naar AM, Shilatifard A, Cooch N, et al. (2005) Integrator, a multiprotein mediator of small nuclear RNA processing, associates with the C-terminal repeat of RNA polymerase II. Cell 123: 265–276.

42. Anamika K, Gyenis A, Poidevin L, Poch O, Tora L (2012) RNA polymerase II pausing downstream of core histone genes is different from genes producing polyadenylated transcripts. PLoS One 7: e38769.

43. Ghazy MA, He X, Singh BN, Hampsey M, Moore C (2009) The essential N terminus of the Pta1 scaffold protein is required for snoRNA transcription termination and Ssu72 function but is dispensable for pre-mRNA 3'-end processing. Mol Cell Biol 29: 2296–2307.

44. Hausmann S, Shuman S (2002) Characterization of the CTD phosphatase Fcp1 from fission yeast. Preferential dephosphorylation of serine 2 versus serine 5. J Biol Chem 277: 21213–21220.

45. Cho EJ, Kobor MS, Kim M, Greenblatt J, Buratowski S (2001) Opposing effects of Ctk1 kinase and Fcp1 phosphatase at Ser 2 of the RNA polymerase II C-terminal domain. Genes Dev 15: 3319–3329.

46. Lin PS, Dubois MF, Dahmus ME (2002) TFIIF-associating carboxyl-terminal domain phosphatase dephosphorylates phosphoserines 2 and 5 of RNA polymerase II. J Biol Chem 277: 45949–45956.

47. Ansari A, Hampsey M (2005) A role for the CPF 3'-end processing machinery in RNAP II-dependent gene looping. Genes Dev 19: 2969–2978.

48. O'Sullivan JM, Tan-Wong SM, Morillon A, Lee B, Coles J, et al. (2004) Gene loops juxtapose promoters and terminators in yeast. Nat Genet 36: 1014–1018.

49. Tan-Wong SM, Wijayatilake HD, Proudfoot NJ (2009) Gene loops function to maintain transcriptional memory through interaction with the nuclear pore complex. Genes Dev 23: 2610–2624.

50. Tan-Wong SM, Zaugg JB, Camblong J, Xu Z, Zhang DW, et al. (2012) Gene loops enhance transcriptional directionality. Science 338: 671–675.

51. Perkins KJ, Lusic M, Mitar I, Giacca M, Proudfoot NJ (2008) Transcription-dependent gene looping of the HIV-1 provirus is dictated by recognition of pre-mRNA processing signals. Mol Cell 29: 56–68.

52. O'Reilly D, Greaves DR (2007) Cell-type-specific expression of the human CD68 gene is associated with changes in Pol II phosphorylation and short-range intrachromosomal gene looping. Genomics 90: 407–415.

53. Tan-Wong SM, French JD, Proudfoot NJ, Brown MA (2008) Dynamic interactions between the promoter and terminator regions of the mammalian BRCA1 gene. Proc Natl Acad Sci U S A 105: 5160–5165.

54. Takao N, Kato H, Mori R, Morrison C, Sonada E, et al. (1999) Disruption of ATM in p53-null cells causes multiple functional abnormalities in cellular response to ionizing radiation. Oncogene 18: 7002–7009.

55. Yunokuchi I, Fan H, Iwamoto Y, Araki C, Yuda M, et al. (2009) Prolyl isomerase Pin1 shares functional similarity with phosphorylated CTD interacting factor PCIF1 in vertebrate cells. Genes Cells 14: 1105–1118.

56. Hirose Y, Manley JL (1998) RNA polymerase II is an essential mRNA polyadenylation factor. Nature 395: 93–96.

11

Avian SERPINB12 Expression in the Avian Oviduct Is Regulated by Estrogen and Up-Regulated in Epithelial Cell-Derived Ovarian Carcinomas of Laying Hens

Gahee Jo[1,9], Whasun Lim[1,9], Seung-Min Bae[1], Fuller W. Bazer[2], Gwonhwa Song[1]*

1 Department of Biotechnology, College of Life Sciences and Biotechnology, Korea University, Seoul, Republic of Korea, 2 Center for Animal Biotechnology and Genomics and Department of Animal Science, Texas A & M University, College Station, Texas, United States of America

Abstract

Serine protease inhibitors (SERPINs) are involved in a variety of biological processes such as blood clotting, angiogenesis, immune system, and embryogenesis. Although, of these, SERPINB12 is identified as the latest member of clade B in humans, little is known of it in chickens. Thus, in this study, we investigated SERPINB12 expression profiles in various tissues of chickens and focused on effects of steroid hormone regulation of its expression. In the chicken oviduct, SERPINB12 mRNA and protein are abundant in the luminal (LE) and glandular (GE) epithelia of the magnum in response to endogenous or exogenous estrogen. Furthermore, SERPINB12 mRNA and protein increase significantly in GE of cancerous ovaries of laying hens with epithelia-derived ovarian cancer. Collectively, these results indicate that SERPINB12 is a novel estrogen-stimulated gene that is up-regulated by estrogen in epithelial cells of the chicken oviduct and that it is a potential biomarker for early detection of ovarian carcinomas in laying hens and women.

Editor: Olivier Gires, Ludwig-Maximilians University, Germany

Funding: This research was funded by the Basic Science Research Program (2013R1A1A2A10005948) through the National Research Foundation of Korea (NRF) funded by the Ministry of Education, Science, and Technology, and also by a grant from the Next-Generation BioGreen 21 Program (No. PJ008142), Rural Development Administration, Republic of Korea. The funders had no role in study design, data collection and analysis, decision to publish, or preparation of the manuscript.

Competing Interests: The authors have declared that no competing interests exist.

* Email: ghsong@korea.ac.kr

9 These authors contributed equally to this work.

Introduction

The chicken oviduct is well known as an excellent model for basic mechanisms of organogenesis during developmental events. In addition, the oviduct of laying hens consists of four anatomical and functional components: the infundibulum is the site of fertilization of the ovum; the magnum produces components of egg-white; the isthmus for provides for formation of the shell membrane in eggs; and the shell gland supports formation of the egg shell [1,2]. Furthermore, the chicken oviduct is very suitable for investigating morphological and physiological effects on development and differentiation of reproductive tissues and organs of vertebrates in response to sex steroid hormones [3]. Estrogen is the primary sex steroid hormone that plays pivotal roles in cell survival and proliferation, synthesis and secretion of egg white proteins and formation and differentiation of tubular glands in the developing oviduct of laying hens [4]. Results of our previous studies indicate that estrogen-stimulated genes such as *AHCYL1*, *AvBD11*, *A2M*, *CTNNB1*, *ERBB1*, *PTN*, *SERPINB3*, *SERPINB11*, and *SPP1* are involved in development of the chicken oviduct due to their activation of various regulatory mechanisms that are mediated via diverse trans-activating transcription factors and complex actions of estrogen to effect physiological changes [5,6,7,8,9,10,11,12,13]. Moreover, changes in circulating concen-

trations of estrogen following cessation and reinitiation of ovarian steroidogenesis in response to induced molting regulates temporal patterns of expression of these genes during remodeling and development of the oviduct in laying hens [14]. However, identification and characterization of additional genes to elucidate mechanisms responsible for development of the chicken oviduct is clearly warranted.

Serpins (serine protease inhibitors) are the largest family of protease inhibitors present in blood and involved in various proteolytic cascades including blood clotting and the complement system in extracellular fluid and blood [15]. A subset of serpins (clade B serpins) are called human ov-serpins because their molecular structures are very similar to egg white proteins such as ovalbumin [16]. Clade B serpins reside primarily within cells and play important roles in various biological processes including protection against certain apoptotic cascades, inhibition of cell migration, and tumor suppression [17]. Of these, *SERPINB12*, also known as yokopin, is located at 18q21.3 in the human genome encoding a protein of 405 amino acids (about 46 kDa) and widely expressed in various human tissues such as brain, heart, lung, liver, spleen, pancreas, kidney, testes, ovary and placenta [18]. However, little is known about the expression and hormonal regulation of SERPINB12 in chickens.

Our research has discovered candidate genes and their expression in the oviduct of immature chicks treated with diethylstilbestrol (DES, a synthetic estrogen agonist) [19]. We have also used DNA microarrays to identify estrogen regulated genes critical to regulatory mechanisms for regression and recrudescence of the oviduct of laying hens during induced molting [14]. Our gene profiling data showed that SERPINB12 increased in the oviduct of chicks treated with DES, but not non-treated chicks; therefore, it is likely to be regulated by changes in secretion of estrogen from the ovary during regression and recrudescence of the oviduct of laying hens during the molting process. Therefore, we hypothesized that SERPINB12 has essential roles in development and differentiation of the oviduct in response to estrogen in laying hens. Hence, the aims of this study were to: 1) determine gene expression patterns for SERPINB12 in the immature chick oviduct treated with estrogen; 2) determine whether expression of SERPINB12 changes in the oviduct of laying hens during regression, remodeling and recrudescence phases following induced molting; and 3) compare differential expression patterns for SERPINB12 in normal and cancerous ovaries of laying hens.

Results

SERPINB12 mRNA expression in various organs from chickens (Study one)

To determine tissue-specific patterns of expression of *SERPINB12* mRNA in chickens, RT-PCR analyses were performed using various organs including brain, heart, liver, kidney, muscle, small intestine, gizzard, testis, ovary, and oviduct of 1- to 2- year-old males (n = 3) and females (n = 3). As shown in Figure 1A and 1B, *SERPINB12* gene is moderately expressed in all organs irrespective of sex. However, there was high expression of *SERPINB12* mRNA in the oviduct of female chickens and in liver and kidney from males. These results are very similar to a previous report that SERPINB12 is detected in many adult and fetal tissues of humans [18]. These results suggest that SERPINB12 plays an important role(s) in various organs.

Estrogen regulates SERPINB12 expression in the developing chicken oviduct (Study two)

To investigate the effects of estrogen on development of the chick oviduct, we first performed quantitative PCR analysis using each segment of the neonatal female chick oviduct treated with DES: infundibulum, magnum, isthmus and shell gland. As illustrated in Figure 2A, *SERPINB12* mRNA increased 3.58-fold ($P<0.001$) in the oviduct of DES-treated chick as compared to non-treated chicks. *SERPINB12* mRNA increased 2.44- ($P<0.01$), 5.17- ($P<0.001$), 3.22- ($P<0.01$) and 3.54-fold ($P<0.001$) in the infundibulum, magnum, isthmus and shell gland segments of the chick oviduct, respectively, in response to DES (Figure 2B). In addition, *in situ* hybridization analysis revealed that *SERPINB12* mRNA was expressed abundantly by luminal (LE) and glandular (GE) epithelia of the magnum and moderately expressed by the infundibulum, isthmus, and shell gland of oviducts from chicks treated with DES (Figure 3A). Consistent with these results, immunoreactive SERPINB12 protein was abundant in LE and GE of the magnum and also expressed in other sections of the oviduct at a low level (Figure 3B). However, the non-specific rabbit IgG, used as a negative control, did not detect immunoreactive SERPINB12 protein in the DES-treated chick oviduct.

Estrogen regulates SERPINB12 expression in the chicken oviduct during the induced molting (Study three)

In this study, we evaluated the effects of estrogen on expression of SERPINB12 during the regression and recrudescence phases of molting in laying hens. As shown in Figure 4A, SERPINB12 expression decreased sharply between Days 6 and 12 of molting (regression phase) and the increased 6.75-fold and 666.83-fold ($P<0.01\sim0.001$) between Days 20 and 30, respectively, of the molting process (recrudescence phase). *In situ* hybridization analysis revealed that *SERPINB12* mRNA is expressed in the LE and GE on Day 0, but expression ceased as concentrations of estrogen in blood decreased from Days 1 to 20 of the molting period (Figure 5A and 5B). However, expression of *SERPINB12* mRNA increased with increasing concentrations of estrogen in blood between Days 21 and 35 of the molting process. Likewise, immunohistochemial analyses revealed that SERPINB12 protein was predominantly localized to LE and GE of the oviduct during the regeneration phase of the oviduct following induced molting.

Distinct expression pattern of SERPINB12 in normal and cancerous ovaries of hens (Study four)

We previously reported that *estrogen receptor alpha (ESR1)* mRNA expression increased about 51-fold ($P<0.001$) and immunoreactive ESR1 protein was most abundant in GE of cancerous ovaries, but not detectable in normal ovaries. Those results indicated that estrogen-stimulated genes in the oviduct may also increase in cancerous ovaries of hens in response to estrogen. Thus, we examined whether SERPINB12 is expressed in cancerous ovaries of laying hen in the present study. As illustrated in Figure 6A, *SERPINB12* mRNA expression increased 206.16-fold ($P<0.01$) in cancerous as compared to normal ovaries of laying hens. To compare cell-specific localization of *SERPINB12* between normal and cancerous ovaries, we performed *in situ* hybridization and immunohistochemical analyses to find abundant expression of SERPINB12 mRNA and protein specifically in GE of cancerous ovaries, but not stromal cells or blood vessels of cancerous ovaries, or in any cells of normal ovaries (Figures 6B and 6C). This result indicates that SERPINB12 expression increases in cancerous ovaries of laying hens as compared with normal ovaries, and that its expression is co-localized to cells in cancerous ovaries expressing ESR1.

Discussion

Results of the present study provide the first evidence of *SERPINB12* gene expression in the reproductive organs of female chickens in response to estrogen. Additionally, the results are the first to demonstrate that SERPINB12 expression is uniquely expressed in cancerous ovaries, but not normal ovaries of laying hens. Thus, results of the present study support our hypothesis that SERPINB12 serves essential roles in development and differentiation of the oviduct, as well as ovarian tumorigenesis in response to estrogen in laying hens.

The serpin superfamily has 16 clades and over 500 members with most involved in proteolytic activity in various biological processes [20]. Of these, the clade B serpins, also known as ovalbumin-related serpins (ov-serpins), are encoded for by 13 genes including *monocyte neutrophil elastase inhibitor (SERPINB1), plasminogen activator inhibitor-2 (SERPINB2), maspin (SERPINB5),* and *plasminogen activator inhibitor-6 (SERPINB6)* [21]. Because SERPINB12 is the latest known member of clade B, little is known about it or its role in the chicken. In this study, we found that chicken SERPINB12 is expressed moderately in various tissues from both sexes as reported for humans [18]. Interestingly,

[A]

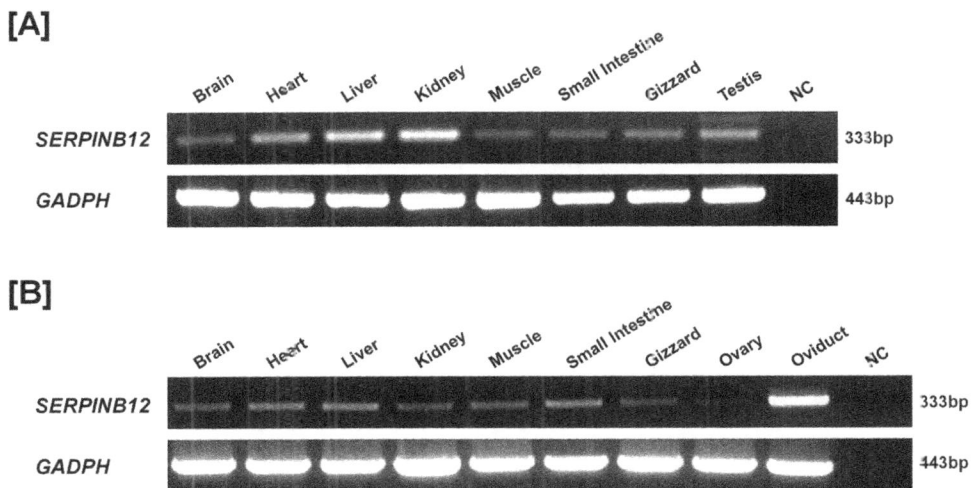

[B]

Figure 1. Expression of SERPINB12 in various organs of both male and female chickens. Results of RT-PCR analysis using cDNA templates from different organs of both male [A] and female [B] chickens with chicken SERPINB12 and chicken GAPDH-specific primers are shown. SERPINB12 gene was generally expressed in all organs from male and female chickens. Its expression was highly abundant in the chicken oviduct.

SERPINB12 mRNA and protein are expressed abundantly in the chicken oviduct, particularly in the LE and GE of the magnum. However, it is also moderately expressed in the infundibulum, isthmus, and shell gland of the chicken oviduct. These results suggest that SERPINB12 plays an essential role in secretion or storage of egg white proteins in the magnum of the chicken oviduct.

Previous gene profiling data indicated that SERPINB12 might be regulated by estrogen in the oviduct of laying hens fed a high zinc diet to induce molting [14]. During molting, the oviduct and ovary of laying hens undergo regression and remodeling and then both undergo recrudescence in response to increases in circulating concentrations of estrogen [22]. Our previous study showed that concentrations of estradiol in serum decreased during the induced molting period (Days 0 to 20) and then increased during the recrudescence period (Days 20 to 35). This result suggested that proliferation of oviductal cells between Days 25 and 30 of the molting period is likely induced by estrogen following resumption of ovarian steroidogenesis. Based on those results and our gene profile data, we selected and examined SERPINB12 expression in

the current study, and we found it to increase significantly in oviductal tissue during the remodeling and recrudescent phases (Day 25 to Day 35) when concentrations of estrogen increase during induced molting. These results indicate that SERPINB12 is a novel gene likely involved in regulatory mechanisms(s) for differentiation of the oviduct of neonatal chicks, as well as tissue regeneration and remodeling of the oviduct of laying hens in response to estrogen.

It is well know that DES implants and endogenous estrogen induce development of the neonatal chick oviduct including differentiation of oviductal tubular glands and ciliated cells, and the synthesis and secretion of egg white proteins by epithelial cells of the magnum [1,23]. Our previous studies showed that SERPINB3 and SERPINB11 are also expressed in a cell-specific manner in the chicken oviduct in response to estrogen [5,8]. Furthermore, using microarray analysis to examine transcript changes in chick oviducts after exposure to DES, we reported that SERPINB12 mRNA expression changed significantly during oviduct development in chickens [19]. Therefore, we examined expression of SERPINB12 in the oviduct of DES treated chicks in

[A]   **[B]**

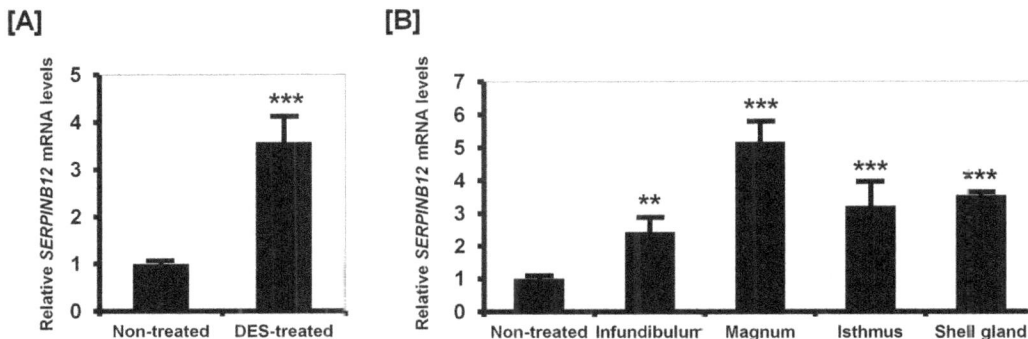

Figure 2. Effect of DES on tissue specificity of expression of chicken SERPINB12 by quantitative PCR analyses. Quantitative PCR analyses showed that SERPINB12 mRNA was more abundant in the oviducts of DES-treated chicks, as compared with oviducts from control chicks [A]. Among the four segments of oviducts from DES-treated chicks, SERPINB12 was most highly expressed in the magnum [3]. The asterisks denote statistically significant differences (mean ± SEM; ** $P<0.01$ or *** $P<0.001$).

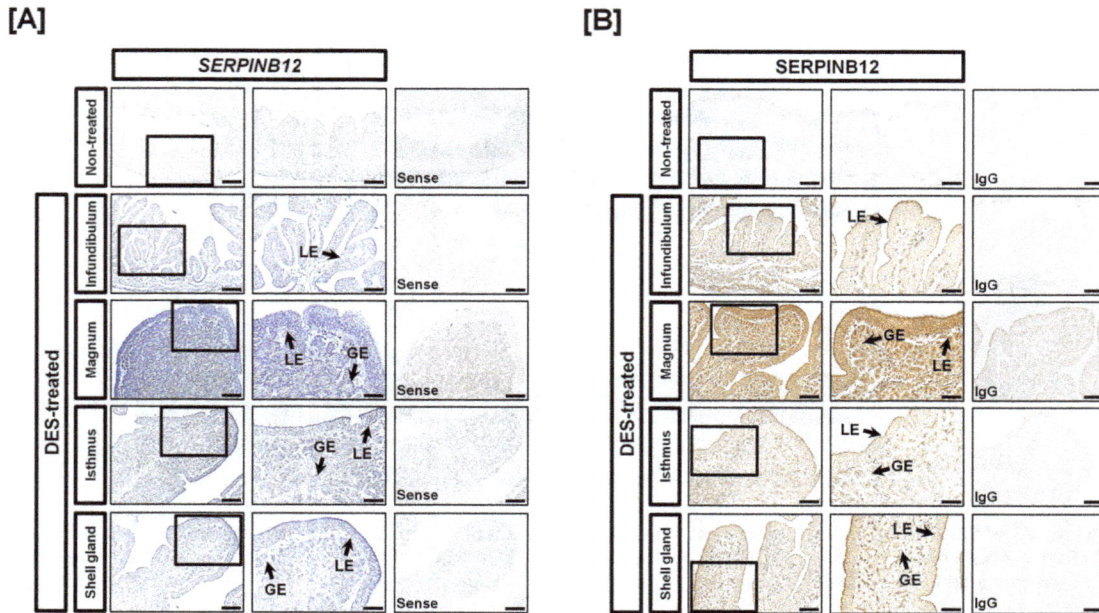

Figure 3. Localization of SERPINB12 mRNA and protein in oviducts of DES-treated and non-treated chicks. [A] *In situ* hybridization analyses indicated cell-specific expression of *SERPINB12* mRNA in GE and LE of the four segments of oviducts from DES-treated chicks. However, there was no expression of *SERPINB12* in oviducts from control chicks. [B] Immunoreactive SERPINB12 protein was localized to LE and GE of oviducts of chicks treated with DES, especially the magnum. For the IgG control, normal rabbit IgG was substituted for the primary antibody. Sections were not counterstained. Legend: LE, luminal epithelium; GE, glandular epithelium; *Scale bar* represents 200 μm (the first columnar panels, sense and IgG) and 50 μm (the second columnar panels).

this study and found that expression of SERPINB12 mRNA and protein is also regulated by estrogen in a tissue- and cell-specific manner during development and differentiation of the chicken oviduct. These results suggest that SERPINB12 may have effects on the process of secretion or storage of egg white proteins.

Recently, Crum and colleagues reported that ovarian cancer (pelvic serous cancer) may originate from the distal end of the oviduct [24]. In fact, epithelial cell-derived ovarian cancer (EOC) is one of the most lethal causes of cancer-related deaths among women due to the fact that there are no or few symptoms and lack of markers for its diagnosis in early stages of the disease [25]. Although several animal models for research to develop novel biomarker for early detection of EOC and appropriate medical

treatment have been developed, most of them are not effective. Interestingly, laying hens are being recognized as a good model because chicken EOC occurs spontaneously, and tumor progression, and histological and pathological types and stages are very similar to that in women [26,27]. Trevino and colleagues reported that EOC may arise from the oviduct and about one-half of up-regulated genes in EOC are oviduct-related genes in response to estrogen stimulation in the laying hen model [28]. Indeed, we reported on several estrogen-stimulated genes including *CTSB, SERPINB11, SERPINB3, AHCYL1, SPP1, A2M, PTN, AvBD-11, CTNNB1* [6,7,9,12,29,30,31,32] and several cell cycle genes [33] which are mainly expressed in the chicken oviduct and expressed abundantly in GE of cancerous ovaries from laying hens.

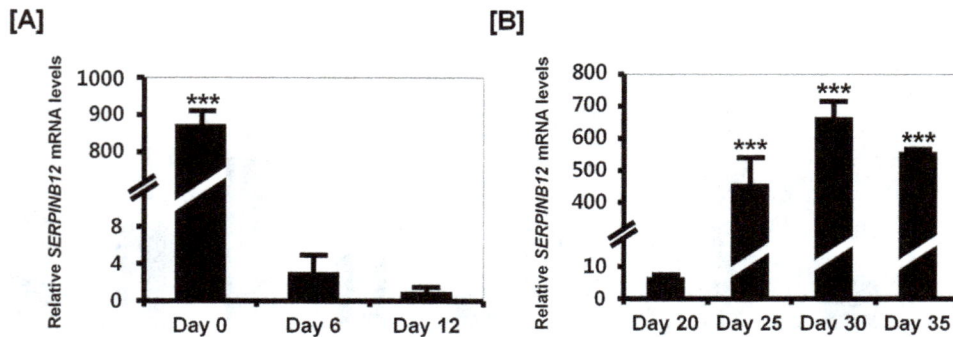

Figure 4. Quantitation of SERPINB12 mRNA during regression and recrudescence of the magnum of laying hens during induced molting. Quantitative PCR was conducted using cDNA templates from the magnum of hens fed a control diet (Day 0), zinc fed hens in which the oviduct was regressing (Day 6 and 12) [A], and hens in which the oviduct was undergoing recrudescence (Day 20, 25, 30, and 35) [B]. These experiments were performed in triplicate and normalized to control *GAPDH* expression. The asterisks denote statistically significant differences (mean ± SEM; ** $P<0.01$ or *** $P<0.001$).

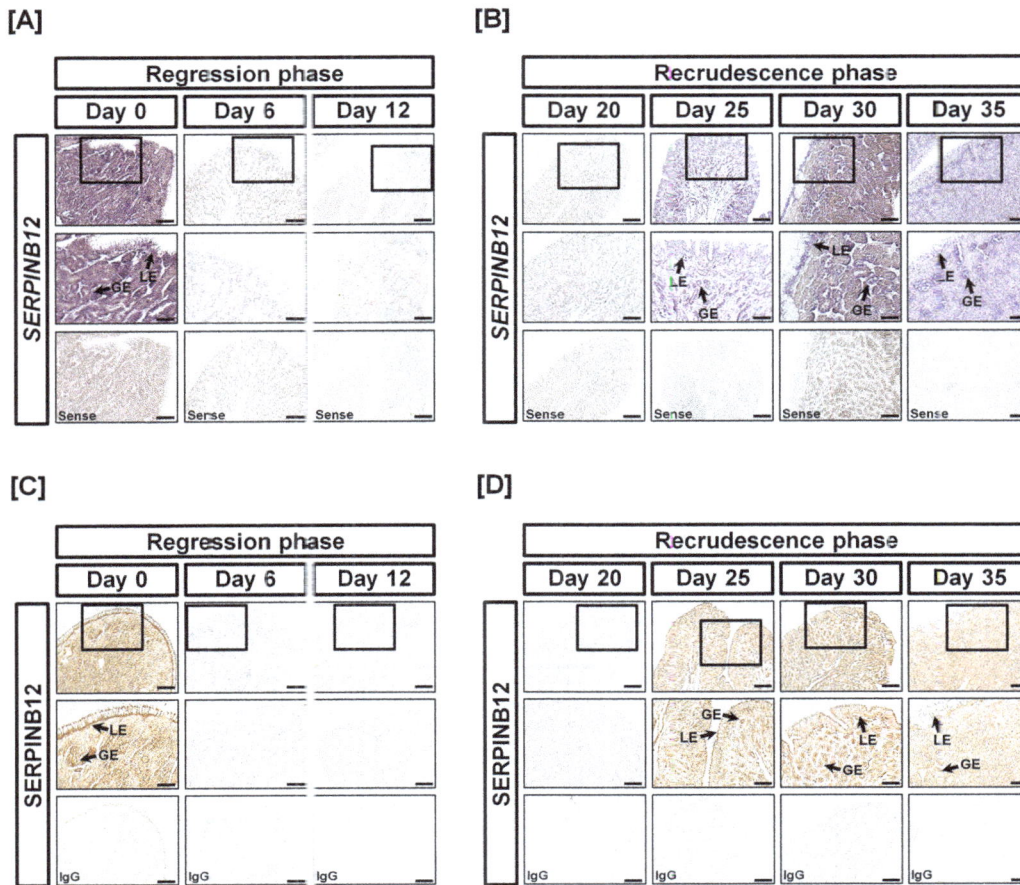

Figure 5. Cell-specific localization of SERPINB12 mRNA and protein during regression and recrudescence of the magnum of oviducts from hens during induced molting. The localization of SERPINB12 mRNA in the magnum of hens on different days of the regression phase [A] and recrudescence phase [B] was analyzed by *in situ* hybridization. It was predominantly expressed in GE and LE of the magnum on Days 0, 25, 30 and 35 of the recrudescence phase of molting. On the other hand, the expression of SERPINB12 was rarely detected on Days 6, 12 and 20 of the regression phase of dietary zinc induced molting. Immunoreactive SERPINB12 protein in the magnum during the regression phase [C] and recrudescence phase [D] was analyzed by immunohistochemistry. Consistent with SERPINB12 mRNA localization. SEPRINB12 protein was mainly detected in GE and LE of the magnum on Days 0, 25, 30 and 35 during the recrudescence phase of the oviduct during molting. Legend: LE, luminal epithelium; GE, glandular epithelium; *Scale bar* represents 200 µm (the first horizontal panels, sense and IgG) and 50 µm (the second horizontal panels).

Similarly, we found that expression of SERPINB12 mRNA and protein increase significantly in GE of in cancerous, but not normal ovaries of laying hens. As a matter of fact, among clade B serpins, SERPINB3 and SERPINB5 are well studied by various cancer researchers. SERPINB3 was first discovered in squamous cell carcinoma of the cervix [34] and it has been investigated in a variety of squamous cell carcinoma types [35,36,37,38]. SERPINB3 is also involved in invasiveness of cells, as well as cell migration, apoptosis, and immune reaction [39,40]. SERPINB5, also known as maspin (mammary serine protease inhibitor) is a class II tumor suppressor gene identified in patients with breast cancer [41]. SERPINB5 promotes apoptotic events of invasive prostate and breast carcinoma cells [41] and down-regulation of its expression causes pancreatic, colorectal, and ovarian adenocarcinomas [42]. On the contrary, over-expression of SERPINB5 was detected in two ovarian cancer cell lines including OVCAR3 and SKOV3 whereas the normal ovarian surface epithelial cells do not express SERPINB5 [43]. In our previous studies, we found increased expression of SERPINB3 in chicken ovarian carcinogenesis and suggested that it as a novel biomarker for predicting

platinum resistance and poor prognosis for survival in patients with EOC [31]. In addition, SERPINB11 expression is dramatically increased in laying hens with progressive endometrioid adenocarcinoma of the ovary [30]. Furthermore, estrogen is a well-known risk factor for gynecological cancer in women [44]. Our previous report showed that *ESR1* mRNA expression is up-regulated 51-fold ($P<0.001$) in ovarian cancer in laying hens while there is little or no detectable ESR1 protein in normal ovaries [12]. Results of the present study indicate that expression of *SERPINB12* mRNA increased about 206-fold ($P<0.001$) only in ovarian carcinoma of laying hens. In addition, we observed that SERPINB12 localized predominantly to GE of cancerous ovaries. This glandular architecture is one of the most common characteristics of ovarian cancer in laying hens and women and it is estrogen-dependent and mainly appears in pre- and peri-menopausal women [26,45]. Therefore, based on our current results, SERPINB12 may play an important role(s) in the development of EOC in laying hens, and be a potential biomarker for early detection and monitoring of therapies used to treat EOC in women.

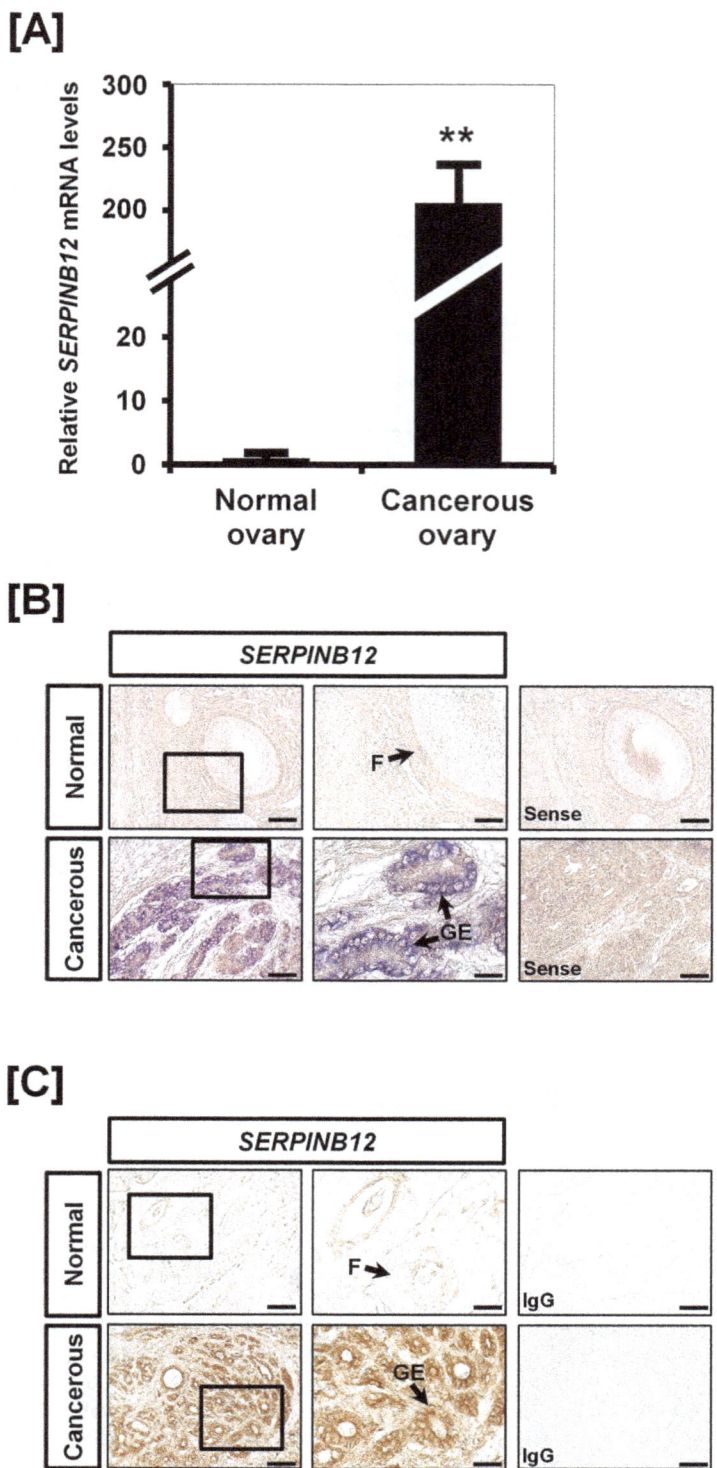

Figure 6. Distribution and localization of SERPINB12 mRNA and protein in normal and cancerous ovaries from laying hens. [A] Quantitative PCR was conducted using cDNA templates from normal and cancerous ovaries of hens. The expression of *SERPINB12* was unique to cancerous ovaries of laying hens as compared with normal ovaries from laying hens. [B] *In situ* hybridization analyses indicated cell-specific localization of *SERPINB12* mRNA in normal and cancerous ovaries from laying hens. *SERPINB12* was strongly expressed in the GE of cancerous ovaries whereas there was no expression in any cells of normal ovaries. [C] Immunoreactive SERPINB12 protein, as for *SERPINB12* mRNA, was also unique to the GE of cancerous ovaries from laying hens. Legend: F, follicle; GE, glandular epithelium. *Scale bar* represents 50 μm (the first columnar panels, sense and IgG) and 200 μm (the second columnar panels).

Collectively, results of the current study demonstrate that *SERPINB12* is a novel estrogen-stimulated gene expressed specifically in GE during development and differentiation of the chicken oviduct. In addition, SERPINB12 is clearly associated with and may be an essential regulatory factor in abnormal growth and dysfunction of ovarian carcinomas. Therefore, our study provides new insights into the use of SERPINB12 as an ethiological and pathogenetical biomarker for chicken ovarian carcinoma and its use as a biomarker for diagnosis and monitoring of the effectiveness of therapies for treatment of EOC in women.

Materials and Methods

Experimental Animals and Animal Care

The experimental use of chickens for this study was approved by the Animal Care and Use Committee of Korea University. All chickens were exposed to a light regimen of 15 h light and 9 h dark with *ad libitum* access to feed and water, and subjected to standard poultry husbandry guidelines.

Tissue Samples

Study one. Following euthanasia of 1- to 2-year-old laying White Leghorn (WL) hens, tissue samples were collected from brain, heart, liver, kidney, small intestine, gizzard, muscle, ovary and oviduct (n = 5). Subsets of these samples were frozen or fixed in 4% paraformaldehyde for further analysis. Frozen tissue samples were cut into 5- to 7-mm pieces, frozen in liquid nitrogen vapor, and stored at $-80°C$. The other samples were cut into 10-mm pieces and fixed in fresh 4% paraformaldehyde in PBS (pH 7.4). After 24 h, fixed tissues were changed to 70% ethanol for 24 h and then dehydrated and embedded in Paraplast-Plus (Leica Microsystems, Wetzlar, Germany). Paraffin-embedded tissues were sectioned at 5 µm.

Study two. Female chicks were identified by PCR analysis using W chromosome-specific primer sets (F: 5′-CTA TGC CTA CCA CAT TCC TAT TTG C-3′ and R: 5′-AGC TGG ACT TCA GAC CAT CTT CT-3′). Treatment with DES and recovery of the oviduct were conducted as reported previously [8]. We implanted a 10 mg DES pellet in the abdominal region of 1-week-old female chicks to release the hormone for 10 days. The DES pellet was removed from all chicks for 10 days, and then a 10 mg daily dose was administered for 10 additional days. Five chicks were assigned to each treatment group. For control chicks, tissues were collected from the middle part of the undifferentiated neonatal oviduct. On the other hand, four segments including infundibulum, magnum, isthmus, and shell gland were collected from the differentiated oviduct of DES-treated neonatal chicks.

Study three. Molting of laying hens was induced by adding 20,000 ppm zinc to the diet to effectively reduce feed-intake and induce molting [46,47]. Briefly, molting was induced by feeding hens a diet containing high zinc (mixed 252 g zinc oxide per 10 kg feed to achieve a final concentration of 20,000 ppm of zinc). Laying hens in the molting group completely ceased egg production within 12 days after feeding the high zinc-diet. The 35 laying hens (47-week-old) were divided into two groups; a molting-progressing group and a post-molting-progressing group, and kept in individual cages. The molting group was further divided into three subgroups based on the number of days of feeding the high zinc diet (normal feeding group, and 6 days and 12 days after onset of zinc feeding). The recrudescence (post-molting) group was divided into four subgroups based on the number of days from complete cessation of egg laying and initiation of feeding a normal commercial diet: 20, 25, 30 or 35 days after onset of zinc feeding or 8, 13, 18 or 23 days on normal

feed after cessation of egg production and removal from the high zinc diet.

Study Four. In this study, a total of 136 laying hens (88 over 36 months of age and 48 over 24 months of age) which had stopped laying eggs were euthanized for collection of normal and cancerous ovaries. From that population of laying hens, we obtained cancerous ovarian tissue from 10 hens and normal ovarian tissues from 10 egg-laying hens of similar age. We evaluated tumor stage of 10 hens with cancerous ovaries according to characteristic features of chicken ovarian cancer [26,30]. Three hens had stage III disease as ovarian tumor cells had metastasized to the gastrointestinal (GI) tract and liver surface with profuse ascites in the abdominal cavity. Five hens had tumor cells spread to distant organs including liver parenchyma, lung, GI tract and oviduct with profuse ascites, indicating stage IV disease. Two hens had stage I disease as tumors were limited to their ovaries. Epithelial ovarian cancers in chickens were classified based on their cellular subtypes and patterns of cellular differentiation with reference to ovarian malignant tumor types in humans [26].

RNA Isolation

Total cellular RNA was isolated from frozen tissues (100 mg) using Trizol reagent (Invitrogen, Carlsbad, CA) according to the manufacturer's recommendations. First, the tissues were homogenized with 1 ml Trizol using a power homogenizer and 0.2 ml of chloroform was added to homogenized samples. Then, the tubes were shaken vigorously by hand for a few seconds. The samples were incubated at room temperature for 3 min and centrifugated at $12,000×g$ for 15 min at $4°C$. The aqueous phase of the sample was placed in a new 1.5 ml tube and 0.5 ml of isopropanol was added. After incubating the mixture for 10 min at room temperature, the samples were centrifuged at $12,000×g$ for 10 min at $4°C$. The resulting RNA pellet was washed using 75% ethanol and dried in air. Finally, the pellet was resuspended in RNase-free water. The quantity and quality of total RNA was determined by spectrometry and denaturing agarose gel electrophoresis, respectively. The quantity and purity of RNA were measured using a Nanodrop 2000 (Thermo Scientific, DE, USA). We confirmed RNA purity in samples with a 260/280 nm ratio greater than 1.8.

Semiquantitative RT-PCR analysis

The cDNA was synthesized from total cellular RNA (2 µg) using random hexamer (Invitrogen, Carlsbad, CA) and oligo (dT) primers and AccuPower RT PreMix (Bioneer, Daejeon, Korea). The cDNA was diluted (1:10) in sterile water before use in PCR. For *SERPINB12*, the forward primer (5′-TCT TAA CCA CGA AGC GTT CC-3′) and reverse primer (5′-AAC TTG TGT TCC CAG GAT GC-3′) amplified a 333-bp product. For *GAPDH* (housekeeping gene), the forward primer (5′-CAC AGC CAC ACA GAA GAC GG-3′) and reverse primer (5′-CCA TCA AGT CCA CAA CAC GG-3′) amplified a 443-bp product. The procedures for generation of primers, PCR amplification and verification of their sequences were as described previously [48,49]. PCR amplification was conducted using approximately 120 ng cDNA as follows: (1) $95°C$ for 3 min; (2) $95°C$ for 20 sec, $60°C$ for 40 sec and $72°C$ for 1 min for 35 cycles for *SERPINB12* and 32 cycles for *GAPDH*; and (3) $72°C$ for 10 min. After PCR, equal amounts of reaction product were analyzed using a 1% agarose gel, and PCR products were visualized using ethidium bromide staining. The amount of DNA present was quantified by measuring the intensity of light emitted from correctly sized bands under ultraviolet light using a Gel Doc XR$^+$system with Image Lab software (Bio-Rad).

Quantitative RT-PCR Analysis

Total RNA was extracted from each tissue sample using Trizol (Invitrogen) and purified using an RNeasy Mini Kit (Qiagen). Complementary DNA was synthesized using a Superscript III First-Strand Synthesis System (Invitrogen). Gene expression levels were determined using SYBR Green (Biotium, Hayward, CA, USA) and a StepOnePlus Real-Time PCR System (Applied Biosystems, Foster City, CA, USA). The *GAPDH* gene was analyzed simultaneously as a control and used for normalization to account for variation in loading. *SERPINB12* and *GAPDH* were analyzed in triplicate. Using the standard curve method, we determined the level of expression of the examined genes using the standard curves and C_T values, and normalized them based on *GAPDH* expression. For *SERPINB12*, the forward primer (5'-GGC TGG AAC AGA CTG AAA GC-3') and reverse primer (5'-TGA GGT TAA AGG TGC CTT CG-3') amplified a 129-bp product. For GAPDH, the forward primer (5'-CAG AAC ATC ATC CCA GCG TC-3') and reverse primer (5'-GGC AGG TCA GGT CAA CAA CA-3') amplified a 133-bp product. The PCR conditions were 94°C for 3 min, followed by 40 cycles at 95°C for 30 sec, 60°C for 30 sec, and 72°C for 30 sec using a melting curve program (increasing the temperature from 55°C to 95°C at 0.5°C per 10 sec) and continuous fluorescence measurement. ROX dye (Invitrogen) was used as a negative control for the fluorescence measurements. Sequence-specific products were identified by generating a melting curve in which the Ct value represented the cycle number at which a fluorescent signal was statistically greater than background, and relative gene expression was quantified using the $2^{-\Delta\Delta Ct}$ method [50]. The relative quantification of gene expression was normalized to the Ct value for the control oviduct.

In Situ Hybridization Analysis

For hybridization probes, PCR products were generated from cDNA using the primers described for RT-PCR analysis. The products were extracted from the gel and cloned into the TOPO Vector (Invitrogen, Carlsbad, CA). After verification of the sequences, plasmids containing gene sequences were amplified with T7- and SP6-specific primers (T7:5'-TGT AAT ACG ACT CAC TAT AGG G-3'; SP6:5'-CTA TTT AGG TGA CAC TAT AGA AT-3') then digoxigenin (DIG)-labeled RNA probes were transcribed using a DIG RNA labeling kit (Roche Applied Science, Indianapolis, IN). Tissues were collected and fixed in fresh 4% paraformaldehyde, embedded in paraffin and sectioned at 5 μm on APES-treated slides. The sections were then deparaffinized in xylene and rehydrated to diethylpyrocarbonate (DEPC)-treated water through a graded series of alcohol. The sections were treated with 1% Triton X-100 in PBS for 20 min, washed two times in DEPC-treated PBS and then digested with 5 μg/ml Proteinase K (Sigma) in TE buffer (100 mM Tris-HCl, 50 mM EDTA, pH 8.0) at 37°C. After post-fixation in 4% paraformaldehyde, sections were incubated twice for 5 min each in DEPC-treated PBS and incubated in TEA buffer (0.1 M triethanolamine) containing 0.25% (v/v) acetic anhydride. The sections were incubated in a prehybridization mixture containing 50% formamide and 4× standard saline citrate (SSC) for at least 10 min at room temperature. After prehybridization, the sections were incubated overnight at 42°C in a humidified chamber in a hybridization mixture containing 40% formamide, 4× SSC, 10% dextran sulfate sodium salt, 10 mM DTT, 1 mg/ml yeast tRNA, 1 mg/ml salmon sperm DNA, 0.02% Ficoll, 0.02% polyvinylpyrrolidone, 0.2 mg/ml RNase-free bovine serum albumin and denatured DIG-labeled cRNA probe. After hybridization, sections were washed for 15 min in 2× SSC at 37°C, 15 min in 1× SSC at 37°C, 30 min in NTE buffer (10 mM Tris, 500 mM NaCl and 1 mM EDTA) at 37°C and 30 min in 0.1× SSC at 37°C. After blocking with 2% normal sheep serum (Santa Cruz Biotechnology, Inc., Santa Cruz, CA), the sections were incubated overnight with sheep anti-DIG antibody conjugated to alkaline phosphatase (Roche, Indianapolis, IN). The signal was visualized following exposure to a solution containing 0.4 mM 5-bromo-4-chloro-3-indolyl phosphate, 0.4 mM nitroblue tetrazolium, and 2 mM levamisole (Sigma Chemical Co., St. Louis, MO).

Immunohistochemistry

Immunocytochemical localization of SERPINB12 protein was performed as described previously using a rabbit polyclonal antibody to SERPINB12 (catalog number: sc-85145; Santa Cruz Biotechnology, Inc., Santa Cruz, CA) at a final dilution of 1:250 (0.4 μg/ml). Antigen retrieval was performed using the boiling citrate method as described previously [51]. Negative controls included substitution of the primary antibody with purified non-immune rabbit IgG at the same final concentration.

Statistical Analyses

Data presented for quantitative PCR are expressed as mean ± SEM unless otherwise stated. Differences in variances between normal and cancerous ovaries were analyzed using the F test, and differences between means were subjected to the Student's t test. Differences with a probability value of $P<0.05$ were considered statistically significant. Excel (Microsoft, Redmond, WA) was used for statistical analyses.

Acknowledgments

The authors thank members of the laboratory of Molecular Development Engineering for their constant support and cooperation.

Author Contributions

Conceived and designed the experiments: GS. Performed the experiments: GJ WL SMB. Analyzed the data: WL FWB GS. Contributed reagents/materials/analysis tools: GS. Contributed to the writing of the manuscript: GJ WL FWB GS.

References

1. Palmiter RD (1972) Regulation of protein synthesis in chick oviduct. II. Modulation of polypeptide elongation and initiation rates by estrogen and progesterone. J Biol Chem 247: 6770–6780.
2. Palmiter RD (1972) Regulation of protein synthesis in chick oviduct. I. Independent regulation of ovalbumin, conalbumin, ovomucoid, and lysozyme induction. J Biol Chem 247: 6450–6461.
3. Dougherty DC, Sanders MM (2005) Estrogen action: revitalization of the chick oviduct model. Trends Endocrinol Metab 16: 414–419.
4. Sanders MM, Mcknight GS (1988) Positive and Negative Regulatory Elements Control the Steroid-Responsive Ovalbumin Promoter. Biochemistry 27: 6550–6557.
5. Lim W, Ahn SE, Jeong W, Kim JH, Kim J, et al. (2012) Tissue specific expression and estrogen regulation of SERPINB3 in the chicken oviduct. Gen Comp Endocrinol 175: 65–73.
6. Lim W, Jeong W, Kim J, Yoshimura Y, Bazer FW, et al. (2013) Expression and regulation of beta-defensin 11 in the oviduct in response to estrogen and in ovarian tumors of chickens. Mol Cell Endocrinol 366: 1–8.
7. Lim W, Jeong W, Kim JH, Lee JY, Kim J, et al. (2011) Differential expression of alpha 2 macroglobulin in response to dietylstilbestrol and in ovarian carcinomas in chickens. Reprod Biol Endocrinol 9: 137.
8. Lim W, Kim JH, Ahn SE, Jeong W, Kim J, et al. (2011) Avian SERPINB11 gene: characteristics, tissue-specific expression, and regulation of expression by estrogen. Biol Reprod 85: 1260–1268.

9. Bae SM, Lim W, Jeong W, Lee JY, Kim J, et al. (2014) Hormonal regulation of beta-catenin during development of the avian oviduct and its expression in epithelial cell-derived ovarian carcinogenesis. Mol Cell Endocrinol 382: 46–54.

10. Ahn SE, Jeong W, Kim JH, Lim W, Kim J, et al. (2012) ERBB receptor feedback inhibitor 1: identification and regulation by estrogen in chickens. Gen Comp Endocrinol 175: 194–205.

11. Jeong W, Kim J, Ahn SE, Lee SI, Bazer FW, et al. (2012) AHCYL1 is mediated by estrogen-induced ERK1/2 MAPK cell signaling and microRNA regulation to effect functional aspects of the avian oviduct. PLoS One 7: e49204.

12. Lee JY, Jeong W, Lim W, Kim J, Bazer FW, et al. (2012) Chicken pleiotrophin: regulation of tissue specific expression by estrogen in the oviduct and distinct expression pattern in the ovarian carcinomas. PLoS One 7: e34215.

13. Lim W, Jeong W, Kim J, Ka H, Bazer FW, et al. (2012) Differential expression of secreted phosphoprotein 1 in response to estradiol-17beta and in ovarian tumors in chickens. Biochem Biophys Res Commun 422: 494–500.

14. Jeong W, Lim W, Ahn SE, Lim CH, Lee JY, et al. (2013) Recrudescence mechanisms and gene expression profile of the reproductive tracts from chickens during the molting period. PLoS One 8: e76784.

15. Silverman GA, Whisstock JC, Askew DJ, Pak SC, Luke CJ, et al. (2004) Human clade B serpins (ov-serpins) belong to a cohort of evolutionarily dispersed intracellular proteinase inhibitor clades that protect cells from promiscuous proteolysis. Cell Mol Life Sci 61: 301–325.

16. Hunt LT, Dayhoff MO (1980) A surprising new protein superfamily containing ovalbumin, antithrombin-III, and alpha 1-proteinase inhibitor. Biochem Biophys Res Commun 95: 864–871.

17. Silverman GA, Bird PI, Carrell RW, Church FC, Coughlin PB, et al. (2001) The serpins are an expanding superfamily of structurally similar but functionally diverse proteins. Evolution, mechanism of inhibition, novel functions, and a revised nomenclature. J Biol Chem 276: 33293–33296

18. Askew YS, Pak SC, Luke CJ, Askew DJ, Cataltepe S, et al. (2001) SERPINB12 is a novel member of the human ov-serpin family that is widely expressed and inhibits trypsin-like serine proteinases. J Biol Chem 276: 49320–49330.

19. Song G, Seo HW, Choi JW, Rengaraj D, Kim TM, et al. (2011) Discovery of candidate genes and pathways regulating oviduct development in chickens. Biol Reprod 85: 306–314.

20. Kaiserman D, Bird PI (2005) Analysis of vertebrate genomes suggests a new model for clade B serpin evolution. BMC Genomics 6: 167.

21. Benarafa C, Remold-O'Donnell E (2005) The ovalbumin serpins revisited: perspective from the chicken genome of clade B serpin evolution in vertebrates. Proc Natl Acad Sci U S A 102: 11367–11372.

22. Berry WD (2003) The physiology of induced molting. Poult Sci 82: 971–980.

23. McKnight GS (1978) The induction of ovalbumin and conalbumin mRNA by estrogen and progesterone in chick oviduct explant cultures. Cell 14: 403–413.

24. Crum CP, McKeon FD, Xian W (2012) The oviduct and ovarian cancer: causality, clinical implications, and "targeted prevention". Clin Obstet Gynecol 55: 24–35.

25. Jemal A, Bray F, Center MM, Ferlay J, Ward E, et al. (2011) Global cancer statistics. CA Cancer J Clin 61: 69–90.

26. Barua A, Bitterman P, Abramowicz JS, Dirks AL, Bahr JM, et al. (2009) Histopathology of ovarian tumors in laying hens: a preclinical model of human ovarian cancer. Int J Gynecol Cancer 19: 531–539.

27. Giles JR, Olson LM, Johnson PA (2006) Characterization of ovarian surface epithelial cells from the hen: a unique model for ovarian cancer. Exp Biol Med (Maywood) 231: 1718–1725.

28. Trevino LS, Giles JR, Wang W, Urick ME, Johnson PA (2010) Gene expression profiling reveals differentially expressed genes in ovarian cancer of the hen: support for oviductal origin? Horm Cancer 1: 177–186.

29. Ahn SE, Choi JW, Rengaraj D, Seo HW, Lim W, et al. (2010) Increased expression of cysteine cathepsins in ovarian tissue from chickens with ovarian cancer. Reprod Biol Endocrinol 8: 100.

30. Lim W, Kim JH, Ahn SE, Jeong W, Kim J, et al. (2012) Avian SERPINB11 gene: a marker for ovarian endometrioid cancer in chickens. Exp Biol Med (Maywood) 237: 150–159.

31. Lim W, Kim HS, Jeong W, Ahn SE, Kim J, et al. (2012) SERPINB3 in the chicken model of ovarian cancer: a prognostic factor for platinum resistance and survival in patients with epithelial ovarian cancer. PLoS One 7: e49869.

32. Jeong W, Kim HS, Kim YB, Kim MA, Lim W, et al. (2012) Paradoxical expression of AHCYL1 affecting ovarian carcinogenesis between chickens and women. Exp Biol Med (Maywood) 237: 758–767.

33. Lee JY, Jeong W, Kim JH, Kim J, Bazer FW, et al. (2012) Distinct expression pattern and post-transcriptional regulation of cell cycle genes in the glandular epithelia of avian ovarian carcinomas. PLoS One 7: e51592.

34. Kato H, Torigoe T (1977) Radioimmunoassay for tumor antigen of human cervical squamous cell carcinoma. Cancer 40: 1621–1628.

35. Chechlinska M, Kowalewska M, Brzoska-Wojtowicz E, Radziszewski J, Ptaszynski K, et al. (2010) Squamous cell carcinoma antigen 1 and 2 expression in cultured normal peripheral blood mononuclear cells and in vulvar squamous cell carcinoma. Tumour Biol 31: 559–567.

36. Shiiba M, Nomura H, Shinozuka K, Saito K, Kouzu Y, et al. (2010) Down-regulated expression of SERPIN genes located on chromosome 18q21 in oral squamous cell carcinomas. Oncol Rep 26: 241–249.

37. Hsu KF, Huang SC, Shiau AL, Cheng YM, Shen MR, et al. (2007) Increased expression level of squamous cell carcinoma antigen 2 and 1 ratio is associated with poor prognosis in early-stage uterine cervical cancer. Int J Gynecol Cancer 17: 174–181.

38. Nakashima T, Yasumatsu R, Kuratomi Y, Masuda M, Kuwano T, et al. (2006) Role of squamous cell carcinoma antigen 1 expression in the invasive potential of head and neck squamous cell carcinoma. Head Neck 28: 24–30.

39. Gettins PG (2002) Serpin structure, mechanism, and function. Chem Rev 102: 4751–4804.

40. Suminami Y, Kishi F, Sekiguchi K, Kato H (1991) Squamous cell carcinoma antigen is a new member of the serine protease inhibitors. Biochem Biophys Res Commun 181: 51–58.

41. Luo JL, Tan W, Ricono JM, Korchynski O, Zhang M, et al. (2007) Nuclear cytokine-activated IKKalpha controls prostate cancer metastasis by repressing Maspin. Nature 446: 690–694.

42. Li HW, Leung SW, Chan CS, Yu MM, Wong YF (2007) Expression of maspin in endometrioid adenocarcinoma of endometrium. Oncol Rep 17: 393–398.

43. Sood AK, Fletcher MS, Gruman LM, Coffin JE, Jabbari S, et al. (2002) The paradoxical expression of maspin in ovarian carcinoma. Clin Cancer Res 8: 2924–2932.

44. Gronroos M, Maenpaa J, Kangas L, Erkkola R, Paul R, et al. (1987) Steroid receptors and response of endometrial cancer to hormones in vitro. Ann Chir Gynaecol Suppl 202: 76–79.

45. Baekelandt MM, Castiglione M (2008) Endometrial carcinoma: ESMO clinical recommendations for diagnosis, treatment and follow-up. Ann Oncol 19 Suppl 2: ii19–20.

46. Creger CR, Scott JT (1977) Dietary Zinc as an Effective Resting Agent for Laying Hen. Poultry Science 56: 1706–1706.

47. Berry WD, Brake J (1985) Comparison of Parameters Associated with Molt Induced by Fasting, Zinc, and Low Dietary-Sodium in Caged Layers. Poultry Science 64: 2027–2036.

48. Stewart MD, Johnson GA, Gray CA, Burghardt RC, Schuler LA, et al. (2000) Prolactin receptor and uterine milk protein expression in the ovine endometrium during the estrous cycle and pregnancy. Biol Reprod 62: 1779–1789.

49. Song G, Bazer FW, Spencer TE (2007) Pregnancy and interferon tau regulate RSAD2 and IFIH1 expression in the ovine uterus. Reproduction 133: 285–295.

50. Livak KJ, Schmittgen TD (2001) Analysis of relative gene expression data using real-time quantitative PCR and the 2(-Delta Delta C(T)) Method. Methods 25: 402–408.

51. Song G, Spencer TE, Bazer FW (2006) Progesterone and interferon-tau regulate cystatin C in the endometrium. Endocrinology 147: 3478–3483.

gga-miR-375 Plays a Key Role in Tumorigenesis Post Subgroup J Avian Leukosis Virus Infection

Hongxin Li[1,2⊙], Huiqing Shang[1⊙], Dingming Shu[3], Huanmin Zhang[4], Jun Ji[1], Baoli Sun[1,2], Hongmei Li[1,2], Qingmei Xie[1,2]*

1 College of Animal Science, South China Agricultural University, Guangzhou, P R China, **2** Key Laboratory of Chicken Genetics, Breeding and Reproduction, Ministry of Agriculture, Guangzhou, P R China, **3** State Key Laboratory of Livestock and Poultry Breeding, Guangzhou, P R China, **4** United States Department of Agriculture (USDA), Agriculture Research Service, Avian Disease and Oncology Laboratory, East Lansing, Michigan, United States of America

Abstract

Avian leukosis is a neoplastic disease caused in part by subgroup J avian leukosis virus J (ALV-J). Micro ribonucleic acids (miRNAs) play pivotal oncogenic and tumour-suppressor roles in tumour development and progression. However, little is known about the potential role of miRNAs in avian leukosis tumours. We have found a novel tumour-suppressor miRNA, gga-miR-375, associated with avian leukosis tumorigenesis by miRNA microarray in a previous report. We have also previously studied the biological function of gga-miR-375; Overexpression of gga-miR-375 significantly inhibited DF-1 cell proliferation, and significantly reduced the expression of yes-associated protein 1 (YAP1) by repressing the activity of a luciferase reporter carrying the 3'-untranslated region of YAP1. This indicates that gga-miR-375 is frequently downregulated in avian leukosis by inhibiting cell proliferation through YAP1 oncogene targeting. Overexpression of gga-miR-375 markedly promoted serum starvation induced apoptosis, and there may be the reason why the tumour cycle is so long in the infected chickens. In vivo assays, gga-miR-375 was significantly downregulated in chicken livers 20 days after infection with ALV-J, and *YAP1* was significantly upregulated 20 days after ALV-J infection ($P<0.05$). We also found that expression of cyclin E, an important regulator of cell cycle progression, was significantly upregulated ($P<0.05$). *Drosophila* inhibitor of apoptosis protein 1 (DIAP1), which is related to caspase-dependent apoptosis, was also significantly upregulated after infection. Our data suggests that gga-miR-375 may function as a tumour suppressor thereby regulating cancer cell proliferation and it plays a key role in avian leukosis tumorigenesis.

Editor: Qiliang Cai, Fudan University, China

Funding: This work was supported by grants from the National Natural Science Foundation of China (31072152) and the Open Project of the State Key Laboratory of Biocontrol (SKLBC2010K07). The funders had no role in study design, data collection and analysis, decision to publish, or preparation of the manuscript.

Competing Interests: The authors have declared that no competing interests exist.

* E-mail: qmx@scau.edu.cn

⊙ These authors contributed equally to this work.

Introduction

Subgroup J avian leukosis virus (ALV-J), belonging to the family *Retroviridae*, subfamily *Orthoretrovirinae* and genus *Alpharetrovirus*, causes a variety of tumours in chickens. ALV-J was first isolated from meat-type chickens in Great Britain in 1988 [1]. In poultry, ALV-J spreads widely and induces myeloid leukosis (ML) and other tumours [2,3,4]. To date, infection of ALV-J in both commercial and meat-type chickens has caused major economic loss and seriously threatened the prosperity of the poultry industry all over the world [5,6,7]. It was reported, for instance, that ALV-J causes up to 60% morbidity and 20% mortality in some Chinese flocks [8]. Emerging ALV-J infections coarsely induce various tumours in both commercial laying hen flocks as well as native Chinese breeds of chickens [8,9], which results in late onset and acute tumours in the field [10,11].

Micro ribonucleic acids (miRNAs) are a class of small, non-coding RNAs consisting of 19–23 nucleotides, which are believed to play important roles in regulating various biological processes including tumorigenesis. It is presumed that miRNAs may eventually be instrumental in the diagnosis and treatment of cancer [12,13,14]. Since the discovery of the first miRNA, lin-4, in *Caenorhabditis elegans* two decades ago [15], there has been over 24,000 curated miRNA entries identified from various species (http://www.mirbase.org/). The expression of miRNAs has been profiled widely and aberrant miRNA expression has been reported to contribute to cancer development and progression [16,17,18,19]. It was estimated that miRNAs are involved in the regulation of more than 30% of all protein-coding genes. It was also suggested that more than 50% of miRNA genes reside in cancer-associated genomic regions or in fragile sites [20,21]. miRNAs offer a fast energy-saving and fine-tuning mechanism for translation control of protein production [22].

miR-375 was originally reported in pancreatic islets of humans and mice to regulate insulin secretion in isolated pancreatic cells [23]. Later, it was reported that miR-375 is commonly downregulated in human tumour tissues, which significantly increased cancer cell development [24,25]. miR-375 was proposed as a candidate tumour suppressor miRNA in gastric carcinoma targeting 14-3-3zeta, Janus kinase 2, and phosphoinositide-dependent kinase 1 [24,25] and was recognized to inhibit neuritis

Table 1. Sequences of RNA oligonucleotides.

Name	Sense Strand/Sense Primer (5'-3')	Antisense Strand/Antisense Primer (5'-3')
Primers for Gene or 3'UTR Cloning		
YAP1	TTCTCGAGGGAGATGGGATGAATATAGAAGG	GGTGTCTAGACCACAGGCAGCAGGAGAC
YAP1-3'UTR	TTATCCCTCCTTTAAGTGAGATTCTCACAATTG	TTAAAGGAGGGATAAAGGAGTTATGGGT
Primers for qRT-PCR		
YAP1 primers	GAACTCAGCATCAGCCATGA	CTACGGAGAGCCAATTCCTG
Cyclin E primers	CACCCTCTCCTGCAACCTAA	TGGTGCAACTTTGGTGGATA
DIAP1 primers	GCCATAACAACTGCTGCTGA	TCTCTTTCAAGGCAGGCAAT
GAPDH primer	AGGCTGAGAACGGGAAACTTG	CACCTGCATCTGCCCATTTG

differentiation by lowering HuD levels [26]. In hepatocellular carcinoma (HCC) research, miR-375 was found as an important regulator of the yes-associated protein (YAP) oncogene with a potential therapeutic role in HCC treatment [27]. miR-375 promotes palmitate-induced lipoapoptosis in insulin-secreting NIT-1 cells through the inhibition of myotrophin (V1) protein expression [28]. Studies show that microRNAs are also involved in various diseases in poultry including avian influenza, avian leukosis, infectious bursal disease, Marek's disease, and ovarian carcinoma [29,30,31,32,33,34].

ALV-J infected broilers are pathologically characterized with clearly visible grey-white nodules on the liver, spleen, and kidney. The nodules range widely in size and can be several times the size of the liver or spleen [35]. Although some signalling molecules have been uncovered that control stem cell proliferation, little is known about the molecular mechanism underlying ALV-J induced tumorigenesis and few prognostic markers have been identified that can predict genetic resistance or susceptibility to ALV-J in poultry. Our previous study has shown that gga-miR-375 to be frequently downregulated in the livers of chickens 10-weeks post ALV-J infection [35]. We also found that gga-miR-375 targets YAP1. Our findings, to some extent, were in agreement with a report on miRNA-375 in which it was shown to target the Hippo signalling effector YAP in human liver cancer and to inhibit tumour propagation [27]. The Hippo pathway was initially identified in flies and was implicated in controlling organ size. Hippo pathway's downstream target genes, such as *cyclin E* and *DIAP1*, are closely related to tumour suppression activities [36,37,38,39].

The antiapoptotic properties and abnormal cell cycle progression are striking features of tumour cells. Overwhelming evidence indicates that aberrant miRNA expression is a cause or indicator of many disease processes. This study was undertaken to explore the roles of gga-miR-375 in chickens with respect to tumour development and progression induced by ALV-J infection. We also intended to elucidate the molecular mechanisms underlying tumorigenesis and to evaluate whether gga-miR-375 expression levels could serve as a biomarker for diagnostic purposes.

Materials and Methods

Virus and cell lines

The NX0101 strain of ALV-J used in all the relevant experiments and was obtained from Professor Cui, Shandong Agricultural University, People's Republic of China. DF-1 was an immortalized chicken embryo fibroblast cell line, and CHO was a continuous cell line of Chinese hamster ovary. DF-1 cell line was cultured in Dulbecco's modified eagle medium (DMEM) supplemented with 10% fetal bovine serum (FBS; Invitrogen Gibco Co, Carlsbad, CA, USA). CHO cell line were cultured in Roswell Park Memorial Institute (RMPI) 1640 supplemented with 10% FBS (Invitrogen Gibco Co).

RNA oligoribonucleotides and cell transfections

The RNA duplex mimic chicken encoded miRNAs (see Table 1 for sequences) were designed as described previously [40]. The control RNA duplex (named gga-miR-NC; sense strand: UU-CUCCGAACGUGUCACGUTT) was nonhomologous to any chicken genome sequence and used for gga-miR-375. All RNA oligoribonucleotides were purchased from Genepharma (Invitrogen Gibco Co). Transfection of RNA oligoribonucleotide(s) was done using X-tremeGENE siRNA Transfection Reagent (Roche Applied Science, Mannheim, Germany) following the manufacturer's protocol. For each transfection, 40 nM of RNA duplex were respectively used in a 6-well plate, unless otherwise indicated.

Cell proliferation assay

Cell proliferation was measured using the WST-1 (Water-soluble tetrazolium, the sodium salt of 4-[3-(4iodophenyl)-2-(4-nitrophenyl)-2H-5-tetrazolio]-1, 3-benzene disulfonate; Roche Applied Science) colorimetric assay. Approximately 24 hours after transfection with gga-miR-375 or negative control oligonucleotides gga-miR-NC (miR-NC), DF-1 cells (1.0×10^5 per millilitre) were seeded, respectively, into a 96-well plate and incubated for another 24, 48, or 72 hours. In addition, a non-transfected (mock) group was used as an additional control. Then, 10 µL of WST-1 reagent was added and incubated for 2 hours at 37°C. Absorbance was subsequently determined at wavelengths of 450 nm using multimode microplate readers (BioTek, Gene Company limited, Hong Kong, People's Republic of China). At least eight replicate wells were included for each experimental group, and all experiments were repeated independently three times. Cell proliferation was calculated by subtracting the absorbance values of the samples from the media alone (background level). The relative cell proliferation was normalized to the respective control.

Colony formation assay

Approximately 24 hours after transfection with gga-miR-375 or mir-NC, 1,000 transfected DF-1 cells were seeded in 6-well plates and maintained in DMEM containing 10% FBS for 2 weeks. Moreover, a mock group was set as another control. Colonies were fixed with methanol and stained with 0.1% crystal violet in 20% methanol for 15 minutes.

Wound healing assay

For the wound healing migration assay, approximately 24 hours after transfection with gga-miR-375 or miR-NC, DF-1 cells (1.6 × 10^5 per millilitre) were seeded on 24-well plates. A mock group was also set. Forty-eight hours after transfection, a scratch wound was made on a confluent monolayer culture of DF-1 cells with a 100 mL pipette tip and fresh media was added for incubation for another 48 hours. The cells were imaged at three different time points (0, 24, and 48 hours after wound induction) using an inverted microscopy system (Leica DM IL LED, Leica Microsystems GmbH, Wetzlar, Germany) equipped with ProgResH MF camera (Jenoptik GmbH, Jena, Germany). The percentage of wound closure (cell migration) was calculated as relative wound area at a given time point normalized to wound area at 0 hours. All experiments were performed independently in triplicate.

Apoptosis assays

Apoptosis was evaluated by apoptotic morphology and Annexin V-fluorescein isothiocyanate/propidium iodide (FITC/PI) assay for which cells were treated in similar ways as for the cell proliferation assay. About 24 hours after transfection with gga-miR-375 or miR-NC, DF-1 cells (1.0 × 10^5 per millilitre) were seeded respectively into a 6-well plate and incubated for another 24, 48, or 72 hours under serum starvation; a blank control was also used. Then, Annexin V-FITC/PI assay (BD Biosciences Pharmingen, Franklin Lakes, NJ, USA) was performed according to the manufacturer's protocol. After staining, cells were analysed by FACS Calibur (Becton Dickinson, San Jose, CA, USA). For morphologic examination, after 48 hour serum starvation treatment, cells were stained with 4'-6'-diamidino-2-phenylindole (DAPI; Sigma-Aldrich Co, St Louis, MO, USA) and those with fragmented or condensed nuclei in deep staining were counted as apoptotic cells. At least 500 cells were counted for each plate. The background luminescence associated with cell culture and assay reagent (blank reaction) was subtracted from the experimental value.

Vector construction

To construct a luciferase reporter vector, pmiRGLO-YAP1-3'UTR-wt, a wild-type 3' UTR fragment of YAP1, was amplified by RT-PCR using the primers 5'-TTCTCGAGGGGAGATGG-GATGAATATAGAAGG-3' and 5'-GGTGTCTAGACCA-CAGGCAGCAGGAGAC-3'. The putative binding sites for gga-miR-375 was inserted downstream of the stop codon of firefly luciferase in pmiRGLO Dual-Luciferase miRNA Target Expression Vector (Ambion, Promega, Beijing, People's Republic of China) as described previously[41] (designated as YAP1'UTR-wt). PmiRGLO-YAP1-3'UTR-mut, which carries a mutated sequence in the complementary site for the seed region of gga-miR-375, was generated using the primers 5'-TTATCCCTCCTTTAAGTGA-GATTCTCACAATTG-3' and 5'-TTAAAGGAGGGATAAAG-GAGTTATGGGT-3' (designated as YAP1-3'UTR-mut).

Dual luciferase reporter assay

Dual luciferase reporter assay was comprised of two reporters; one is Renilla luciferase expression construct, the other is a firefly luciferase expression construct in pmiRGLO containing the assayed 3'UTR sequences. For luciferase reporter assay, CHO cells (3.4 × 10^5) were plated in a 24-well plate and then co-transfected with 10 nmol/L gga-miR-375 or miR-NC, 20 ng YAP1-3'UTR-wt, or YAP1-3'UTR-mut, and 4 ng pRL-TK (Promega) using X-tremeGENE siRNA Transfection Reagent (Invitrogen Roche Applied Science) following the manufacturer's

protocol. Cells were collected 48 hours after transfection and analysed using the Dual-Luciferase Reporter Assay System (Promega). Luciferase activity was detected by Lumat LB 9507 Ultra Sensitive Tube Luminometer (Titertek Berthold, Nanjing, People's Republic of China). Firefly luciferase activity of each sample was normalized by Renilla luciferase activity. Transfections were done in duplicates and repeated independently at least three times.

Western blotting

At 48 or 72 hours after transfection with gga-miR-375 or miR-NC, DF-1 or CHO cells were subjected to Western blot analysis as described previously [42]. In addition, a non-transfected (mock) group was set. The primary antibodies used for Western blot analysis were polyclonal rabbit anti anti-YAP1 (1:600; predicted molecular weight: 65 kDa; Bioss Inc, Wobourn, MA, USA) and β-actin (1:600; predicted molecular weight: 42 kDa; Bioss Inc) which served as a protein loading control. Secondary antibody was goat polyclonal anti rabbit IgG (H+L)-horseradish peroxidase (HRP; Bioss Inc).

Animal experiment

Specific pathogen-free (SPF) chickens were purchased from Guangdong Wen's Foodstuffs Group Co Ltd (Yunfu, People's Republic of China), housed in negatively-pressured biosecurity isolators under quarantine conditions, and provided with water and commercial feed ad libitum. One hundred and one day old SPF chickens were randomly divided into two groups of fifty chickens each. The first group (NX0101) was inoculated intra-abdominally at 1 day of age with $10^{3.7}$ $TCID_{50}$/0.2 mL virulent NX0101 strain. The other group (NC) was inoculated with the same volume of nutrient solution. The second group was used as the control group. Three chickens from each of the two groups were euthanized for necropsy every 10 days post infection. Tissues samples were collected from each chicken at necropsy and snap frozen in liquid nitrogen. The tissue samples were stored at −70 °C until subsequent analysis. Institutional and national guidelines for the use and care of experimental animals were closely followed. Use of animals in this study was approved by the South China Agricultural University Committee of Animal Experiments (approval ID 201004152).

Extraction of total RNA and miRNA

Total RNA was extracted from tissue samples with TRIzol reagent (Invitrogen); miRNA was extracted using the mirVana miRNA Isolation Kit (Life Technologies, Calrsbad, CA, USA) following the manufacturer's instructions.

miRNA microarray

Microarray analysis was performed as described previously [24]. High-quality total RNAs, isolated from three chickens for each of the two groups (infection group and control group) at 10 weeks of age and three normal liver tissues using TRIzol reagent according to the manufacturer's instructions (Invitrogen), was carried out using the μParaflo microfluidic technology according to the manufacturer's protocol (LC Sciences, Houston, TX, USA).

Real-time quantitative RT-PCR

gga-miR-375 and related gene expression was evaluated for absolute quantification using real-time quantitative reverse transcriptase polymerase chain reaction (RT-PCR) assays. gga-miR-375 and reference 5S rRNA, or the target genes and the reference gene glyceraldehyde-3-phosphate dehydrogenase (GAPDH) were

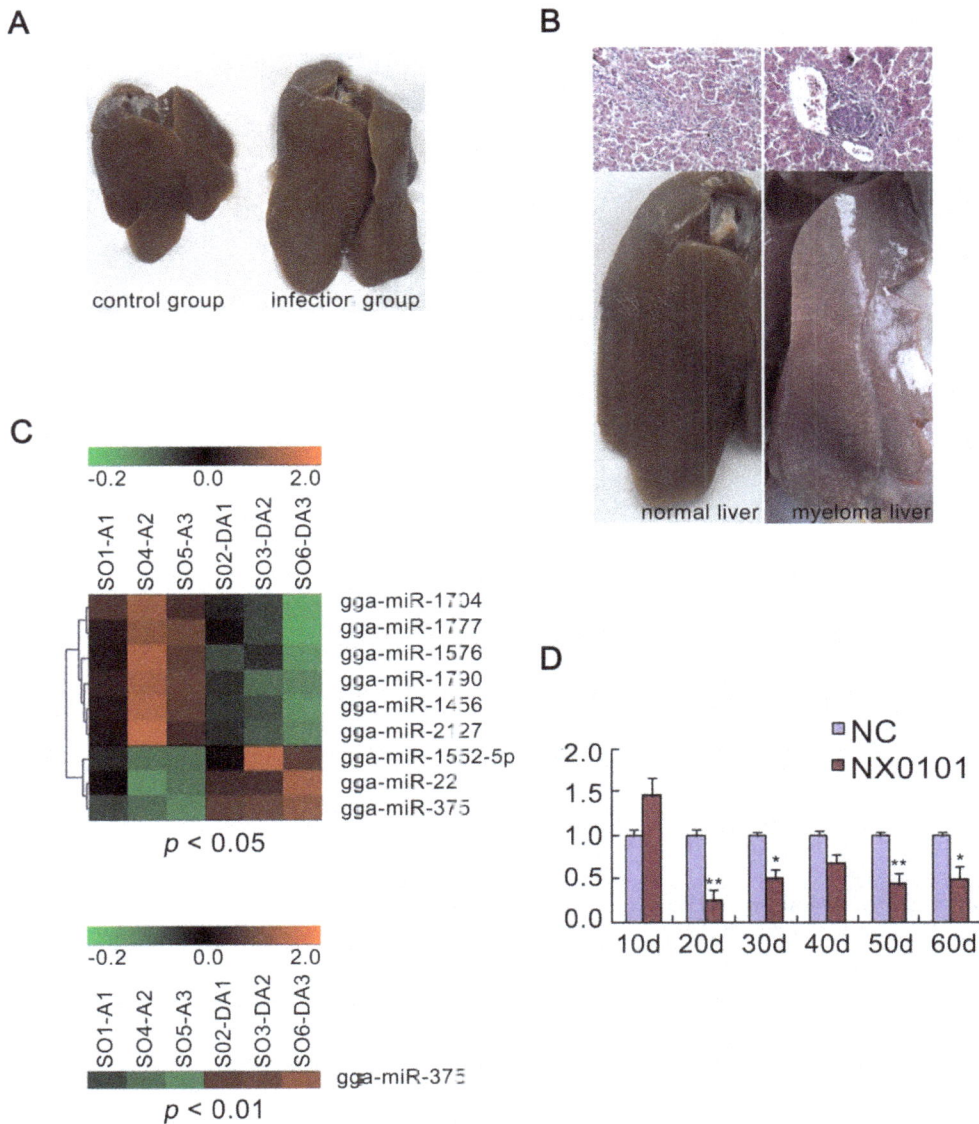

Figure 1. gga-miR-375 expression was frequently downregulated in ALV-J induced cancer. Liver lesions induced by viral infection in SPF white leghorn chickens at 70 days (**A**) and later (**B**). Representative histological features of nontumour liver and myeloma liver are shown with hematoxylin and eosin staining, 400× (**B**). (**C**) The miRNAs significantly associated with ALV-J by significance analysis of microarrays are listed. gga-miR-375 is most significantly associated with ALV-J infected liver tissue, as determined by significance analysis of microarrays. (**D**) Quantitative real-time PCR quantification of gga-miR-375 expression in the liver of ALV-J infected chickens every 10 days between 10 and 60 days post transfection (**P < 0.01, *p<0.05).

amplified, cloned, and used as standard controls to generate standard curves following a previously described protocol [42]. The data from the real-time quantitative RT-PCR were analysed as relative miRNA expression using the $2^{-\triangle\triangle Ct}$ method. The 5S rRNA was used as an internal control.

Statistical analysis

Fixed effect was assessed by one-way analysis of variance (ANOVA). Unless otherwise noted, pairwise comparisons were done using Student's two-tailed t-test, and the differences were assessed by one-way analysis of variance (ANOVA) when more than two groups were compared. Results are presented as mean ± standard error of the mean (SEM) unless otherwise noted. The

differences between groups were analysed when two or more groups were compared. Differences were considered statistically significant when $P<0.05$.

Accession number

The microarray data were MIAME compliant and our data have been deposited in a MIAME compliant database (ArrayExpress, GEO ID: GSE28434). The sequences of gga-miR-375, hsa-miR-375 and rno-miR-3375 (MI0003705, MI0000783, MI0006140) described in this paper have been deposited in miRBase (http://www.mirbase.org/).

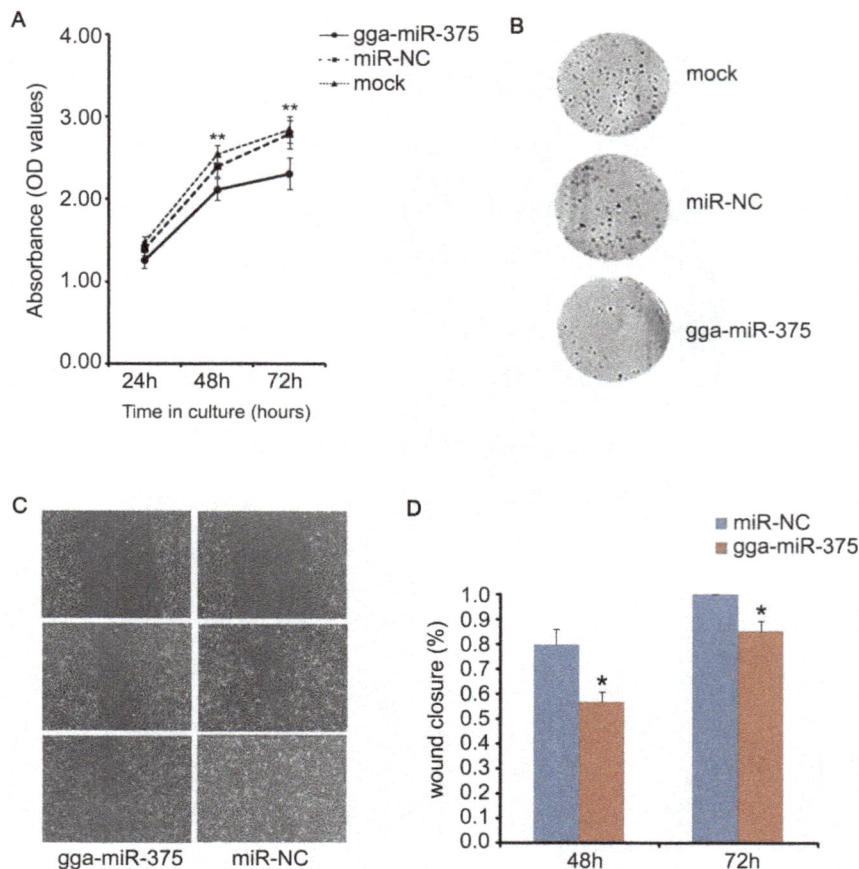

Figure 2. gga-miR-375 inhibited DF-1 cell proliferation and invasion. The cells transfected with gga-miR-375, miR-NC, or mock were subjected toWST-1 analysis, colony formation, and wound healing assay. (**A**) Effects of gga-miR-375 on proliferation over different time periods. Plotted means and standard errors were computed from data of three independent experiments; bars, SEM. **$P<0.01$. (**B**) Effects of gga-miR-375 on colony formation of DF-1 cells. (**C**) Images of cell migration from wound healing assay. Scratch wounds were made on confluent monolayer cultures 48 hours post transfection. Images of wound repair were taken at 0, 24, and 48 hours after wound. (D)The percentage of wound closure was normalized to the wound area at hour 0 (above panel). Plotted means and standard errors were computed from data of three independent experiments. The comparisons were evaluated using t-test; bars, SEM. *$P<0.05$.

Results

Expression of gga-miR-375 in the liver of ALV-J infected chickens

Compared to control chickens, most chickens in the ALV-J infected group showed gradual emaciation. Livers of the infected chickens were evidently bigger than the control group at 10 weeks (Figure 1A), and some developed tumour formations (Figure 1B). miRNA microarray profiling was performed in SPF chicken livers of controls and animals infected with ALV-J NX0101 strain, and the results showed that gga-miR-375 was significantly downregulated in SPF chicken livers of infected chickens at 10 weeks ($P<0.01$; Figure 1 C). In Animal experiments, the gga-miR-375 was significantly downregulated in liver tissue from the ALV-J infected chickens from 20 days post infection (Figure 1D), which may serve as a biomarker for diagnostic purposes.

Overexpression of gga-miR-375 inhibited DF-1 cell proliferation and invasion

To explore the role of gga-miR-375 in ALV-J carcinogenesis, we examined the effect of gga-miR-375 overexpression on the proliferation of DF-1 cell lines. The cells were transfected with either gga-miR-375 (gga-miR-375) or negative control oligonucle-

otides gga-miR-NC (miR-NC), and then cultured for various periods of time (24, 48, or 72 hours). In addition, a NT (mock) group was set as another control. Cell proliferation reagent WST-1 assays showed that all three groups (mock, miR-NC, and gga-miR-375) displayed fewer cells and overexpression of gga-miR-375 significantly inhibited the proliferation of DF-1 cells from 48 hours after transfection (Figure 2A) compared to the NC (miR-NC) or the mock group. Colony formation assay confirmed this inhibition (Figure 2B). To determine the effect of gga-miR-375 on the invasion of DF-1 cells, we conducted a wound healing assay. This assay showed that the invasion of the gga-miR-375 transfected cells was slower than the NC and non-transfected (NT) treated cells (Figure 2C, 2D). These results suggested that gga-miR-375 inhibits cell proliferation and invasion.

gga-miR-375 promotes serum starvation induced apoptosis

Approximately 24, 48 and 72 hours after transfection, apoptosis was assessed by morphological examination and Annexin V-FITC/PI staining. The DAPI staining data suggests that gga-miR-375 overexpression remarkably increased serum starvation induced apoptosis in DF-1 cells ($P<0.001$; Figure 3A, 3B) at 48 and 72 hours. The analysis of Annexin V-FITC/PI staining

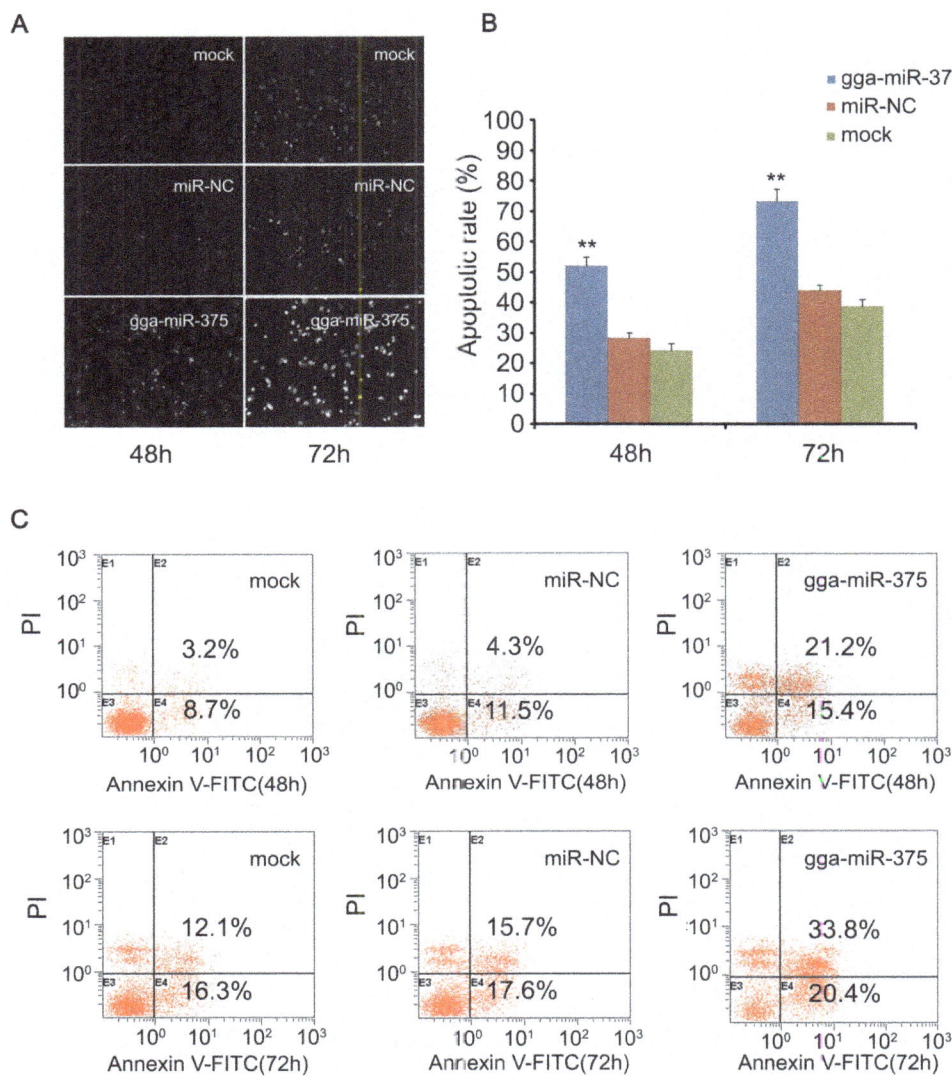

Figure 3. gga-miR-375 promoted serum starvation induced apoptosis. The cells transfected with gga-miR-375, miR-NC, or mock were subjected to DAPI and Annexin V-FITC/PI staining. (**A**) Apoptotic rates of DF-1 cells were evaluated by apoptotic morphology examination; (**C**) Apoptotic rates of DF-1 cells evaluated by Annexin V-FITC/PI staining during 48 and 72 hours post-transfection. (**B**) Apoptotic rate plot showing differences between gg-miR-375, NC, and mock treatment groups. Plotted means and standard errors were computed from data of three independent experiments; bars, SEM. **$P<0.01$.

confirmed the gga-miR-375 increased serum starvation induced apoptosis from 54.2% to 36.6% (Figure 3C). These results collectively demonstrate that gga-miR-375 may inhibit cell proliferation and invasion by increasing apoptosis under serum starvation.

gga-miR-375 represses YAP1 protein production through 3′-UTR binding

To explore the role that gga-miR-375 plays in ALV-J carcinogenesis, TargetScan, miRBase, and RNAhybrid algorithms were employed to search for putative cellular protein-coding gene targets of gga-miR-375. Based on TargetScan and miRBase search, YAP1 was predicted as a potential target gene of gga-miR-375 (Figure 4A). The gga-miR-375 differs from homo sapiens miR-375 and rattus norvegicus miR-375 by a single base (Figure 4A). To test whether the predicted gga-miR-375-binding sites in the 3′-UTR of YAP1 mRNA were responsible for its

regulatory role, the 3′-UTR region of YAP1 was cloned downstream of a luciferase reporter gene (YAP1-3′UTR-wildtype), and co-transfected DF-1 cells with gga-miR-375 precursor, miR-NC, or NT cells. The luciferase activity of cells transfected with a gga-miR-375 precursor was significantly decreased compared to the NC ($P<0.01$; Figure 4B), indicating the mutation within the putative gga-miR-375-binding site clearly abrogated the repression of luciferase activity caused by gga-miR-375 overexpression. To further confirm YAP1 as a direct target of gga-miR-375, YAP1 protein expression was assayed 48 and 72 hours after transfection with gga-miR-375, miR-NC, or NT in DF-1 or CHO cells. The gga-miR-375 significantly suppressed the expression of YAP1 compared to miR-NC and NT (Figure 4C). These data suggested that gga-miR-375 may directly inhibit YAP1 protein production through binding to the 3′-UTR of YAP1.

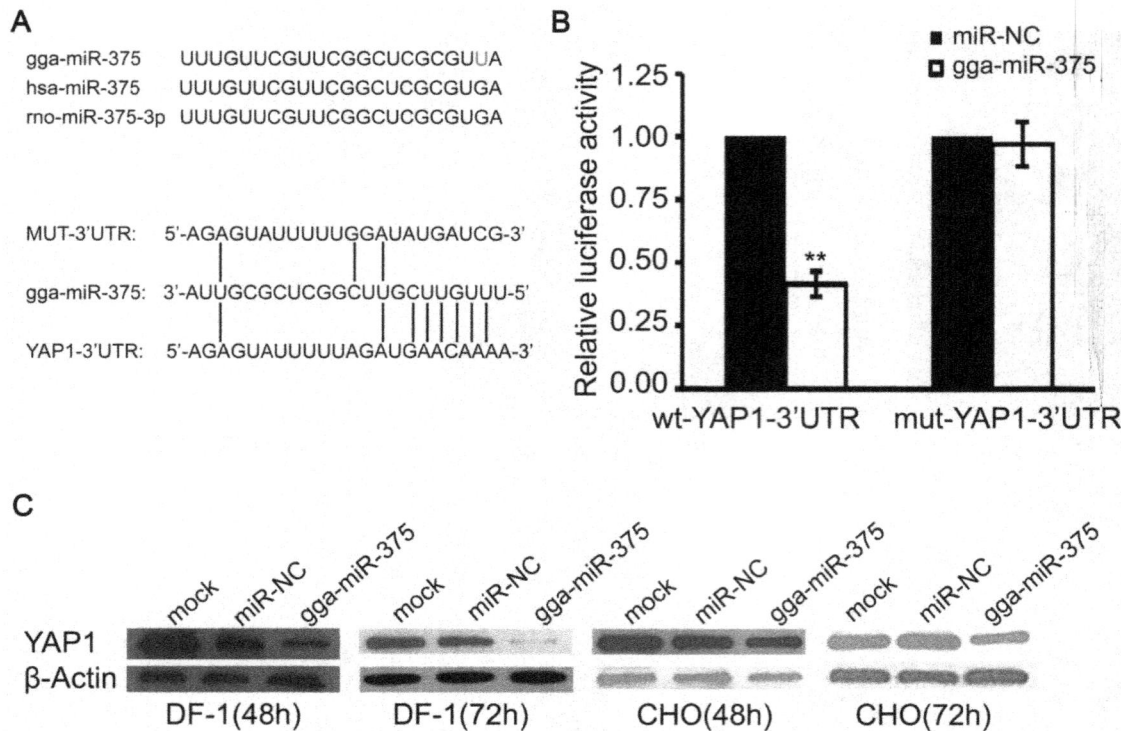

Figure 4. YAP1 is a direct gga-miR-375 target. (**A**) Differences in gga-miR-375, homo sapiens miR-375, and rattus norvegicus miR-375. Alignment of YAP1-3'UTR, gga-miR-375, and MUT-3'UTR, where the complementary site for the seed region of gga-miR-375 is indicated. (**B**) The regulation of luciferase activity by YAP1-3'UTR is dependent on gga-miR-375. CHO cells were co-transfected with YAP1-3'UTR-wt with either gga-miR-375 or miR-NC (left), and YAP1-3'UTR-mut with either gga-miR-375 or miR-NC (right). Columns, mean of at least three independent experiments done in duplicate; bars, SEM. **$P<0.01$, compared to miR-NC-transfected cells. (**C**) Ectopic expression of gga-miR-375 reduced YAP1 protein production in both DF-1 and CHO cells. β-actin levels were used as a control. Each experiment was repeated three times, and each sample was assayed in triplicate.

mRNA expression of *YAP1*, *cyclin E*, and *DIAP1* in the liver, blood, bone marrow, and spleen of ALV-J infected chickens

Because gga-miR-375 inhibited cell proliferation and invasion and suppressed YAP1 protein production at the cellular level, we checked whether gga-miR-375 targeted Hippo signalling effector YAP1 in ALV-J infected chickens at intervals of 10 days up to 60 days. Cyclin E, a prognostic marker in other tumours, was tested in this study. DIAP1, which is associated with apoptosis, was also detection in this study. The mRNA expression of *YAP1*, *cyclin E*, and *DIAP1* during 50–60 days post infection was significantly upregulated (Figure 5A) suggesting that this period might be important for tumour formation and development. *YAP1*, *cyclin E*, and *DIAP1* expression were also significantly upregulated in livers but bone marrow and spleen at 20–30 days post-infection (Figure 5B, 5C). *YAP1* was also upregulated in blood during this period (Figure 5B). No significant differences were observed at other time points during the testing period.

Discussion

There is substantial literature on miR-375 documenting this microRNA as a tumour suppressor in humans. However, such a role for gga-miR-375 has not been investigated to date. The data from this study showed that gga-miR-375 was significantly downregulated in liver tissue of chickens 10 weeks post ALV-J infection, which inhibited cell proliferation and promoted cell apoptosis under serum starvation. This finding suggests that YAP1

is a direct target gene of gga-miR-375. This also suggests that gga-miR-375 in chickens and miR-375 in humans are consistent on the function [40,43], implicating mechanisms in different species and cancer types may reveal many similarities. Moreover, by directly targeting Hippo signalling effector YAP1, gga-miR-375 may directly or indirectly affect cyclin E and DIAP1 during the early stages of ALV-J infection, resulting in a range of effects on tumour development.

The role of the Hippo pathway originally defined in *Drosophila melanogaster* was to restrain cell proliferation and to promote apoptosis affecting normal cell fate and tumorigenesis [44,45]. YAP, a transcriptional co-activator amplifier, is a pivotal effector of the Hippo pathway in mouse and human cancers; YAP1 and YAP2 are potent oncogenic drivers and independent prognostic risk factors for HCC [27,38,46,47]. The importance of Hippo signalling pathway in mammalian growth control is supported by reports that transgenic overexpression of YAP, or loss of Mst1/2, leads to massive hepatomegaly and rapid progression to HCC and that YAP is amplified in some tumours and may transform immortalized mammary epithelial cells in vitro [38,46,48,49,50]. We know the size of liver and spleen in dead or sick ALV-J infected birds are enlarged to several times their normal size. However, little is known about the role of YAP1 in ALV-J induced tumours. YAP1 has several domains containing a TEAD binding region and 2 WW domains, which are DNA binding domains that function as transcriptional coactivators through interactions with DNA binding transcription factors [51,52,53]. YAP1 can transactivate growth-promoting genes and enhance p73-dependent

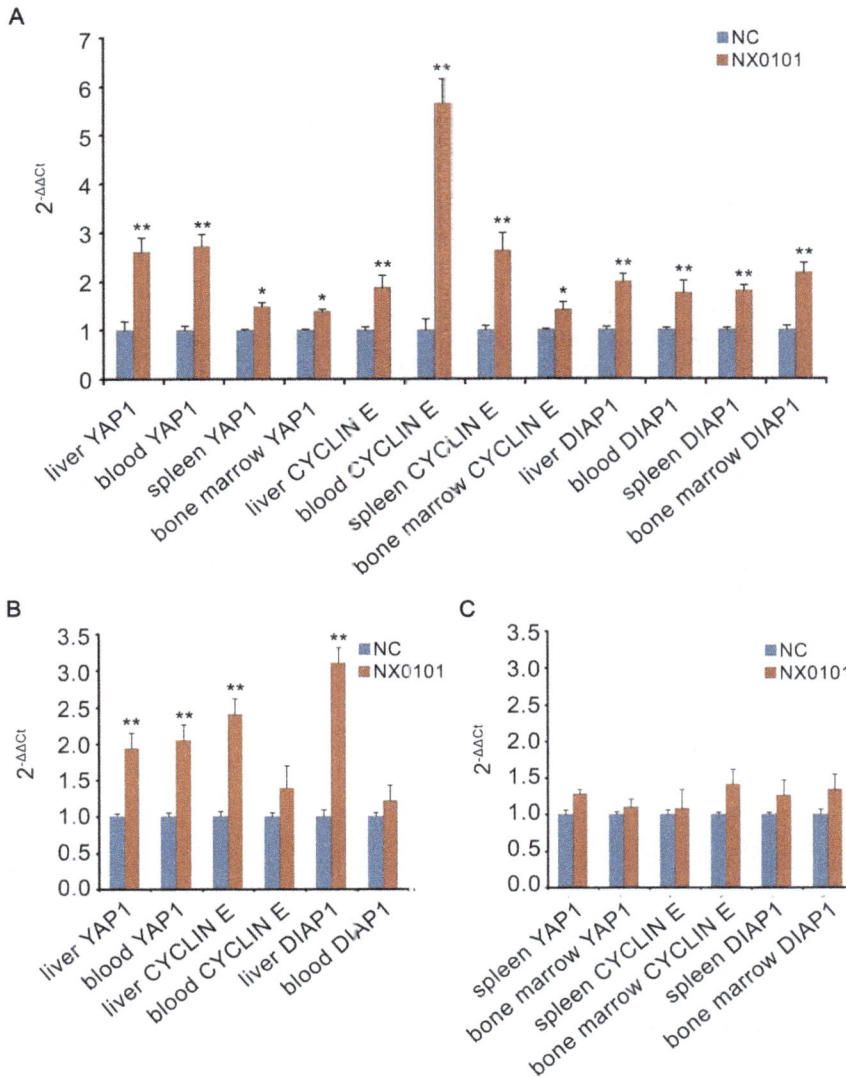

Figure 5. *YAP1, cyclin E,* **and** *DIAP1* **gene expression in the liver, bone marrow, blood, and spleen of chickens infected with ALV-J quantified by real-time RT-PCR.** (A) *YAP1, cyclin E,* and *DIAP1* gene expression at 50–60 days post-infection; *YAP1, cyclin E,* and *DIAP1* gene expression in the liver and blood (B) and in the spleen and bone marrow (C) 20–30 days post infection (**$P < 0.01$, *$p < 0.05$).

apoptosis in response to DNA damage by binding to specific domains [54,55,56]. Here, for the first time, we show that YAP1 is a direct target of gga-miR-375. The growth of DF-1 cells was suppressed along with YAP1 expression and significantly reduced when gga-miR-375 was overexpressed, and *YAP1* appeared highly expressed in infected chickens, suggesting that YAP1 may be an oncogenic gene involved in ALV-J infection.

Organisms remove damaged or unwanted cells by an evolutionarily conserved process called programmed cell death or apoptosis [57,58,59]. For tumour-inducing viruses, apoptosis is a major obstacle for virus survival and the malignant transformation of host cells [60]. Overexpression of gga-miR-375 sufficiently enhanced serum starvation induced apoptosis, implying gga-miR-375 may also activate the Hippo pathway to augment apoptosis by transactivating growth-promoting genes through the TEAD binding domain of YAP1. The reason why there was different degrees of inhibition of YAP1 in DF-1 or CHO cells may be related to the mutation base (U) (Figure 4A), suggesting that for the

mature RNA the miRNAs 3' end is important and provides evidence of an evolutionary relationship between the different species studies.

DIAP1 functions in the early embryo was to inhibit apoptosis [61]. In the absence of DIAP1, most cells undergo caspase-dependent apoptosis [62]. Increased DIAP1 levels are suspected to facilitate survival, as cells are very sensitive to even low levels of apoptotic inhibitors in the presence of pro-apoptotic stimuli [63,64,65]. As per a previous report [66], the Hippo pathway may signal through Warts to promote apoptosis by decreasing levels of the caspase inhibitor, DIAP1. Cyclin E was discovered by screening human cDNA for a rescue deficiency in G1 cyclin function in budding yeast [67]. Cyclin E is an important regulator of cell cycle progression and it reaches maximal levels of expression during the G1-to-S phase transition. This protein also exhibits specific properties that together indicate that it has an essential and rate-limiting function for allowing cells to enter into the S phase of the cell cycle [67,68,69,70]. Altered expression of the cyclin E protein was reported in most breast tumour tissues and leukemia solid tumours examined to date, and aberrant levels increase with increases in tumour grade and stage [36,71], which makes it a potential prognostic marker for some tumours. Between 50–60 days, the significant increase in levels of *DIAP1* and *cyclin E* seen in this study may serve to resist apoptosis and affect cell cycle, supporting tumour formation.

Yorkie, a Drosophila homolog of the YAP, is required for the transcription of the DIAP1 and cyclin E genes and its inactivation leads to growth arrest and apoptosis [52,72]. As downstream genes of the Hippo pathway, *cyclin E* and *DIAP1* in mammals are significantly upregulated in the liver following the significant downregulation of gga-miR-375 in the liver, and *YAP1* is significantly upregulated. There may be a similar Hippo pathway operating in chickens. From a previous report, we know that avian leukosis infection is age-dependent; chicken resistance noticeably strengthens following growth in the first 3 weeks [73,74,75]. *Cyclin E* may be a prognostic marker that sharply augments, as it is similar to human *cyclin E* [71].

Together, these data show that the ALV-J virus may inhibit gga-miR-375 thus blocking the Hippo pathway and facilitate tumour progression. Additionally, the expression of related genes differed among organs; for instance, compared to the bone marrow and spleen, which are immune organs, liver and blood had higher levels of *YAP1*, *cyclin E*, and *DIAP1* expression from 20–30 days post infection. During the first 8 weeks post ALV-J infection, transformed follicles and histopathological changes are detected in 82% of the susceptible animals [76]. In this study, all tested mRNA expression was significantly upregulated during days 50–60, suggesting that this period might be a key time-point for tumour formation and development.

In poultry, ALV is the most common naturally occurring avian retroviral infection, and it causes neoplastic diseases and other production problems. As the virus infection spreads by both vertical and horizontal transmission that can have a long latency period and cause fatal damage, studies investigating the control of avian leukosis are difficult but also necessary. Our data shows that gga-miR-375 directly targets YAP1, induces cell apoptosis, and weakens the Hippo pathway, suggesting that ALV-J viruses might inhibit gga-miR-375 to influence cell proliferation, invasion, and apoptosis and subsequently affect normal cell fate and tumorigenesis. Although there is no available ALV cancer cell line, transfection with synthetic gga-miR-375 oligonucleotides as well as other approaches designed to increase endogenous gga-miR-375 together with follow-up tests of long term animal infection are needed in further studies.

Acknowledgements

We thank Zhizhong Cui at Shandong Agricultural University for kindly providing the NX0101 strain of ALV-J for this study.

Author Contributions

Conceived and designed the experiments: Hongxin Li HS QX. Performed the experiments: Hongxin Li HS. Analyzed the data: Hongxin Li JJ Hongmei Li. Contributed reagents/materials/analysis tools: HZ DS BS. Wrote the paper: Hongxin Li QX.

References

1. Payne LN, Howes K, Gillespie AM, Smith LM (1992) Host range of Rous sarcoma virus pseudotype RSV(HPRS-103) in 12 avian species: support for a new avian retrovirus envelope subgroup, designated J. J Gen Virol73 (Pt11): 2995–2997.
2. Payne LN, Gillespie AM, Howes K (1991) Induction of myeloid leukosis and other tumours with the HPRS-103 strain of ALV. Vet Rec 129: 447–448.
3. Payne LN (1998) Retrovirus-induced disease in poultry. Poult Sci 77: 1204–1212.
4. Payne LN, Gillespie AM, Howes K (1992) Myeloid leukaemogenicity and transmission of the HPRS-103 strain of avian leukosis virus. Leukemia 6: 1167–1176.
5. Gao Y, Yun B, Qin L, Pan W, Qu Y, et al. (2012) Molecular epidemiology of avian leukosis virus subgroup J in layer flocks in China. J Clin Microbiol 50: 953–960.
6. Cui Z, Sun S, Zhang Z, Meng S (2009) Simultaneous endemic infections with subgroup J avian leukosis virus and reticuloendotheliosis virus in commercial and local breeds of chickens. Avian Pathol 38: 443–448.
7. Payne LN, Nair V (2012) The long view: 40 years of avian leukosis research. Avian Pathol 41: 11–19.
8. Gao YL, Qin LT, Pan W, Wang YQ, Le Qi X, et al. (2010) Avian leukosis virus subgroup J in layer chickens, China. Emerg Infect Dis 16: 1637–1638.
9. Cheng ZQ, Zhang L, Liu SD, Zhang LJ, Cui ZZ (2005) [Emerging of avian leukosis virus subgroup J in a flock of Chinese local breed]. Wei Sheng Wu Xue Bao 45: 584–587.
10. Fadly AM, Smith EJ (1999) Isolation and some characteristics of a subgroup J-like avian leukosis virus associated with myeloid leukosis in meat-type chickens in the United States. Avian Dis 43: 391–400.
11. Payne LN, Gillespie AM, Howes K (1993) Recovery of acutely transforming viruses from myeloid leukosis induced by the HPRS-103 strain of avian leukosis virus. Avian Dis 37: 438–450.
12. Volinia S, Calin GA, Liu CG, Ambs S, Cimmino A, et al. (2006) A microRNA expression signature of human solid tumors defines cancer gene targets. Proc Natl Acad Sci U S A 103: 2257–2261.
13. Rosenfeld N, Aharonov R, Meiri E, Rosenwald S, Spector Y, et al. (2008) MicroRNAs accurately identify cancer tissue origin. Nat Biotechnol 26: 462–469.
14. Zhang X, Yan Z, Zhang J, Gong L, Li W, et al. (2011) Combination of hsa-miR-375 and hsa-miR-142-5p as a predictor for recurrence risk in gastric cancer patients following surgical resection. Ann Oncol 22: 2257–2266.
15. Lee RC, Feinbaum RL, Ambros V (1993) The C. elegans heterochronic gene lin-4 encodes small RNAs with antisense complementarity to lin-14. Cell 75: 843–854.
16. Xia L, Zhang D, Du R, Pan Y, Zhao L, et al. (2008) miR-15b and miR-16 modulate multidrug resistance by targeting BCL2 in human gastric cancer cells. Int J Cancer 123: 372–379.
17. Fassan M, Sachsenmeir K, Rugge M, Baffa R (2011) Role of miRNA in distinguishing primary brain tumors from secondary tumors metastatic to the brain. Front Biosci (Schol Ed) 3: 970–979.
18. Lotterman CD, Kent OA, Mendell JT (2008) Functional integration of microRNAs into oncogenic and tumor suppressor pathways. Cell Cycle 7: 2493–2499.
19. Yang Y, Li X, Yang Q, Wang X, Zhou Y, et al. (2010) The role of microRNA in human lung squamous cell carcinoma. Cancer Genet Cytogenet 200: 127–133.
20. Calin GA, Ferracin M, Cimmino A, Di Leva G, Shimizu M, et al. (2005) A MicroRNA signature associated with prognosis and progression in chronic lymphocytic leukemia. N Engl J Med 353: 1793–1801.
21. Filipowicz W, Bhattacharyya SN, Sonenberg N (2008) Mechanisms of post-transcriptional regulation by microRNAs: are the answers in sight? Nat Rev Genet 9: 102–114.
22. Bartel DP (2004) MicroRNAs: genomics, biogenesis, mechanism, and function. Cell 116: 281–297.
23. Poy MN, Eliasson L, Krutzfeldt J, Kuwajima S, Ma X, et al. (2004) A pancreatic islet-specific microRNA regulates insulin secretion. Nature 432: 226–230.
24. Ding L, Xu Y, Zhang W, Deng Y, Si M, et al. (2010) MiR-375 frequently downregulated in gastric cancer inhibits cell proliferation by targeting JAK2. Cell Res 20: 784–793.
25. Tsukamoto Y, Nakada C, Noguchi T, Tanigawa M, Nguyen LT, et al. (2010) MicroRNA-375 is downregulated in gastric carcinomas and regulates cell survival by targeting PDK1 and 14-3-3zeta. Cancer Res 70: 2339–2349.
26. Abdelmohsen K, Hutchison ER, Lee EK, Kuwano Y, Kim MM, et al. (2010) miR-375 inhibits differentiation of neurites by lowering HuD levels. Mol Cell Biol 30: 4197–4210.
27. Liu AM, Poon RT, Luk JM (2010) MicroRNA-375 targets Hippo-signaling effector YAP in liver cancer and inhibits tumor properties. Biochem Biophys Res Commun 394: 623–627.
28. Li Y, Xu X, Liang Y, Liu S, Xiao H, et al. (2010) miR-375 enhances palmitate-induced lipoapoptosis in insulin-secreting NIT-1 cells by repressing myotrophin (V1) protein expression. Int J Clin Exp Pathol 3: 254–264.
29. Wang Y, Brahmakshatriya V, Zhu H, Lupiani B, Reddy SM, et al. (2009) Identification of differentially expressed miRNAs in chicken lung and trachea with avian influenza virus infection by a deep sequencing approach. BMC Genomics 10: 512.

30. Lee JY, Jeong W, Kim JH, Kim J, Bazer FW, et al. (2012) Distinct expression pattern and post-transcriptional regulation of cell cycle genes in the glandular epithelia of avian ovarian carcinomas. PLoS One 7: e51592.

31. Lian L, Qu L, Chen Y, Lamont SJ, Yang N (2012) A systematic analysis of miRNA transcriptome in Marek's disease virus-induced lymphoma reveals novel and differentially expressed miRNAs. PLoS One 7: e51003.

32. Stik G, Dambrine G, Pfeffer S, Rasschaert D (2013) The oncogenic microRNA OncomiR-21 overexpressed during Marek's disease lymphomagenesis is transactivated by the viral oncoprotein Meq. J Virol 87: 80–93.

33. Wang Q, Gao Y, Ji X, Qi X, Qin L, et al. (2013) Differential expression of microRNAs in avian leukosis virus subgroup J-induced tumors. Vet Microbiol 162: 232–238.

34. Yao Y, Zhao Y, Xu H, Smith LP, Lawrie CH, et al. (2008) MicroRNA profile of Marek's disease virus-transformed T-cell line MSB-1: predominance of virus-encoded microRNAs. J Virol 82: 4007–4015.

35. Li H, Ji J, Xie Q, Shang H, Zhang H, et al. (2012) Aberrant expression of liver microRNAs in chickens infected with subgroup J avian leukosis virus. Virus Res 169: 268–271.

36. Keyomarsi K, Tucker SL, Buchholz TA, Callister M, Ding Y, et al. (2002) Cyclin E and survival in patients with breast cancer. N Engl J Med 347: 1566–1575.

37. Nolo R, Morrison CM, Tao C, Zhang X, Halder G (2006) The bantam microRNA is a target of the hippo tumor-suppressor pathway. Curr Biol 16: 1895–1904.

38. Overholtzer M, Zhang J, Smolen GA, Muir B, Li W, et al. (2006) Transforming properties of YAP, a candidate oncogene on the chromosome 11q22 amplicon. Proc Natl Acad Sci U S A 103: 12405–12410.

39. Wang SL, Hawkins CJ, Yoo SJ, Muller HA, Hay BA (1999) The Drosophila caspase inhibitor DIAP1 is essential for cell survival and is negatively regulated by HID. Cell 98: 453–463.

40. Lim LP, Lau NC, Garrett-Engele P, Grimson A, Schelter JM, et al. (2005) Microarray analysis shows that some microRNAs downregulate large numbers of target mRNAs. Nature 433: 769–773.

41. Su H, Yang JR, Xu T, Huang J, Xu L, et al. (2009) MicroRNA-101, down-regulated in hepatocellular carcinoma, promotes apoptosis and suppresses tumorigenicity. Cancer Res 69: 1135–1142.

42. Chen CY, Xie QM, Xue Y, Ji J, Chang S, et al. (2012) Characterization of cytotoxicity-related gene expression in response to virulent Marek's disease virus infection in the bursa of Fabricius. Res Vet Sci.

43. Nishikawa E, Osada H, Okazaki Y, Arima C, Tomida S, et al. (2011) miR-375 is activated by ASH1 and inhibits YAP1 in a lineage-dependent manner in lung cancer. Cancer Res 71: 6165–6173.

44. Harvey K, Tapon N (2007) The Salvador-Warts-Hippo pathway - an emerging tumour-suppressor network. Nat Rev Cancer 7: 182–191.

45. Saucedo LJ, Edgar BA (2007) Filling out the Hippo pathway. Nat Rev Mol Cell Biol 8: 613–621.

46. Camargo FD, Gokhale S, Johnnidis JB, Fu D, Bell GW, et al. (2007) YAP1 increases organ size and expands undifferentiated progenitor cells. Curr Biol 17: 2054–2060.

47. Zhao B, Wei X, Li W, Udan RS, Yang Q, et al. (2007) Inactivation of YAP oncoprotein by the Hippo pathway is involved in cell contact inhibition and tissue growth control. Genes Dev 21: 2747–2761.

48. Zender L, Spector MS, Xue W, Flemming P, Cordon-Cardo C, et al. (2006) Identification and validation of oncogenes in liver cancer using an integrative oncogenomic approach. Cell 125: 1253–1267.

49. Song H, Mak KK, Topol L, Yun K, Hu J, et al. (2010) Mammalian Mst1 and Mst2 kinases play essential roles in organ size control and tumor suppression. Proc Natl Acad Sci U S A 107: 1431–1436.

50. Zhou D, Conrad C, Xia F, Park JS, Payer B, et al. (2009) Mst1 and Mst2 maintain hepatocyte quiescence and suppress hepatocellular carcinoma development through inactivation of the Yap1 oncogene. Cancer Cell 16: 425–438.

51. Wang K, Degerny C, Xu M, Yang XJ (2009) YAP, TAZ, and Yorkie: a conserved family of signal-responsive transcriptional coregulators in animal development and human disease. Biochem Cell Biol 87: 77–91.

52. Bertini E, Oka T, Sudol M, Strano S, Blandino G (2009) YAP: at the crossroad between transformation and tumor suppression. Cell Cycle 8: 49–57.

53. Zhao B, Li L, Lei Q, Guan KL (2010) The Hippo-YAP pathway in organ size control and tumorigenesis: an updated version. Genes Dev 24: 862–874.

54. Oka T, Sudol M (2009) Nuclear localization and pro-apoptotic signaling of YAP2 require intact PDZ-binding motif. Genes Cells 14: 607–615.

55. Oka T, Mazack V, Sudol M (2008) Mst2 and Lats kinases regulate apoptotic function of Yes kinase-associated protein (YAP). J Biol Chem 283: 27534–27546.

56. Strano S, Munarriz E, Rossi M, Castagnoli L, Shaul Y, et al. (2001) Physical interaction with Yes-associated protein enhances p73 transcriptional activity. J Biol Chem 276: 15164–15173.

57. Vaux DL, Korsmeyer SJ (1999) Cell death in development. Cell 96: 245–254.

58. Raff MC (1992) Social controls on cell survival and cell death. Nature 356: 397–400.

59. Wyllie AH, Kerr JF, Currie AR (1980) Cell death: the significance of apoptosis. Int Rev Cytol 68: 251–306.

60. Xu S, Xue C, Li J, Bi Y, Cao Y (2011) Marek's disease virus type 1 microRNA miR-M3 suppresses cisplatin-induced apoptosis by targeting Smad2 of the transforming growth factor beta signal pathway. J Virol 85: 276–285.

61. Yoo SJ, Huh JR, Muro I, Yu H, Wang L, et al. (2002) Hid, Rpr and Grim negatively regulate DIAP1 levels through distinct mechanisms. Nat Cell Biol 4: 416–424.

62. Yokokura T, Dresnek D, Huseinovic N, Lisi S, Abdelwahid E, et al. (2004) Dissection of DIAP1 functional domains via a mutant replacement strategy. J Biol Chem 279: 52603–52612.

63. Hay BA, Wassarman DA, Rubin GM (1995) Drosophila homologs of baculovirus inhibitor of apoptosis proteins function to block cell death. Cell 83: 1253–1262.

64. Goyal L, McCall K, Agapite J, Hartwieg E, Steller H (2000) Induction of apoptosis by Drosophila reaper, hid and grim through inhibition of IAP function. EMBO J 19: 589–597.

65. Lisi S, Mazzon I, White K (2000) Diverse domains of THREAD/DIAP1 are required to inhibit apoptosis induced by REAPER and HID in Drosophila. Genetics 154: 669–678.

66. Edgar BA (2006) From cell structure to transcription: Hippo forges a new path. Cell 124: 267–273.

67. Dulic V, Lees E, Reed SI (1992) Association of human cyclin E with a periodic G1-S phase protein kinase. Science 257: 1958–1961.

68. Koff A, Giordano A, Desai D, Yamashita K, Harper JW, et al. (1992) Formation and activation of a cyclin E-cdk2 complex during the G1 phase of the human cell cycle. Science 257: 1689–1694.

69. Dulic V, Drullinger LF, Lees E, Reed SI, Stein GH (1993) Altered regulation of G1 cyclins in senescent human diploid fibroblasts: accumulation of inactive cyclin E-Cdk2 and cyclin D1-Cdk2 complexes. Proc Natl Acad Sci U S A 90: 11034–11038.

70. Ohtsubo M, Theodoras AM, Schumacher J, Roberts JM, Pagano M (1995) Human cyclin E, a nuclear protein essential for the G1-to-S phase transition. Mol Cell Biol 15: 2612–2624.

71. Keyomarsi K, O'Leary N, Molnar G, Lees E, Fingert HJ, et al. (1994) Cyclin E, a potential prognostic marker for breast cancer. Cancer Res 54: 380–385.

72. Huang J, Wu S, Barrera J, Matthews K, Pan D (2005) The Hippo signaling pathway coordinately regulates cell proliferation and apoptosis by inactivating Yorkie, the Drosophila Homolog of YAP. Cell 122: 421–434.

73. Mays JK, Pandiri AR, Fadly AM (2006) Susceptibility of various parental lines of commercial white leghorn layers to infection with a naturally occurring recombinant avian leukosis virus containing subgroup B envelope and subgroup J long terminal repeat. Avian Dis 50: 342–347

74. Maas HJ, De Boer GF, Groenendal JE (1982) Age related resistance to avian leukosis virus. III. Infectious virus, neutralising antibody and tumours in chickens inoculated at various ages. Avian Pathol 11: 309–327.

75. Rubin H, Fanshier L, Cornelius A, Hughes WF (1962) Tolerance and immunity in chickens after congenital and contact infection with an avian leukosis virus. Virology 17: 143–156.

76. Baba TW, Humphries EH (1985) Formation of a transformed follicle is necessary but not sufficient for development of an avian leukosis virus-induced lymphoma. Proc Natl Acad Sci U S A 82: 213–216.

13

Evolution and Development of Ventricular Septation in the Amniote Heart

Robert E. Poelmann[1,2,9]*, **Adriana C. Gittenberger-de Groot**[2], **Rebecca Vicente-Steijn**[1], **Lambertus J. Wisse**[1], **Margot M. Bartelings**[1], **Sonja Everts**[1], **Tamara Hoppenbrouwers**[9], **Boudewijn P. T. Kruithof**[3], **Bjarke Jensen**[4,5], **Paul W. de Bruin**[6], **Tatsuya Hirasawa**[7], **Shigeru Kuratani**[7], **Freek Vonk**[8], **Jeanne M. M. S. van de Put**[9], **Merijn A. de Bakker**[9], **Michael K. Richardson**[9]

1 Department of Anatomy and Embryology, Leiden University Medical Center, Leiden, The Netherlands, 2 Department of Cardiology, Leiden University Medical Center, Leiden, The Netherlands, 3 Department of Molecular Cell Biology, Leiden University Medical Center, Leiden, The Netherlands, 4 Department of Anatomy, Embryology and Physiology, AMC Amsterdam, Amsterdam, The Netherlands, 5 Department of Bioscience-Zoophysiology, Aarhus University, Aarhus, Denmark, 6 Department of Radiology, Leiden University Medical Center, Leiden, The Netherlands, 7 Laboratory for Evolutionary Morphology, RIKEN Center for Developmental Biology, Kobe, Japan, 8 Naturalis Biodiversity Center, Darwinweg 2, Leiden, The Netherlands, 9 Institute of Biology Leiden (IBL), Leiden University, Sylvius Laboratory, Leiden, The Netherlands

Abstract

During cardiogenesis the epicardium, covering the surface of the myocardial tube, has been ascribed several functions essential for normal heart development of vertebrates from lampreys to mammals. We investigated a novel function of the epicardium in ventricular development in species with partial and complete septation. These species include reptiles, birds and mammals. Adult turtles, lizards and snakes have a complex ventricle with three cava, partially separated by the horizontal and vertical septa. The crocodilians, birds and mammals with origins some 100 million years apart, however, have a left and right ventricle that are completely separated, being a clear example of convergent evolution. In specific embryonic stages these species show similarities in development, prompting us to investigate the mechanisms underlying epicardial involvement. The primitive ventricle of early embryos becomes septated by folding and fusion of the anterior ventricular wall, trapping epicardium in its core. This folding septum develops as the horizontal septum in reptiles and the anterior part of the interventricular septum in the other taxa. The mechanism of folding is confirmed using DiI tattoos of the ventricular surface. Trapping of epicardium-derived cells is studied by transplanting embryonic quail pro-epicardial organ into chicken hosts. The effect of decreased epicardium involvement is studied in knock-out mice, and pro-epicardium ablated chicken, resulting in diminished and even absent septum formation. Proper folding followed by diminished ventricular fusion may explain the deep interventricular cleft observed in elephants. The vertical septum, although indistinct in most reptiles except in crocodilians and pythonidsis apparently homologous to the inlet septum. Eventually the various septal components merge to form the completely septated heart. In our attempt to discover homologies between the various septum components we aim to elucidate the evolution and development of this part of the vertebrate heart as well as understand the etiology of septal defects in human congenital heart malformations.

Editor: Leonard Eisenberg, New York Medical College, United States of America

Funding: B.J. is supported by The Danish Council for Independent Research Natural|Sciences. B.P.T.K. is supported by the Netherlands Institute for Regenerative Medicine and the BioMedical Materials Institute, M.K.R. is supported by AgentschapNL, Smartmix SSM06010, R.V.-S. receives support from the Netherlands Heart Foundation, 2012T71. The funders had no role in study design, data collection and analysis, decision to publish, or preparation of the manuscript.

Competing Interests: The authors have declared that no competing interests exist.

* Email: r.e.poelmann@lumc.nl

Introduction

During evolution the ventricle of the heart became divided into left and right chambers by a complete septum. Interestingly, ventricular septation evolved independently in mammals and in the archosaurs, comprising birds and crocodilians [1] [2] (**Fig. 1A**). It is therefore, a textbook case of convergent evolution. Not only is cardiac ventricular septation of great intrinsic interest to evolutionary biologists, it is also crucial for the understanding of many types of heart defects in humans [3]. Our studies show that part of the septum is critically dependent on interactions between myocardium and the epicardium including the epicardium-derived cells (EPDCs) [4,5] for its development, and will develop abnormally if the epicardium is disturbed [6,7].

Cardiac ventricular development in vertebrates, starting in an embryonic state with a primitive common ventricular tube, leading to separation into the left (LV) and right ventricle (RV), is a complex phenomenon. The mechanisms involve both the ventricular inflow and outflow compartment [8–10]. Complete cardiac septation is found in mammals as well as in crocodilians and birds. Other extant reptile groups including the squamates (lizards and snakes) and chelonians (turtles) show what is assumed to be a primitive pattern. They show partial septation by a horizontal and a vertical septum leading to a ventricle that is

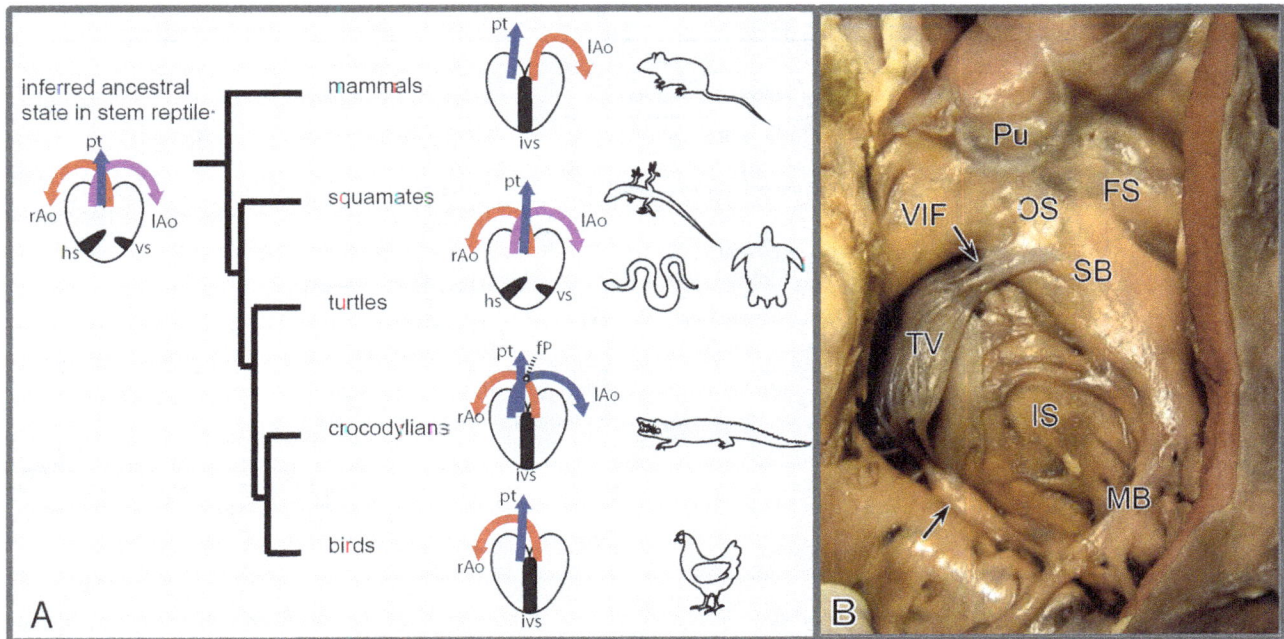

Figure 1. Evolution and septation of the heart. A. Evolution of hearts in higher vertebrates. Archosaurs (crocodilians, birds) and mammals independently evolved complete ventricular septation. Birds and mammals have lost either a left (lAo) or right (rAc) aorta. The horizontal (hs) and vertical septum (vs) are schematically indicated, together with the pulmonary trunk (Pt). The evolutionary tree s based on ref (2). B. Septum components in the human heart. Right face of the septum in a human heart after opening the right ventricle (RV), with inlet and folding components. Dissection line of the RV free wall in pink. Abbreviations: FS folding septum, IS inlet septum; MB moderator band; Pu pulmonary semilunar valve leaflets; SB septal band; TV anterior tricuspid valve leaflet with chordae tendineae (arrows) connected to SB and IS.; VIF ventriculo-infundibular fold. Fig. courtesy dr. L. Houyel.

divided into 3 interconnected cavities. The cavities are located sinistro-dorsal (the cavum arteriosum and cavum venosum unified as cavum dorsale) and dextro-ventral (the cavum ventrale, more often named the cavum pulmonale). Controversy exists on the homology of the cardiac chambers and the septal components and how the primitive reptilian pattern was modified into the complete septum in mammals, birds and crocodiles. Reptilian cardiac development has been addressed [2,11–13], with regard to the horizontal and vertical septum [14]. It is not possible to reconstruct the evolutionary history of ventricular septation as fossil records of embryonic soft tissues are non-existent.

Our strategy was to investigate the pattern and mechanisms of ventricular septation across the higher vertebrates using multiple lines of evidence. We focused on the functional role of the epicardium and EPDCs using various approaches: 1. immunohistochemistry of epicardium and EPDCs in embryos of different vertebrate species (lizard, snake, turtle, chicken, mouse, human), 2. quail-chicken chimeras by transplantation of early quail proepicardial organ (PEO) into the pericardial cavity of chicken to label subsets of epicardial and endothelial cells. 3. epicardium-deficient animal models such as the podoplanin knockout mouse, and epicardial ablation experiments in chicken embryos, 4. DiI-labelling experiments of the myocardial surface in chicken embryos to analyze outgrowth of the cardiac compartments. 5. expression of Tbx5 in embryos of several species, paying attention to the different septal components. *Tbx5* is reported to show gradients along the cardiac tube for various amniote embryos [14], and to be highly enriched in the left ventricle and the left face of the interventricular septum [15], 6. dissection of the extremely deep interventricular sulcus in adult elephant hearts [16].

To visualize the developmental anatomy of the different septal components we made animated 3D reconstructions of embryonic hearts from a range of species, including humans.

Results

Septum components in the completely septated heart

Very heterogeneous terminology is used for components of the ventricular septum and their respective boundaries and we adapted the following(**Table 1, Fig. 1B**). The *inlet septum* is the posterior (or dorsal in prone animals) component of the interventricular septum between the left and right atrioventricular junctions. The *folding septum* (a new term introduced in this paper) is the anterior (or ventral in prone animals) component. The septal band is a muscular profile on the RV septal surface situated between the inlet and folding septa. The outflow tract (OFT) or aorto-pulmonary septum depending highly on neural crest contribution, differs considerably among species. In the completely septated heart it is the last component that seals the interventricular communication. Development of the aorto-pulmonary component of the ventricular septum [8] has not been specifically studied here.

The presence of the epicardium in the various species

To infer primitive conditions in amniotes we examined embryos of the copperhead rat snake (*Coelognathus radiates*), Macklot's python (*Liasis mackloti*) (see also **Figure S1**), bearded dragon (*Pogona vitticeps*), European pond turtle *(Emys orbicularis)*, and Chinese soft-shell turtle (*Pelodiscus sinensis*). The presence of epicardium, covering the outer face of the myocardial hart tube is confirmed. A pronounced subepicardium in the inner curvature of

Table 1. Index for the terminology used.

Level of septation	Non-crocodilian reptiles	Avian, mouse, human
Outflow tract[1] at arterial, semilunar valve and intracardiac levels	Aorto-pulmonary septum	Aorto-pulmonary septum/conotruncal septum/OFT septum
Primary or bulboventricular fold[2]	Primary or bulboventricular fold	Primary or bulboventricular fold
Ventricular levels		
Anterior[3]	Horizontal/folding septum	Anterior/primary/folding septum
Apical[4]	When present: trabeculations	Apical trabecular septum
Posterior[5]	When present: Vertical septum	Inlet septum
Atrioventricular canal[6]		Membranous atrio- and interventricular septum

The numbers refer to the superscripts in the Table.

[1]. Combinations of aortic sac, truncus, conus and bulbus have been used to describe this segment. Conus and bulbus are usually myocardial, whereas aortic sac and truncus refer mostly to the vascular part. Septation from the vascular, semilunar valve and intracardiac levels is interchangeably referred to as either aorto-pulmonary septum or outflow tract (OFT) septum. The endocardial cushions in the proximal intracardiac part myocardialize through induction by neural crest cells forming the aorto-pulmonary or OFT septum. The distal part of the cushions is remodelled into semilunar valves that are separated by fibrous tissue between the orifices of the great arteries. In reptiles the aorto-pulmonary septum is branched and separates the two aortae and the pulmonary trunk. In mammals the distinction between proximal and distal endocardial cushions is inconspicuous.

[2]. Bulboventricular fold, synonymous with the primary fold, between outflow and inlet portion of the primitive ventricle.

[3]. Anterior (positional), primary (time-related) and folding (mechanistic, new in this paper) septum are synonymously used. The bulboventricular fold extends apically over the anterior surface of the heart and deepens to enclose epicardium and subepicardial tissue, thus forming an anteriorly located folding septum. The folding septum is considered to be homologous to the reptilian horizontal septum, which is also called the muscular ridge (see for further synonyms ref 13).

[4]. The apical trabecular septum develops from the coalescence of many trabeculations and does not show a clear demarcation with the folding septum or the inlet septum.

[5]. The inlet septum in early stages of eventually completely septated hearts and in some reptiles (presence is species-dependent) is a dense muscular structure on the posterior wall of the ventricle without an infolding mechanism. In the current study we have clearly shown that the anterior margin of the inlet septum with the folding septum is formed by the septal band or trabecula septomarginalis. In earlier literature the septal band has been described as the posterior margin of the primary (or folding) septum.

[6]. The superior and inferior atrioventricular endocardial cushions fuse in the midline. In the central part the cushions are remodelled into fibrous (membranous) tissue that becomes part of the fibrous heart skeleton. Part of this forms the membranous septum which is located between right atrium and outflow of the left ventricle (atrio-ventricular component) and the remainder between RV and LV (interventricular component). This tissue is obliquely embedded in both the atrial and ventricular septa and as such is sometimes referred to as atrioventricular septum.

the looping heart tube at the site of the bulboventricular fold is noted (**Fig. 2A-D**) while in e.g. the turtle *Emys* (**Fig. 2A, B**), harboring an extensive loop of the OFT [17], the subepicardium, here referred to as epicardial cushion, is almost as elaborate as the flanking endocardial atrioventricular (AV) and OFT cushions.

In chicken embryos the presence of the epicardium during folding (**Fig. 3A**) is demonstrated (see also **Figure S2**), although less extensive compared to the turtle.

Development of the epicardium in mice has been extensively described [18–21], and here the presence of an epicardial epithelium in the folding zone (**Fig. 4A, B**) is shown (**Fig. 4M-P, Figure S3**).

To examine the wider implications of our model, we examined human embryos at Carnegie stages 11–15 (3.6–7 mm, **Figure S4**). The surface of the heart is covered by the epicardial epithelium while the inner curvature harbors an extensive epicardial cushion comparable to the turtle. The anterior component of the septum seems to form by folding (**Fig. 5A-D**) as in the other species (**Fig. 5F**).

In contrast, in adult elephants the folding septum is poorly developed and lacks a solid muscular core, as is evident from the deep anterior interventricular sulcus. We examined hearts with computed tomography and magnetic resonance imaging and found a deep epicardial fat pad separating the two ventricles over more than 50% of their antero-posterior extent (**see Fig. 5G, H**). Examination of the internal right septal surface revealed the septal band with tricuspid valve chordae tendineae attached, continuing as the moderator band to the free RV wall as in the human heart (**Fig. 1B**). Using these structures as landmarks, we concluded that

the septum consists of an inlet component comparable to human, but the muscular walls have not fused in the folding component.

The epicardium in the avian heart

In the inner curvature of the early looping heart tube the bulboventricular fold is positioned between the AV canal and the OFT with, as yet, no sign of ventricular septation. Between HH22–27 by outgrowth of the left and right chambers the interventricular sulcus (**Figs. 3A, D-G**) can be appreciated as extension of the bulboventricular fold. The chambers identified in this project are depicted in a serially sectioned HH27 chicken embryo (**Fig. 6A-F**) immune-incubated for cardiac troponin I. Moving from outflow to apex, the inner curvature is present in the first section (**Fig. 6A**) containing the epicardial cushion between the OFT and AV cushions. More apically the folding septum appears (**Fig. 6B**), as well as the interventricular communication between left and right sided chambers with the tip of the septal OFT cushion adjacent to the folding septum (**Fig. 6D**). This is followed by the appearance of the posteriorly located inlet septum, that fuses with the anteriorly located folding septum (**Fig. 6E**). Note the hinge point between the two septal components (arrow) and note that the septal OFT cushion ends here. Closest to the apex (**Fig. 6F**) the LV is characteristically circular in shape, and the RV crescent-like. Apically, the inlet and folding septum components become inconspicuous as the number of trabecula-tions increase and we refer to this component as the apical trabecular septum (Table 1). Close to the inner curvature the folding septum is trapping the epicardial cushion containing mesenchymal subepicardial cells, or more apically only an epithelial sheet of epicardium. At HH 31 this has been embedded

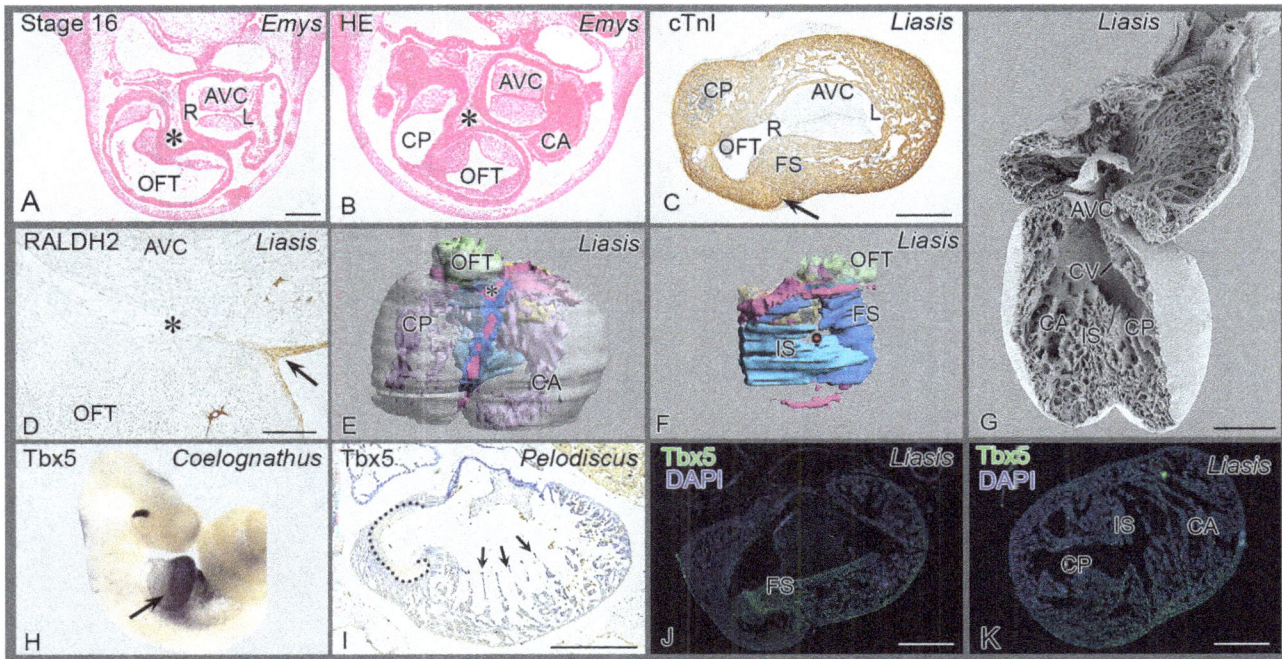

Figure 2. Reptile cardiac development. (A, B) The epicardial cushion (*) is located between OFT and AV cushions. **(C, D)** cardiac troponin I (cardiac muscle) and RALDH2 (epicardial cells) stainings show folding septum (arrow, asterisk). **(E)** 3D reconstruction in an anterior view, the epicardial patches are depicted in pink. See also Figure S1 1 for full animation. **(F)** right sided view of the septum, folding (FS) and inlet (IS) septum are depicted in shades of blue. For further colors see legend to Fig. 5E. **(G)** Scanning electron microscopy of anterior inner face, note communication between the three cava. The folding (syn. horizontal) septum is cut of view. **(H)** A sharp decline of Tbx5 mRNA expression (arrow) between cavum dorsale and OFT. **(I)** Sharp boundary at muscular OFT (inside of dotted line) and wall of cavum pulmonale (outside dotted line), but the tip of trabeculations in the cavum dorsale stain strongly (arrows). **(J)** Section downstream of Fig C, showing sharp decline of Tbx5 protein expression at folding septum. **(K)** Section more to the apex of J, showing uniform immunostaining for Tbx5. Abbreviations as in Fig 1, others: AVC: AV cushions; ca, cp, cv: cavum arteriosum, pulmonale and venosum; L left AV orifice; OFT outflow tract cushions; R right AV orifice; →: infolding; * epicarcium and EPDCs; ● position of cavum venosum in 3D reconstruction of Fig F.

in the heart and is no longer visible as an epithelium. The accompanying subepicardial cells or EPDCs apparently have been dispersed in the cardiac wall (see below).

To investigate the fate of the epicardium during folding we constructed quail-chicken chimeras. An isochronic quail PEO including a small piece of adjacent liver tissue to provide endothelial cells (as explained in the Materials and Methods section) was transplanted into the pericardial cavity of HH15–17 chick embryos in an anterior position, relative to the inner curvature. Using quail-specific antibodies quail EPDCs and endothelial cells were demonstrated in the folding septum, but not in the inlet septum. In later stages the quail epicardial sheet dispersed into individual cells that became distributed between the cardiomyocytes, mostly in the core of the folding septum **(Fig. 3H, I)**. Both quail EPDCs (stained with the nuclear QCPN antibody) and adjacent co-transplanted endothelial cells (stained with the cytoplasmatic QH1 antibody) were encountered. They occupy the same regions in the chimera although the area occupied in sister sections by endothelial cells is slightly more restricted. Nevertheless, we have chosen to present the latter **(Fig. 3H-L)** as these are better visualized in the low magnifications needed.

In a second set of chimeras, we positioned the quail PEO dorsally to the inner curvature, leading to the presence of quail EPDCs and endothelial cells on the posterior ventricular surface, subsequently migrating into the ventricular inlet septum including the septal band, but not in the anterior folding septum. This indicates the development of the inlet septum as a separate

component **(Fig. 3J-L)**. An epicardial sheet, reminiscent of the anterior folding septum was not encountered, therefore, we concluded that expansion of the ventricles immediately downstream of the AV canal occurs in a ventral direction, resulting in anterior folding of the ventricular wall, but not of the posterior wall, probably because of the physical constraints imposed by the dorsal body wall. In HH30 the right (tricuspid) opening in the AV canal is visible and separated from the mitral orifice. The right AV orifice is flanked on the left-side by the inlet septum and on the right-side by the RV wall. After chimerization both sides contain quail cells that have migrated into these parts of the cardiac wall **(Fig. 3L)**.

To study the folding mechanism fluorescent DiI was applied to the surface of the myocardium [22] ventrally to the inner curvature in HH 15–17 chicken embryos before epicardial covering, and embryos were sacrificed between HH22–33 **(Fig. 3M-P)**. Fluorescent patches positioned exactly in the future fold separated during further development into a left-sided fragment, incorporated in the left side of the folding septum **(Fig. 3M)** and a right-sided fragment **(Fig. 3N)** extending to the surface towards the apex **(Fig. 3O)**. This indicates a longitudinally directed morphogenetic expansion of the right ventricular wall [23] compared to a transverse expansion of the left wall [22]. At stage 31 the DiI labelled myocardial cells were completely embedded in the folding septum as visualized by a narrow fluorescent strip located close to the right ventricular face of the septum **(Fig. 3P)**, indicative of a more massive contribution of the left ventricular wall to the septum.

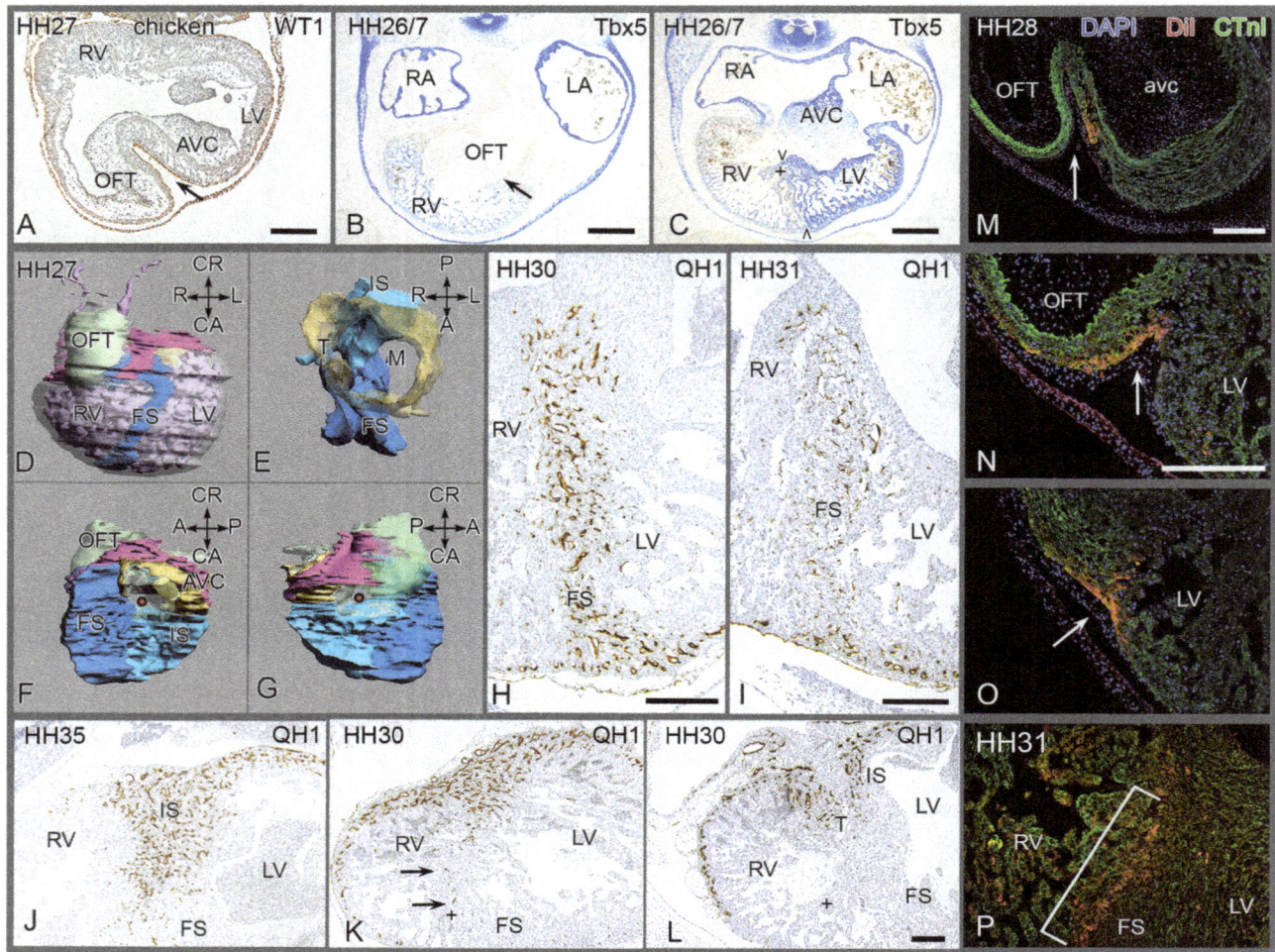

Figure 3. Development of chicken septum. (A) epicardium (→) infolding, located between OFT and AVcanal. **(B)** In situ hybridisation showing weakly positive Tbx5 of the RV and negative OFT with boundary (arrow). The atria are strongly positive. **(C)** more posterior section of the same embryo through folding septum (FS), the stronger left sided expression is evident, as is the septal band (+); boundary (> <) indicates FS. **(D-G)** 3D reconstruction with septum components and epicardial cushion. See also figure S2 for full animation and Fig. 6 for underlying sections, explaining the various components. **(H-L)** PEO quail-chicken chimeras. **(H, I)** anterior quail PEO(+liver) transplant, quail endothelial cells are exclusively present in FS and anterior free wall **(J-L)** posterior PEO (+liver) transplant with quail vascular profiles in IS **(J, K)** and right face of tricuspid orifice **(L)**, but not in FS. **(K)** Several quail cells (arrows) in septal band (+), but FS does not harbor quail cells and remains negative **(K, L)**. **(M-P)** DiI marking at HH17 of anterior myocardium surviving until HH28 and 31. **(M)** parts of the DiI patch (arrow) after survival to HH28 on left, **(N)** DiI on the right face and **(O)** DiI near the apex. **(P)** DiI inside the septum at HH31. Abbrev. as in Fig 2. Others: AVC atrioventricular cushions; LA/LV left atrium and ventricle; RA/RV right atrium and ventricle; + septal band.

Disturbance of the epicardium

Development of the epicardium in mice has been analyzed in recent years [18–21]. As the folding and epicardial incorporation is very similar to avian development it will not be treated separately. Deficient septation, however, is known in several animal models including PEO ablation in chicken embryos [25] and mouse mutants and we have chosen to analyze the podoplanin mutant mouse. Podoplanin is expressed in the lining of the body cavities including the epicardium and pericardium. The podoplanin mutant mouse has been morphometrically analysed [7] and presents with an underdeveloped PEO (40% compared to wildtype) and abnormal epicardial covering with hardly epithelial to mesenchymal transition resulting in only a few EPDCs. Analysis of this PEO-deficient mouse shows multiple malformations including an atrioventricular septum defect and a thin myocardium. At embryonic day (ED) 12.5 the folding septum is very thin, and the inlet septum spongy. The diminutive septum is nearly

devoid of EPDCs (**Fig. 4C-H**), suggestive of an instructive role for these cells in completion of the septum.

The 3-D reconstruction of a wild type mouse embryo is provided in **Fig. 4M-P**.

Septum components in reptilian hearts

The horizontal septum (syn. muscular ridge, "Muskelleiste") is found in similar developmental stages, in the same location, separating ventricular cavities, and harboring an epithelial epicardial sheet much the same as the folding septum in mammals and birds. The mentioned names have been used interchangeably but we prefer to address this structure as 'folding septum' (**Table 1**).

The presence and extent of the vertical septum (homologous to inlet septum) differs among turtles and squamates, being virtually absent in turtles (**Fig. 2I**), but being more prominent in varanids and pythonidae (**Figs. 2E-G, K**). Myocardial apical trabecula-

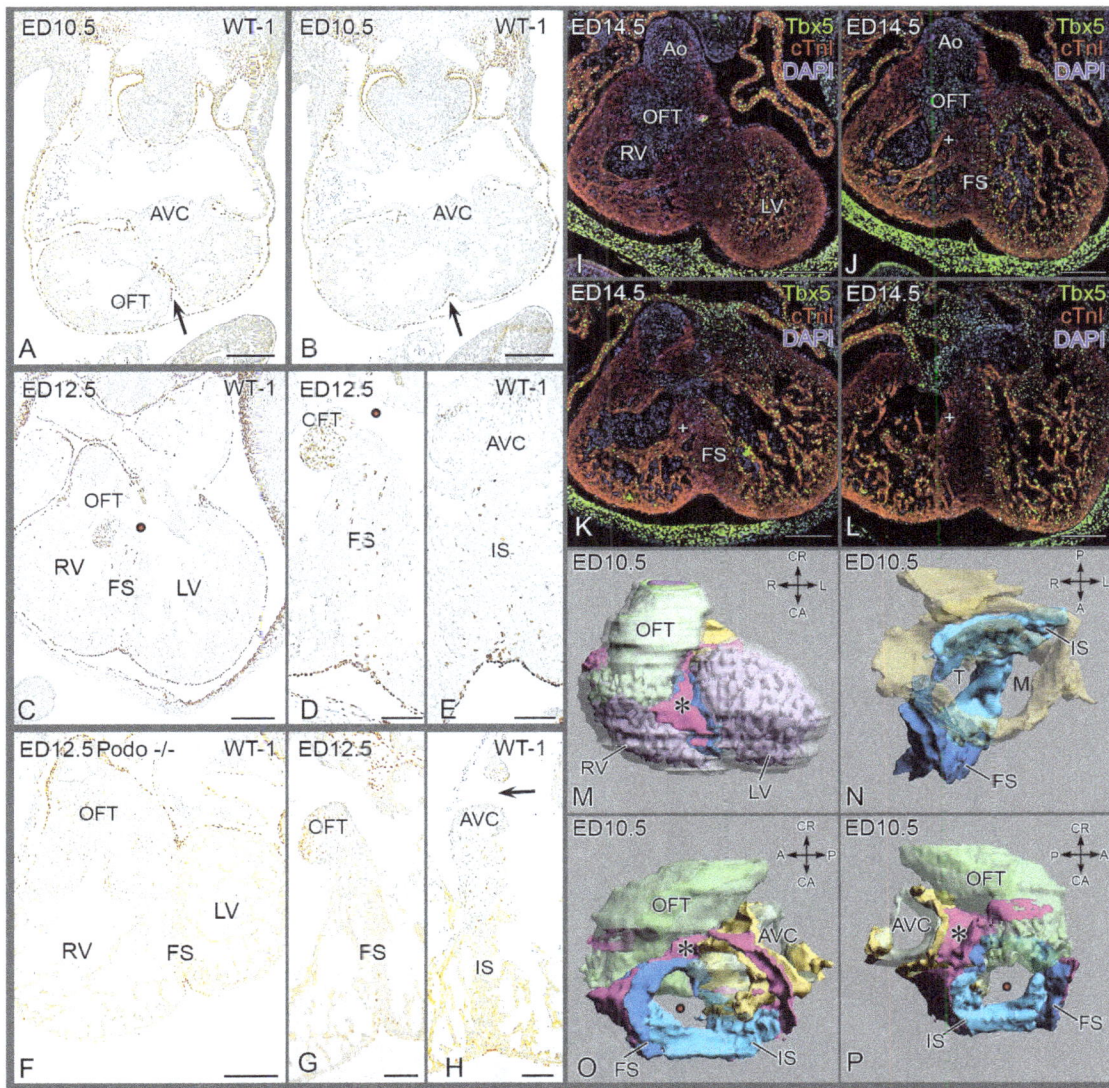

Figure 4. Septum formation in the mouse. (A, B) Almost transverse sections of the same embryo showing WT1+ epicardial cells in the folding septum (FS,→) at ED 10.5, Fig. B is more apically located. **(C, D)** epicardial cells in FS of wildtype mouse at ED 12.5 and **(E)** present in the inlet septum underneath the posterior AV cushion. Note: WT1 staining of mesenchyme in septal OFT cushion is unrelated to epicardial cells. **(F-H)** Podoplanin mutant with diminutive PEO, presents with sparse epicardium lining the pericardial cavity **(F, G)** and with an underdeveloped septum lacking EPDCs in both FS **(G)** and inlet septum (IS) **(H)**. **(I-L)** Immunostained for Tbx5 in a wild type mouse ED 14.5, four levels from anterior-posterior. **(I)** Tbx5 in LV trabeculations but not in the RV close to the outflow tract; core of septum is negative. **(J-L)** More posteriorly located sections, trabeculations in RV belonging to the inlet part become positive for Tbx5. **(M-P)** Four positions of a 3D Amira reconstruction of ED 10.5. The epicardial cushion in pink (*), the folding septum in dark blue and the inlet septum in light blue. Endocardial cushions in green and the AVC myocardium in yellow. See also Fig. S3 for animated 3D. Abbrev. AVC atrioventricular cushions; FS folding septum; IS inlet septum; LV left ventricle; M mitral orifice; RV right ventricle; OFT outflow tract cushion; T tricuspid orifice, ● interventricular communication, + septal band.

tions (**Fig. 2G**) traversed the cavum dorsale connecting the anterior and posterior myocardial walls, partly separating the cavum dorsale into the cavum venosum and arteriosum, but at a different position compared to the folding septum.

Tbx5 expression patterns

The T-box transcription factor Tbx5 has been reported to be expressed in the heart from left to right in a gradient that declines towards the right side [14]. Tbx5 expression in the early embryonic copperhead rat snake, *Coelognathus* (**Fig. 2H**) was evident in the cavum dorsale and absent in the OFT. In the embryonic turtle (**Fig. 2I**) as well as in the python (**Fig. 2J, K**)

there was a boundary at the folding septum and ventricle. Tbx5 thus exhibited a distinct decline in expression only at the cavum dorsale/OFT boundary (**Fig. 2J**). Its expression was uniform over the three cava in the direction of the apex (**Fig. 2K**). It is obvious that in the turtle the tip of the trabeculations show a stronger Tbx5 expression than the adjacent tissues (**Fig. 2I**).

In the chicken the Tbx5 mRNA gradient identifies the RV(weak expression) and the Tbx5-negative OFT (**Fig. 3B**). Additionally, a second Tbx5 gradient is found at the folding septum showing strong expression in the LV but weak in the RV (**Fig. 3C**), whereas the expression is present on both sides of the inlet septum including the septal band. Thus, the two components of the

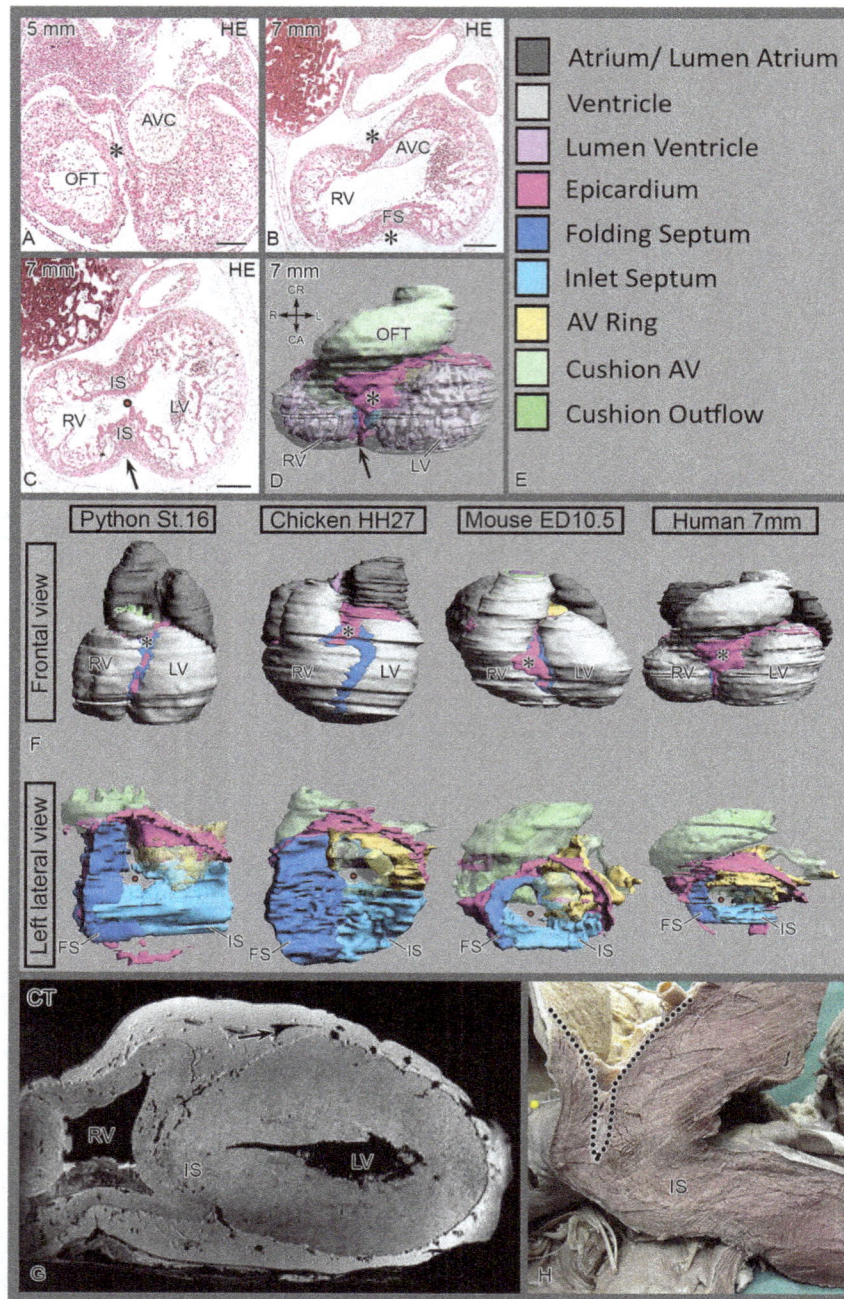

Figure 5. Embryonic human and adult elephant hearts. (Fig. A, B) Epicardial cushion (*) in inner curvature between OFT and AVC. The inlet septum becomes apparent more apically **(Fig. C)**. **(D)** and **Fig S4** represent a reconstructed 7 mm embryo with folding and inlet septum formation. **(F)** 3D Comparison of development. Top row, 4 species showing relation of epicardial cushion (pink) with folding septum (dark blue). Bottom row, left lateral view, showing connection of folding septum with the inlet septum (light blue). The interventricular foramen connects left and right ventricle. In python this connection is represented by the cavum venosum. The AV myocardium is depicted in yellow and the various cushion tissues in green. **(G, H) adult elephant heart (G)** CT-image of first bifid elephant heart. **(H)** Anatomical dissection of the interventricular septum of the second heart. Left descending coronary artery (arrow) and the deep epicardial fat pad are outlined. Note the relative absence of a folding component. The dissected chordae tendineae of the tricuspid valve are visible in the lower part of Fig 5H. Abbrev. AVC atrioventricular cushions; FS folding septum; IS inlet septum; LV left ventricle; RV right ventricle; OFT outflow tract cushion.

ventricular septum in chicken, namely folding and inlet septa **(Fig. 3D-G)** are differentially identified by Tbx5 gradients.

In the mouse similar to the chicken, two Tbx5 gradients identify the inlet and folding septum as exemplified by the protein pattern **(Fig. 4I-L)**. Protein expression is strongest in the trabeculations of the LV **(Fig. 4I, J)**, weaker in the RV inlet including septal band, **(Fig. 4 J, K)** and weakest to negative in the RV OFT **(Fig 4I)**.

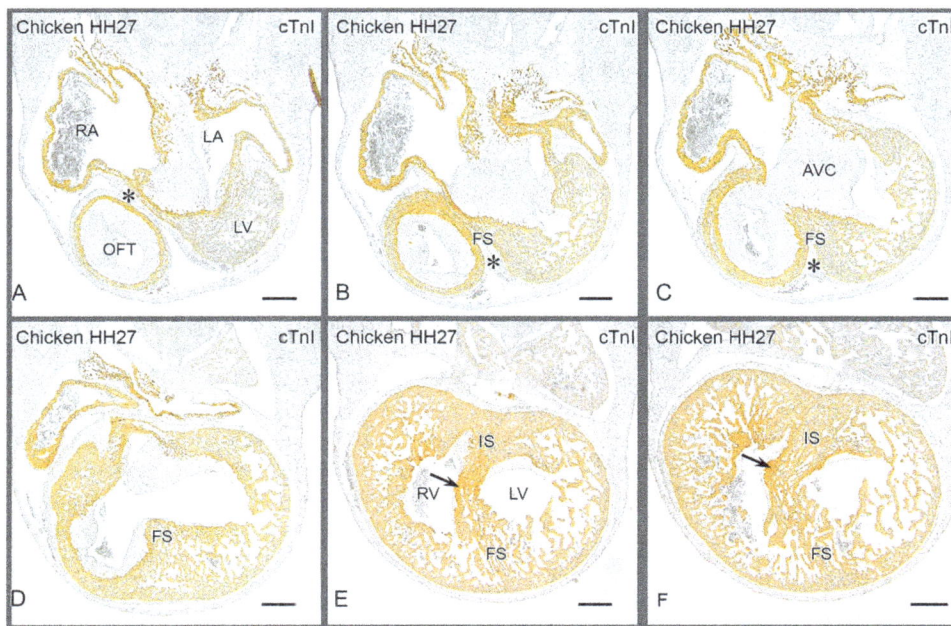

Figure 6. Chicken embryo HH27. Provides 6 sections of a serially sectioned chicken embryo (HH27) to demonstrate the merging of the folding and inlet components before septation is finished. From this embryo Fig 3D and Fig. S2 have been reconstructed. Similar series served as basis for the other species depicted in Fig S1, S3 and S4. **Fig 6.A** is most cranial, showing the epicardial cushion (*) at a level between outflow tract (OFT) and the right (RA) and left atria (LA) with the AV cushions in between. The cranial cap of the left ventricle (LV) is grazed in the section. **Fig 6.B** and **C** give the cranial extension of the folding septum (FS) with the epicardial cushion (*) located between the outflow tract and the fused AV cushions (AVC). **Fig 6.D** shows the FS bordering the interventricular foramen. It is evident that the core of the folding septum is lined on the left and right side by many trabeculations. **Fig. 6.E** The AV cushions are attached to the flanks of the inlet septum where also the tip of the septal OFT cushion is found (arrow). The inlet (IS) and folding components have fused and constitute the floor of the interventricular foramen. **Fig 6.E, F** The right AV junction is present in the RV immediately above the arrow and can be traced upstream in Fig F and downstream in Fig. D. Note the close relationship to the IS. The folding septum becomes less compact and the trabeculations become more conspicuous. Abbrev. AVC atrioventricular cushions; FS folding septum; IS inlet septum; LA left atrium; LV left ventricle; RA right atrium; RV right ventricle; OFT outflow tract cushion with its proximal tip indicated by arrow in Fig. F.

Discussion

The evolution from the single fish ventricle to the two ventricles of mammals and archosaurs is the result of ventricular septation and has been studied and discussed for more than a century. It is tempting to apply to intermediate stages the partially septated ventricles encountered in turtles, lizards and snakes. Indeed, all amniotes exhibit a horizontal or folding septum and here we provided a mechanism for its formation involving the epicardium in fusion of two opposing myocardial walls. In the completely septated ventricle, the folding septum forms the anterior part of the septum. The epicardium originating from the PEO [24–26], important for multiple aspects of heart development, is already present in lampreys [27]. Reducing the size of the PEO is exemplified by the podoplanin mouse mutant [7] and inhibits its outgrowth in chicken [6]. Both lead to diminished or retarded covering of the myocardium and lack of epithelial-mesenchymal transition producing a decreased number of EPDCs (this paper). Many genes are expressed in the epicardium lining the myocardium of the heart. They can be divided in 'epithelial genes' including integrins and b-catenin, but also others like RALDH2, and several transcription factors [28] including Tbx18, Tbx5, Tcf21, NFATc1, and WT1. A subset of these genes is important for epithelial to mesenchymal transition resulting in the formation of EPDCs. The latter can differentiate into smooth muscle cells of the coronary vessels and into perivascular and interstitial fibroblasts. It is evident that mechanical or genetical interference with the epicardium or EPDCs not only disturbs coronary vascularisation but can strongly influence cardiomyocyte differentiation and ventricular septation [6,7]

Different septal components have been identified in the completely septated hearts but the origin and subdivision of the primary interventricular septum has been the subject of continued debate [8–10,29], with varying implications for the explanation of the position of central muscular ventricular septal defects, the origin of the left and right bundle branch and the connection of the tricuspid valve tendinous cords related to either septal components. It was generally agreed that the septal band belonged to an anterior component of the septum, the primary septum, but we describe it as a derivative of the posterior inlet septum. In normal hearts but also in patients with septal defects [3] the position of the AV valve leaflets and the connection of the tension apparatus to the septal band and the inlet septum is consistent with our evolutionary and developmental concept. We could identify the border between inlet and folding component according to the combination of the following criteria: (i) the AV cushions are connected to the inlet component including the septal band; (ii) the proximal tip of the septal OFT cushion is incorporated in the zone where the inlet septum merges with the folding septum; (iii) quail cells derived from anterior PEO chimeras do not crossover from folding septum to inlet septum (including septal band) and, likewise, there is no crossover in posterior chimeras from inlet septum (including the septal band) to folding septum; (iv) Tbx5 expression is strong in the complete inlet septum (both left and right side) including the septal band, whereas the folding septum only shows left-sided positivity consistent with this part belonging to the primitive embryonic ventricle. These combined characteristics support the concept that the early embryonic ventricle in birds and mammals (homologous to the cavum dorsale in reptiles) gives rise to both the LV and inlet of the RV, also implicating that the inlet septum originates in its entirety from the wall of the primitive ventricle (the cavum dorsale) and becomes partitioned over the LV and the RV inlet. The origin of the trabeculated apical portion of the interventricular septum has not been approached by our experiments but it can be traced back in the primary heart tube of chicken [30]. In the RV the antero-posterior boundary is determined by the septal band, which is lacking in the LV, leaving the LV boundary less well determined.

Our new model for partitioning of the ventricles comprising more than one component (note: the membranous septum related to the AV cushions, and the muscular OFT septum related to the neural crest, have not been specifically studied here) has consequences for the development of AV and muscular ventricular septal defects at the border of septal components.

Interestingly, the curious 'bifid' heart of elephants, seacows [31] and some other marine mammals may be interpreted as the retention of an embryonic feature resulting from non-progressed fusion of the ventricular walls of the folding septum, rather than regression from a well-developed septum. Obviously, the small coney or rock hyrax (*Procavia capensis*), a close extant relative to the elephant and the seacows, does not present a bifid heart (unpublished).

The development of both the folding and inlet components with differently incorporated EPDCs is severely hampered in mutant mice deficient for epicardially expressed genes.

This may lead to an abnormal septum [7,32,33] and to multiple muscular ventricular septal defects, reported in human congenital heart disease [3] as a result of deficient myocardial compaction [19].

Strengths and weaknesses

An important novelty of this study is found in distinguishing several septal components each with their own characteristic developmental history and mechanistic program. The exact boundaries in the fully septated heart, however, remain difficult to define. It is a limitation of the study that no specific tissue markers are available, implying that we rely on the combination of various approaches, each with their own strength and weakness. We are of the opinion that combining the different techniques and referring to data in which their combination supports the conclusions, reduces substantially erroneous interpretations and allows us to add substantial new evidence to earlier studies that postulated the existence of more than one ventricular septal component.

In conclusion, we have added in an evo-devo context a hitherto overlooked, but important role for the epicardium in ventricular septation and have clarified complex homologies of the ventricular septum in amniotes aiding to understand clinical disorders of the heart in humans.

Materials and Methods

Animal material

Animal material was obtained as follows. Mouse embryos were harvested between embryonic day 12.5–15.5 from strains housed in the fully licensed Animal Facility of the Leiden University Medical Center. Pregnant dams were killed by cervical dislocation. The podoplanin mouse material was obtained in a collaboration and provided by P. Uhrin (Vienna). Animal care and all experimental procedures were approved by the Animal Experimental Committee of the Medical University of Vienna, and by the Austrian Ministry of Science (License No. 1321/115, and 66.009/0103-C/GI/2007). Fertilized chicken eggs were obtained from a commercial breeder. Chicken and reptile eggs are not considered experimental animals under the Dutch law. The oldest chicken stages harvested after manipulation are Hamburger Hamilton stage 34. All reptile embryos were obtained legally from general-purpose captive-bred populations and were harvested and fixed in conformity with local and international regulations. Embryos were euthanized instantly by fixation to cause minimum suffering. Their use was approved according to the Regulations of the Animal Experimental Committee of the LUMC and Leiden University based legally on the national "Wet op de Dierproeven" (Article 9). This regulation serves as the implementation of Guidelines on the protection of experimental animals by the Council of Europe, Directive 86/609/EEC. Stages used were before 35% of embryonic development prior to hatching, which is in conformity with the local, national and European Union regulations. The two available elephant hearts were obtained after euthanasia of the animals for unrelated metabolic or arthritic disease and made available by the dept. Veterinary Pathology (M. Kik, University of Utrecht) with a licence for the study of protected animal material. The anonymized human embryos belong to the collection of the department of Anatomy (Vienna) and have been photographed over 3 decades ago for peer reviewed publications on development.

Quail-chicken chimeras

To study EPDCs and endothelial (precursor) cell migration in more detail, chimeras were generated using quail (*Coturnix coturnix*) embryos as donors and White Leghorn chicken embryos as hosts. Embryos were staged [34]. Quail-chicken chimeras were made as described [35]. In brief, the PEO of a HH15–18 quail embryo was isolated together with a tiny piece of liver tissue to provide endothelial precursor cells to the PEO transplant and transplanted into the pericardial cavity of a HH15–18 chicken host embryo through the naturally occurring hiatus in the body wall that exists until HH18. The transplant was positioned along the developing heart tube, either dorsal or ventral in the inner curvature. Chimeras (n = 31) were harvested between stage HH19 and HH32.

Histological Procedures

Normal and chimeric embryos were fixed by overnight immersion in 4% paraformaldehyde (PFA) in 0.1 M phosphate buffer (pH 7.4) at 4°C. Serial sections were immunostained. Tissue processing and immunohistochemistry were described [7].

DiI labelling in chicken

White Leghorn chicken (*Gallus gallus domesticus*) eggs (n = 36 analysed) were incubated at 38°C on stationary shelves until stage 18–19 [36]. The outer face of the looping heart tube was tattooed with a 1.25 mg/ml red fluorescent mix of 1,1'-dioctade dioctadecyl-3,3,3'-tetramethylindocarbocyanine perchlorate (DiI, Invitrogen, D-282) and 5-carboxytetramethylrhodamine, succinimidyl ester (TAMRA SE, Invitrogen, C2211) as described [36]. Localized microinjection, using an ultrathin glass needle (inner diameter 0.01 mm) and a Narishige IM-300 microinjector, of a minimal fluorophore volume at the anterior surface near the inner curvature resulted in labelling mostly myocardial but also epicardial cells. After microinjection, the eggs were re-incubated to allow further development to stages 24–35.

Mouse podoplanin

Mouse embryos were generated and obtained in collaboration with P.Uhrin (Vienna). After staging of pregnancy the dams were killed and embryos harvested and fixed. Serial sections were immunostained for the myocardial marker MLC2-a (1/6000, kindly provided by S.W.Kubalak, Charleston, SC, USA) and WT-1 (1/1000, Santa Cruz Biotechnology, CA, USA) as previously described [7].

Mouse Tbx5 immunochemistry

For immunofluorescent detection of Tbx5, 6 µm serial sections of wild type embryos were deparaffinized and rehydrated. After boiling in Antigen Unmasking Solution (H-3300, Vector Laboratories) in a pressure cooker for antigen retrieval, sections were blocked in 1% bovine serum albumin (Sigma) in 0,1% Tween-PBS for 30 minutes and incubated O/N at 4°C. Next, the sections were incubated with Tbx5 antibody (a kind gift from C.J.Hatcher, Ithaca, NY) and subsequently with horseradish peroxidase-conjugated secondary antibody, followed by Tyramide Signal Amplification (Perkin Elmer Life Science; #NEL700A). Signal was visualized using Alexa 488-conjugated streptavidin (Invitrogen, S-11223). Subsequently, (sister) sections were incubated with antibodies directed against cardiac Troponin I (cTNI, HyTest Ltd, CN-4T21_2) followed by Alexa 555-conjugated secondary antibodies (Invitrogen, A-21432) to visualize the myocardium. Sections were mounted using ProLong Gold antifade reagent (Invitrogen, CN-P36930) with DAPI.

Whole mount chicken and reptile Tbx5 mRNA

This has been performed as described [37,38].

CT scanning

This was performed on two adult Asian elephant (*Elephas maximus*) hearts using a 64 slice CT scanner (Toshiba Aquilion, Toshiba Medical Systems, Otawara, Japan). Helical acquisition settings were tube voltage 100 kV, tube charge 500 mAs, and pitch 0.83. Reconstruction with slice thickness 0.5 mm, FOV 500 mm, 512×512 matrix and a soft tissue reconstruction filter (FC03). 3D Visualisation by using volume rendering (Osirix v5.6). We examined the hearts with computed tomography and magnetic resonance imaging and found a deep epicardial fat pad separating the two ventricles over more than 50% of their antero-posterior extent.

SEM protocol

Hearts of embryos of the pythonid snake *Liasis mackloti* were fixed in 70% ethanol. The vessels at the venous pole of the heart were removed and hearts dissected prior to critical point drying with a BAL-TEC Critical Point Drier 030. The samples were sputter coated with a gold layer by using a Polaron E5100 and were analysed with a JEOL scanning electron microscope.

Amira protocol

Three-dimensional reconstructions of snake, chicken, mouse and human hearts were made. Photographs of serial sections stained for cTnI (snake, chicken mouse) or HE (human) were taken with an Olympus Provis AC70 microscope fitted with Olympus UPlan Apo-objectives, using an Olympus XC50 camera. Every 5th section of 5 µm was used. The sections in between were stained with various other antibodies such as WT1. Pictures were optimized for Amira version 5.4.2 with Adobe Photoshop CS6 Extended. See **Figure 6** as example using the chicken. Sections were both automatically and manually aligned and labels were added to the different structures, based on morphology and stains. Surface views were executed to PDF formats by using Adobe Acrobat 9.5 Extended. The animation of Amira files with help of Adobe pdf (version 11 or higher is needed for full functionality) has been described [39].

Supporting Information

Text S1 Figs. S1–S4 present the fully animated reconstructions of python, chicken, mouse and human hearts, respectively. Comparison between the four species is provided in main text figure 5F from a frontal and left lateral view. A short manual how to use the animated Adobe pdfs (version 11 or higher is needed). The color legend provided (Fig 5E) is useful to recognize the various structures. Moreover, several structures in the animated reconstructions have been annotated for your convenience.

Figure S1 Animated pdf of an embryonic python heart (Liasis mackloti). The inlet septum is indicated in light blue and the folding septum in dark blue. In this stage, the epicardium (pink) is mainly associated with the AV-ring (yellow) and the folding septum. Endocardial cushion sets (OFT and AV) are represented in shades of green.

Figure S2 Chicken embryo HH27. Animated pdf of the same chicken embryo of Fig. S2. The inlet septum is indicated in light blue and the folding septum in dark blue. In this stage, the epicardium (pink) is mainly associated with the AV-ring (yellow) and the folding septum. Endocardial cushion sets (OFT and AV) are represented in shades of green.

Figure S3 Animated pdf of a mouse embryo. The inlet septum is indicated in light blue and the folding septum in dark blue. In this stage, the epicardium (pink) is mainly associated with the AV-ring (yellow) and the folding septum. Endocardial cushion sets (OFT and AV) are represented in shades of green.

Figure S4 Animated pdf of a human embryo. The inlet septum is indicated in light blue and the folding septum in dark blue. In this stage, the epicardium (pink) is mainly associated with the AV-ring (yellow) and the folding septum. Endocardial cushion sets (OFT and AV) are represented in shades of green.

Acknowledgments

M.R.M. Jongbloed and M.C. DeRuiter are acknowledged for discussions on various aspects of the work. C. Mummery critically commented on the manuscript. L. de Graaf and V. de Winter assisted with obtaining reptile embryos. H. Lie-Venema initially performed chicken-quail chimeras. Podoplanin mouse material was provided by P. Uhrin (Vienna). J. den Boeft assisted with preparing the elephant hearts, obtained via M. Kik (University of Utrecht). CT scans were performed in the Department of Radiology (LUMC) by J. Geleijns, and W.M. Teeuwisse. Sections of human embryos from the collection of the Dept. Anatomy and Embryology, Vienna were photographed by A.C.G. Wenink. M. van der Heiden assisted with Fig. 1a. The photograph in Fig. 1B is kindly provided by L. Houyel (Paris, France). The polyclonal Tbx5 antibody is a gift from C.J. Hatcher (Cornell, Ithaca, USA).

Author Contributions

Conceived and designed the experiments: REP ACGdG MKR. Performed the experiments: REP RVS SE T. Hoppenbrouwers BPTK BJ JMMSvdP.

Analyzed the data: REP ACGdG MMB BJ PWdB MKR. Contributed reagents/materials/analysis tools: LJW T. Hoppenbrouwers BPTK T. Hirasawa SK FV MAdB MKR. Wrote the paper: REP ACGdG BJ MKR.

References

1. Holmes EB (1975) A reconsideration of the phylogeny of the tetrapod heart. J Morphol 147: 209–228.

2. Wang Z, Pascual-Anaya J, Zadissa A, Li W, Niimura Y, et al. (2013) The draft genomes of soft-shell turtle and green sea turtle yield insights into the development and evolution of the turtle-specific body plan. Nat Genet 45: 701–706. doi: 10.1038/ng.2615.

3. Jacobs JP, Quintessenza JA, Burke RP, Mavroudis C (2000) Congenital Heart Surgery Nomenclature and Database Project: atrial septal defect. Ann Thorac Surg 69: S18–S24.

4. Gittenberger-de Groot AC, Vrancken Peeters M-PFM, Mentink MMT, Gourdie RG, Poelmann RE (1998) Epicardium-derived cells contribute a novel population to the myocardial wall and the atrioventricular cushions. Circ Res 82: 1043–1052.

5. Von Gise A, Pu WT (2012) Endocardial and epicardial epithelial to mesenchymal transitions in heart development and disease. Circ Res 110: 1628–1645. doi: 10.1161/CIRCRESAHA.111.259960

6. Lie-Venema H, van den Akker NMS, Bax NAM, Winter EM, Maas S, et al. (2007) Origin, fate, and function of epicardium-derived cells (EPCDs) in normal and abnormal cardiac development. Scientific World Journal 7: 1777–1798.

7. Mahtab EAF, Wijffels MCEF, van den Akker NMS, Hahurij ND, Lie-Venema H, et al. (2008) Cardiac malformations and myocardial abnormalities in podoplanin knockout mouse embryos: correlation with abnormal epicardial development. Dev Dyn 237: 847–857. doi: 10.1002/dvdy.21463

8. Wenink ACG (1981) Embryology of the ventricular septum. Separate origin of its components. Virchows Arch 390: 71–79.

9. van Mierop LH, Kutsche LM (1985) Development of the ventricular septum of the heart. Heart Vessels 1: 114–119.

10. Lamers WH, Wessels A, Verbeek FJ, Moorman AF, Virágh Sz, et al. (1992) New findings concerning ventricular septation in the human heart. Implications for maldevelopment. Circulation 86: 1194–1205.

11. Goodrich ES (1919) Note on the Reptilian Heart. J Anat 53: 298–304.

12. Jensen B, van den Berg G, van den Doel R, Oostra RJ, Wang T, et al. (2013) Development of the hearts of lizards and snakes and perspectives to cardiac evolution. PLoS One 8: e63651. doi: 10.1371/journal.pone.0063651

13. Jensen B, Moorman AF, Wang T (2014) Structure and function of the hearts of lizards and snakes. Biol Rev Camb Philos Soc 89: 302–336. 10.1111/brv.12056 [doi].

14. Koshiba-Takeuchi K, Mori AD, Kaynak BL, Cebra-Thomas J, Sukonnik T, et al. (2009) Reptilian heart development and the molecular basis of cardiac chamber evolution. Nature 461: 95–98. doi: 10.1038/nature08324

15. Greulich F, Rudat C, Kispert A (2011) Mechanisms of T-box function in the developing heart. Cardiovasc Research 91: 212–222. doi: 10.1093/cvr/cvr112

16. Robb JS (1965) Comparative basic Cardiology. Grune and Stratton, New York, London. p.212–213.

17. Bertens LM, Richardson MK, Verbeek FJ (2010) Analysis of cardiac development in the turtle Emys orbicularis (Testudines: Emidydae) using 3-D computer modeling from histological sections. Anat Rec (Hoboken) 293: 1101–1114. doi: 10.1002/ar.21162

18. Smart N, Dube KN, Riley PR (2013) Epicardial progenitor cells in cardiac regeneration and neovascularisation. Vascul Pharmacol 58: 164–173. doi: 10.1016/j.vph.2012.08.001

19. Gittenberger-de Groot AC, Winter EM, Bartelings MM, Goumans MJ, DeRuiter MC, et al. (2012) The arterial and cardiac epicardium in development, disease and repair. Differentiation 84: 41–53. doi: 10.1016/j.diff.2012.05.002

20. Zhou B, Pu WT (2012) Genetic Cre-loxP assessment of epicardial cell fate using Wt1-driven Cre alleles. Circ Res 111: e276–e280. doi: 10.1161/CIRCRESAHA.112.275784

21. Huang GN, Thatcher JE, McAnally J, Kong Y, Qi X, et al. (2012) C/EBP transcription factors mediate epicardial activation during heart development and injury. Science 338: 1599–1603. doi: 10.1126/science.1229765.

22. Rana MS, Horsten NC, Tesink-Taekema S, Lamers WH, Moorman AF, et al. (2007) Trabeculated right ventricular free wall in the chicken heart forms by ventricularization of the myocardium initially forming the outflow tract. Circ Res 100: 1000–1007.

23. Meilhac SM, Esner M, Kelly RG, Nicolas JF, Buckingham ME (2004) The clonal origin of myocardial cells in different regions of the embryonic mouse heart. Dev Cell 6: 685–698.

24. Schlueter J, Brand T (2009) A right-sided pathway involving FGF8/Snail controls asymmetric development of the proepicardium in the chick embryo. Proc Natl Acad Sci U S A 106: 7485–7490. doi: 10.1073/pnas.0811944106

25. Gittenberger-de Groot AC, Vrancken Peeters M-PFM, Bergwerff M, Mentink MMT, Poelmann RE (2000) Epicardial outgrowth inhibition leads to compensatory mesothelial outflow tract collar and abnormal cardiac septation and coronary formation. Circ Res 87: 969–971.

26. Mikawa T, Gourdie RG (1996) Pericardial mesoderm generates a population of coronary smooth muscle cells migrating into the heart along with ingrowth of the epicardial organ. Dev Biol 174: 221–232.

27. Pombal MA, Carmona R, Megias M, Ruiz A, Perez-Pomares JM, et al. (2008) Epicardial development in lamprey supports an evolutionary origin of the vertebrate epicardium from an ancestral pronephric external glomerulus. Evol Dev 10: 210–216 doi: 10.1111/j.1525-142X.2008.00228.x.

28. Braitsch CM, Combs MD, Quaggin SE, Yutzey KE (2012) Pod1/Tcf21 is regulated by retinoic acid signaling and inhibits differentiation of epicardium-derived cells into smooth muscle in the developing heart. Dev Biol 368: 345–357. doi: 10.1016/j.ydbio.2012.06.002.

29. Van Praagh R (2011) The cardiovascular keys to air-breathing and permanent land-living in vertebrates. Kardiochir Thorakochir Polska 8: 1–22.

30. De la Cruz MV, Moreno-Rodriguez R (1998) Embryological development of the apical trabeculated region of both ventricles. The contribution of the primitive interventricular septum in the ventricular septation. In: De la Cruz MV, Markwald RR, editors. Living morphogenesis of the hearts. Birkhauser. pp.121–130.

31. Rowlatt U (1990) Comparative anatomy of the heart of mammals. Zool Journal Linnean Society 98: 73–110.

32. Bax NA, Bleyl SB, Gallini R, Wisse LJ, Hunter J, et al. (2010) Cardiac malformations in Pdgfralpha mutant embryos are associated with increased expression of WT1 and Nkx2.5 in the second heart field. Dev Dyn 239: 2307–2317. doi: 10.1002/dvdy.22363

33. Hsu WH, Yu YR, Hsu SH, Yu WC, Chu YH, et al. (2013) The Wilms' tumor suppressor Wt1 regulates Coronin 1B expression in the epicardium. Exp Cell Res 319: 1365–1381. doi: 10.1016/j.yexcr.2013.03.027

34. Hamburger V, Hamilton HL (1951) A series of normal stages in the development of the chick embryo. J Morphol 88: 49–92.

35. Poelmann RE, Gittenberger-de Groot AC, Mentink MMT, Bökenkamp R, Hogers B (1993) Development of the cardiac coronary vascular endothelium, studied with antiendothelial antibodies, in chicken-quail chimeras. Circ Res 73: 559–568.

36. Darnell DK, Garcia-Martinez V, Lopez-Sanchez C, Yuan S, Schoenwolf GC (2000) Dynamic labeling techniques for fate mapping, testing cell commitment, and following living cells in avian embryos. Methods Mol Biol 135: 305–321.

37. Moorman AF, Houweling AC, de Boer PA, Christoffels VM (2001) Sensitive nonradioactive detection of mRNA in tissue sections: novel application of the whole-mount in situ hybridization protocol. J Histochem Cytochem 49: 1–8.

38. de Bakker MA, Fowler DA, den OK, Dondorp EM, Navas MC, et al. (2013) Digit loss in archosaur evolution and the interplay between selection and constraints. Nature 500: 445–448. doi: 10.1038/nature12336

39. de Boer BA, van den Berg G, de Boer PA, Moorman AF, Ruijter JM (2012) Growth of the developing mouse heart: an interactive qualitative and quantitative 3D atlas. Dev Biol 368: 203–213. doi: 10.1016/j.ydbio.2012.05.001

14

Systematic Characterization and Comparative Analysis of the Rabbit Immunoglobulin Repertoire

Jason J. Lavinder[1,2], **Kam Hon Hoi**[3], **Sai T. Reddy**[1,3¤], **Yariv Wine**[1,2], **George Georgiou**[1,2,3,4]*

1 Department of Chemical Engineering, University of Texas at Austin, Austin, Texas, United States of America, **2** Institute for Cellular and Molecular Biology, University of Texas at Austin, Austin, Texas, United States of America, **3** Department of Biomedical Engineering, University of Texas at Austin, Austin, Texas, United States of America, **4** Section of Molecular Genetics and Microbiology, University of Texas at Austin, Austin, Texas, United States of America

Abstract

Rabbits have been used extensively as a model system for the elucidation of the mechanism of immunoglobulin diversification and for the production of antibodies. We employed Next Generation Sequencing to analyze Ig germline V and J gene usage, CDR3 length and amino acid composition, and gene conversion frequencies within the functional (transcribed) IgG repertoire of the New Zealand white rabbit (*Oryctolagus cuniculus*). Several previously unannotated rabbit heavy chain variable (VH) and light chain variable (VL) germline elements were deduced bioinformatically using multidimensional scaling and k-means clustering methods. We estimated the gene conversion frequency in the rabbit at 23% of IgG sequences with a mean gene conversion tract length of 59 ± 36 bp. Sequencing and gene conversion analysis of the chicken, human, and mouse repertoires revealed that gene conversion occurs much more extensively in the chicken (frequency 70%, tract length 79 ± 57 bp), was observed to a small, yet statistically significant extent in humans, but was virtually absent in mice.

Editor: Javier Marcelo Di Noia, Institut de Recherches Cliniques de Montréal (IRCM), Canada

Funding: This work was funded by the Defense Advanced Research Projects Agency (DARPA, www.darpa.mil) grant HR0011-10-0052 and the Defense Threat Reduction Agency (DTRA, www.dtra.mil) grant HDTRA1-12-COO07. JJL was supported by a postdoctoral fellowship from the Cancer Prevention Research Institute of Texas (CPRIT, www.cprit.state.tx.us). The funders had no role in study design, data collection and analysis, decision to publish, or preparation of the manuscript.

Competing Interests: The authors have declared that no competing interests exist.

* Email: gg@che.utexas.edu

¤ Current address: Department of Biosystems Science and Engineering, Swiss Federal Institute of Technology (ETH) Zürich, Basel, Switzerland

Introduction

B cell development and repertoire diversification vary significantly among vertebrate species [1]. Diversification of the Ig repertoire occurs through the combinatorial joining of numerous V, D, and J gene segments for the Ig heavy chain (or just V and J gene segments in the case of Ig light chains) through several mechanisms collectively referred to as VDJ recombination, followed by somatic mutagenesis upon subsequent B-cell encounter with foreign antigen. Compared to humans and mice, which use a diverse assortment of germline VH gene segments during VDJ recombination of the heavy chain, the rabbit IgH repertoire displays highly restricted VH gene segment usage. Earlier studies had indicated that the majority of B cells in the rabbit utilize the VH1 gene, the most D-proximal VH locus [2]. VH1 Ig are serotypically VHa-positive, and there are three distinct VHa allotypic lineages (a1, a2, and a3) [3,4]. In addition, approximately 10–20% of expressed Ig in rabbits are serotypically VHa-negative (VHn) [4,5]. The VHn Ig genes that have been annotated in rabbits (VHx, VHy, and VHz) are encoded by loci significantly upstream (>100 kb) of the VH1 gene locus [6]. Recently, sequencing of the rabbit genome has enabled the identification of germline Ig elements in a Thorbecke inbred rabbit [7]. Overall, >300 VH-like gene sequences were identified within 79 unplaced genomic scaffolds (i.e. unknown chromosomal locations). The large number of previously unannotated VH-like sequences identified within the a1/a2 Thorbecke rabbit, as well as previously identified sequences from latent heavy chain allotypes [2,8], clearly demonstrate the complexity of the germline Ig repertoire. However, because the sequenced Thorbecke rabbit was heterozygous at the IgH locus (a1/a2 based on mapping of the VH1 gene), the actual number of distinct VH gene elements in the haploid genome is unclear.

Another major source of Ig repertoire diversity derives from the somatic introduction of non-templated nucleotides into the imprecise junctions formed by the variable ligation of recombining V-D and D-J gene segments—a process known as N-nucleotide addition. This hypervariable V-*N*-D-*N*-J interval defines CDR3 of the heavy chain (CDRH3). Species such as cattle have extremely long CDRH3s [9] as a result of increased levels of N-nucleotide addition. Longer CDRH3s not only create a more expansive and diverse sequence space in the Ig repertoire, but may also hold unique functional relevance in protection against disease [10]. For most mammalian species, N-nucleotide addition during VJ recombination of the light chain is limited and therefore junctional diversity in the light chain is much less pronounced compared to the heavy chain; however, rabbits have been shown to have light chain CDR3s (CDRL3s) that are unusually longer and more diverse, indicating significant N-nucleotide addition during light chain VJ recombination [11].

After VDJ recombination, the naïve Ig repertoire in rabbits is further diversified in the first 2 months of age by extensive somatic mutagenesis in the gut-associated lymphoid tissue (GALT) [12], through both somatic hypermutation (SHM) and gene conversion events [13], both of which have been shown to be dependent upon the exposure of the naïve B cell repertoire to the gut microflora [14]. Ig gene conversion is employed not only by rabbits, but also by other species including chickens and involves the non-reciprocal homologous recombination of upstream donor V gene loci into the recombined VDJ (and VJ) locus. Like SHM, Ig gene conversion is mediated through the enzyme activation induced cytidine deaminase (AID) [15] and thus is often found to occur proximal to hotspot AID motifs conserved within germline V genes. In chickens, gene conversion has been shown to be the dominant mechanism of AID-mediated mutagenesis [16] and involves a single functional VH and VL gene undergoing gene conversion with numerous upstream VH and VL pseudogenes, respectively [17]. In rabbits, however, the upstream loci are a mix of functional V genes and pseudogenes that can serve as potential donor sequences in gene conversion events. The fundamental properties of gene conversion events and the relative extent to which gene conversion plays a role in rabbit Ig diversification is not entirely clear, mostly due to limitations in sampling and difficulty in precise, automated identification of gene conversion events in highly mutated Ig sequences.

Here, we present a thorough characterization of the expressed rabbit IgG repertoire. We identify several unannotated functional rabbit germline VH and VL germline gene sequences and provide a comprehensive survey of the salient features of the rabbit Ig repertoire. We estimate the gene conversion frequency in the rabbit and demonstrate that it is significantly less than that observed in the chicken repertoire and, not surprisingly, much greater than that observed in humans and mice.

Materials and Methods

Ethics Statement

Three New Zealand white (NZW) rabbits and one white leghorn chicken were used for this work, as approved through the Institutional Animal Care and Use Committee (IACUC) of the University of Texas at Austin (protocol AUP-2011-00016). All efforts were made to ensure animal welfare and minimize suffering in accordance with the United States Department of Agriculture (USDA) Animal and Plant Health Inspection Service (APHIS) Guidelines for animal care and husbandry.

Isolation of B cells from immunized rabbits, chicken, mouse, and human

At sacrifice, rabbit femoral bone marrow (BM) cells were isolated and approximately 100 ml blood was collected into heparin tubes. Blood aliquots of 20 ml were gently layered over 20 ml of Histopaque 1077 (Sigma, MO, USA) and centrifuged in a swinging bucket rotor at 400 g, 45 min at 25°C (Beckman Coulter). The serum was removed from the top of the gradient and stored at −20°C. PBMCs were isolated from the intermediate layer. Each collected tissue (BM and PBMC) was processed as previously described [18], with the exception that the PBMCs did not require red blood cell lysis after gradient centrifugation. CD138[+] cells were isolated as previously described [19]. PBMCs or CD138[+] BM plasma cells (PCs) were centrifuged at 930×g, 5 min at 4°C. Cells were then lysed with TRI reagent (Ambion, TX, USA) and total RNA was isolated according to the

manufacturer's protocol in the Ribopure RNA isolation kit (Ambion). RNA concentrations were measured with an ND-1000 spectrophotometer (Nanodrop, DE, USA).

For the chicken, total RNA was prepared from splenic tissue of a white leghorn chicken using TRIzol reagent (Life technologies) and purified with RNeasy Micro Kit (Qiagen, CA). cDNA was generated from total RNA using oligo(dt) according to the manufacturer's protocol (Superscript II First strand Synthesis kit, Life Technologies), PCR-amplified as described previously [20] using chicken IgY-specific primers listed in Table S1, and sequenced using the 2×250 paired end MiSeq Next Generation Sequencing (NGS) platform (Illumina, San Diego, CA). The two Illumina 2×250 output files were aligned using FLASH [21] and CDRH3 and full-length VH sequences were determined using in-house probabilistic model [18] for delimiting the CDRH3 regions based on *Gallus gallus* Ig sequences found in NCBI Genbank.

Amplification and high-throughput sequencing of rabbit VH and VL gene repertoires

Approximately 0.5 μg of ethanol precipitated RNA was used for first-strand cDNA synthesis according to the manufacturer's protocol for 5′ RACE using the SMARTer RACE cDNA Amplification kit (Clontech, CA, USA). The cDNA reaction was diluted into 100 μl of Tris-EDTA buffer and stored at −20°C. 5′ RACE PCR amplification was performed on the first strand cDNA to amplify the VH repertoire with the kit-provided, 5′ primer mix and 3′ rabbit IgG-specific primers RIGHC1 and RIGHC2 (Table S1). The rabbit VL repertoire was amplified via 5′ RACE, using a 3′ primer mix specific for both the Vκ and Vλ rabbit constant regions. The VL primers comprised 90% RIGκC mix and 10% RIGλC mix (Table S1) to approximate known ratios of light chain isotypes in rabbits. Reactions were carried out in a 50 μl volume by mixing 35.25 μl H$_2$O, 5 μl 10X Advantage-2 PCR buffer (Clontech), 5 μl 10X Universal Primer A mix (Clontech), 0.75 μl Advantage-2 polymerase mix (Clontech), 2 μl cDNA, 200 nM V$_H$ or V$_L$ primer mix, and 200 μM dNTP mix. PCR conditions were: 95°C for 5 min, followed by 30 cycles of amplification (95°C for 30 sec, 60°C for 30 sec, 72°C for 2 min), and a final 72°C extension for 7 min. The PCR products were gel-purified to isolate the amplified VH or VL DNA (~500 bp). 100 ng of each 5′ RACE amplified VH or VL DNA was processed for Roche GS-FLX 454 DNA sequencing according to the manufacturer's protocol. The 454 dataset has been deposited at the NIH SRA (Sequence Read Archive) under accession number SRP042296.

All 454 data were first processed using the sequence quality and signal filters of the 454 Roche pipeline and then subjected to bioinformatics analysis that relied on homologies to conserved framework regions using IMGT/HighV-Quest Tool [22]. Additional filters were applied for full repertoire database construction as follows: (i) Length cutoff: full-length sequences were filtered by aligned amino acid lengths >70 residues and aligned framework 4 region lengths >2 residues; (ii) Stop codons: aligned amino acid sequences containing stop codons were removed.

IgBLAST alignment, Multidimensional scaling (MDS), and k-means analysis

An IgBLAST database for germline annotation of the rabbit IgG sequences was constructed using the following sequences: the IMGT rabbit V germline reference set that includes the allotypic a2 sequences in BAC clones AY386694 and AY386697 [23], allotypic a2 sequences from an Alicia rabbit (AF176997 through AF177016) [24], potentially latent IGHV (M12180, M60121,

M60336) [8,25,26], allotypic a1 sequences VH1-a1 (M93171), VH3-a1 (M93177), and VH4-a1 (M93181) [27], and the allotypic a3 sequences VH1-a3 through VH7-a3 (M93173, M93176, M93179, M93183, M93184, M93185, M93186) [13,27]. In addition to the IMGT rabbit reference set, initial IgBLAST database included VH8-a3 through VH11-a3 (L27311, L27312, L27313, L27314) [28], VHx (L03846) [29], and VHy (L03890) [29]. For light chain, the IMGT database was used without addition. IgBLAST alignments against the database were analyzed by bit score (and equivalently the number of called nucleotide mutations per sequence). Aligned (annotated to a certain germline) sequences with greater than 30 called mutations were extracted from this initial IgBLAST alignment and these poorly aligned sequences were aligned using MUSCLE [30] multiple sequence alignment (BLOSUM80 substitution matrix, gap open penalty -15, gap extend penalty -3). For calculating distance matrices and performing MDS, the package bios2mds [31] in the R environment was used. The MUSCLE alignment was imported into R and the pairwise distance matrix calculation using the 'mat.dif' function, which computes a distance matrix based on pairwise differences between each sequence was performed. Metric MDS analysis of the pairwise distance matrix was performed using the function 'mmds', which reduces the dimensionality of the distance matrix into Euclidean space. These Euclidean values are analyzed by k-means silhouette scoring (function 'sil.score) and k-mean clustering (function 'Kmeans') to identify distinct sets of sequences that each derive from an unannotated germline Ig sequence. The sequences from each cluster are extracted and aligned in MUSCLE. For each derived cluster alignment, the consensus sequence was searched by BLASTn against the non-redundant nucleotide collection and the rabbit genome.

IMGT and IgBLAST repertoire analyses

Germline V gene assignments were derived from IgBLAST alignments against the database described above. Germline J gene assignments and CDR3 sequences (rabbit, mouse, and human) were derived from IMGT HighV-Quest alignments. Chicken CDR3 sequences were derived from a position weight matrix motif search of the FR3 and J region in chickens.

Gene conversion analysis

For rabbits, IgBLAST alignments of the NGS data sets was performed using custom BLAST databases for rabbit, as detailed above. For the chicken, the IgBLAST database included the functional VH1 sequence, along with 18 known VH pseudogenes [17]. For mouse and human, the IgBLAST-provided database was used. IgBLAST was used to assign the best-scoring germline VH reference sequence for each query sequence. To detect gene conversion events in the query, the assigned germline reference sequence was then scored against all other germline reference sequences in the IgBLAST alignment as follows: 1) For each VH germline in the alignment (each a possible donor VH sequence) except the assigned one, we used a scoring function that assigns a '1' at each position only if the putative donor VH matches and the assigned reference VH germline mismatches, a '0' at each position that both references either match or both mismatch, and a '−1' at each position that the assigned reference VH matches and the putative donor VH mismatches. 2) Search each scored putative donor VH for stretches of positions that score as '1', with a putative gene conversion event called only if three positions scoring '1' are uninterrupted by positions scoring −1. The gene conversion event boundaries were defined by positions scoring '−1' (long tract boundary) or by the most distal positions of the tract that score '1' (short tract boundary). Adjacent long tracts from the same donor VH are automatically combined by allowing long tracts with a shared boundary to connect. Positions of the alignment that have gaps in the query are scored as '0' in all putative donor VH scored positions. To exclude PCR crossover products or gene replacement events (single crossover events), all gene conversion events that start within the first 15 positions or end with the last 15 positions of the aligned VH gene are excluded (e.g. the gene conversion must be an internal double crossover event with sufficient sequence from the assigned VH on each side). The donor VH selected represents the germline VH with the highest scoring tract (sum of the tract positional scores). P-values for the gene conversion events are scored as described [32], with the exception that all polymorphic sites are permuted during the permutation test. The p-values described here are local p-values calculated via 1000 iterations of positional permutation of the assigned and donor VH germlines. Only gene conversion events with a p-value below 0.05 (95% confidence interval) and a minimum tract score >4 (to avoid effects of high SHM) are considered as high confidence events.

Results

Identification of putative rabbit VH germline elements using multidimensional scaling of high throughput sequencing data

Total RNA was isolated from BM PCs and total PBMCs of three adult NZW rabbits. IgG heavy chain and Igκ/Igλ light chain cDNAs were amplified by 5′ RACE using primers that annealed respectively to the CH1 or Cκ/Cλ constant region directly 3′ of the J segment (Table S1), and the resulting amplicons were sequenced by Roche 454 sequencing. 172,126 high quality reads corresponding to 88,830 unique heavy chain sequences across the three rabbits were obtained (Table 1). Germline VH usage was determined with IgBLAST [33] alignments using a custom database that included NZW rabbit germline sequences compiled from a number of sources [8,13,23,24,25,26,27,28,29] (see Materials and Methods). For the VHa sequences in all three rabbits, >99% were of the a3 allotype, strongly indicating that the cohort of NZW rabbits examined here is homozygous a3/a3 at the IgH locus. However, the IgBLAST alignments revealed a non-normal distribution of VH germline alignment scores (Figure S1A). Based on an analysis by Gertz et al. [7] revealing a number of unannotated germline elements in an a1/a2 Thorbecke rabbit, we hypothesized that the NZW rabbit germline database may be incomplete and thus lack the germline V gene sequences for these poorly scoring Ig alignments. MDS [34], a space-based method that has been used to identify patterns in distance matrices derived from multiple sequence alignments (MSAs) of large biological sequence data sets [31,35,36], was employed to deduce putative germline V gene segments. MDS allows MSA distance matrices to be analyzed in Euclidean space, facilitating k-means clustering [37] of the sequences. In the case of somatically mutated Ig V gene sequences, the consensus sequence of each of these k-means defined clusters represents a putative germline V gene sequence. Figure 1 shows the MDS and k-means clustering of the poorly aligned VH gene sequences (higher than 30 nt differences from the nearest VH germline) in the NZW rabbit repertoire. For each of the three rabbits, four distinct VH clusters were identified. Each cluster of VH sequences was extracted and aligned, and the consensus sequence for each of the four clusters was compared across the three rabbits. Each of the four VH consensus sequences (Table S2) matched identically across all three rabbits, strongly supporting our hypothesis that the poorly aligned sequences are

Table 1. Summary of 454 NGS data.

Sample	reads	unique VH/VL amino acid sequences	unique CDRH3/CDRL3
Rabbit rab1 PBMC VH	16102	9447	5525
Rabbit rab1 Bone marrow PC VH	31136	19044	5954
Rabbit rab2 PBMC VH	24251	13459	7220
Rabbit rab2 Bone marrow PC VH	76510	34762	11564
Rabbit rab3 Bone marrow PC VH	24127	12118	5958
Rabbit rab1 Bone marrow PC VL	24489	10446	5629
Rabbit rab2 Bone marrow PC VL	17155	7487	4465
Rabbit rab3 Bone marrow PC VL	23761	12581	7139

derived from unannotated germline VH elements encoded in the NZW rabbit genome.

The four putative germline sequences identified by MDS and k-means clustering were searched by BLASTn to identify homology to publicly available rabbit genomic and transcript sequences (Table 2). For three of the four putative VH germline sequences, NZW rabbit genomic or transcript sequence matches were found that were identical or within 1–3 nucleotide differences. The closely matching transcript sequences (AY676808, AF264452, and AF264440) were derived from rabbits that have a ligated appendix (LigApx) [14,38], which effectively eliminates SHM and gene conversion. Three of the four putative germline sequences

contained a $^{70}WVN^{72}$ motif, consistent with VHa-negative (VHn) immunoglobulins (VHa sequences have a $^{70}WAK^{72}$ motif), while one sequence (VHs1) had a $^{70}SVK^{72}$ motif, which is predominant in VHs immunoglobulins (which are also VHa-negative) and ancestral to hares [39]. VHx2 was highly identical (281/288 nt) to the VHx32 allele previously annotated [29] and may represent a distinct VHx allele (hence its designated ID). These four new putative germline sequences in the NZW rabbit were added to our existing NZW rabbit germline database (see Materials and Methods for full description) and using this updated database, IgBLAST was used to assign VH and JH germline usage (Figure 2). Consistent with earlier observations [2,19], the VH1

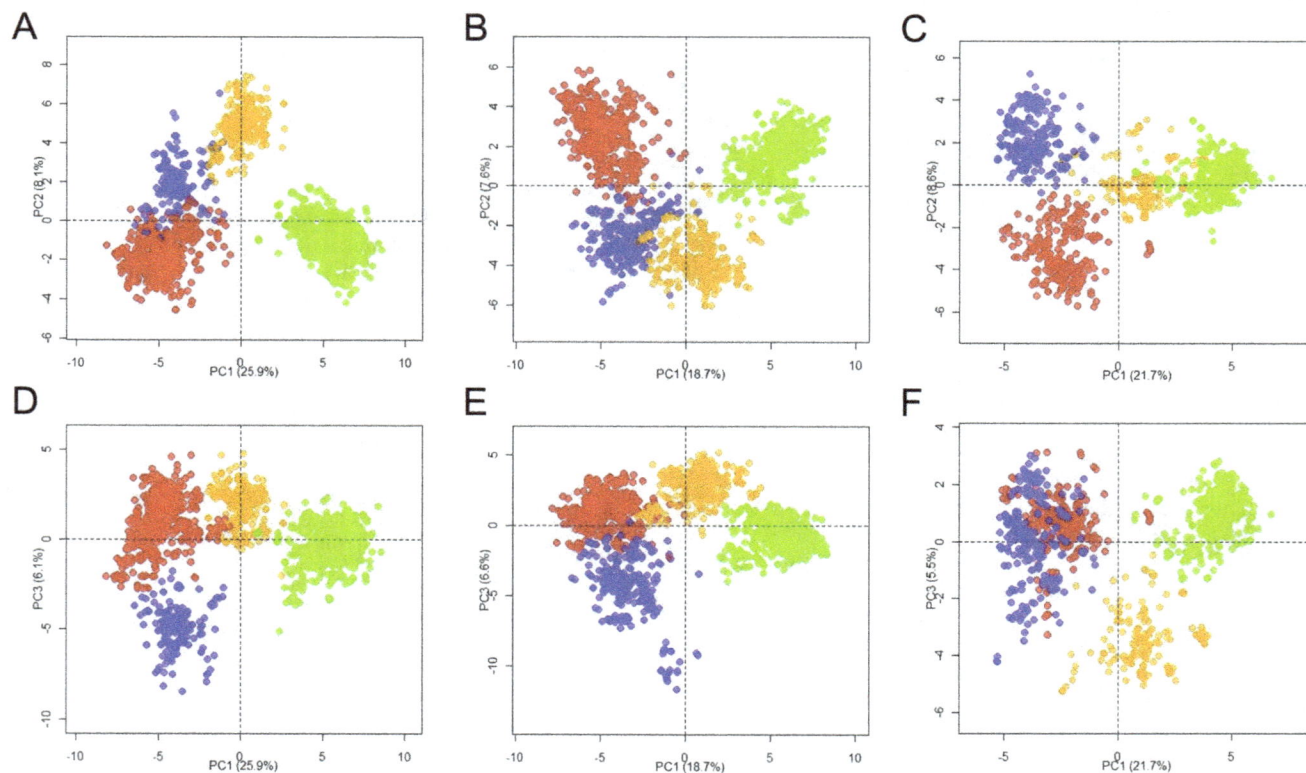

Figure 1. MDS and k-means clustering of low scoring alignments for VH sequences. The first three principle components of the MDS are shown here, with k-means defined clusters colored differently. PC1 (principle component 1) v. PC2 (principle component 2) for (A) rab1, (B) rab2, and (C) rab3, and PC1 v. PC3 (principle component 3) for (D) rab1, (E) rab2, and (F) rab3. Each color represents a cluster of sequences as determined by k-means clustering of the Euclidean MDS-derived values.

Table 2. BLASTn results of the four putative VH germline sequences identified by MDS and k-means clustering.

VH germline (query sequence)	NCBI GenBank accession number	mismatches/total length (nt)	Source of highly identical BLASTn sequence match
VHn3	M12386	3/288	Putative VH from liver genomic DNA
	NW_003160066	4/288	Unplaced genomic scaffold in Thorbecke rabbit
VHs1	JN131896	2/285	Genomic VDJ recombinant from immature B cell
	AY676808	1/285	mRNA from peripheral B cell in LigApx rabbit
	NW_003161050	3/285	Unplaced genomic scaffold in Thorbecke rabbit
VHx2	AF264452	0/288	mRNA from peripheral B cell in LigApx rabbit
	AF264440	0/288	mRNA from peripheral B cell in LigApx rabbit
	NW_003159519	3/288	Unplaced genomic scaffold in Thorbecke rabbit
VHn2	AF245499	10/288	mRNA from rabbit bone marrow and spleen

Each NCBI GenBank accession number is a sequence identified by BLASTn as a highly homologous sequence to the query putative VH germline sequence.

gene is heavily utilized in all three rabbits, as is the VH4 gene, which is >97% identical to VH1. The VHa-negative sequences (combined) account for 12%, 22%, and 11% of the total IgG sequences in rab1, rab2 and rab3 respectively. All three rabbits also exhibit highly restricted JH usage, with JH4 accounting for 60–70% of the IgG repertoire.

Vκ and Jκ usage in the rabbit

Similar to mice, rabbits utilize the kappa light chain isotype at a much higher frequency than the lambda isotype [40]. We amplified the light chain repertoire from BM PCs in all three rabbits using 5′ RACE and sequenced the VL region using Cκ and Cλ specific primers. A total of 65,405 high quality reads and 30,514 unique sequences across the three rabbits were obtained (Table 1). As expected, the utilization of lambda light chain sequences sets was very low (<1%). Rabbit immunoglobulin kappa light chains have four allotypes: b4, b5, b6, and b9 [41]. For each of the three rabbits examined here, >98% of the unique VL sequences were of the b4 allotype, indicating this cohort of NZW

rabbits was b4/b4 homozygous. Similar to the results of the VH IgBLAST alignments, Vκ gene alignment scores also revealed a non-normal distribution, with a group of sequences exhibiting significantly lower alignment scores as compared to the bulk of the Vκ sequences (Figure S1B). These poorly aligned sequences were examined more closely by MDS and k-means clustering as described above and in the Materials and Methods, and four new Vκ clusters were identified (Figure S2). Two of the four putative Vκ germline sequences, NZWk57r and NZWk155g (Table S2), were utilized in all three rabbits. NZWk57r and NZWk155g have also been detected in non-functional light chain sequences (VJ junction out-of-frame) in the bone marrow of a 1 day old b5/b5 NZW rabbit (i.e. early development when naive, unmutated Ig sequences are common in the rabbit) [11]. For the other two putative Vκ germlines (Table S2), one was identified only in the rab2 and rab3 rabbits (NZWk807y), while the other was identified only in the rab2 rabbit (NZWk529g). Nonetheless, all four cluster consensus sequences were also found by BLASTn analysis as either exact matches or differing by only 1 nt (NZWk807y) from

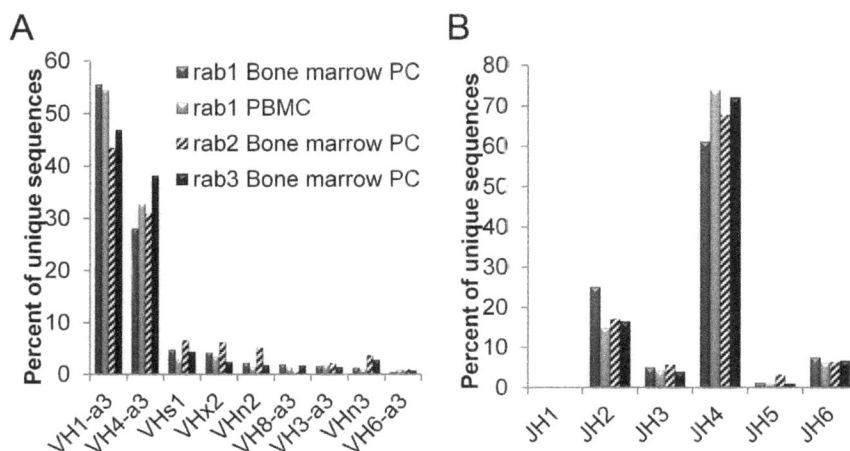

Figure 2. Heavy chain germline gene usage. (A) VH and (B) JH germline usage of unique IgG sequences in rabbits. Sample sizes of unique sequences: rab1 Bone marrow PC, N = 19,291; rab1 PBMC, N = 9,235; rab2 Bone marrow PC, N = 10,148; rab3 Bone marrow PC, N = 12,107.

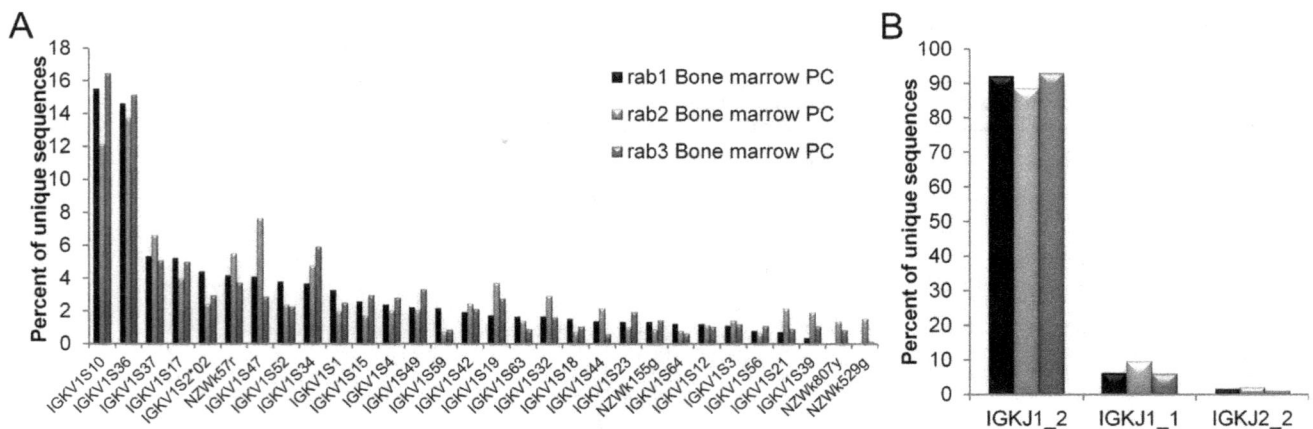

Figure 3. Light chain germline gene usage. (A) Vκ and (B) Jκ germline usage of unique kappa light chain sequences in rabbits. Sample sizes of unique sequences: rab1 Bone marrow PC, N = 10,446; rab2 Bone marrow PC, N = 7,481; rab3 Bone marrow PC, N = 12,580. Germline gene IDs are as listed in the IMGT database.

previously identified germline genes in the Thorbecke inbred rabbit.

The four putative Vκ sequences were added to our existing NZW IgBLAST database, which was then used to assign germline Vκ usage (Figure 3). Contrary to the sharp germline restriction seen in the VH gene repertoire, Vκ gene usage is very diverse, with the top germline gene segment used at ~10–20% and 30 Vκ germlines utilized at least >1% (of total unique Vκ sequences) across the three rabbits. Jκ germline usage, on the other hand, is mostly restricted to the IGKJ1_2 gene (~90%) and to a very small extent IGKJ1_1 and IGKJ2_2.

Characterization of the CDRH3 and CDRL3 in the rabbit IgG repertoire as compared to other species

In addition to the rabbit NGS data set, we also analyzed human [42], mouse [18], and chicken NGS data sets to compare and contrast repertoire characteristics across species. For the chicken, we obtained 320,468 high quality VH sequence reads (231,165 unique VH amino acid sequences) from the splenic B cell repertoire of a white leghorn chicken using the Illumina MiSeq 2×250 NGS platform. A comparison of the CDRH3 length distribution is shown in Figure 4. Rabbit IgG CDRH3 lengths are intermediate (mean = 14.8±3.6 aa, mode = 13 aa) relative to mice (mean = 11.1±2.0 aa, mode = 10 aa), humans (mean = 15.3±4.0 aa, mode = 15 aa), and chickens (mean = 17.9±2.8 aa, mode = 16 aa). The length distribution of the CDRH3 for all unique IgG sequences was similar across all three rabbits (Figure S3). For CDRL3, mice and humans both exhibit very little junctional diversity and are severely restricted in length, with the vast majority of CDRL3s for both species being 9±1 amino acids (Figure 4); However, the rabbit exhibits significant junctional diversity in the CDRL3, with a wide distribution of CDRL3 lengths (range: 5aa–16aa) and a much greater mean length, equal to 12±1.6 aa.

The amino acid composition of the rabbit Ig CDRH3 is dominated by tyrosine (Y), glycine (G), and aspartate (D) which together represent half (49%) of the amino acid usage in the CDRH3 loop (Figure 4), while the top five amino acids used (GYDAS) represent a full two-thirds (66%) of the amino acid usage. In that regard, the overall amino acid utilization in the rabbit is highly similar to the other species, consistent with earlier observations [43] that the average hydrophobicity of CDRH3—

and, hence, the center of the antigen binding site—is conserved across evolution to be slightly hydrophilic and enriched for glycine, serine and tyrosine. Nevertheless, when compared to other species, the CDRH3 amino acid composition in rabbits does show some distinct features. Human CDRH3s use glycine and tyrosine at a much lower frequency than that seen in rabbits. Chicken CDRH3s have less tyrosine (~2-fold less than rabbits) but utilize much higher cysteine content (~5–10-fold higher than humans or rabbits). The higher utilization of Cys residues in the chicken CDRH3 repertoire has previously been shown to be important for stabilizing (by disulfide bonds) the longer CDRH3 loops seen in chickens [44]. The amino acid utilization of the rabbit CDRL3 is also shown in Figure 4 for comparative purposes.

Diversification of the rabbit IgG repertoire by SHM and gene conversion

The rabbit Ig repertoire is known to undergo extensive AID-mediated mutagenesis (via both SHM and gene conversion) early on in development when the antigen-inexperienced naïve B cell repertoire migrates from the bone marrow to the GALT [6]. Earlier studies with rabbits lacking an established gut microflora demonstrated significantly reduced levels of AID-mediated diversification of the repertoire, with most Ig having sequences that approximate the germline elements from which they are derived [14,38].

We compared the overall level of mutation (combined SHM and gene conversion) within the IgG repertoires of rabbits, chicken, mice and humans (Figure 5). The mutational load varied as follows: chicken>rabbit ≈ human>mouse. It should be noted that the reported mutational load is a combination of both biological processes mediated by AID and inherent PCR/ sequencing error, which has been reported to be approximately 1% for both 454 GS-FLX [45] and Illumina MiSeq sequencing [46]. To determine the relative contribution of gene conversion to the diversification of the primary repertoire, we developed a script that searches Ig sequences for tracts of putative gene conversion events. Gene conversion tracts are detected as a contiguous block of nucleotides within a query Ig sequence that closely matches a different germline element (e.g. not the query's assigned germline element) in the IgBLAST database. Additionally, to rule out possible PCR template switching artifacts, the gene conversion tracts were required to be bound on each end by positions (tracts)

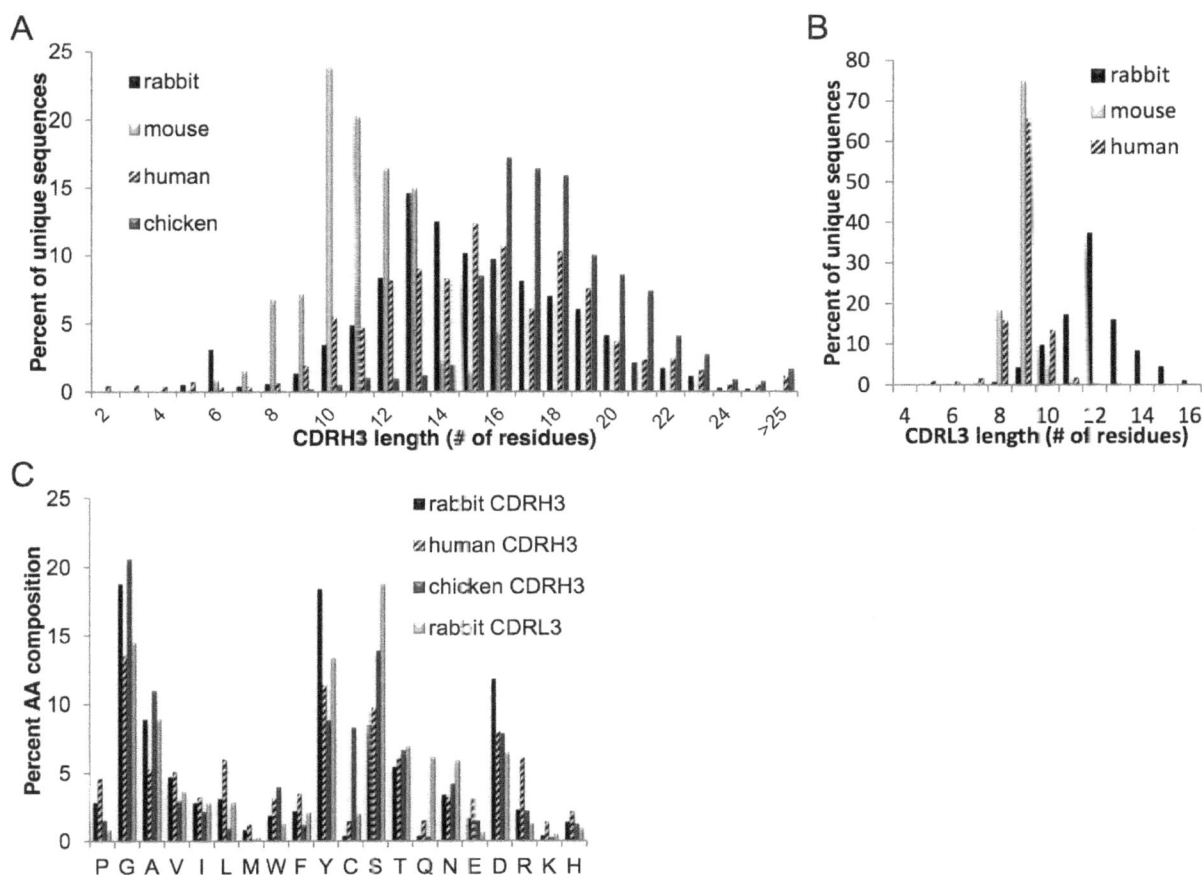

Figure 4. Characterization of the CDRH3 and CDRL3 in rabbit as compared to other species. (A) CDRH3 lengths, (B) CDRL3 lengths, and (C) CDRH3 and CDRL3 amino acid composition. All data shown here is derived from unique heavy chain or light chain sequences. Sample sizes of unique IgG/IgY sequences: mouse heavy chain, N = 2,762; rabbit heavy chain, N = 29,439; rabbit light chain, N = 10,446; human heavy chain, N = 2,948; chicken heavy chain, N = 231,165.

that match the query's assigned VH germline sequence (i.e. the gene conversion event was not contiguous with the 5′ or 3′ ends of the sequence). Additionally, minimum scoring and p-value thresholds were applied as described in the methods. Strict statistical thresholds were set to ensure that the identified gene conversion events were highly significant and not attributed to high loads of point mutation. For these reasons, the reported frequencies of gene conversion events should be considered as a lower bound of the actual biological frequencies (Table 3).

The vast majority of unique chicken IgY sequences examined (70%) display evidence of gene conversion events. In rabbits, 23% of IgG sequences and 32% of Igκ sequences were the products of gene conversion. There have been previous, although somewhat controversial, indications suggesting gene conversion occurs in humans and mice as well, albeit at a much lower frequency [47,48,49]. We find that, in the mouse, putative gene conversion events are nearly absent, with an estimated frequency of 0.1% of all unique IgG sequences. Whereas an earlier analysis of gene conversion in a small set of human IgG sequences indicated that ~7% (8 out of 121) display evidence of having undergone gene conversion [49], our present analysis of a much larger data set revealed a lower frequency of 2.5%. We note that, in humans and mice, the low p-values (p<0.05) in the detection of gene conversion events suggest that these are high confidence identifications despite the fact that the average tract lengths detected

were significantly lower than those in the rabbit and chicken (Table 3).

The frequencies of donor germline VH usage for gene conversion in the rabbit are largely unknown. Figure 6 shows the donor germline VH usage for query sequences that were assigned by IgBLAST to one of three heavily utilized germline VH gene segments in the rabbit (VH1, VHs1, and VHn3). Because gene conversion occurs through homologous recombination, the frequency is heavily dependent on donor VH sequence homology and proximity. High homology donor VH genes directly upstream of the assigned VH reference (e.g. the VH germline originally used during VDJ recombination) are expected to be used in gene conversion more frequently than donor genes that are more distal or less homologous. The donor germline usage for VH1 is consistent with this expectation, with the genes directly upstream being used as donors for gene conversion more frequently than those more distal to VH1. The two VHa-negative sequences (VHs1 and VHn3) have very different patterns of germline VH donor usage. The genomic location and organization of these two VHa-negative elements are not known, but it is clear that VHs1 must be downstream of VHn3 as it heavily utilizes VHn3 as a donor sequence for gene conversion.

The tract lengths and start/end residue numbers of the gene conversion events for assigned VH1 sequences are shown in Figure 6B and 6C. The majority of gene conversion tracts in rabbit

Table 3. Gene conversion analysis: comparison of the frequencies, average scores, and average GC tract lengths in four different species.

Organism/sample	unique seqs	% with GC	max GC score	avg tract (nt)
rabbit BM PC IgG	10680	23	17	59±36
rabbit BM PC Igκ	2062	32	19	86±48
chicken Spleen IgY	10000	70	23	79±57
mouse BM PC IgG	946	0.1	5	6
human PBMC IgG	1028	2.5	7	39±27

IgG are under 30 bp in length, although some identified tracts are much longer (>120 bp). As expected for AID-mediated events, the gene conversion tracts have start and end positions that mostly localize to the CDRH1 and CDRH2 regions of the V genes, where a number of conserved AID hotspot motifs are located. These CDRs, along with CDRH3, constitute a large amount of the paratope in antibodies and thus are strongly mutated and selected during the affinity maturation process.

Discussion

The vertebrate adaptive immune system is unparalleled in its ability to sample the depths of protein sequence space for the production of high-affinity antibodies endowed with exquisite specificity. Not only are antibodies extremely useful in the lab as affinity reagents, but they also represent the fastest growing sector of the biologics drug market, with annual global sales for monoclonal antibodies approaching $50 billion [50]. This has resulted in an increased interest for mining the antibody repertoires within vertebrates in a systematic, high resolution manner, something afforded by increasingly economical NGS technologies that enable the collection of thousands to millions of DNA sequences in a single sequencing run. Several species' Ig repertoires have been characterized by NGS to date [9,18,51,52,53,54]. In this report, we used 5′ RACE-amplification of rabbit IgG and Igκ/Igλ transcripts, followed by NGS and bioinformatics analyses, to elucidate key features of the repertoire. We provide evidence that the existing rabbit germline VH gene database, as annotated from a number of sources [8,13,23,24,25,26,27,28,29] (see Materials and Methods), is incomplete. This was not surprising based on previous estimations of the number of Ig germline elements in the rabbit and also a very recent survey of Ig germline elements detected in the genome of a Thorbecke inbred rabbit [7].

There are typically two types of approaches for examining sequence relationships in the multiple alignments of homologous sequences: (1) tree-based methods (e.g. phylogenetics) and (2) space-based methods that, unlike phylogenetics, do not infer a hierarchical or a specific structure within the sequence alignment. For the assignment of germline sequences, space-based methods provide a statistical framework for comparing and clustering the sequences based on pairwise identities or similarities. MDS is a space-based method that allows the pairwise distances in the multiple sequence alignment to be reduced to a small number of principle components that aid in clustering the data within Euclidean space. This type of analysis applied to large Ig sequence data sets allows accurate genotyping of the germline elements within the species simply based upon the detection of highly frequent shared polymorphisms observed across individuals [55]. We show that MDS combined with k-means clustering provides an efficient approach towards discovery of new Ig germline elements in NGS data sets, even with repertoires that exhibit high loads of mutations, as is the case with the rabbit IgG repertoire where a large fraction of Ig sequences deviate significantly from the germline due to gene conversion events. MDS combined with k-means clustering could be successfully applied to a multitude of species for which the germline Ig loci are poorly annotated.

The large sample size provided by NGS also allows the diversification mechanism of Ig repertoires to be analyzed in great

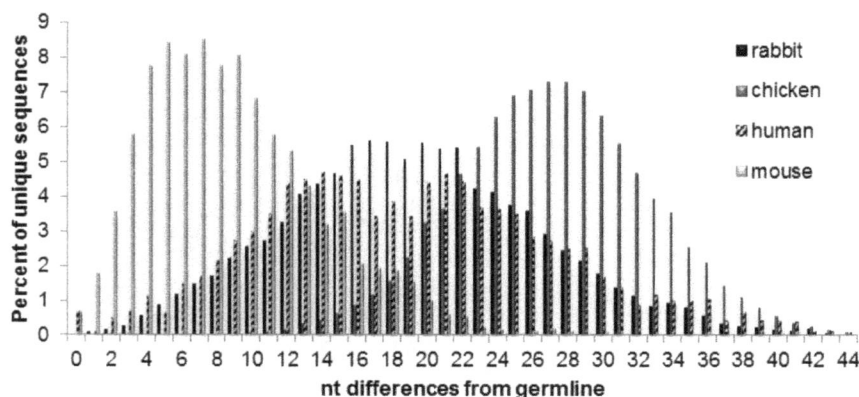

Figure 5. Diversification of the immunoglobulin repertoire – comparison of overall levels of nucleotide changes in the VH sequences across four species. The number of nucleotide differences from the nearest (assigned) VH germline derived from IgBLAST alignments as discussed in the text for rabbits.

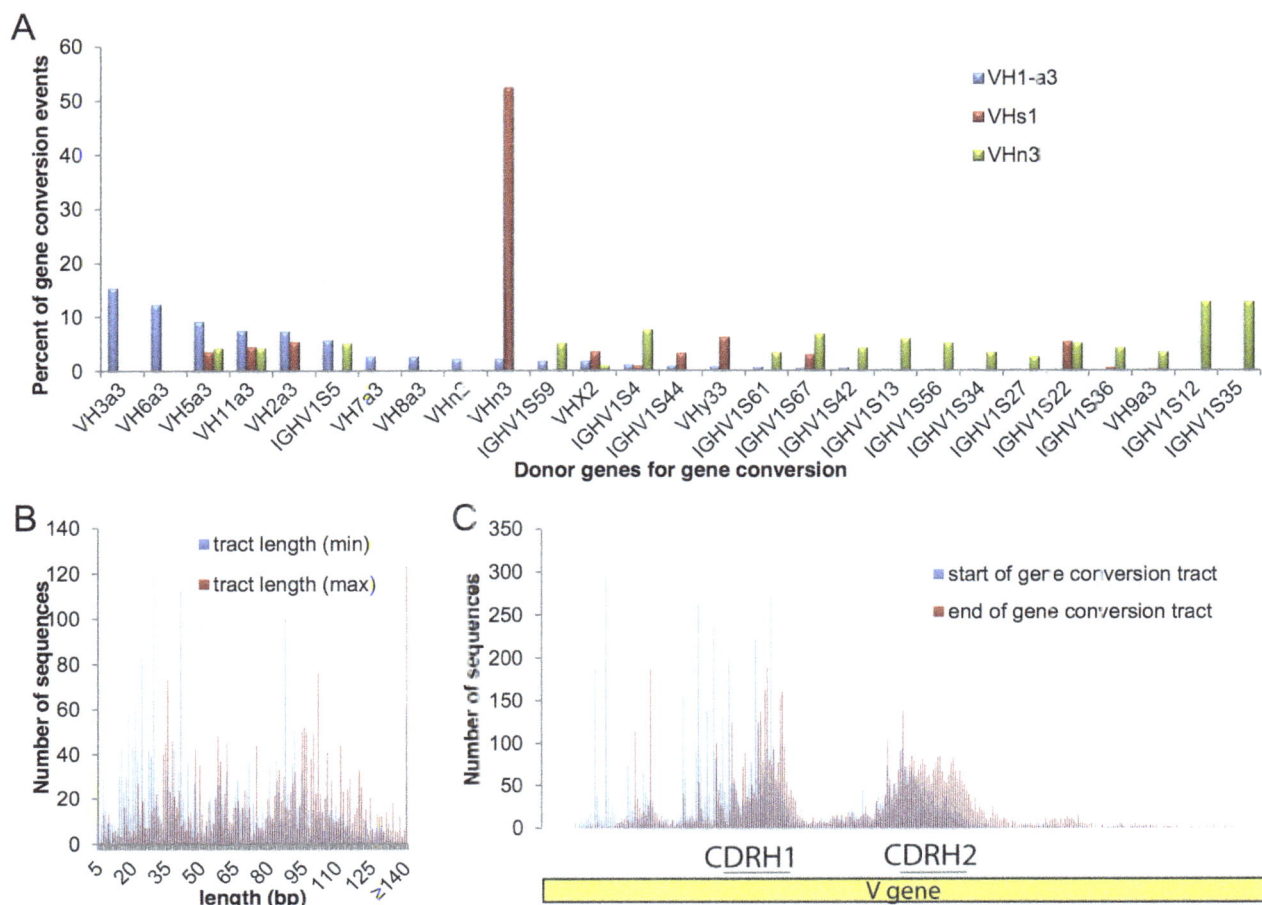

Figure 6. Gene conversion analysis. (A) Examination of gene conversion donor VH germline usage for recipient query sequences. For the recipient query sequences examined here, the germline VH usage (i.e. the original recombined V gene) was either VH-1-a3, VHs1, or VHn3. (B) Gene conversion tract lengths (i.e. lengths of recombined fragments). See Materials and Methods for description of short (min) versus long (max) tract lengths. (C) The nucleotide positions along the VH gene sequence where the gene conversion recombination events start and stop. Start and end positions are based upon the short (min) tract as described in the Materials and Methods.

detail. We show that in the rabbit, the frequency of gene conversion is significantly lower than in the chicken. Consistent with this finding, it had been previously reported that chickens depend on gene conversion as the primary mechanism of Ig diversification and that SHM play a smaller role [56]. In rabbits, the chromosomal organization of VH gene elements is quite complex, with many VH germline genes located in genomic regions far removed from the commonly utilized VH1 germline gene. This may effectually limit the relative frequency of gene conversion, as gene conversion of VH1 is limited mostly to those donor genes directly upstream. Further, several of these upstream donor genes are functional, whereas in chickens there exists a single functional germline VH and a pool of upstream pseudogenes that are used exclusively as donor genes for gene conversion. Interestingly, and consistent with earlier data [47], we report a detectable amount of gene conversion in the human IgG repertoire, but not in the mouse. The gene conversion tract lengths are significantly lower in the expressed human IgG repertoire as compared to the rabbit and chicken, but nonetheless are of high statistical confidence (p<0.05). This finding argues that gene conversion needs to be explicitly taken into account in the analysis of the antibody repertoire.

Supporting Information

Figure S1 IgBLAST database alignment performance. Comparison of IgBLAST alignment performance before and after addition of putative (A) VH and (B) Vκ germline sequences identified by MDS and k-means clustering. Before addition of the newly annotated germline sequences (IgBLAST database v.1), a large shoulder of very high 'mutation' load is evident in the IgBLAST alignments. After addition of the germline sequences identified by MDS and k-means clustering (IgBLAST database v.2), the vast majority of the sequences with high 'mutation' load now align to one of the new germline annotations and thus have a lower amount of nt differences from the nearest VH germline sequence.

Figure S2 MDS and k-means clustering of low scoring Vκ-aligned sequences in rab2 rabbit bone marrow PC IgG. The first three components of the MDS are shown here, comparing (A) PC1 v. PC2 and (B) PC1 v. PC3. Included in yellow are all the germline Vκ sequences in the original IgBLAST database (v.1). The population in blue represents light chain sequences that cluster with germline Vκ already existing in the original IgBLAST database (v.1). rab1 and rab3 MDS and k-means clustering produced similar results, but unlike with VH clusters, not all four identified clusters were observed across all three rabbits (as detailed in the main text). Here, for example, the rab2 rabbit only has three new k-means clusters (in red, green, and gray). Due to the high identity (94%) between NZWk155g and NZWk57r (both part of the red cluster here), k-means was unable to separate these into two distinct clusters for the rab2 analysis.

Table S1 Primers used to amplify IgH and Igκ/Igλ repertoires.

Table S2 NZW rabbit VH and Vκ germline sequences identified by MDS and k-means clustering.

Acknowledgments

We are extremely grateful to Dr. Scott Hunicke-Smith for assistance with NGS, Constantine Chrysostomou for assistance in data analysis, Bob Glass for assistance with rabbit immunization and bone marrow isolation, Prof. Gregory Ippolito for reading the manuscript, and Prof. Andrew Ellington and Brent L. Iverson for useful discussions and comments.

Author Contributions

Conceived and designed the experiments: JJL GG. Performed the experiments: JJL KHH STR YW. Analyzed the data: JJL GG. Contributed reagents/materials/analysis tools: KHH. Contributed to the writing of the manuscript: JJL KHH GG.

References

1. Lanning DK, Rhee KJ, Knight KL (2005) Intestinal bacteria and development of the B-lymphocyte repertoire. Trends Immunol 26: 419–425.
2. Knight KL (1992) Restricted VH gene usage and generation of antibody diversity in rabbit. Annu Rev Immunol 10: 593–616.
3. Dray S, Lennox ES, Oudin J, Dubiski S, Kelus A (1962) A notation for allotypy. Nature 195: 785–&.
4. Kim BS, Dray S (1973) Expression of A, X, and Y variable region genes of heavy-chains among IgG, IgM, and IgA molecules of normal and a locus allotype-suppressed rabbits. J Immunol 111: 750–760.
5. Dray S, Young GO, Nisonoff A (1963) distribution of allotypic specificities among rabbit gamma-globulin molecules genetically defined at 2 loci. Nature 199: 52–&.
6. Mage RG, Lanning D, Knight KL (2006) B cell and antibody repertoire development in rabbits: The requirement of gut-associated lymphoid tissues. Dev Comp Immunol 30: 137–153.
7. Gertz EM, Schaffer AA, Agarwala R, Bonnet-Garnier A, Rogel-Gaillard C, et al. (2013) Accuracy and coverage assessment of Oryctolagus cuniculus (rabbit) genes encoding immunoglobulins in the whole genome sequence assembly (OryCun2.0) and localization of the IGH locus to chromosome 20. Immunogenetics 65: 749–762.
8. Roux KH, Dhanarajan P, Gottschalk V, Mccormack WT, Renshaw RW (1991) Latent A1 VH germline genes in an alpha-2-alpha-2 rabbit evidence for gene conversion at both the germline and somatic levels. J Immunol 146: 2027–2036.
9. Larsen PA, Smith TPL (2012) Application of circular consensus sequencing and network analysis to characterize the bovine IgG repertoire. BMC Immunol 13.
10. Wang F, Ekiert DC, Ahmad I, Yu WL, Zhang Y, et al. (2013) Reshaping antibody diversity. Cell 153: 1379–1393.
11. Sehgal D, Johnson G, Wu TT, Mage RG (1999) Generation of the primary antibody repertoire in rabbits: expression of a diverse set of Igk-V genes may compensate for limited combinatorial diversity at the heavy chain locus. Immunogenetics 50: 31–42.
12. Weinstein PD, Anderson AO, Mage RG (1994) Rabbit IGH sequences in appendix germinal-centers - VH diversification by gene conversion-like and hypermutation mechanisms. Immunity 1: 647–659.
13. Becker RS, Knight KL (1990) Somatic diversification of immunoglobulin heavy-chain VDJ genes - evidence for somatic gene conversion in rabbits. Cell 63: 987–997.
14. Lanning D, Sethupathi P, Rhee KJ, Zhai SK, Knight KL (2000) Intestinal microflora and diversification of the rabbit antibody repertoire. J Immunol 165: 2012–2019.
15. Harris RS, Sale JE, Petersen-Mahrt SK, Neuberger MS (2002) AID is essential for immunoglobulin V gene conversion in a cultured B cell line. Curr Biol 12: 435–438.
16. Arakawa H, Buerstedde JM (2004) Immunoglobulin gene conversion: Insights from bursal B cells and the DT40 cell line. Dev Dynam 229: 458–464.
17. Reynaud CA, Dahan A, Anquez V, Weill JC (1989) Somatic hyperconversion diversifies the single VH-gene of the chicken with a high-incidence in the D-region. Cell 59: 171–183.
18. Reddy ST, Ge X, Miklos AE, Hughes RA, Kang SH, et al. (2010) Monoclonal antibodies isolated without screening by analyzing the variable-gene repertoire of plasma cells. Nat Biotech 28: 965–U920.
19. Wine Y, Boutz DR, Lavinder JJ, Miklos AE, Hughes RA, et al. (2013) Molecular deconvolution of the monoclonal antibodies that comprise the polyclonal serum response. Proc Natl Acad Sci USA 110: 2993–2998.
20. Finlay WJ, Bloom L, Cunningham O (2011) Optimized generation of high-affinity, high-specificity single-chain Fv antibodies from multiantigen immunized chickens. Methods Mol Biol 681: 383–401.
21. Magoc T, Salzberg SL (2011) FLASH: fast length adjustment of short reads to improve genome assemblies. Bioinformatics 27: 2957–2963.
22. Lefranc MP, Giudicelli V, Ginestoux C, Bodmer J, Muller W, et al. (1999) IMGT, the international ImMunoGeneTics database. Nucleic Acids Res 27: 209–212.
23. Ros F, Puels J, Reichenberger N, van Schooten W, Buelow R, et al. (2004) Sequence analysis of 0.5 Mb of the rabbit germline immunoglobulin heavy chain locus. Gene 330: 49–59.
24. Zhu X, Boonthum A, Zhai SK, Knight KL (1999) B lymphocyte selection and age-related changes in VH gene usage in mutant Alicia rabbits. J Immunol 163: 3313–3320.
25. Bernstein KE, Alexander CB, Mage RG (1985) Germline VH genes in an a3 rabbit not typical of any one VHa allotype. J Immunol 134: 3480–3488.
26. Fitts MG, Metzger DW (1990) Identification of rabbit genomic Ig-VH pseudogenes that could serve as donor sequences for latent allotype expression. J Immunol 145: 2713–2717.
27. Knight KL, Becker RS (1990) Molecular basis of the allelic inheritance of rabbit immunoglobulin VH allotypes: implications for the generation of antibody diversity. Cell 60: 963–970.
28. Raman C, Spieker-Polet H, Yam PC, Knight KL (1994) Preferential VH gene usage in rabbit Ig-secreting heterohybridomas. J Immunol 152: 3935–3945.
29. Friedman ML, Tunyaplin C, Zhai SK, Knight KL (1994) Neonatal VH, D, and JH gene usage in rabbit B lineage cells. J Immunol 152: 632–641.
30. Edgar RC (2004) MUSCLE: multiple sequence alignment with high accuracy and high throughput. Nucleic Acids Res 32: 1792–1797.
31. Pele J, Becu JM, Abdi H, Chabbert M (2012) Bios2mds: an R package for comparing orthologous protein families by metric multidimensional scaling. BMC Bioinformatics 13.
32. Sawyer S (1989) Statistical tests for detecting gene conversion. Mol Biol Evol 6: 526–538.
33. Ye J, Ma N, Madden TL, Ostell JM (2013) IgBLAST: an immunoglobulin variable domain sequence analysis tool. Nucleic Acids Res 41: W34–W40.
34. Torgerson WS (1952) Multidimensional scaling: I. Theory and method. Psychometrika 17: 401–419.
35. Pele J, Abdi H, Moreau M, Thybert D, Chabbert M (2011) Multidimensional scaling reveals the main evolutionary pathways of class A G-protein-coupled receptors. Plos One 6.
36. Higgins DG (1992) Sequence ordinations: a multivariate analysis approach to analysing large sequence data sets. Comp Appl Biosci: CABIOS 8: 15–22.
37. Jain AK (2010) Data clustering: 50 years beyond K-means. Pattern Recogn Lett 31: 651–666.
38. Rhee KJ, Jasper PJ, Sethupathi P, Shanmugam M, Lanning D, et al. (2005) Positive selection of the peripheral B cell repertoire in gut-associated lymphoid tissues. J Exp Med 201: 55–62.
39. Esteves PJ, Lanning D, Ferrand N, Knight KL, Zhai SK, et al. (2005) The evolution of the immunoglobulin heavy chain variable region (IgV(H)) in Leporids: an unusual case of transspecies polymorphism. Immunogenetics 57: 874–882.
40. Appella E, Chersi A, Rejnek J, Reisfeld R, Mage R (1974) Rabbit immunoglobulin lambda chains: isolation and amino acid sequence of cysteine-containing peptides. Immunochemistry 11: 395–402.
41. Dubiski S, Muller PJ (1967) A "new" allotypic specificity (A9) of rabbit immunoglobulin. Nature 214: 696–697.
42. Lavinder JJ, Wine Y, Giesecke C, Ippolito GC, Horton AP, et al. (2014) Identification and characterization of the constituent human serum antibodies elicited by vaccination. Proc Natl Acad Sci USA 111: 2259–2264.
43. Schroeder HW, Ippolito GC, Shiokawa S (1998) Regulation of the antibody repertoire through control of HCDR3 diversity. Vaccine 16: 1383–1390.
44. Wu LY, Oficjalska K, Lambert M, Fennell BJ, Darmanin-Sheehan A, et al. (2012) Fundamental characteristics of the immunoglobulin VH repertoire of chickens in comparison with those of humans, mice, and camelids. J Immunol 188: 322–333.

45. Gilles A, Meglecz E, Pech N, Ferreira S, Malausa T, et al. (2011) Accuracy and quality assessment of 454 GS-FLX Titanium pyrosequencing. BMC Genomics 12: 245.

46. Quail MA, Smith M, Coupland P, Otto TD, Harris SR, et al. (2012) A tale of three next generation sequencing platforms: comparison of Ion Torrent, Pacific Biosciences and Illumina MiSeq sequencers. BMC Genomics 13: 341.

47. Darlow JM, Stott DI (2006) Gene conversion in human rearranged immunoglobulin genes. Immunogenetics 58: 511–522

48. D'Avirro N, Truong D, Xu B, Selsing E (2005) Sequence transfers between variable regions in a mouse antibody transgene can occur by gene conversion. J Immunol 175: 8133–8137.

49. Duvvuri B, Wu GE (2012) Gene conversion-like events in the diversification of human rearranged IGHV3-23*01 gene sequences. Front Immunol 3: 158.

50. Huston JS (2012) Engineering antibodies for the 21st century. Protein Eng Des Sel 25: 483–484.

51. Weinstein JA, Jiang N, White RA, Fisher DS, Quake SR (2009) High-throughput sequencing of the zebrafish antibody repertoire. Science 324: 807–810.

52. DeKosky BJ, Ippolito GC, Deschner RP, Lavinder JJ, Wine Y, et al. (2013) High-throughput sequencing of the paired human immunoglobulin heavy and light chain repertoire. Nat Biotech 31: 166–169.

53. Boyd SD, Marshall EL, Merker JD, Maniar JM, Zhang LN, et al. (2009) Measurement and clinical monitoring of human lymphocyte clonality by massively parallel V-D-J pyrosequencing. Sci Transl Med 1.

54. Castro R, Jouneau L, Pham HP, Bouchez O, Giudicelli V, et al. (2013) Teleost fish mount complex clonal IgM and IgT responses in spleen upon systemic viral infection. Plos Pathogens 9.

55. Boyd SD, Gaeta BA, Jackson KJ, Fire AZ, Marshall EL, et al. (2010) Individual variation in the germline Ig gene repertoire inferred from variable region gene rearrangements. J Immunol 184: 6986–6992.

56. Arakawa H, Kuma K, Yasuda M, Ekino S, Shimizu A, et al. (2002) Effect of environmental antigens on the Ig diversification and the selection of productive V-J joints in the bursa. J Immunol 169: 818–828.

15

Construction of Recombinant Marek's Disease Virus (rMDV) Co-Expressing AIV-H9N2-NA and NDV-F Genes under Control of MDV's Own Bi-Directional Promoter

Zhenjie Zhang[1,2◐], **Chengtai Ma**[1,2◐], **Peng Zhao**[1,2], **Luntao Duan**[1,2], **Wenqing Chen**[1,2], **Fushou Zhang**[1,2], **Zhizhong Cui**[1,2]*

1 College of Veterinary Medicine, Shandong Agricultural University, Taian, China, 2 Animal Disease Prevention Technology and Research Center of Shandong Province, Taian, China

Abstract

To qualitatively analyze and evaluate a bi-directional promoter transcriptional function in both transient and transgenic systems, several different plasmids were constructed and recombinant MDV type 1 strain GX0101 was developed to co-express a Neuraminidase (NA) gene from Avian Influenza Virus H9N2 strain and a Fusion (F) gene from the Newcastle disease virus (NDV). The two foreign genes, NDV-F gene and AIV-NA gene, were inserted in the plasmid driven in each direction by the bi-directional promoter. To test whether the expression of pp38/pp24 heterodimers are the required activators for the expression of the foreign genes, the recombinant plasmid pPpp38-NA/1.8kb-F containing expression cassette for the two foreign genes was co-transfected with a pp38/pp24 expression plasmid, pBud-pp38-pp24, in chicken embryo fibroblast (CEF) cells. Alternatively, plasmid pPpp38-NA/1.8kb-F was transfected in GX0101-infected CEFs where the viral endogenous pp38/pp24 were expressed via virus infection. The expression of both foreign genes was activated by pp38/pp24 dimers either via virus infection, or co-expression. The CEFs transfected with pPpp38-NA/1.8kb-F alone had no expression. We chose to insert the expression cassette of Ppp38-NA/1.8kb-F in the non-essential region of GX0101ΔMeq US2 gene, and formed a new rMDV named MZC13NA/F through homologous recombination. Indirect fluorescence antibody (IFA) test, ELISA and Western blot analyses indicated that F and NA genes were expressed simultaneously under control of the bi-directional promoter, but in opposite directions. The data also indicated the activity of the promoter in the 1.8-kb mRNA transcript direction was higher than that in the direction for the pp38 gene. The expression of pp38/pp24 dimers either via co-tranfection of the pBud-pp38-pp24 plasmid, or by GX0101 virus infection were critical to activate the bi-directional promoter for expression of two foreign genes in both directions. Therefore, the confirmed function of the bi-directional promoter provides better feasibilities to insert multiple foreign genes in MDV genome based vectors.

Editor: Huaijun Zhou, University of California, Davis, United States of America

Funding: The study was supported by a grant from National Natural Science Foundation Project (31072149). The funders had no role in study design, data collection and analysis, decision to publish, or preparation of the manuscript.

Competing Interests: The authors have declared that no competing interests exist.

* E-mail: zzcui@sdau.edu.cn

◐ These authors contributed equally to this work.

Introduction

Marek's disease viruses (MDV) belong to a subgroup of the alphaherpesviridae [1]. Serotype 1 MDV is the prototype virus for this group of avian viruses, it contains a double strand DNA genome of about 178kb. The genome has a unique long (U_L) and unique short (U_S) sequences flanked with terminal repeat long (T_{RL}) and internal repeat long (I_{RL}) or internal repeat short (I_{RS}) and terminal repeat short (T_{RS}) [2]. So far, more than 100 genes or ORFs have been identified in the MDV genome. The 1.8-kb mRNA transcript family [3,4] and the 38kd phosphorylated protein gene (pp38) [5,6] are two of many MDV-specific genes. They were located on I_{RL} region and separated by a short sequence of only about 305 bp but with several enhancer motifs such as TATA-box, CAAT-box, Oct-1, and Sp1, and as a bi-directional promoter to initiate transcription of two genes in opposite directions [7,8,9]. By use of *GFP* gene and chloramphenicol acetyltransferase (*CAT*) as reporter genes, it was demonstrated that the activity of the bi-directional promoter could be strongly increased by the expression of MDV pp38/pp24 dimers as a trans-acting transcriptional factor [10]. However, it is still unclear whether this bi-directional promoter is able to simultaneously drive two genes expression in both directions.

In the past decades, development of recombinant vaccines has been one of the most active aspects of molecular virology. Several different attenuated viruses were used as vectors to express foreign antigens from other viruses. The viruses have large genomes were more favored as vectors, such as fowl pox virus (FPV) and MDV. Successful examples have been reported in expressing viral genes such as F gene of Newcastle disease virus (NDV) [11,12] or hemagglutinin gene of avian influenza virus (AIV) [13,14,15]. However, with the widely existing maternal antibodies against FPV, these recombinant FPVs were strongly denied for commercial use. In contrast to FPV, MDVs are less influenced by maternal

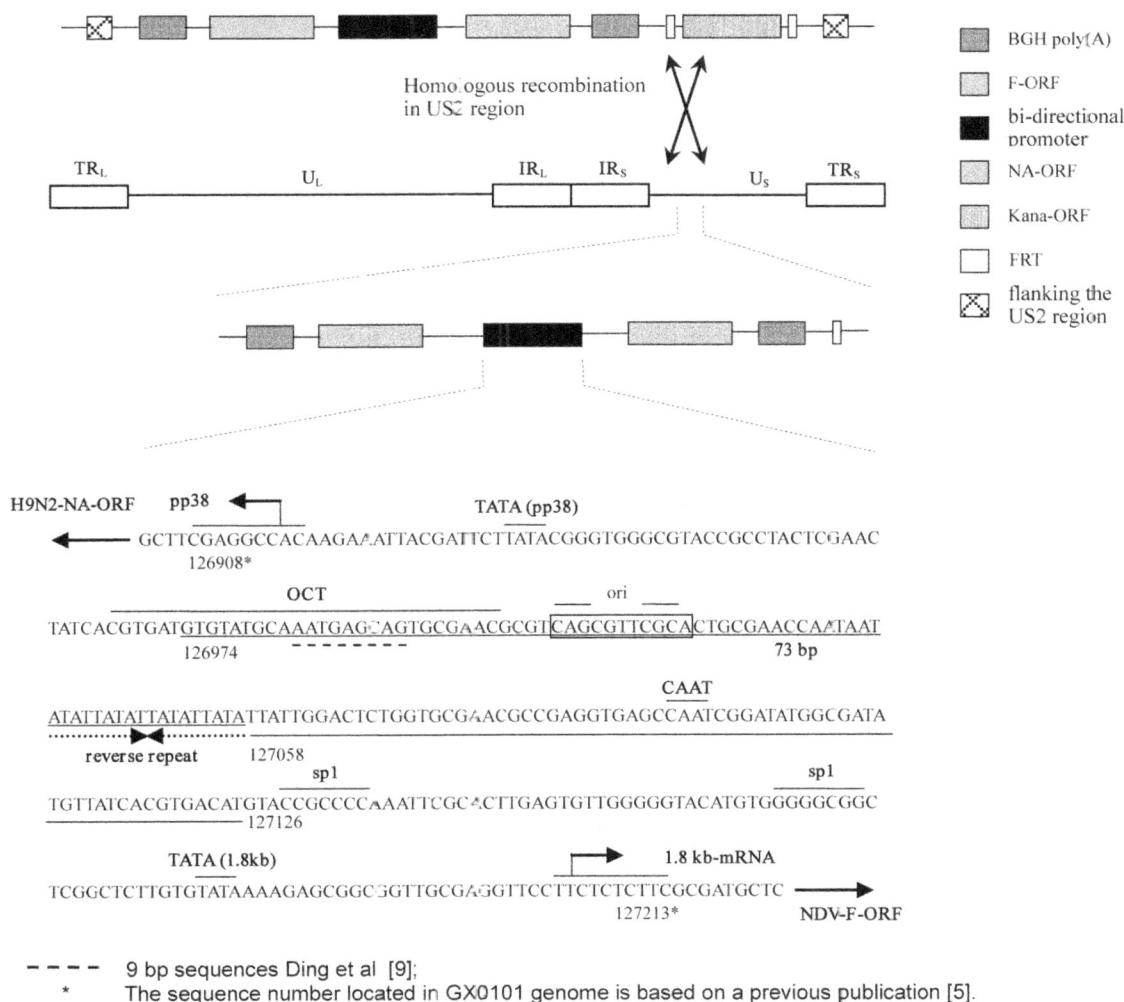

Figure 1. The structure of insertion sites of the H9N2-AIV-NA and NDV-F gene under the control of the bi-directional promoters on MZC13NA/F genome (GX0101/Δmeq/Kan⁻/gpt⁻/NA/F).

antibodies, and a recombinant Herpesvirus of turkeys (HVT) expressing VP2 gene of infectious bursal disease virus (IBDV) has been successfully used as a commercial vaccine to control IBDV in chickens. In recent years, recombinant MDV vaccines expressing other foreign genes were studied and reported, and some rMDV vaccines have demonstrated good protective immunity in SPF or commercial chickens [16,17,18]. In the most studies, foreign genes were expressed under control of some common promoters such as CMV, SV40, or β-actin, which are also foreign sequences to MDV. An exception was using MDV endogenous promoter for gB gene to drive the express of NDV-F gene, and this rMDV vaccine was demonstrated as an effective and stable polyvalent vaccine against both vvMDV and NDV even in the presence of maternal antibodies [16]. In this study, we designed experiments to test if the MDV endogenous bi-directional promoter would drive the expression of two foreign genes in recombinant MDV.

Bacterial artificial clone (BAC) technology is a very powerful technique in molecular virology. With the BAC vector system, many strains MDV BAC infectious clones were constructed and used to study gene function or construction of recombinant vaccines [19,20,21]. Recently, we successfully constructed and rescued recombinant MDVs with a very virulent MDV field

strain, GX0101, as well as its meq-deletion mutant, GX0101ΔMeq [22,23]. Our studies indicated that GX0101Δmeq not only lost its pathogenicity but also could provide better protective immunity than the classical vaccine strain CVI988/Rispens against the challenge of very virulent MDV [24]. Similar results were also demonstrated by others in the field with a meq-deleted Md5 strain of MDV constructed based on cosmid system [25,26,27,28].

In this study, we attempt to insert and express the NDV-F gene and AIV-H9N2-NA gene in the GX0101Δmeq, under the control of MDV endogenous bi-directional promoter (Fig. 1). It will be more interesting to evaluate the protective immunity against MDV, NDV and AIV in the future studies.

Materials and Methods

Viruses and cell culture

MDV GX0101 is a field strain isolated from a layer farm in Guangxi Province of China [29]. Infectious meq-deleted BAC-GX0101 virus(GX0101/Δmeq/Kan⁻/gpt⁺) was previously constructed and rescued in our studies [22,23]. NDV TZ060107 strain (GenBank Accession No. FJ011448) was originally isolated

Table 1. List of primers used for construction of different recombinant plasmids.

No.	Primers	Sequence of primers(5'-3')	Related genes and sizes of the expected PCR products
1	F-P1*	CTGGCTAGCGTTAGCATGGACCGCGCGGTTAAC	NDV-F gene (1692 bp)
	F-P2*	CTCGGTACCAAGCTATTAAACTCTATCATCCTTG	
2	NA-P1*	CCCTCTAGATCAGCATGGACCCAAATCAGAAGAT	AIV-NA gene (1431 bp)
	NA-P2*	GCAGAATTCCACCTATTATATAGGCATGAAGTTGA	
3	pp38-P1*	CGAGCGGCCGCCACCTGGCTAGCGTTGAGCATCGCG	The bi-directional promoter
		AAGAGAGA	(392 bp)
	pp38-P2*	ATGCCTGCAGGTCGGACTCTAGAGGATCCGTCGACAA	
		GGCTT CGAGGCCACAAGAAATT	
4	F-pol(A)-p1*	CTGGCTAGCGTTAGCATGGACCGCGCGGTTAAC	NDV-F gene and BGH-pol(A)
	F-pol(A)-p2*	CGAGCGGCCGCCACTGGGGATACCCCCTAGAG	(1932 bp)
5	NA-pol(A)-p1*	TCCGTCGACAAGAGCATGGACCCAAATCAGAAGAT	AIV-NA gene and BGH-pol(A)
	NA-pol(A)-p2*	GACTCTAGAGGATGGGGATACCCCCTAGAG	(1671 bp)
6	KanR-F*	GACTCTAGAGGATCGTGTAGGCTGGAGCTGCTTC	KanR cassette flanked by FRT
	KanR-R*	ATGCCTGCAGGTCGCATTCCGGGGATCCGTCGAC	sites (1332 bp)
7	US2-F/NA-Ka-F**	**ATGGGTGTGTCCATGATAACTATAGTCACACTTCTAG**	NA-F-KanR cassette gene with
		ATGAATGCGATCGCACTGGGGATACCCCCTAGAG	MDV sequence flanking the US2
	US2-F/NA-Ka-R**	**CTACTCATTTGAGGTGGTTCGATTTCCGGAGGTTTTA**	arm (5435 bp)
		GAGGATTGGGTGGCATTCCGGGGATCCGTCGAC	
8	US2-F	GGTTTTAGAGGATTGGGTGG	The junction between MDV-US2
	US2-R	CTTCTAGATGAATGCGATCG	and Kan-cassette (600bp)
9	F-pro-F	TGCTCACTCCTCTTGGCGAC	Labeling NDV F probes with
	F-pro-R	GCTGCATCTTCCCAACTGCC	digoxigenin (300 bp)
10	NA-pro-F	AGTTGGGTGTCCCGTTTCAT	Labeling H9N2 NA probes with
	NA-pro-R	CACTT CCTGACAATGGGCTA	digoxigenin (330 bp)
11	pp38-pro-F	GACGCGTTCGCACTGCTCATTTG	Labeling H9N2 F probes with
	pp38-pro-R	CGTTGCCGTTCGATCCAGGTCTC	digoxigenin (190 bp)

* For the convenience of plasmid construction, Restriction Endonuclease sites introduced with the primers were underlined. ** For primers US2-F/NA-Ka-F and US2-F/NA-Ka-R, underlined sequences in the upper case indicate the sequences from template constructed used to amplify co-expression cassette, and sequences in bold indicate MDV sequence flanking the US2 region.

from a poultry farm [30]; AIV-H9N2 LG1 strain was kindly provided by Qilu Animal Health Products co (Shandong, China). All cell cultures were maintained in Dulbecco's minimum essential medium (DMEM) supplemented with 5% fetal bovine serum (FBS), 100 U/ml penicillin and 100 ug/ml streptomycin. Cells were incubated at 37°C in an atmosphere containing 5% CO_2. All culture reagents were purchased from Gibco (Grand Island, USA).

Antibodies and reagents

Chicken polyclonal anti-F serum raised against NDV TZ060107, mouse anti-NA polyclonal serum and mouse anti-pp38 monoclonal antibody H19 were all obtained from Avian Disease and Oncology diagnostic Laboratory (Taian, China); The pp38 and pp24 genes co-expression plasmid pBud-pp38-pp24 was constructed previously [31]; Platinum high fidelity Taq DNA polymerase, were purchased from Invitrogen (CA, USA); Gel Extraction Kit and Plasmid Maxi Kit were purchased from QIAGEN (Hilden, Germany).

RT-PCR

RNA were extracted using Trizol reagent (TransGen Biotech, Beijing, China) as described by the manufacturer. The RT-PCR

Access kit (Promega, Wisconsin, USA) was used to detect the transcription of F and NA genes. The method was essentially the same as that described previously using primer pairs #1 or #2 (Table 1) [3]. To include the pol(A) at 3' end of expressed genes, the PCR products of NA gene (1431bp), and F gene (1692bp), were cloned directly into the pcDNA3.1(-) expressing vector (Invitrogen, California, USA) to form a new vector named as pcDNA-NA or pcDNA-F, and sequences were confirmed by automated sequencing (BGI, Beijing, China).

Plasmids construction

MDV viral DNA was prepared from the GX0101-infected CEFs as previously described [32]. To obtain the bi-directional promoter element, PCR were carried out using GX0101 DNA as the template with the primer pair #3 (Table 1). The *NotI* and *NheI* sites were introduced to the 5' end of the pp38-P1 primer and the *XbaI* and *SalI* sites were introduced to the 5' end of the pp38-P2 primer (underlined positions of primers, Table 1) for the convenience of plasmid construction. The 392 bp PCR product was purified with the Gel Extraction kit (Qiagen) and sequenced (BGI, Beijing, China). The recovered DNA fragment was then subcloned into pMD18-T vector to form a new transfer vector named as pBiP.

The NA or F gene, includes the pol(A) at 3′ end, were amplified with gene-specific primer pairs #4 or #5 (Table 1) from plasmid pcDNA-NA or pcDNA-F. Different constructions were designed to express NA or F separately or in combo. For individual gene expression, the NA-pol(A) fragment was inserted into the plasmid pBiP at the downstream of pp38 bi-directional promoter between the SalI and XbaI sites, result in plasmid, pPpp38-NA. Similarly, the F-pol(A) was amplified and inserted into the plasmid pBiP at the upstream of pp38 bi-directional promoter, between the NheI and NotI sites and the outcome recombinant plasmid was named pP1.8kb-F. To express both foreign genes in combo, the NA-pol(A) fragment was cloned between the SalI and XbaI sites of pP1.8kb-F resulted in plasmid pPpp38-NA/1.8kb-F. The Kan^R cassette flanked by FRT sites was amplified using primer pair #6 (Table 1) from pKD13 [33], then the Kan fragment was cloned between XbaI and Sse8387I sites of pPpp38-NA/1.8kb-F creating plasmid pPpp38- NA/1.8kb-F-Kan.

Plasmid transfection

CEF cells were trypsinized, propagated from primary CEFs. About 6×10^5 cells were plated in each well of 24-well plates with DMEM medium supplemented with 5% FCS (fetal calf serum) and free of antibiotics. Cells were incubated at 37°C for 18–24 hrs till 90–95% confluent. Plasmid transfection was performed by using LipofectAMI-NETM reagent (Gibco, BRL) according to the manufacturer's instructions. Briefly, 0.8 μg plasmid DNA and 2 μL LipofectAMI-NETM were added into two polypropylene tubes separately with 50 μL of OPTI-MEM I medium free of serum and antibiotic. Two solutions were then mixed and incubated for 20 min at room temperature and then added into another 400 μL plain DMEM. Total of 0.5 ml of transfection solution was carefully dropped onto the cell monolayers in each well. 4 hrs after incubating at 37°C CO_2 incubator, 0.5 ml of complete medium with 5% FCS was added to each transfected cell monolayers. All plates were maintained at 37°C in a CO_2 incubator.

When pBud-pp38-pp24 was co-transfected with pPpp38-NA, pP1.8kb-F or pPpp38-NA/1.8kb-F,0.8 μg of each plasmid DNA was used and accordingly 4 μL of LipofectAMI-NETM reagent was used to prepare the transfection mixture for each well. The transfection was performed following the steps described above.

To express pPpp38-NA, pP1.8kb-F or pPpp38-NA/1.8kb-F in GX0101-infected CEFs, primary CEFs were seeded in a 60 cm^2 flask and inoculated with GX0101 stocks of about 1×10^5 plaque form unit (PFU). Around 3–4 days post infection, the cell pathogenic effects (CPE) were visualized in about a quarter of cells in the monolayers. The GX0101-infected CEF monolayers

were trypsinized and resuspended in DMEM medium supplemented with 5% FCS. Un-infected primary CEF cells were also trypsinized and resuspended as well. Both types of cells were then counted and then mixed (infected cells:uninfected cells = 1:2) and seeded into 24-well plate (6×10^5 cells per well). Transfections were then carried out when the monolayers formed about 18 hrs later.

Indirect immunofluorescence assay (IFA)

IFA assays were conducted To check the expression of genes introduced to the cells following standard procedures, Briefly, two days after transfection, cells were fixed with acetone and ethanol solution (3:2) for 5 min at room temperature, followed by blocking with 10% normal goat serum in PBS for 30 min at room temperature. The cells were then incubated with the chicken polyclonal anti-F serum or mouse anti-NA polyclonal serum at a dilution of 1:100 in blocking buffer for 1 hr at 37°C. Lastly, the cells were incubated with corresponding FITC-conjugated secondary antibodies, anti-chicken IgG (Fc) (1:500 dilution with the diluent; Bethyl Laboratories, Inc., Montgomery, TX) to detect signal of F gene expression and anti-mouse IgG to detect signal of NA gene expression. Cells were washed with PBS (0.15 M NaCl, 15 mM Na_3PO_4, pH 7.4) twice between each step. At the end the NA and F co-expressed cells were evaluated by fluorescent microscopy.

The CEF cells in each group were trypsinized and washed 3 times with PBS, and then fixed with acetone and ethanol solution for 5 min at room temperature. The cells were suspended in 1% FBS (prepared with calcium/magnesium free PBS) to the density of about 1×10^6 cells/ml. IFA assay was performed following the steps described above. The cell pellet was resuspended in PBS with 1% FBS for analysis on flow cytometer after centrifugation at $2000\times g$.

Construction of MZC13NA/F

GX0101-BAC DNA containing the whole genome of GX0101 were transformed into Escherichia coli (E. coli) EL250 cells. A single clone was picked and grown in Luria-Bertani (LB) medium containing chloramphenicol (25 μg/ml) at 37°C overnight. The overnight culture was then inoculated into 10 ml of LB medium (with chloramphenicol) and grown at 32°C until an optical density at 600 nm of 0.5 was reached. The cultures were then induced to express the recE, recT, and λ gam proteins at 42°C, followed by chilling on ice for 15 min. The cells were then collected to be prepared as electrocompetent cells by a standard protocol [34,35,36]. Co-expression cassette with Kan^R cassette flanked by FRT sites at 3′ end of the cassette (Fig. 1) was amplified from template DNA, pPpp38-NA/1.8kb-F-Kan, using primer pair #6 (Table 1) and the PCR products were purified. About 300 ng of

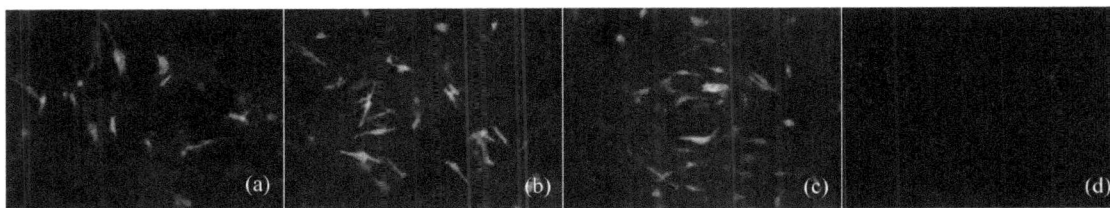

Figure 2. Demonstration of H9N2-NA or NDV-F expressing cells in IFA with monospecific sera in CEF transfected with different plasmids or their combinations (×200). (a) IFA with mouse anti-NA polyclonal serum to NA in CEF co-transfected with pPpp38-NA and pBud-pp38-pp24; (b) IFA with chicken polyclonal anti-F serum to F in CEF co-transfected with pP1.8kb-F and pBud-pp38-pp24; (c) IFA with mouse anti-NA polyclonal serum and chicken polyclonal anti-F serum to NA and F in CEF co-transfected with pPpp38-NA/1.8kb-F and pBud-pp38-pp24; (d) Detection of NA and F in CEF transfected with pPpp38-NA/1.8kb-F only.

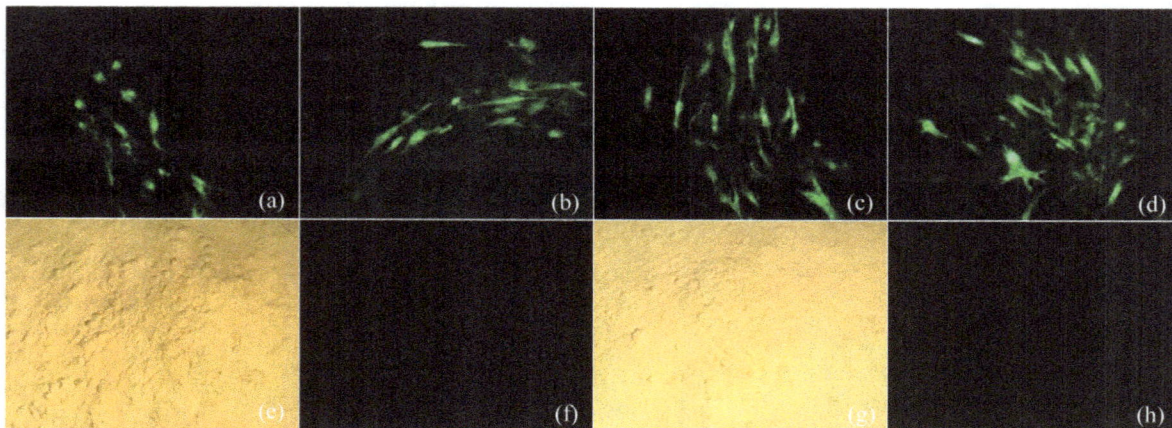

Figure 3. Demonstration of H9N2-NA or NDV-F expressing cells in IFA with specific antibodies in GX0101-CEF transfected with different plasmids (×200). (a) IFA with mouse anti-NA polyclonal serum to NA in GX0101-CEF transfected with pPpp38-NA; (b) IFA with chicken polyclonal anti-F serum to F in GX0101-CEF transfected with pP1.8kb-F; (c) IFA with mouse anti-NA polyclonal serum to NA in GX0101-CEF transfected with pPpp38-NA/1.8kb-F; (d) IFA with chicken polyclonal anti-F serum to F in GX0101-CEF transfected with pPpp38-NA/1.8kb-F; (e,f) IFA with mouse anti-NA polyclonal antibody against NA in GX0101-CEF; (g,h) IFA with chicken polyclonal anti-F serum against F protein in GX0101-CEF.

the PCR products were electroporated into 50 μL of electro-competent EL250 cells harboring the GX0101-BAC using standard electroporation parameters (2.0 kV, 100 Ω and 25 μF). After electroporation, the cells were grown in 1 ml of SOC medium (2% Trypton, Oxoid; 0.5% yeast extract, Oxoid; 0.05% NaCl; 2.5 mM KCl; 10 mM MgCl₂; 20 mM glucose) for 2 hrs and spread onto LB agar plates containing chloramphenicol (Cm, 25 μg/ml) and kanamycin (Kan, 50 μg/ml).Colonies were picked and grown in liquid LB medium with proper antibiotics. Excision of the Kan^R cassette was carried out by addition 0.1% arabinose (Sigma Aldrich, St. Louis, MO) into the medium to induce expression of FLPe recombinase. Then the cultures were plated on LB plates that contained Cm. After overnight incubation, colonies were picked and transferred to plates that contained either Cm or Kan. Recombinant clones, which were selected for growth on the plate with Cm, were confirmed by PCR. The EL250 cells harboring MZC13NA/F were grown up and the BAC DNA was prepared using commercially available kits (Qiagen) according to the standard protocols [37].

The MZC13NA/F virus infected cell culture characterization

The growth kinetics and the size of vial plaques of MZC13NA/F were compared with those of wild type virus, GX0101. One-step growth kinetics of MZC13NA/F were performed as described previously [38,39]. Briefly, for each virus MZC13NA/F or GX0101, 100 plaque-forming units (PFU) of virus were plated to infect fresh CEFs separately. At 0, 24, 48, 72, 96, 120, 144 and 170 hrs postinfection, virus-infected CEFs were harvested and a serial of 2-fold dilutions were prepared and distributed in triplicate onto the 96-well plates of CEFs. The titers of the virus at each time point were calculated based on the number of PFUs formed at each dilution and the growth curves of MZC13NA/F and GX0101 were determined.

Southern blot hybridization

The recombinant virus MZC13NA/F was generated based on homologous recombination of plasmid DNA and wild type virus GX0101at the insertion site of US2 region, therefore the genetic purity of the MZC13NA/F clone needs to be verified. To do this, a PCR amplification targeted the US2 region with primer pair, #8 (Table 1), was performed using DNA prepared from MDV infected CEFs. The amplified PCR products were hybridized with

Table 2. Comparison of the expression of NDV-F or AIV-NA in the presence or absence of pp38/pp24 dimers via IFA.

Cells	Transfected with plasmids[a]					
	pPpp38-NA	pP1.8kb-F	pPpp38-NA + pBud-pp38-pp24	pP1.8kb-F + pBud-pp38-pp24	pPpp38-NA/1.8kb-F	pPpp38-NA/1.8kb-F+ pBud-pp38-pp24
CEF	−	−	38%	57%	NA(−); F(−)	NA(29%); F(32%)
GX0101-infected CEF	65%	80%	ND[b]	ND	NA(51%); F(100%)[c]	ND

[a]The intensity of protein expression is based on the detection rate of positive cells by flow cytometry. When the plasmids constructed by the bi-directional promoter transfect CEF, the molar ratio of different plasmids is basically the same, which could enter the cells to minimize the influence of transcriptional activity due to the different molar ratios, while there is no correlation between the molar ratio and transcription activity of bi-directional promoter transfecting CEF in two directions.
[b]ND = not determined.
[c]It is taken the ratio of positive cells appeared in the MDV-infected CEF which was transfected with the pPpp38-NA/1.8kb-F plasmid as 100%, and the data in the table was the relative expression intensity 48 hrs post different plasmids transfection.

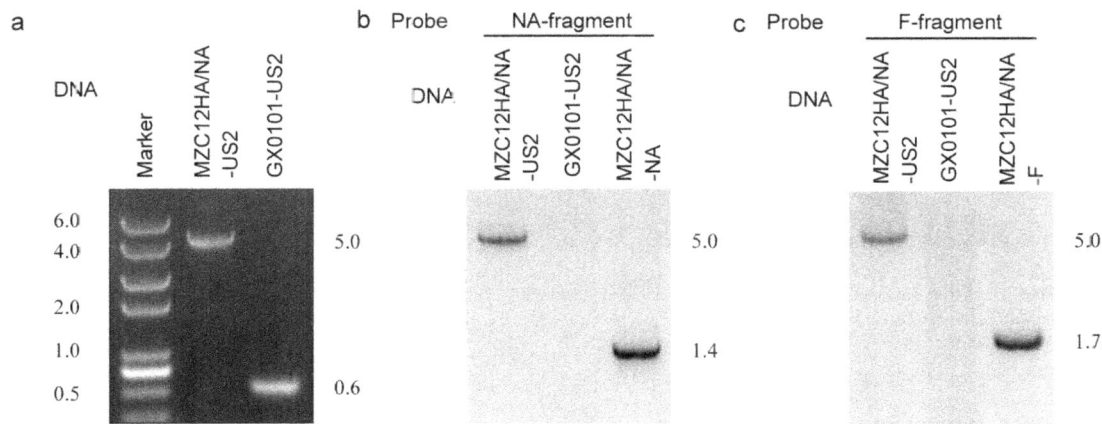

Figure 4. Southern blot of PCR products amplified from GX0101 and its recombinant genomes to demonstrate foreign genes. (a) Amplification of MZC13NA/F and GX0101 US2 gene by PCR. The US2 gene was amplified from DNAs prepared from the MZC13NA/F-infected CEF cells, or GX0101-infected CEF cells. The reaction products were submitted to agarose gel electrophoresis and stained with ethidium bromide; (b) The same DNAs and the NA ORF amplified from DNAs prepared from MZC13NA/F-infected CEF cells were processed for Southern blot analysis with the NA probe; (c) The same DNAs and the F ORF amplified from DNAs prepared from MZC13NA/F-infected CEF cells were processed for Southern blot analysis with the F probe.

NA or F probes. The NA or F probe DNA were amplified using primer pairs #9 or #10 (Table 1), respectively. and were labeled with digoxigenin (DIG) according to the manufacturer's manual (Boehringer Mannheim, Mannheim, Germany). The DIG-labeled hybridized DNA was detected by enzyme-linked immunoassay using anti-DIG antibodies conjugated with alkaline phosphatase and subsequent enzyme-catalyzed color reaction with 5-bromo-4-chloro-3-indolyl phosphate and nitro blue tetrazolium salt.

NA, F and pp38 mRNA analysis

In order to quantitate the transcription level of NA, F and pp38 mRNA driven by different promoters in MZC13NA/F, a highly sensitive real-time relative quantification of reverse transcription and polymerase chain reaction (RT-qPCR) was used for quantifying the NA, F and pp38 mRNA in the infected CEF cells. The NA, F or pp38 DNA fragment was each amplified and

constructed in the pMD18-T vector in order to generate the plasmid DNA standards in the quantitative assay. DNA concentration was determined by spectrophotometry at 260 nm. Serial dilutions ranging from 10^9 copies/μL to 10^1 copies/μL were used as standard controls. To generate standard curves, the crossing point (Cp) values in every dilution were measured in triplicates and then plotted against the logarithm of the corresponding template copy numbers. Each standard curve was generated by linear regression of the plotted points.

The primer pairs (#9, 10 and 11) and probes for the detection of NA, F and pp38 genes were designed at the highly conserved region and each amplify a fragment around 300 bp in length. The secondary CEF cells were cultured in one 6-well tissue culture plates (1×10^6 cells each well) in order to have six repeats for each testing sample. When the CEF cells form monolayer, one 6-well tissue culture plate was inoculated with 100 PFU of MZC13NA/F

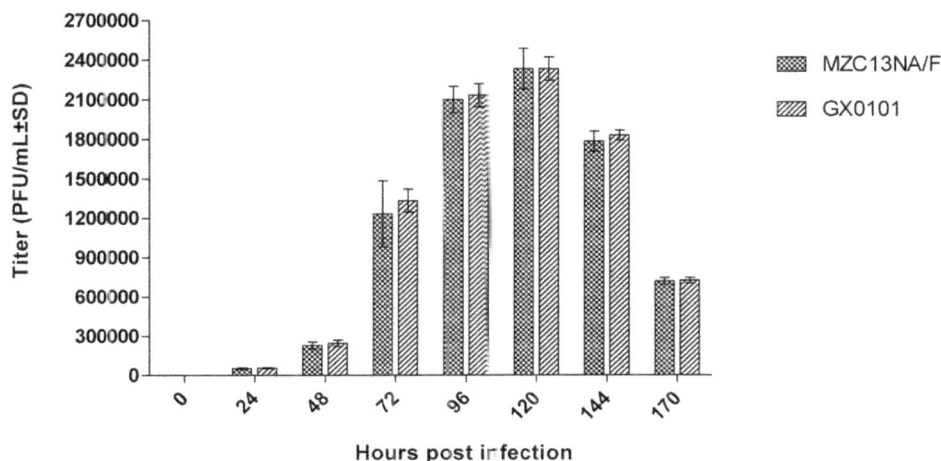

Figure 5. Growth curves (one-step growth kinetics) of GX0101 and its recombinant MZC13NA/F. After inoculation of CEFs with 100 PFU of each virus, virus titers were determined at different times and steadily increased from 48 to 170 hrs post-infection until maximal titers were reached. The data of the growth curve of MZC13NA/F were cited from reference [40]. The results represent the mean±standard deviations of three independent replicates.

Figure 6. Plaque sizes of Meq-deleted MDV GX0101 and its recombinant MZC13NA/F in infected CEF cells at second passage (×200). The mouse anti-pp38 monoclonal antibody H19 bound to the MZC13NA/F (a) or its parent GX0101 (b) infected CEF cells by IFA, or to the CEF cells (c). MZC13NA/F plaque size was similar to GX0101 at some level, 96 hours post-infection showing bulging and aggregation of rounded cells.

and cultured for 4 days. RNA was extracted from the infected cells in every well according to the product manual of the OMEGA Viral RNA Kit (Omega Bio-Tek, Inc., Georgia, USA). Then DNA fragments were synthetized by reverse transcription with TaKaRa RNA PCR Kit (AMV) Ver.3.0 (TaKaRa, Dalian, China). 20 µl the PCR mixture was prepared according to the product manual of TaKaRa SYBR Premix ExTaqTM (TaKaRa, Dalian, China) and the data was analyzed using Applied Biosystems 7500 Fast Real-Time PCR System (ABI, Foster City, USA).

Indirect immunofluorescence assay (IFA) on virus infected cells

About 1×10^5 PFU of MZC13NA/F virus were inoculated in a 24-well plate of secondary CEFs and incubated for 3–4 days until CPE can be visualized. The cells were fixed with 70% acetone at room temperature. IFA assay was performed following the steps described above to confirm NA and F protein co-expression in cultured CEF cells infected with the MZC13NA/F.

ELISA

To quantitate the expression of NA, F and MDV-pp38 protein, CEFs transfected with each transfer plasmid were harvested, washed with PBS, and added to the wells of three 96-well microtiter plates at identical cell concentrations, respectively. These plates were dried overnight at 37°C. Then, mouse anti-NA polyclonal serum (diluted 1:200), chicken antiserum raised against

NDV (diluted 1:200) and mouse anti-pp38 monoclonal antibody H19 (diluted 1:1,000) in PBS containing 5% fetal bovine serum was added to each well respectively, and the plates were incubated overnight at 4°C. After extensive washing of the plates with PBS containing 0.05% Tween 20, anti-mouse or chicken IgG labeled with HRP (Bio-Rad Laboratories) at a 1:300 dilution in the same buffer was added, and incubation continued for 1 hr at 37°C. After another wash as before, the wells were developed by adding 0.1 ml of the ABTS (2,2′-azino-bis-[3-ethylbenzthiazoline-6-sulfonic acid]; Sigma Chemical Co., St. Louis, Mo.) solution (0.5 mg/ml) and incubating for 30 min at room temperature. The absorbance at 405 and 490 nm was read with a spectrophotometer.

Western blot

Western blot analysis was carried out following the established procedures. Briefly, after 3–4 days post infection, MZC13NA/F-infected CEFs were washed twice with PBS and directed lysed with 700 ul of SDS-PAGE loading buffer (8 M urea, 10 mM Tris-HCl (pH 6.8), 2% SDS, 2% 2-mercaptoethanol, and 0.1% bromophenol blue) via boiling and vertexing. A 20 ul of the cell lysates was subjected to 10% SDS-polyacrylamide gels, followed by electrophoretically transferring to a nitrocellulose membrane of 0.45 µm pore size (Millipore, USA). The membranes were blocked with 1% bovine serum albumin (BSA) in PBS followed by washing with PBS containing 0.1% Tween-20 (PBS-T). The membranes were cut into strips to incubate with different antibodies individually at 37°C for 2 hrs. The following primary antibodies were properly diluted and applied to each strip with labeling: mouse anti-NA polyclonal antibody at a dilution of 1:200, chicken antiserum against NDV (TZ060107) at a dilution of 1:200 and mouse anti-pp38 monoclonal antibodies H19 at a dilution of 1:1000. All strips of membrane were well washed and then incubated with horseradish peroxidase (HRP)-conjugated goat anti-mouse or anti-chicken IgG (Southern Biotech, USA). After washing 5 times the signals protein image were visualized by chemiluminescence with LAS-1000 (DAB) according to the manufacturer's manual.

Results

The expression of AIV-NA or NDV-F is dependent on the presence of pp38/pp24 dimers

As it is demonstrated in Fig 2, in the presence of pp38/pp24 dimer expression in the CEF cells, the expression of AIV-NA (Fig. 2 a and c) and NDV-F(Fig. 2 b and c) genes can be detected by specific anti-sera via IFA. No NA or F protein expression was detected in CEFs transfected with pPpp38-NA/1.8kb-F only (Fig. 2 d). Expression of AIV-NA and NDV-F were also detected when

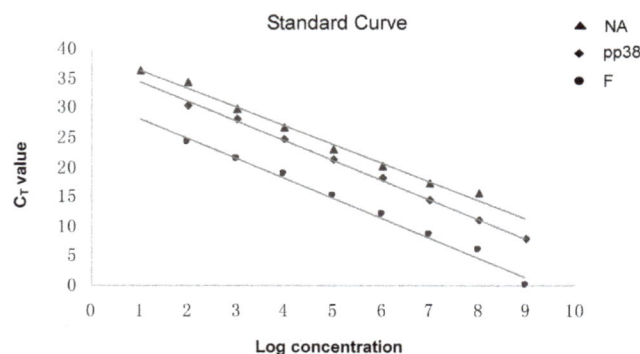

Figure 7. Standard curve of NA, F and pp38 assay. Copy number for the plasmid pMD18-NA, -F or -pp38 were respectively determined spectrophotometrically and diluted serially from 10^9 copies/µL to 10^1 copies/µL for use as standard controls. Standard curve ($y = 39.369 - 3.122x$, $y = 31.533 - 3.332x$ and $y = 37.614 - 3.299x$) for NA, F and pp38 genes quantification ($R^2 = 0.9918$, 0.9906 and 0.9978; Efficiency $= 100 \pm 1.5$%) was analyzed with the GraphPad Software 5.0®. The concentration refers to the template copy number per reaction.

Table 3. The absolute quantification (AQ) values of cDNAs of different genes (copies/ul).

	H9N2-NA	NDV-F	MDV-pp38	Ratio of NA/F/pp38 mRNA
RT-PCR C_T value Average ± standard deviation	19.0±0.0 (6*)	21.2±0.1 (6)	20.9±0.3 (6)	—
AQ of cDNAs	$10^{5.64}$	10^7	$10^{5.89}$	1 : 23 : 1.8

* The numbers of sample.

the plasmid DNA, pPpp38-NA/1.8kb-F were transiently transfected with GX0101 infected CEFs, Moreover, the positive fluorescence signal were co-localized with typical MDV infection plaques (Fig. 3 a, b, c and d). These results suggested that the bi-directional promoter could co-transcript foreign genes NA and F in two directions and the pp38/pp24 dimers expressed by pBud-pp38-pp24 in co-transfected cells or endogenously by GX0101 infection were critical to activate the bi-directional promoter for expression of two foreign genes in two directions at the same time. The results were also summarized in Table 2.

Generation, Purification and Verification of MZC13NA/F

To rescue MZC13NA/F viruses, the BAC DNA was transfected into CEF using Lipofectamine reagent according to manufacturer's instructions (Invitrogen). The transfected CEFs exhibited the CPE at around 5–6 d post transfection of MZC13NA/F BAC DNA. Highly purified MZC13NA/F was obtained after four rounds of plaque purification and proliferation. Schematic presentation of the recombinant virus genome of MZC13NA/F (Fig.1) identified the insertion site of the NA and F gene driven by the bi-directional promoters at the US2 gene. DNA prepared from MZC13NA/F infected CEFs and GX0101 virus infected CEFs were analyzed with PCR and Southern blot assays (Fig. 4 A, B and C). A 0.6 kb US2 fragment was amplified from GX0101infected cells, while the US2 fragment amplified from MZC13NA/F infected cells was about 5.0 kb, demonstrating the insertion of DNA sequences (Fig. 4 A). The PCR products were then probed with NA fragment and F fragment in southern blot, and confirmed presence of NA gene and F gene in MZC13NA/F virus infected cells, while the DNA from GX0101 virus infected cells clearly showed negative with NA and F probe(Fig. 4 B and C).

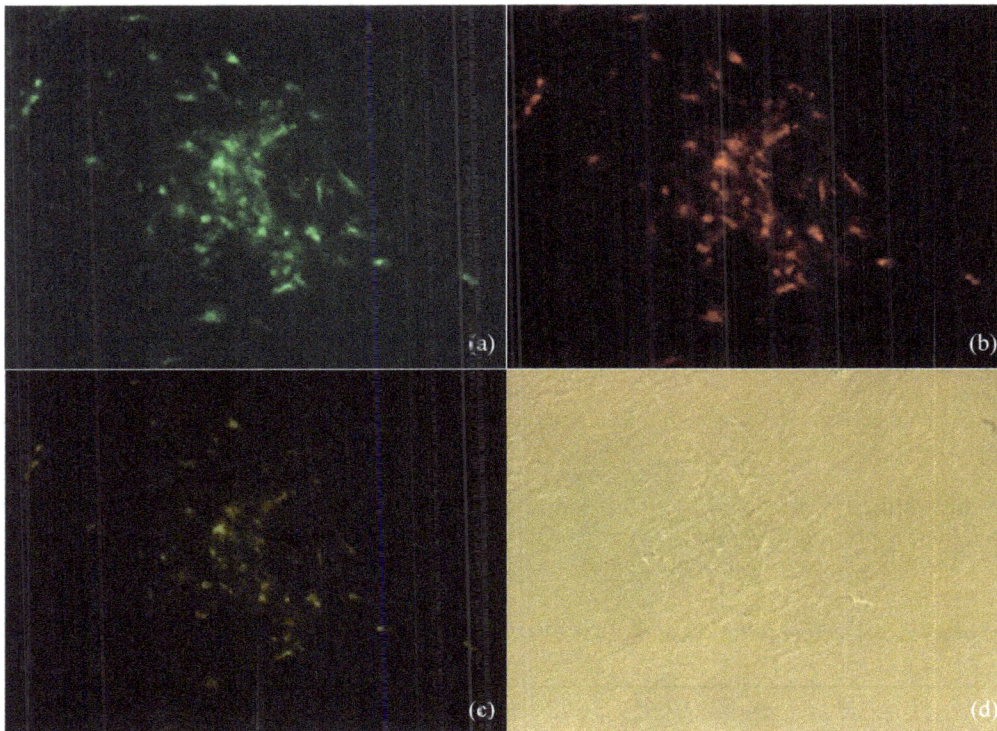

Figure 8. Demonstration of H9N2-NA or NDV-F expressing cells in IFA with monospecific sera in the MZC13NA/F-infected CEF (×200). (a) Distribution of NA was detected by staining with mouse anti-NA polyclonal antibody and Fluorescein Isothiocyanate-labelled anti-mouse secondary antibody (green fluorescence); (b) Distribution of F expression was detected using NDV virus-specific chicken antiserum and phycoerythrin-conjugated goat anti-chicken IgG (red fluorescence); (c) The merge including the visualization of red and green images showed that the co-expression proteins were visible in the cytoplasm or on the cell surface; (d) Photo was taken under regular light from the same plaque in the same visual field.

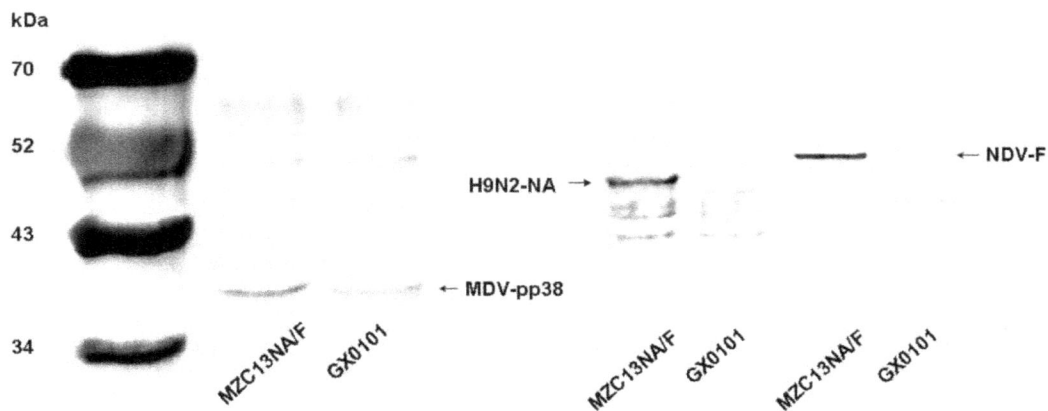

Figure 9. Western blot analyses of recombinant NA and F proteins co-expressed in MZC13NA/F infected CEF cells. CEF lysates prepared from the MZC13NA/F-infected CEF cells were subjected to SDS-polyacrylamide gel electrophoresis and transferred to a nitrocellulose membrane. The blotted membrane was blocked and reacted with mouse anti-pp38 monoclonal antibodies H19, mouse anti-NA polyclonal serum and chicken antiserum raised against NDV followed by HRPO-conjugated goat anti-mouse or chicken IgG antibodies and ECL Western blotting detection reagents. The prestained molecular size marker was included in the same gel and three experiments were accomplished independently.

Characterization of Recombinant MZC13NA/F

The growth curves (one-step growth kinetics) were compared between recombinant MZC13NA/F virus and the wild type MDV1 strain GX0101. Similar growth kinetics were demonstrated between these two viruses (Fig. 5). The titers for both viruses started increasing steadily from 48 hrs post infection, both reached the highest titer at 120 hrs. The plaque size for each virus were also compared at 120 hrs post infection, and no difference was identified (Fig. 6).

Comparison of promoter activity in real time quantification RT-PCR for NA and F mRNA in CEF infected with different rMDVs

The RT-qPCR was done to determine the transcription efficacy of NA, F and pp38 gene driven by different promoters. In order to quantitate each cDNA, all the threshold cycle (CT) values were compared to a standard curve generated using the plasmid DNA containing the same gene of interest (Fig. 7). As shown in Table 3, the absolute quantification (AQ) value of F gene cDNA was the highest and the AQ of NA gene was the lowest. Therefore, the mRNA level of F gene in 1.8kb mRNA transcript direction was much higher than that in pp38 direction for the NA gene. The mRNA of pp38 gene was slightly higher than that of NA gene though driven by the same pp38 direction.

Detection of Recombinant NA or F Protein Expression

To confirm the expression of the recombinant NA or F proteins, CEF cells infected with purified MZC13NA/F virus were analyzed by IFA. Fluorescence signals were detected in the MZC13NA/F-infectious plaques with both mouse anti-H9 AIV-NA serum and chicken antiserum raised against NDV (Fig. 8). CEFs infected with MDV1 strain GX0101 were set as negative controls, and no detectable NA or F protein signal was observed in the cytoplasm or on the cell surface of GX0101-infected CEFs (Figure not shown).

Comparison of the expression of F, NA and pp38 protein in MZC13NA/F-infected cells

For the NA and F genes expressed in cells infected with MZC13NA/F virus, the expression level was compared with the endogenous viral pp38 protein by ELISA assays. As presented in Table 4, the NA gene expressed under the pp38 promoter and the F gene under the 1.8kb mRNA promoter were expressed equally in cells infected with MZC13NA/F, and the level of NA or F protein expression were nearly detectable in GX0101-infected cells. It is also noticed that the expression level of pp38 protein was almost equal (1.344 ± 0.101 vs 1.363 ± 0.096) in MZC13NA/F or parental GX0101 infection CEF, which indicated that the introduction of bi-directional promoter did not affect the expression of viral protein pp38 itself. However, in MZC13NA/F infected cells, the foreign genes (NA or F) expressed much less than pp38 (0.445 ± 0.031 or 1.096 ± 0.040 vs 1.344 ± 0.101). Even though the NA gene and pp38 gene were expressed under the same promoter in the same direction, NA was expressed much less than virus endogenous pp38 (0.445 ± 0.031 vs 1.344 ± 0.101). The expression of NA, F and pp38 were also demonstrated by Western Blot assay (Fig. 9).

Table 4. Comparison of relative expression levels of H9N2-NA and NDV-F to MDV-pp38 in MDV- GX0101 and its recombinant with H9N2-NA and NDV-F genes.

Virus	ELISA value			Ratio of NA/F/pp38
	H9N2-NA	**NDV-F**	**MDV-pp38**	
MZC13NA/F	0.445 ± 0.031(8*)	1.096 ± 0.040(8)	1.344 ± 0.101(8)	1: 1.34: 3.02
GX0101	0.021 ± 0.002(8)	0.015 ± 0.002(8)	1.363 ± 0.096(8)	—

* The numbers of sample.

Discussion

Previous studies have proved the existence of a bi-directional promoter between pp38 gene and 1.8-kb mRNA transcripts of MDV genome. The sequence of this bi-directional promoter is only about 305 bp but contains several enhancer motifs [3,5]. More *in vitro* studies from our laboratory demonstrated that the promoter could drive the expression of *GFP* or *CAT* as reporter genes in each direction with the presence of MDV pp24/pp38 hetero-dimer. The pp24/38 dimer was able to bind to the promoter and worked in-trans as a transcriptional factor to enhance the transcription activity [9,41]. In this study, use recombinant plasmid transient transfected CEF as well as in recombinant MDV infected CEF, it was demonstrated that different foreign genes could be expressed simultaneously under control of the bi-directional promoter. As it is summarized in Table 2 and Fig.2, both AIV-NA and NDV-F genes can be expressed whether or not they were constructed in separate plasmid or in the same recombinant plasmid. Moreover NA or F gene expression in the transfected plasmid was all dependent on the presence of pp24/pp38 dimer. It is also noticed that the F gene driven by promoter p1.8kb has a higher expression level than the NA gene which is driven by promoter Ppp38. All these imply that two foreign genes can be expressed simultaneously under this bi-directional promoter in opposite directions, and the activity of the promoter in the 1.8-kb mRNA transcript direction was higher than that in the direction for the pp38 gene. The critical dependence of the pp24/pp38 dimer had been proved previously as a critical transcriptional factor to successfully express the reporter genes such as *GFP* or *CAT* driven by the bi-directional promoter [42,43]. In this study, the pp24/pp38 dimer was either introduced by transient expression plasmid or expressed endogenously via MDV virus infection. Surprisingly, we found the foreign genes NA and F expressed much higher (Fig. 3) with the viral endogenous pp24/pp38 dimer, than transient expressed pp24/38. We speculate this was due to a high level of expression of pp24/pp38 in MDV infected cells. This may be one of the limit of this bi-birectional promoter that it would be best used in the MDV vector system, but not in other virus vectors.

Extended study were carried out by constructing recombinant MDV virus, MZC13NA/F, that contains the NA and F foreign genes driven by the same bi-directional promoter. As shown in Fig. 6 that both NDV-F and H9N2 AIV-NA antigens were successfully expressed and detected in IFA with monoclonal antibody against NDV-F or AIV-NA proteins. The results clearly indicated that the MDV endogenous bi-directional promoter of only 305 bp could effectively drive two foreign genes expressed in two directions. As mentioned above, the bi-directional promoter is located between pp38 and 1.8kb mRNA family transcription start sites and pp38/pp24 dimer could work as trans-acting transcriptional factor to strongly enhance its promoter activity [8,9,41], the bi-directional promoter may have some advantages in expression of two foreign genes in the early stage of MDV infection. This is because pp38 is an early gene related to MDV early B-lymphocyte cytolytic infection [44], and pp38/pp24 dimers should be available in the early stage of infection. Such suggestion is to be further confirmed in comparative studies with other common promoters such as CMV, SV40, even MDV gB promoter.

With the cosmid vector system, meq-deleted Md5 strain of MDV was constructed and confirmed to provide better protective immunity than CVI988/Rispens in chickens against very virulent

plus MDV challenges [26,27,28]. We constructed a BAC-vectored infectious clone of very virulent MDV strain GX0101 [22], a series of meq-deleted GX0101 mutants were constructed and some of them demonstrated better protective immunity against MDV in SPF chickens than CVI988/Rispens vaccine as well [24,45]. One of protective Meq-deleted mutant virus is SC9-1, which was fully attenuated and able to provide better protective immunity than CVI988/Rispens [45]. In fact, In this study, the recombinant MDV virus, MZC13NA/F, was derived from SC9-1 virus by inserting NDV-F and H9N2 AIV-NA expression cassette under control of MDV's own bi-directional promoter. It will be more interesting to investigate whether protective efficacy can be demonstrated when challenged with Avian Influenza virus or Newcastle Disease virus. These investigation are currently ongoing in the lab.

In construction of recombinant MDV as vectors to express foreign genes, two of the key questions are where to insert and which promoter to choose. Should a strong promoter result in higher protein expression, and will thereafter provide better protective efficacy? So far, heterogenous promoters such as CMV and SV40 promoters or homogenous promoter such as MDV gB promoter were used to drive expression of foreign genes [16,17,46]. In one study, the NDV-F gene was expressed in rMDV either driven by the MDV gB promoter, or driven by non-MDV promoters, such as SV40 and β-actin promoters. However, with MDV gB promoter, rMDV-F induced lower antibody titers based on ELISA assay, but it provided better protective efficacy then the F gene driven by SV40 or β-actin promoters [16]. The bi-directional promoter used in this study was also a MDV's own promoter as gB promoter, but it had unique characteristics different from gB promoter. It was able to drive two genes express simultaneously in opposite direction and its activity was strongly depended on a trans-acting transcriptional factor pp38/pp24 dimer [8].

Experiment results in both recombinant expression plasmid-transfected CEF and recombinant MZC13NA/F-infected CEF suggest that the MDV's own bi-directional promoter may have some potential use in construction of recombinant MDV vaccines expressing foreign genes of other viruses, especially for construction of recombinant MDV vaccine expressing two foreign genes. By using the bi-directional promoter, recombinant expression cassettes for two genes could be inserted into suitable sites of MDV genomes in one step, and it helps to simplify the procedure and reduce troublesome and difficulties to insert different expression cassette plasmid DNA into MDV genomes again. The strong dependence on pp38/pp24 dimer as a trans-acting transcriptional factor may have some limitation in the use of the bi-directional promoter in construction of recombinant viruses other than use MDV as viral vectors.

Acknowledgments

We are grateful to Dr. Xiaoping Cui for critical review of this manuscript. We would also like to acknowledge the staff of College of Veterinary Medicine, Shandong Agricultural University; Taian, Shandong, China.

Author Contributions

Conceived and designed the experiments: ZC ZZ CM. Performed the experiments: ZZ CM LD WC FZ. Analyzed the data: ZZ CM PZ. Contributed reagents/materials/analysis tools: CM LD PZ. Wrote the paper: ZC ZZ CM PZ.

References

1. McGeoch DJ, Dolan A, Ralph AC (2000) Toward a comprehensive phylogeny for mammalian and avian herpesviruses. Journal of Virology 74: 10401–10406.

2. Su S, Cui N, Cui ZZ, Zhao P, Li YP, et al. (2012) Complete Genome Sequence of a Recombinant Marek's Disease Virus Field Strain with One Reticuloendotheliosis Virus Long Terminal Repeat Insert. Journal of Virology 86(24): 13818.

3. Barlic-Maganja D, Grom J (2001) Highly sensitive one-tube RT-PCR and microplate hybridisation assay for the detection and for the discrimination of classical swine fever virus from other pestiviruses. Journal of virological methods 95: 101–110.

4. Bradley G, Hayashi M, Lancz G, Tanaka A, Nonoyama M (1989) Structure of the Marek's disease virus BamHI-H gene family: Genes of putative importance for tumor induction. Journal of Virology 63: 2534–2542.

5. Bradley G, Lancz G, Tanaka A, Nonoyama M (1989) Loss of Marek's disease virus tumorigenicity is associated with truncation of RNAs transcribed within BamHI-H. Journal of Virology 63: 4129–4135.

6. Cui ZZ, Lee LF, Liu JL, Kung HJ (1991) Structural analysis and transcriptional mapping of the Marek's disease virus gene encoding pp38, an antigen associated with transformed cells. Journal of Virology 65: 6509–6515.

7. Cui ZZ, Qin AJ, Lee LF (1992) Expression and processing of Marek' s disease virus pp38 gene insert cells and immunological characterization of the gene product. Proceedings of 19th World' s Poulty Congress: 123–126.

8. Ding JB, Cui ZZ, Jiang SJ, Sun AJ, Sun SH (2005) Study on the characterization of the bi-directional promoter between pp38 gene and 1.8kb mRNA transcripts of Marek's disease viruses. Acta Microbidogica Sinica 45(3): 363–367.

9. Ding JB, Cui ZZ, Jiang SJ, Reddy SJ (2006) The enhancement effect of pp38 gene product on the activity of its upstream bi-directional promoter in Marek's disease virus. Science in China: Series C Life Sciences 49: 1–10.

10. Ding JB, Cui ZZ, Lee LF (2007) Marek's disease virus unique genes pp38 and pp24 are essential for transactivating the bi-directional promoters for the 1.8-kb mRNA transcripts. Virus Genes 35(3): 643–650.

11. Taylor J, Edbauer C, Rey-Senelonge A, Bouquet JF, Norton E, et al. (1990) Newcastle disease virus fusion protein expressed in a fowl pox virus recombinant confers protection in chickens. Journal of Virology 64(4): 1441–1450.

12. Morgan RW, Gelb J, Schreurs CS, Lütticken D, Rosenberger JK, et al. (1992) Protection of chickens from Newcastle and Mark's disease with a recombinant herpserivus of turkey's vaccine expressing the Newcastl e disease virus fusion protein. Avian Dis 36: 858–870.

13. Webster RG, Kawaoka Y, Taylor J, Weinberg R, Paoletti E (1991) Efficacy of nucleo protein and haemagglutinin antigens expressed in fowl pox virus as vaccine for influenza in chickens. Vaccine 9: 303–308.

14. Swayne DE, Garcia M, Beck JR, Kinney N, Suarez DL (2000) Protection against diverse highly pathogenic H5 avian influenza viruses in chickens immunized with recombinant fowl pox vaccine containing an H5 avian influenza hemagglutinin gene insert. Vaccine 18: 1088–1095.

15. Ma MX, Jin NY, Wang ZG, Wang RL, Fei DL, et al. (2006) Construction and immunogenicity of recombinant fowlpox vaccines coexpressing HA of AIV H5N1 and chicken IL18. Vaccine 24(20): 4304–431.

16. Sonoda K, Sakaguchi M, Okamura H, Yokogawa K, Tokunaga E, et al. (2000) Development of an Effective Polyvalent Vaccine against both Marek's and Newcastle Diseases Based on Recombinant Marek's Disease Virus Type 1 in Commercial Chickens with Maternal Antibodies. Journal of Virology 74(7): 3217–3226.

17. Tsukamoto K, Kojima C, Komori Y, Tanimura N, Mase M, et al. (1999) Protection of Chickens against Very Virulent Infectious Bursal Disease Virus (IBDV) and Marek's Disease Virus (MDV) with a Recombinant MDV Expressing IBDV VP2. Virology 257: 352–362.

18. Sakaguchi M, Nakamura H, Sonoda K, Okamura H, Yokogawa K, et al. (1998) Protection of chickens with or without maternal antibodies against both Marek's and Newcastle diseases by one-time vaccination with recombinant vaccine of Marek's disease virus type 1. Vaccine 16(5): 472–479.

19. Petherbridge L, Brown AC, Baigent SJ, Howes K, Sacco MA, et al. (2004) Oncogenicity of virulent Marek's disease virus cloned as bacterial artificial chromo-somes. J Virol 78: 13376–13380.

20. Baigent SJ, Petherbridge LJ, Smith LP, Zhao Y, Chesters PM, et al. (2006) Herpesvirus of turkey reconstituted from bacterial artificial chromosome clones induces protection against Marek's disease. J Gen Virol 87: 769–776.

21. Cui HY, Wang YF, Shi XM, Tong GZ, Lan DS, et al. (2008) Construction of Marek's disease virus serotype 814 strain as an infectioous bacterial artificial chromosome. Chinese journal of biotechnology 24: 569–575.

22. Sun AJ, Petherbridge LP, Zhao Y, Li YP, Cui ZZ, et al. (2009) A BAC clone of MDV strain GX0101 with REV-LTR integration retained its pathogenicity. Chinese Science Bulletin 54 (15): 2641–2647.

23. Li YP, Sun AJ, Su S, Zhao P, Cui ZZ, et al. (2011) Deletion of the Meq gene significantly decreases immunosuppression in chickens caused by pathogenic Marek's disease virus. Virology journal: 2–8.

24. Su S, Li YP, Sun AJ, Zhao P, Cui ZZ, et al. (2010) Protective immunity of a meqΔdeleted Marek's disease virus against very virulent virus challenge in chickens. Acta Microbiologica Sinica 50(3): 380–386.

25. Reddy SM, Lupiani B, Gimeno IM, Silva RF, Lee LF, et al. (2002) Rescue of a pathogenic Marek's disease virus with overlapping cosmid DNAs: use of a pp38 mutant to validate the technology for the study of gene function. Proc Natl Acad Sci USA 99(May (10)): 7054–7059.

26. Lee LF, Kreager KS, Arango J, Paraguassu A, Beckman B, et al. (2010) Comparative evaluation of vaccine dfficacy of recombinant Marek's disease virus vaccine lacking Meq oncogene in commercial chickens. Vaccine 28(5): 1294–1299.

27. Lee LF, Zhang HM, Heidari M, Lupiani B, Reddy SM (2011) Evaluation of factors affecting vaccine dfficacy of recombinant Marek's disease virus lacking the Meq oncogene in chickens. Avian Dis.55(2): 172–179.

28. Lee LF, Heidari M, Zhang HM, Lupiani B, Reddy SM, et al. (2012) Cell culture attenuation eliminates rMd5Meq-induced bursal and thymic atrophy and renders the mutant virus as an effective and safe vaccine against Marek's disease. Vaccine 30(34): 5151–5158.

29. Zhang Z, Cui ZZ (2005) Isolation of recombinant field strains of Marek's disease virus integration with reticuloendotheliosis virus genome fragments. Sci. China C. Life Sci 48 (1): 81–88.

30. Sun SH, Cui ZZ (2007) Biological Identification of Newcastle Disease Viruses Isolated from Eggs of a Parent Breeder Farm. Acta Veterinaria et Zootechnica Sinica 38(7): 741–743.

31. Jiang SJ, Ding JB, Meng SS, Cui ZZ, Yang HC (2005) Co-expression and Construction of Eukaryote Plasmid of pp38 and pp24 of Marekcs Disease Virus.Virologica sinica 20(4): 404–407.

32. Morgan RW, Cantello JL, McDermott CH (1990) Transfection of chicken embryo fibroblasts with Marek's disease virus DNA. Avian Diseases 34: 345–351.

33. Datsenko KA, Wanner BL (2000) One-step inactivation of chromosomal genes in Escherichia coli K-12 using PCR products. Proc Natl Acad Sci U S A 97 (12): 6640–6645.

34. Muyrers JP, Zhang Y, Testa G, Stewart AF (1999) Rapid modification of bacterial artificial chromosomes by ET-recombination. Nucleic Acids Res 27 (6): 1555–1557.

35. Narayanan K, Williamson R, Zhang Y, Stewart AF, Ioannou PA (1999) Efficient and precise engineering of a 200 kb beta-globin human bacterial artificial chromosome in E. coli DH10B using an inducible homologous recombination system. Gene Ther 6: 442–447.

36. Yu D, Ellis HM, Lee EC, Jenkins NA, Copeland NG, et al. (2000) An efficient recombination system for chromosome engineering in Escherichia coli. Proc Natl Acad Sci U S A 97: 5978–5983.

37. Sun AJ, Xu XY, Lawrence PB, Zhao YG, Cui ZZ (2010) Functional evaluation of the role of reticuloendotheliosis virus long terminal repeat (LTR) integrated into the genome of a field strain of Marek's disease virus. Virology 397: 270–276.

38. Cui HY, Wang YF, Shi XM, An TQ, Tong GZ, et al. (2009) Construction of an infectious Marek's disease virus bacterial artificial chromosome and characterizationof protection induced in chickens. Journal of Virological Methods 156: 66–72.

39. Schumacher D, Tischer BK, Fuchs W, Osterrieder N (2000) Reconstitution of Marek's disease virus serotype 1 (MDV-1) from DNA cloned as a bacterial artificial chromosome and characterization of a glycoprotein B-negative MDV-1 mutant. Journal of Virology 74: 11088–11098.

40. Li Y, Reddy K, Reid SM, Cox WJ, Brown IH, et al. (2011) Recombinant herpesvirus of turkeys as a vector-based vaccine against highly pathogenic H7N1 avian influenza and Marek's disease. Vaccine 29: 8257–8266.

41. Ding JB, Cui ZZ, Jiang SJ, Li YP (2008) The structure of unique genes pp38 and pp24 and the effect on the activity of its upstream bi-directional promoters in Marek's disease. Science in China: Series C Life Sciences 38: 760–765.

42. Shigekane H, Kawaguchi Y, Shirakata M, Sakaguchi M, Hirai K (1999) The bi-directional transcriptional promoters for the latency-relating transcripts of the pp38/pp24 mRNAs and the 1.8 kb-mRNA in the long inverted repeats of Marek's disease virus serotype 1 DNA are regulated by common promoter-specific enhancers. Archives of Virology 144: 1893–1907.

43. Ding JB, Cui ZZ, Lee LF, Cui XP, Reddy SM (2006) The role of pp38 in regulation of Marek's disease virus bi-directional promoter between pp38 and 1.8-kb mRNA. Virus Genes 32(2): 193–201.

44. Gimeno IM, Witter RL, Hunt HD, Reddy SM, Lee LF, et al. (2005) The pp38 gene of Marek's disease virus (MDV) is necessary for cytolytic infection of B cells and maintenance of the transformed state but not for cytolytic infection of the feather follicle epithelium and horizontal spread of MDV. Journal of Virology, 79(7): 4545–4549.

45. Su S (2013) Comparisons of Biological Characteristic between Recombinant Marek's disease virus and its Mutant Strains. Tai'an: Shandong Agricultural university.

46. Cui HY, Gao HB, Cui XL, Zhao Y, Wang YF, et al. (2013) Avirulent Marek's Disease Virus Type 1 Strain 814 Vectored Vaccine Expressing Avian Influenza (AI) Virus H5 Haemagglutinin Induced Better Protection Than Turkey Herpesvirus Vectored AI Vaccine. PLoS ONE 8(1): e53340.

16

Factors Associated with the Seroprevalence of Leishmaniasis in Dogs Living around Atlantic Forest Fragments

Nelson Henrique de Almeida Curi[1][x], **Ana Maria de Oliveira Paschoal**[2], **Rodrigo Lima Massara**[2], **Andreza Pain Marcelino**[3], **Adriana Aparecida Ribeiro**[3], **Marcelo Passamani**[1], **Guilherme Ramos Demétrio**[1], **Adriano Garcia Chiarello**[4]

[1] Postgraduate program in Applied Ecology, Department of Biology, Federal University of Lavras, Lavras, Brazil, [2] Postgraduate program in Ecology, Conservation and Management of Wildlife, Department of Biology, Institute of Biological Sciences, Federal University of Minas Gerais, Belo Horizonte, Brazil, [3] Laboratory of Leishmaniasis, Ezequiel Dias Foundation-FUNED, Belo Horizonte, Brazil, [4] Department of Biology, University of São Paulo, Ribeirão Preto, Brazil

Abstract

Canine visceral leishmaniasis is an important zoonosis in Brazil. However, infection patterns are unknown in some scenarios such as rural settlements around Atlantic Forest fragments. Additionally, controversy remains over risk factors, and most identified patterns of infection in dogs have been found in urban areas. We conducted a cross-sectional epidemiological survey to assess the prevalence of leishmaniasis in dogs through three different serological tests, and interviews with owners to assess features of dogs and households around five Atlantic Forest remnants in southeastern Brazil. We used Generalized Linear Mixed Models and Chi-square tests to detect associations between prevalence and variables that might influence *Leishmania* infection, and a nearest neighbor dispersion analysis to assess clustering in the spatial distribution of seropositive dogs. Our findings showed an average prevalence of 20% (ranging from 10 to 32%) in dogs. Nearly 40% (ranging from 22 to 55%) of households had at least one seropositive dog. Some individual traits of dogs (height, sterilization, long fur, age class) were found to positively influence the prevalence, while some had negative influence (weight, body score, presence of ectoparasites). Environmental and management features (number of cats in the households, dogs with free-ranging behavior) also entered models as negative associations with seropositivity. Strong and consistent negative (protective) influences of the presence of chickens and pigs in dog seropositivity were detected. Spatial clustering of cases was detected in only one of the five study sites. The results showed that different risk factors than those found in urban areas may drive the prevalence of canine leishmaniasis in farm/forest interfaces, and that humans and wildlife risk infection in these areas. Domestic dog population limitation by gonadectomy, legal restriction of dog numbers per household and owner education are of the greatest importance for the control of visceral leishmaniasis in rural zones near forest fragments.

Editor: Luzia H. Carvalho, Centro de Pesquisa Rene Rachou/Fundação Oswaldo Cruz (Fiocruz-Minas), Brazil

Funding: Thanks to CAPES, who provided the scholarship for the first author. The Brazilian Science Council (CNPq 472802/2010-0) and Minas Gerais Science Foundation (FAPEMIG APQ 01145-10) also provided funds. The funders had no role in study design, data collection and analysis, decision to publish, or preparation of the manuscript.

Competing Interests: The authors have declared that no competing interests exist.

* Email: nelsoncuri@hotmail.com

Introduction

Landscape changes such as urbanization and human encroachment are among the main drivers of the alteration of disease dynamics, e.g., the increased or altered prevalence and incidence of disease in humans, domestic animals, and wildlife [1–4]. The introduction of exotic domestic species often accompanies human movements during such changes and poses a threat to both wildlife and human health. Since their domestication, pet animals have been closely associated with humans, and dogs (*Canis familiaris*) are the most common and distributed companion animal worldwide [5–6]. Unfortunately, this ubiquitous human-dog bond also brings many host species into contact with their pathogens because dogs occupy both natural and human-modified areas and may therefore enhance disease transmission and persistence in

humans and wildlife [7]. But because of this close bilateral interaction, domestic dogs may also be used as sentinels of disease for both human and wildlife populations [8–10].

In Brazil, there are about 40 million dogs, of which five million are represented by rural dogs. Most of these live unrestricted, exhibiting free-ranging behavior, and move in both urban and natural areas [6]. Accordingly, recent studies have shown that the domestic dog has become increasingly common in several Brazilian protected areas [11–12], but the ecological and epidemiological impact of this invasion generally remains unknown. In a study conducted in India, Vanak and Gompper [13] have shown that dogs interfere with the spatial distribution of sympatric native carnivore species. Therefore, they also disturb the spatial distribution of hosts and parasites, affecting disease dynamics and the resulting impact on wildlife and human populations that have contact with these dogs. The contact events and the presence of parasites in domestic dogs indeed increase the risk of disease for both humans and wildlife [7,14–15] and must be investigated if the aim is to minimize risk and to understand the dynamics of the systems into which dogs are introduced and with which they interfere. Human behavior also has the potential to alter parasite dynamics in wildlife-human-domestic animal inter-faces [16]. For instance, wild carnivores are more exposed to pathogens in places where they face more frequently their domestic counterparts [15], and dog ownership is itself an important risk factor for human leishmaniasis [14,17].

Visceral leishmaniasis is a dangerous systemic disease among the most significant zoonosis in Brazil, affecting both dogs and humans. Brazil holds the higher number of cases in South America and is one of the six most affected countries worldwide. The disease is caused by parasites of the species *Leishmania infantum*, whose vectors are phlebotomine sand flies of the genus *Lutzomyia* (Psychodidae) [18–20]. The main reservoir of *L. infantum* is the domestic dog, although the possible participation of asymptomatic infected persons is currently been suggested [21–23]. Other wild mammal species may be infected and may develop clinical signs, but their role as reservoirs remains to be clarified [22,24–26]. One of the few well studied species is the widely distributed and relatively abundant South American wild canid crab-eating fox *Cerdocyon thous*, a host with low infectiveness unable to sustain *Leishmania* cycles without the presence of sympatric dogs [21].

Recent studies have considered the surrounding environment and its relation to the epidemiology of human and canine visceral leishmaniasis (CVL). Their results are mixed, although several interesting patterns have arisen, e.g., the influence of other domestic animals as attractors for the vector, which ultimately produces an increased risk of infection in dogs and humans [27–30]. Furthermore, according to a topical review, there is still controversy over risk factors associated with infection in dogs, and surveillance and information is scarce in some areas in Brazil [31]. A recently published paper has identified peridomestic risk factors for both canine and human cutaneous leishmaniasis in an agricultural area of southern Brazil [32].

Visceral leishmaniasis affects mostly poor communities in remote rural areas [19]. However, for CVL, many areas and contexts such as rural settlements around forest fragments and other human-wildlife-domestic animal interface zones have been poorly evaluated. The control and elimination of leishmaniasis is far from realistic in Latin America because it is a zoonosis with a very large domestic reservoir and probably a substantial sylvatic reservoir (though this is a point which still needs further investigation), and the existence of gaps in knowledge and surveillance along with a lack of political involvement [33]. Thus, the goals of this study are to evaluate the seroprevalence of CVL, a neglected but important zoonosis in Brazil, in areas of unknown epidemiological status in the Atlantic Forest domain and to correlate this presence with dog individual traits, animal manage-ment and environmental factors. In this way, the patterns of infection detected here can ultimately be targeted or managed by programs for the control of the disease.

Materials and Methods

Ethics statement

Sampling and interviewing were performed under consent obtained from the household head or other responsible individual. Licenses from the State Forest Institute – IEF (UC: 080/10, 081/10 and 082/10) and approval from the Ethics Commission on the Use of Animals of the Pontiphical Catholic University of Minas

Gerais (CEUA, PUC Minas 037/2010) were obtained prior to the initiation of the field work. Regarding the collection of data from human participants, our project was examined by the Ethics Research Committee (Comitê de Ética em Pesquisa) of the Pontiphical Catholic University of Minas Gerais (PUC-Minas). We did collect some information on the number of people inhabiting the house with the approved consent of the household head. A Consent Term about the confidential character of the records was read to every interviewed person. Animal manipula-tion procedures adhered to the recommendations from the COBEA (Brazilian College of Animal Experimentation) and the Animal Ethics Committee of FIOCRUZ (Oswaldo Cruz Institute Foundation) of the Brazilian Ministry of Health.

Study sites

Rural settlements surrounding five protected areas in the Atlantic Forest domain of the state of Minas Gerais, southeastern Brazil, were selected for this study. These areas comprise two state parks, Serra do Brigadeiro (PESB, municipality of Araponga) and Sete Salões (PESS, municipality of Santa Rita do Itueto), and three private reserves, Fazenda Macedônia (RPPNFM, municipality of Ipaba), Feliciano Miguel Abdala (RPPNFMA, municipality of Caratinga), and Mata do Sossego (RPPNMS, municipality of Simonésia) (Figure 1, table 1). All of the areas had humans living in their vicinity and various degrees of domestic dog occupancy recorded within their borders [12]. The landscapes around the protected areas are mostly composed of a mosaic of forest borders, small rural properties, their legal reserves and small human settlements. Households were mostly located near forests, water bodies, and had vegetation in their vicinities (Figure 2), which are considered risk factors for *Leishmania* infection [31]. According to the official Brazilian health services, these areas are characterized by an absence of recorded human leishmaniasis cases except for Ipaba municipality, where a few records have been obtained in recent years (Table 1). Several species of the genus *Lutzomyia* occurs at the Atlantic Forest in both peridomiciliary and forest environments [34–35]. All households were located near potential breeding sites for the vectors (forested areas, water bodies, peridomiciliary microhabitats and plantations). Sand flies are indeed abundant in human-disturbed open areas such as plantations and secondary forest and homesteads with the presence of dogs [36]. Thus, our sampling sites located in rural/forest interfaces are likely not free of the presence of vector species.

Sampling

The study was conducted between January 2011 and August 2012. Overall, 291 dogs older than two months were sampled in 124 rural households located up to two kilometers from protected area boundaries around the study areas, and this was the sole eligibility criteria used for this study. After physical restraint, blood was collected from the jugular vein and a complete clinical examination of the dogs was performed (focusing on clinical alterations of visceral leishmaniasis such as weight loss, skin lesions, nail overgrowth and increased volume of the liver and spleen). A standardized questionnaire survey was administered to the owners. Factors related to animal management and behavior (number of dogs, mobility of dogs, access of dogs to the forest and villages, observed interactions between dogs and wildlife, ectoparasite treatment), the presence of vector attractors in peridomestic dwellings (i.e., other domestic species), number of people and geographic coordinates were recorded for each household. The individual and clinical features of the dogs (sex, age, height, weight, fur type, breed, sterilization, body condition, clinical alterations, and the presence of ectoparasites such as fleas and

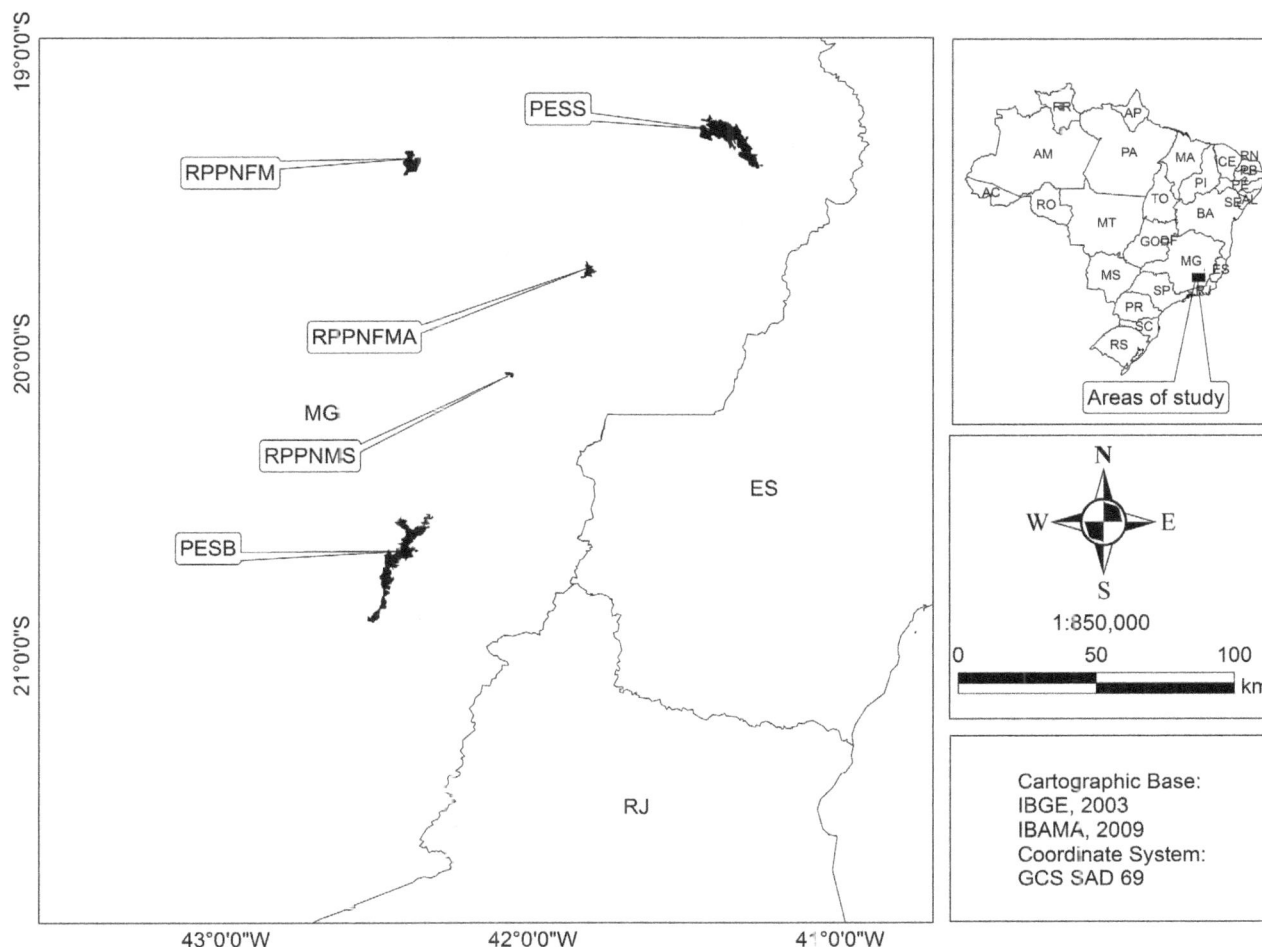

Figure 1. Study areas location in the Atlantic Forest domain, Minas Gerais state, southeastern Brazil.

ticks) were recorded in individual files. Weight was measured with a precision scale (Pesola, 50 kg capacity), and height was measured from the footpad to the top of the scapulae of standing dogs. Body condition of dogs was scored from 0 (extreme emaciation) to 5 (extreme obesity). Refusals to the survey occurred in four cases because the responsible were absent from the households at the time of collection. There were no other refusals, and we believe

that the houses that were not surveyed did not affect the overall results.

Laboratory analysis

Blood samples were allowed to clot for 4 h at room temperature and then centrifuged for serum extraction. Serum samples were initially stored at $-20°C$, and sent later to be stored at $-80°C$ at Fundação Ezequiel Dias, Belo Horizonte, prior to analysis.

Table 1. Epidemiological features of five protected areas in the Atlantic Forest of the state of Minas Gerais, southeastern Brazil.

Study site	Distance from nearest city (km)	Altitude (m)	Area size (ha)	Transmission status[1]	Human cases[1]/population[2]	Human:dog ratio
RPPNFM	0.3	320	3,343	Sporadic	2/16,708	1.2
PESB	3.3	1,437	15,015	Silent	0/8,152	1.9
PESS	4.7	687	13,370	Silent	0/5,697	1.8
RPPNFMA	10.5	430	1,312	Silent	0/22,242	1.1
RPPNMS	7.7	1,340	392	Silent	0/18,298	2.9
Total	-	-	-	-	2/71,097	1.8 (0.2–8)

[1]Data from 2010–2012 (Brazilian Ministry of Health).
[2]Data from the 2010 census (Brazilian Institute of Geography and Statistics, www.ibge.gov.br).

Figures 2. Typical households and peridomestic scenarios of rural areas surrounding Atlantic Forest fragments in Minas Gerais State, southeast Brazil.

Immune enzyme assays (ELISA), indirect immunofluorescence reaction (IFI), and dual path platform immunochromatographic rapid test (DPP) analyses were performed using Biomanguinhos kits (Fiocruz, Manguinhos, Rio de Janeiro, Brazil). These tests are

Table 2. Seroprevalence of canine leishmaniasis in rural dogs sampled around five protected areas of the Atlantic Forest.

Study site	Number of dogs	Number sampled (%)	Dogs/house	Dog prevalence	P value	Household prevalence	P value
RPPNFM	98	84 (85)	3.9	13.1% (11/84)	<0.0001	40% (10/25)	0.4233
PESB	86	67 (77)	2.7	32.8% (22/67)	0.0072	54.8% (17/31)	0.4723
PESS	53	48 (90)	2.1	14.6% (7/48)	<0.0001	24% (6/25)	0.0163
RPPNFMA	60	50 (83)	3.3	10% (5/50)	<0.0001	22.2% (4/18)	0.0184
RPPNMS	49	42 (85)	1.9	30.9% (13/42)	0.0136	44% (11/25)	0.6889
Total	346	291 (84)	2.8 (1–15)	19.9% (58/291)	<0.0001	38.7% (48/124)	0.0270

currently used for the diagnosis of CVL in endemic areas by the laboratories of public health [37–39]. IFI tests were performed with a cut-off point at the dilution of 1:40. The ELISA results are expressed in absorbance values and the DPP test provide visual interpretation of seropositivity.

Statistical analysis

Spearman correlation matrices were built in order to test correlations and assess the level of agreement between the three serological tests, as well as to assess correlation between the ectoparasite presence and previous insecticide treatment in dogs. Dog individual traits, ecological (presence of animals attractive for the sand flies), and management factors (level of dog's restriction, access to forest and urban areas, and ectoparasite treatment) that may be linked to CVL transmission according to previous literature (see [31] and related papers) were used as explanatory or independent variables for different scenarios of seropositivity (positives for at least one test, ELISA, IFI, and DPP positives, and paired tests) for *Leishmania* in dogs, the binary response (dependent) variables. Households were considered positive if they had at least one seropositive dog. At the individual level, sex, age class (younger or older than 12 months), fur type (fur less than 3 cm long was considered short), sterilization, breed (purebred and mixed bred), and the presence of ectoparasites were used as the independent binary variables. Age, weight, height and body condition were included as quantitative variables. For the households, the continuous variables were the numbers of dogs, people, and cats. The presence of chickens, livestock mammals (cattle, horses and pigs), small pets (e.g. rabbits and birds), whether dogs were kept free or not, the access of dogs to the nearest cities and to the protected areas, whether owners observed interactions with wildlife, and ectoparasite treatment, were included as binary factors. Generalized linear mixed models (GLMMs) adjusted with a binomial distribution for the response data and controlling for households and areas as random effects (all other variables were set as fixed effects in the models), were used to select the most important factors or combinations of factors associated with seropositivity. This type of model is considered suitable for cross-sectional epidemiological studies [40]. The variables were subsequently removed from the complete model (significantly different from a null model) by a backward stepwise approach according to their level of significance, until the difference between subsequent models was significant (p<0.05). Comparisons of prevalence ratios among the study areas, and for binary variables of dog individual traits (gender, sterilization, age class: young (< 1 yr) versus adult (>1 yr), pure breed versus mixed breed, short fur versus long fur dogs, presence of ectoparasites), and management and environmental features (mobility, access to forests and villages, presence of other domestic animal species, interactions with wildlife and previous ectoparasitic treatment) were performed with multiple and two proportion Yates-corrected Chi-square tests. We did not applied Chi-square or similar tests with the prevalence ratios of continuous variables to avoid unnecessary data categorization and redundancy with the GLMM tests. A threshold of p<0.05 was used to determine statistical significance. The GLMM tests were run in package lme4 of R software, and the other analyses were performed in BioEstat 5.0 [41–42]. To assess spatial clustering of seropositive dogs, we used a nearest neighbor dispersion analysis of dog locations with the software BIOTAS version 2.0a 3.8. We based on the STROBE statement [43] as a guide for the reporting of our observational results.

Table 3. Prevalence ratios for *Leishmania* seropositive dogs (for at least one test) in rural areas around Atlantic Forest fragments, and Chi-square tests results for binary variables.

Variable	Category	Number	Positives	Prevalence ratio	Z	P value
Gender	Males	193	37	19.2		
	Females	98	21	21.4	0.45	0.64
Sterilized	Yes	19	6	31.6		
	No	272	52	19.1	1.32	0.18
Breed	Mixed bred	245	51	20.8		
	Purebred	46	7	15.2	0.87	0.38
Hair	Short	266	52	19.5		
	Long	25	6	24.0	0.53	0.59
Age class	Young	64	9	14.1		
	Adult	227	49	21.6	−1.33	0.18
Ectoparasites	Yes	255	50	19.6		
	No	36	8	22.2	0.36	0.71
Mobility	Free	278	54	19.4		
	Restrained	13	4	30.8	1	0.31
Access to forest	Yes	239	46	19.2		
	No	52	12	23.1	0.62	0.53
Access to villages	Yes	75	11	14.7		
	No	216	47	21.8	−1.32	0.18
Presence of chickens	Yes	271	46	17.0		
	No	20	12	60.0	4.64	<0.0001
Presence of cattle	Yes	180	30	16.7		
	No	111	28	25.2	1.77	0.07
Presence of horses	Yes	153	27	17.6		
	No	138	31	22.5	−1.02	0.3
Presence of pigs	Yes	155	19	12.3		
	No	136	39	28.7	−3.49	0.0005
Presence of small pets*	Yes	59	9	15.3		
	No	232	49	21.1	1	0.31
Interaction with wildlife	Yes	161	32	19.9		
	No	130	26	20.0	−0.02	0.97
Ectoparasite treatment	Yes	226	41	18.1		
	No	65	17	26.2	−1.42	0.15

*Rabbits and cage birds.

Results

The sex ratio of the dogs was 2:1 (193 males: 98 females), the average age of the dogs was 3.3 yr (ranging from 3 months to 18 yr), and adult dogs (>1 yr old) represented 78% of the total (227/291). Only 8.6% (25/291) of the dogs had long fur, and purebred dogs represented 15.8% (46/291). The mean body condition score was 2.2 (ranging from 0.5 to 3.5). Low body scores (up to 2) were detected in 170 (58.4%) dogs. Ectoparasites (fleas or ticks) were found in 86% of the dogs, and 77% (226/291) were submitted to previous ectoparasite treatment, and infestation were inversely but weakly correlated to previous treatment ($r = -0.12$; $p = 0.032$). Only nineteen dogs (6.5%) had been sterilized. The mean number of dogs per household was 2.8 (including dogs that could not be sampled, maximum number = 15). Ninety-five percent (278/291) of dogs were kept without space restriction. The mean number of people was 3.6 per household, with a maximum of eight. Average human to dog ratio was approximately 2:1. In 80% of the households, the dogs had access to the forest, and they had access to the nearest cities in 36.5% of the households. Chickens were present in 90%, cattle in 55%, horses in 46%, pigs in 38%, cats in 48%, and small pets (rabbits and cage birds) in 14.5% of the households.

There was low correlations between the serological tests used ($r = 0.42$, $p<0.0001$ for IFI and ELISA; $r = 0.23$, $p<0.0001$ for IFI and DDP; $r = 0.05$, $p = 0.3131$ for ELISA and DPP). The ELISA test revealed 13.7% (40/291) of positive samples (39% of positive samples had absorbance values above the cut-off point, including those from symptomatic dogs). Only 9.6% (28/291) of the dogs were seropositive for *Leishmania* sp. according to the IFI test. In the DPP test, eleven samples (3.8%) were positive. When tests were combined, 5.5% of the samples (16/291) were positive for ELISA and IFI. Three samples (1%) were positive for ELISA and DPP. Five samples (1.7%) tested positive for IFI and DPP, and only

Table 4. Best supported GLMMs analyzing associations for leishmaniasis-seropositive rural dogs living around Atlantic Forest fragments.

Scenario/Variables	Estimate (SE*)	Z	P value
+ in at least one test			
Sterilized	1.196 (0.569)	2.1	0.03558
Weight	−0.130 (0.044)	−2.9	0.00341
Height	0.139 (0.036)	3.7	0.00016
Presence of chickens	−1.778 (0.530)	−3.3	0.00079
Presence of pigs	−1.084(0.347)	−3.1	0.001804
+ ELISA			
Height	0.043 (0.020)	2.09	0.03663
Presence of chickens	−1.411 (0.557)	−2.5	0.01136
Presence of pigs	−1.144 (0.417)	−2.7	0.00616
+ IFI			
Sterilized	2.294 (0.766)	2.9	0.002739
Body score	−1.132 (0.501)	−2.2	0.024009
Weight	−0.205 (0.083)	−2.4	0.013545
Height	0.142 (0.052)	2.7	0.006397
Presence of ectoparasites	−1.582 (0.659)	−2.4	0.016469
Number of cats	−0.453 (0.218)	−2.07	0.038373
Mobility of dogs	−2.976 (0.823)	−3.6	0.000301
Presence of pigs	−0.992 (0.480)	−2.06	0.039026
+ ELISA/+ IFI			
Sterilized	1.307 (0.618)	2.1	0.034550
Long fur	1.375 (0.574)	2.4	0.016681
Age class	1.130 (0.597)	1.89	0.058377
Body score	−0.824 (0.344)	−2.4	0.016719
Height	0.048 (0.020)	2.4	0.015223
Presence of chickens	−1.919 (0.546)	−3.5	0.000442
Presence of pigs	−1.343 (0.384)	−3.5	0.000481

*Standard error.

three samples (1%) were positive for all tests. Because of the low level of agreement among the diagnostic methods used, we calculated prevalence data based on the number of dogs seropositive for at least one test.

Overall seropositivity was 19.9% (58/291). Ten of the 58 positive dogs (17%) were symptomatic for leishmaniasis, showing clinical signs such as weight loss, skin lesions, and nail overgrowth. Forty eight of 124 (38.7%) households had at least one seropositive

dog. If the protected areas were considered separately, seroprevalence ranged from 10 to 32% in dogs and from 22 to 55% in households, with significant differences in the prevalence between the areas. Dog and household prevalence were significantly higher in PESB and RPPNMS (Table 2). Differences in prevalence ratios regarding binary variables were detected by the Chi-square tests for the cohabitation of dogs with chickens and pigs (Table 3).

Table 5. Nearest neighbor dispersion analysis results for leishmaniasis seropositive rural dogs around five protected fragments of the Atlantic Forest in the State of Minas Gerais, Brazil.

Study site	Mean distance between seropositive dogs (m)	Distance standard deviation	Z score	Spatial pattern
RPPNFM	951.4	166.2	0.12	Random
PESB	160.8	47.5	−4.48	Clustered
PESS	1351.7	162.5	3.9	Uniform
RPPNFMA	874.3	112.4	4.04	Uniform
RPPNMS	298.6	64.7	−1.38	Random

The results of the GLMM modeling are summarized in table 4. Models for four of eight possible scenarios (DPP, DPP+IFI, DPP+ELISA, DPP+ELISA+IFI) could not be built due to the small number of positive outputs. In the four viable final models, eleven of 23 entered variables remained in at least one model. The presence of pigs entered all models as a negative association, while the presence of chickens featured in three models, also negatively associated with prevalence. Weight and body score entered two models with negative relationships to infection. The presence of ectoparasites, number of cats per household and mobility of dogs figured in one of the four final models showing negative relationships with seropositivity.

Height of dogs appeared in all models as a positive association with CVL. Sterilization was positively associated with infection in three scenarios. Long fur entered one model with a positive association. Age class was positively associated with infection in one model. The correlation matrices provided contained no value above 0.6, thus no colinearity was found that would have prevented the variables to be included in the same model.

Spatial clustering of seropositive dogs was detected only in PESB (Table 5), and seropositive dogs were randomly or uniformly distributed in the other four sites.

Discussion

Because the dog is the primary reservoir and the infection in dogs generally precedes human cases [22], more attention should be given to the disease in dogs wherever they occur, i.e., all human-occupied areas. Even though relatively few humans live in our study areas and have access to these dogs, and the ecological impact of leishmaniasis may be greater than the public health impact, rural families' welfare should never be neglected. Additionally, there is ecotourism activity inside and around parks, and human encroachment is ongoing at these sites. Consequently, dogs may be useful as sentinels for zoonotic leishmaniasis in areas with uncertain epidemiological status, and efforts to reveal their patterns of infection are of the highest importance for control and prevention.

We acknowledge that the low accuracy of the serological tests used is a limitation of our study and without a molecular test is not possible to rule out cross-reactions with other protozoans, such as *Trypanosoma* sp., in a proportion of dogs sampled. The same serum samples were tested for *Babesia canis* (Curi et al., unpublished data), and only four (1.3%) were positive for both *Leishmania* and *Babesia*. Therefore, the occurrence of this cross reaction may be considered low or nonexistent in this study. Instead, coinfection by both agents is possible. Our results show a low level of agreement between the serological tests used which may be related, among other factors, to the relatively low indirectly estimated (through ELISA) antibody concentrations detected in most samples. Other studies have reported discrepancies in serologic tests, such as differences in sensitivity and specificity [37]. This is of great concern because tests such as ELISA and DPP are currently employed for epidemiological screening and control of CVL in Brazil [38–39], and such inconsistency may hamper any research or control efforts. Therefore, our strategy to use concomitantly different serologic tests is recommended, preferably along with molecular diagnostic methods [44].

Many studies have identified risk factors for zoonotic human and CVL. However, most studies on dogs were primarily concerned with urban zones [28–32,44–45,47]. In our study, seven individual traits of dogs were associated with seropositivity. Height was positively associated with seropositivity in all four models. This factor is possibly linked to a target size effect or differences in heat and CO_2 irradiation between small and large sized dogs, enhancing the finding of larger hosts by the vectors. Weight and body score were negatively associated in two scenarios of seropositivity, and this can be explained by the fact that low body condition animals may have impaired immune function and higher susceptibility to infection. However, dog size was not associated with infection in previous studies [31].

The literature shows that ectoparasites may be positively, negatively or neutrally associated with dog infection [31]. However, despite some controversy, other authors claim that ticks may be able to transmit the parasite [22,46]. In our study, the presence of ectoparasites in dogs has entered one final model, but with a negative association with seropositivity. This finding do not corroborate with studies from urban areas [31], but the work of Dantas-Torres and colleagues [45] with dogs from a rural community in northeastern Brazil have showed that ticks are not relevant as vectors of *Leishmania*. Our analysis revealed a weak negative correlation between the presence of ectoparasites and previous ectoparasite treatment, meaning that this intervention has been ineffectively performed in the study areas, and is probably either ineffective against sand flies.

Surprisingly, long fur was positively associated with dog seropositivity in our study by one of the models, because, according to the literature, short fur is considered as a strong predictor of canine leishmaniasis infection in Brazilian cities [28,31,47]. However this relationship did not hold in our data set. Possibly, the lower densities of rural dogs when compared to urban dogs [6] balance the detectability of shorthaired and longhaired dogs by sand flies. Thus, control measures in rural zones should not target any particular dog phenotype, contrary to the focus on shorthaired dogs proposed for urban populations [31].

Dogs older than one year were more likely to be infected, according to one GLMM scenario. Conversely, age did not enter the models and there was no difference in prevalence between young and adult dogs according to the Chi-square tests. Thus, we believe that age is not a strong predictor for *Leishmania* infection and dogs of all ages may be reservoirs in the study areas, and this is in general agreement with previous literature [31].

Sterilized dogs were found to be seropositive more frequently according to three scenarios. This is expected since gonadectomized dogs tend to roam or escape less and spend more time quiet [48–49] being more easily found by the vectors. Conversely, this would depend very much on sand fly density at different sites and peak times of sand fly feeding and of canine resting habits, since sand flies could easily feed on immobile dogs whether they sometimes roam or not.

Four other significant variables linked to dog management (dogs kept free) and vector attractiveness (presence of chickens, pigs and number of cats) entered final models as negatively associated with seropositivity. In the same way as aforementioned about gonadectomized dogs, free-roaming dogs are less sedentary and more difficult targets to vectors, whilst dogs living in restrict spaces spend more time quiet being more easily found, bitten and infected in these rural scenarios. Additionally, the negative association with dog mobility in one of the models indicates that being kept near a human dwelling is associated with increased risk for dog infection. However, in the review of Belo and coauthors [31] is mentioned that the general relationship is the inverse. Perhaps the detectability of dogs by the sand flies varies in some ways between cities and rural areas. Moreover, a purely peridomestic cycle of CVL may be happening in these scenarios, and warrants interesting future investigation.

Negative associations of dog seropositivity and the presence of chickens and pigs were revealed both by the GLMM models and the Chi-square tests. The strongly negative association between positive dogs and the presence of pigs in the households do not agree with most of the past findings. Previous studies have highlighted the presence of large domestic mammals as a positive influence on infection rates in dogs and humans [28–29,31,50]. Our data show that in these rural sites, the presence of large mammalian livestock (cattle and horses) did not influenced *Leishmania* seroprevalence in dogs, but the presence of pigs may be diverting sand fly bites away from dogs, and then reducing their infection rates. The pig is one of the preferred species as blood sources for the phlebotomines [51], but is apparently an incompetent reservoir [52]. This may facilitate the pig's zooprophylactic effect against CVL in rural zones, what seemingly happened in our case.

The negative association between the presence of chickens and seroprevalence reveals another evidence of the protective effect of some domestic species against leishmaniasis. This result is also quite controversial because some studies have also identified chickens as attractors for sand flies, implying that the presence of chickens ultimately produces increased infection rates in dogs and humans [27,28]. Nonetheless, a recent review of risk factors for visceral leishmaniasis in Brazil shows both positive and negative associations of chickens for canine infection [31]. Our results are pointed at the same direction that those aforementioned for pigs. Because chickens are the preferred vertebrate target for the vectors [27,53] but not suitable hosts for *Leishmania* parasites [54], they also divert the attention of the vectors from the dogs, thus reducing the bite rates and, consequently, the infection rates in dogs. The role of chickens as food sources, vector attractors, and zooprophylactic agents for leishmaniasis has previously been discussed [27,54–55], but only in the context of human infection. The number of cats followed the same pattern, being negatively associated with dog seropositivity (more cats per household are associated with less positive dogs). Cats have been found to be infected with *Leishmania*, can infect sand flies, but do not seem to develop high parasite burdens [22], and may also turn infection away from dogs when in high numbers and densities. Animal sheds and animals on which sand flies feed can increase sand fly density [36] but may also decrease infection prevalence and feeding on dogs and humans, so that the net impact on VL transmission depends on the balance of these outcomes. In our rural context, the balance appears to be favoring a zooprophylactic function of domestic fowl, swine and cats against CVL.

Since there was weak evidence of spatial clustering of seropositive dogs (exclusively for one study site), we believe that the disease is not being maintained in focal points throughout the study areas. Thus, control efforts must be equally employed and cover all properties in these scenarios. One possible explanation for the clustering at PESB is that its higher altitude and the steeper topography drives most human settlements to be located at some of the few valleys and flat areas in the region, resulting in spatial aggregation of households, and consequently, of their dogs.

The Atlantic Forest is a highly diverse and fragmented ecosystem located at the most developed region in Brazil [56]. Therefore, a strong presence of drivers of the dynamic alterations of disease, such as anthropogenic environmental change and increased contact between humans, wildlife, and domestic animals, is expected [4]. However, although governmental prevention programs exist for rural areas, interface areas such as rural zones around forest fragments have received little scientific or government attention in terms of health issues. Our findings show that the study areas should be considered endemic for canine leishmaniasis and that despite the recent trend toward urbanization of the disease [57], it is advisable that government health agencies return to look at rural zones beyond Brazilian urban areas if the aim is to widely control zoonotic leishmaniasis and other tropical diseases. Specifically, in our case, the study areas deserve more attention and thorough investigation through surveys of leishmaniasis in humans, reservoir dogs, wildlife and vectors. Additionally, higher prevalence areas such as PESB and RPPNMS should be prioritized by control programs. The Brazilian visceral leishmaniasis control program should expand the focus to embrace rural and ecosystem health in a holistic view of the problem, and the data presented here should be used as a reference for research and intervention in Brazilian human/wildlife interface areas.

Habitat loss and fragmentation and the subsequent decrease in biodiversity may cause, among many other effects, alterations in parasite ecology that result in increased rates of infection in wildlife [58–60]. Although we have no data on wildlife prevalence, the scenario of infected dogs living around and actually entering important biodiversity sites such as Atlantic Forest remnants [12] raises concerns about possible transmission to and from wild animals. Wild mammals can develop clinical signs of leishmaniasis, especially in stressful situations such as captivity [25], and the prevalence of the disease in many captive and free-ranging populations has been reported [22,24–26]. Therefore, the presence of infected reservoir dog populations around small forest fragments under strong human pressure may warrant persistence, circulation, and the possible, yet unknown, deleterious effects of leishmaniasis on the health and fitness of wild animals. Control programs should primarily involve a reduction in the dog population size and density, e.g., by sterilization (not culling), owner education, and legally limiting the number of dogs per rural household in settlements close to wildlife refuges and by restricting the access of dogs to protected areas, thus reducing the probability of disease transmission to and from humans and wildlife. Other measures that reduce attractiveness for sand flies, e.g. application of insecticides and keeping zooprophylactic species such as pigs or chickens around the house may be also recommendable in rural areas. Of course, the latter needs more investigation to detect general patterns before being adopted. Cats are especially not recommended because they cause great damage to wildlife species [61].

Finally, the results presented here suggest another important reason for controlling and monitoring dog populations around protected areas: the risk of visceral leishmaniasis for humans and wildlife. Our findings also highlight the need for additional surveys to detect epidemiological patterns of leishmaniasis in Brazilian rural zones, especially around wildlife-rich protected areas. Another noteworthy aspect of the results is the difference between the profile of risk factors and the results of most previous studies from urban areas. These differences are crucial for planning thoughtful and effective management initiatives that will protect the interdependent health of humans, domestic animals, and wildlife.

Acknowledgments

The authors wish to thank the people who helped in the field work, especially the workers from protected areas and the land owners whose health is in the aim of our effort. The anonymous reviewers also deserve our gratitude for valuable comments on earlier versions of the manuscript. Finally, we are grateful to IEF, for licenses conceded and field support, and to Ludimilla Zambaldi and Nathália Carvalho for assistance in the spatial analysis.

Author Contributions

Conceived and designed the experiments: NHAC AMOP RLM AGC MP. Performed the experiments: NHAC AMOP RLM APM AAR. Analyzed

the data: NHAC GRD AGC. Contributed reagents/materials/analysis tools: APM AAR. Wrote the paper: NHAC MP.

References

1. Daszak P, Cunningham AA, Hyatt AD (2000) Emerging infectious diseases of wildlife - threats to biodiversity and human health. Science 287: 443–449.

2. Patz JA, Daszak P, Tabor GM, Aguirre AA, Pearl M, et al. (2004) Unhealthy landscapes: policy recommendations on land use change and infectious disease emergence. Environ Health Perspect 112: 1092–1098.

3. Bradley CA, Altizer S (2006) Urbanization and the ecology of wildlife diseases. Trends Ecol Evol 22: 95–102.

4. Brearley G, Rhodes J, Bradley A, Baxter G, Seabrook L, et al. (2012) Wildlife disease prevalence in human-modified landscapes. Biol Rev 88: 427–442.

5. Young JK, Olson KA, Reading RP, Amgalanbaatar S, Berger J (2011) Is wildlife going to the dogs? Impacts of feral and free-roaming dogs on wildlife populations. Bioscience 61: 125–132.

6. Gompper ME (2014) Free-ranging Dogs & Wildlife Conservation. Oxford: Oxford University Press. 312 p.

7. Cleaveland S, Laurenson MK, Taylor LH (2001) Diseases of humans and their domestic mammals: pathogen characteristics, host range and the risk of emergence. Phil Trans R Soc Lond B 356: 991–999.

8. Rabinowitz PM, Gordon Z, Holmes R, Taylor B, Wilcox M, et al. (2005) Animals as sentinels of human environmental health hazards: an evidence-based analysis. EcoHealth 2: 26–37.

9. Cleaveland S, Meslin FX, Breiman R (2006) Dogs can play useful role as sentinel hosts for disease. Nature 440: 605.

10. Halliday JEB, Meredith AL, Knobel DL, Shaw DJ, Bronsvoort BMC, et al. (2007) A framework for evaluating animals as sentinels for infectious disease surveillance. J R Soc Interface 4: 973–984.

11. Lacerda ACR, Tomas WM, Marinho-Filho J (2009) Domestic dogs as an edge effect in the Brasília National Park, Brazil: interactions with native mammals. Anim Conserv 12: 477–487.

12. Paschoal AM, Massara RL, Santos JL, Chiarello AG (2012) Is the domestic dog becoming an abundant species in the Atlantic Forest? A study case in southeastern Brazil. Mammalia 76: 67–76.

13. Vanak AT, Gompper ME (2010) Interference competition at the landscape level: the effect of free-ranging dogs on a native mesocarnivore. J Appl Ecol 47: 1225–1232.

14. Gavgani SM, Mohite H, Edrissian GH, Mohebali M, Davies CR (2002) Domestic dog ownership in Iran is a risk factor for human infection with *Leishmania infantum*. Am J Trop Med Hyg 67: 511–515.

15. Woodroffe R, Prager KC, Munson L, Conrad PA, Dubovi EJ, et al. (2012) Contact with domestic dogs increases pathogen exposure in endangered African wild dogs (Lycaon pictus). PLoS One 7(1): e30099.

16. Alexander KA, McNutt JW (2010) Human behavior influences infectious disease emergence at the human–animal interface. Front Ecol Environ 8: 522–526.

17. Reithinger R, Espinoza JC, Llanos-Cuentas A, Davies CR (2003) Domestic dog ownership: a risk factor for human infection with *Leishmania (Viannia)* species. Trans R Soc Trop Med Hyg 97: 141–145.

18. Grimaldi GJ, Tesh R (1993) Leishmaniases of the New World: current concepts and implications for future research. Clin Microb Rev 6: 230–250.

19. Chappuis F, Sundar S, Hailu A, Ghalib H, Rijal S, et al. (2007) Visceral leishmaniasis: what are the needs for diagnosis, treatment and control? Nature Rev 5: 873–882.

20. Otranto D, Dantas-Torres F, Breitschwerdt EB (2009) Managing canine vector-borne diseases of zoonotic concern: part one. Trends Parasitol 25: 157–163.

21. Courtenay O, Quinnell RJ, Garcez LM, Dye C (2002) Low infectiousness of a wildlife host of *Leishmania infantum*: the crab-eating fox is not important for transmission. Parasitology 125: 407–414.

22. Quinnell RJ, Courtenay O (2009) Transmission, reservoir hosts and control of zoonotic visceral leishmaniasis. Parasitology 136: 1915–1934.

23. Dantas-Torres F (2007) The role of dogs as reservoirs of *Leishmania* parasites, with emphasis on *Leishmania (Leishmania) infantum* and *Leishmania (Viannia) braziliensis*. Vet Parasitol 149: 139–146.

24. Curi NHA, Miranda I, Talamoni SA (2006) Serologic evidence of *Leishmania* infection in free-ranging wild and domestic canids around a Brazilian National Park. Mem Inst Oswaldo Cruz 101: 99–101.

25. Luppi MM, Malta MC, Silva T, Silva FL, Motta RO, et al. (2008) Visceral leishmaniasis in captive wild canids in Brazil. Vet Parasitol 155: 146–151.

26. Souza TD, Turchetti AP, Fujiwara RT, Paixão TA, Santos RL (2014) Visceral leishmaniasis in zoo and wildlife. Vet Parasitol. 200: 233–241.

27. Alexander B, Carvalho RL, McCallum H, Pereira MH (2002) Role of the domestic chicken (*Gallus gallus*) in the epidemiology of urban visceral leishmaniasis in Brazil. Emerg Infect Dis 8: 1480–1485.

28. Moreira ED, Souza VM, Sreenivasan M, Lopes NL, Barreto RB, et al. (2003) Peridomestic risk factors for canine leishmaniasis in urban dwellings: new findings from a prospective study in Brazil. Am J Trop Med Hyg 69: 393–397.

29. Faye B, Bucheton B, Bañuls AL, Senghor MW, Niang AA, et al. (2011) Seroprevalence of Leishmania infantum in a rural area of Senegal: analysis of

risk factors involved in transmission to humans. Trans R Soc Trop Med Hyg 105: 333–340.

30. Almeida ABPF, Sousa VRF, Cruz FACS, Dahroug MAA, Figueiredo FB, et al. (2012) Canine visceral leishmaniasis: seroprevalence and risk factors in Cuiabá, Mato Grosso, Brazil. Rev Bras Parasitol Vet 21: 359–365.

31. Belo VS, Struchiner CJ, Werneck GL, Barbosa DS, Oliveira RB, et al. (2013) A systematic review and meta-analysis of the factors associated with Leishmania infantum infection in dogs in Brazil. Vet Parasitol 195: 1–13.

32. Membrive NA, Rodrigues G, Gualda KP, Bernal MVZ, Oliveira DM, et al. (2012) Environmental and animal characteristics as factors associated with American cutaneous leishmaniasis in rural locations with presence of dogs, Brazil. PLoS One 7: e47050.

33. Romero GAS, Boelaert M (2010) Control of visceral leishmaniasis in Latin America - A systematic review. PLoS Negl Trop Dis 4: e584.

34. Donalisio MR, Peterson T, Costa PL, Silva FJ, Valença HF, et al. (2012) Microspatial distributional patterns of vectors of cutaneous leishmaniasis in Pernambuco, northeastern Brazil. J Trop Med doi:10.1155/2012/642910.

35. Pinto IS, Ferreira AL, Valim V, Carvalho FS, Silva GM, et al. (2012) Sand fly vectors (Diptera, Psychodidae) of American visceral leishmaniasis areas in the Atlantic Forest, state of Espírito Santo, southeastern Brazil. J Vector Ecol 37: 90–96.

36. Quinnell RJ, Dye C (1994) Correlates of the peridomestic abundance of *Lutzomyia longipalpis* (Diptera: Psychodidae) in Amazonian Brazil. Med Vet Entomol 8: 219–224.

37. Rodríguez-Cortés A, Ojeda A, Todolí F, Alberola J (2013) Performance of commercially available serological diagnostic tests to detect *Leishmania infantum* infection on experimentally infected dogs. Vet Parasitol 191: 363–366.

38. Coura-Vital W, Ker HG, Roatt BM, Aguiar-Soares RDO, Leal GGA, et al. (2014) Evaluation of change in canine diagnosis protocol adopted by the Visceral Leishmaniasis Control Program in Brazil and a new proposal for diagnosis. PLoS One 9: e91009.

39. Grimaldi G, Teva A, Ferreira AL, dos Santos CB, Pinto IDS, et al. (2012) Evaluation of a novel chromatographic immunoassay based on Dual-Path Platform technology (DPP CVL rapid test) for the serodiagnosis of canine visceral leishmaniasis. Trans R Soc Trop Med Hyg 106: 54–59.

40. Skov T, Deddens G, Petersen MR, Endahl L (1998) Prevalence proportion ratios: estimation and hypothesis testing. Int J Epidemiol 27: 91–95.

41. R Core Team (2012) R: A language and environment for statistical computing. R Foundation for Statistical Computing, Vienna, Austria. Available: http://www.R-project.org. Accessed 20 December 2012.

42. Ayres M, Ayres Jr M, Ayres DL, Santos AAS (2007) Bioestat 5.0 - Aplicações estatísticas nas áreas das ciências bio-médicas. Publicações avulsas do Mamirauá. Available: http://www.mamiraua.org.br/downloads/programas. Accessed 2013 March 15.

43. von Elm E, Altman DG, Egger M, Pocock SJ, Gøtzsche PC, et al. (2007) The Strengthening the Reporting of Observational Studies in Epidemiology (STROBE) Statement: guidelines for reporting observational studies. Bull WHO 85: 867–872.

44. Coura-Vital W, Marques MJ, Veloso VM, Roatt BM, Aguiar-Soares RDO, et al. (2011) Prevalence and factors associated with Leishmania infantum infection of dogs from an urban area of Brazil as identified by molecular methods. PLoS Negl Trop Dis 5: e1291.

45. Dantas-Torres F, de Paiva-Cavalcanti M, Figueredo LA, Melo MF, Silva FJ, et al. (2010) Cutaneous and visceral leishmaniosis in dogs from a rural community in northeastern Brazil. Vet Parasitol 170: 313–317.

46. Dantas-Torres F (2011) Ticks as vectors of *Leishmania* parasites. Trends Parasitol 27: 155–159.

47. Coura-Vital W, Reis AB, Fausto MA, Leal GGA, Marques MJ, et al. (2013) Risk factors for seroconversion by Leishmania infantum in a cohort of dogs from an endemic area of Brazil. PloS One 8: e71833.

48. Maarschalkerweerd RJ, Endenburg N, Kirpensteijn J, Knol BW (1997) Influence of orchiectomy on canine behaviour. Vet Rec 140: 617–619.

49. Spain CV, Scarlett JM, Houpt KA (2004) Long-term risks and benefits of early-age gonadectomy in dogs. J Am Vet Med Assoc 224: 380–387.

50. Bern C, Courtenay O, Alvar J (2010) Of cattle, sand flies and men: a systematic review of risk factor analyses for South Asian visceral leishmaniasis and implications for elimination. PLoS Negl Trop Dis 4: e599.

51. Noguera P, Rondón M, Nieves E (2006) Effect of blood source on the survival and fecundity of the sandfly *Lutzomyia ovallesi* Ortiz (Diptera: Psychodidae), vector of Leishmania. Biomedica 26(Supplement 1): 57–63.

52. Moraes-Silva E, Antunes FR, Rodrigues MS, Julião FS, Dias-Lima AG, et al. (2006) Domestic swine in a visceral leishmaniasis endemic area produce antibodies against multiple Leishmania infantum antigens but apparently resist to L. infantum infection. Acta Trop 98: 176–182.

53. Dias FOP, Lorosa ES, Rebêlo JMM (2003) Blood feeding sources and peridomiciliation of *Lutzomyia longipalpis* (Lutz & Neiva, 1912) (Psychodidae, Phlebotominae). Cad Saúde Púb 19: 1373–1380.

54. Otranto D, Testini G, Buonavoglia C, Parisi A, Brandonisio O, et al. (2010) Experimental and field investigations on the role of birds as hosts of Leishmania infantum, with emphasis on the domestic chicken. Acta Trop 113: 80–83.

55. Caldas AJM, Costa JML, Silva AAM, Vinhas V, Barral A (2002) Risk factors associated with asymptomatic infection by *Leishmania chagasi* in north-east Brazil. Trans R Soc Trop Med Hyg 96: 21–28.

56. Ribeiro MC, Metzger JP, Martensen AC, Ponzoni FJ, Hirota MM (2009) The Brazilian Atlantic Forest: how much is left, and how is the remaining forest distributed? Implications for conservation. Biol Conserv 142: 1141–1153.

57. Arias JR, Monteiro PS, Zicker F (1996) The reemergence of visceral leishmaniasis in Brazil. Emerg Infect Dis 2: 145–146.

58. Cottontail VM, Wellinghausen N, Kalko EKV (2009) Habitat fragmentation and haemoparasites in the common fruit bat, *Artibeus jamaicensis* (Phyllostomidae) in a tropical lowland forest in Panama. Parasitology 136: 1133–1145.

59. Susán G, Esponda F, Carrasco-Hernández R, Aguirre AA (2012) Habitat fragmentation and infectious disease ecology. In: Aguirre AA, Ostfeld RS, Daszak P, editors. New directions in Conservation Medicine: applied cases of ecological health. New York: Oxford University Press. 135–150.

60. Xavier SCC, Roque ALR, Lima VdS, Monteiro KJL, Otaviano JCR, et al. (2012) Lower richness of small wild mammal species and Chagas disease risk. PLoS Negl Trop Dis 6: e1647.

61. Loss SR, Will T, Marra PP (2013) The impact of free-ranging domestic cats on wildlife of the United States. Nature Comm 4: 1396–1402.

17

Subcellular Localization of ENS-1/ERNI in Chick Embryonic Stem Cells

Sophie Blanc[1], Florence Ruggiero[1], Anne-Marie Birot[1], Hervé Acloque[1,2], Didier Décimo[3], Emmanuelle Lerat[4], Théophile Ohlmann[3], Jacques Samarut[1]*, Anne Mey[1]*¤

1 Institut de Génomique Fonctionnelle de Lyon, Université de Lyon, Université Lyon 1, CNRS UMR 5242, INRA USC 1370, Ecole Normale Supérieure de Lyon, Lyon, France, 2 Laboratoire de Génétique Cellulaire-INRA, ENVT, Castanet Tolosan, France, 3 CIRI, International Center for Infectiology Research, Université de Lyon, INSERM U1111, Ecole Normale Supérieure de Lyon, Lyon, France, 4 Université de Lyon, Lyon, France; Université Lyon 1, Villeurbanne, France; CNRS, UMR5558, Laboratoire de Biométrie et Biologie Evolutive, Villeurbanne, France

Abstract

The protein of retroviral origin ENS-1/ERNI plays a major role during neural plate development in chick embryos by controlling the activity of the epigenetic regulator HP1γ, but its function in the earlier developmental stages is still unknown. ENS-1/ERNI promoter activity is down-regulated upon differentiation but the resulting protein expression has never been examined. In this study, we present the results obtained with custom-made antibodies to gain further insights into ENS-1 protein expression in Chicken embryonic stem cells (CES) and during their differentiation. First, we show that ENS-1 controls the activity of HP1γ in CES and we examined the context of its interaction with HP1γ. By combining immunofluorescence and western blot analysis we show that ENS-1 is localized in the cytoplasm and in the nucleus, in agreement with its role on gene's promoter activity. During differentiation, ENS-1 decreases in the cytoplasm but not in the nucleus. More precisely, three distinct forms of the ENS-1 protein co-exist in the nucleus and are differently regulated during differentiation, revealing a new level of control of the protein ENS-1. In silico analysis of the Ens-1 gene copies and the sequence of their corresponding proteins indicate that this pattern is compatible with at least three potential regulation mechanisms, each accounting only partially. The results obtained with the anti-ENS-1 antibodies presented here reveal that the regulation of ENS-1 expression in CES is more complex than expected, providing new tracks to explore the integration of ENS-1 in CES cells regulatory networks.

Editor: Anton Wutz, Wellcome Trust Centre for Stem Cell Research, United Kingdom

Funding: This work was supported by grants from the ANR (Agence Nationale pour la Recherche) Genanimal (ANR-06-GANI-007-01), from the ANR Blanc (ANR-09-BLAN-0141) http://www.agence-nationale-recherche.fr/. The funders had no role in study design, data collection and analysis, decision to publish, or preparation of the manuscript.

Competing Interests: The authors have declared that no competing interests exist.

* E-mail: jacques.samarut@ens-lyon.fr (JS); anne.mey@lyon.inra.fr (AM)

¤ Current address: Laboratoire CarMeN, INSERM1060, Université Lyon 1, INRA 1235, INSA Lyon, Hospices Civils Lyon, Faculté de Médecine Lyon-Sud, Oullins France

Introduction

Endogenous retroviruses (ERVs) are known to play an important role in the expression of their host genome, notably during the first developmental stages when totipotent [1] or pluripotent cells [2] adopt new cell fates. ERV's promoters act as enhancers in different cellular models and lineages [3]. More rarely, ERVs introduce coding sequences that are adopted by the host genome to play a major role in host survival. This is the case for Syncitin, involved in formation of the mammalian placenta [4], and for ERNI, also called ENS-1 [5], that controls the timing of the neural plate emergence during chick embryonic development [6]. More precisely, ENS-1/ERNI acts as a boundary element between the epigenetic regulator HP1γ and the protein complex that is recruited to the promoter of the neural plate inducer *Sox2* before its expression. This linker property involves two distinct motifs in the protein, the HP1box engaged with HP1γ and the coiled-coil domain interacting with other proteins recruited to the promoter. The repression, mediated by HP1γ, on the promoter of *Sox2* is released in the prospective neural plate due to competition

between ERNI and the newly synthesized protein BERT, another coiled-coil domain protein that does not bind HP1 [7].

In addition, *Erni* is expressed earlier during the chick developmental process notably in the hypoblast, in the pluripotent epiblast [8] and in its derived embryonic stem cells, cultivated *in vitro* [9], where it was called *Ens-1*. Silencing of the gene occurs later, as final differentiation is achieved [8,9]. In the epiblast, this expression pattern is managed by the pluripotency transcription factor Nanog and by a combination of Gata and Ets transcription factors that are expressed in the epiblast, in the hypoblast and in the prospective neural plate [8]. It is probable that the ERV *Ens-1/Erni* controls transcription of the host's genome in pluripotent cells either directly with its promoter sequences spread over the genome [5], or indirectly by controlling HP1γ on its target genes. Among them *Sox2* is also known as a key-player in the maintenance of pluripotent cells in mammals [10,11], but other genes, mainly involved in cell proliferation control, have also been described in mouse ES cells as HP1γ target genes [12].

Despite its important role during chick development, demonstrated using transient transfections, the expression of the *Ens-1* gene has only been followed at the transcriptional level and the endogenous protein ENS-1 has never been observed.

In this study, we raised ENS-1 specific antibodies in order to follow ENS-1 expression in CES cells and during their differentiation. Our results reveal that the distribution of ENS-1 is a new level of its regulation.

Materials and Methods

Cell culture and DNA transfection

CES are pluripotent stem cells that were isolated from chick embryonic epiblast [13] and expressing the pluripotency supporting genes [14]. CES were cultivated as indicated previously [15]. Plasmid DNA transfections were performed using Lipofectamine 2000 (Invitrogen) in cell cultures at 30% confluency according to the manufacturer's instruction. Stable overexpression was obtained using linearized DNA plasmids, and stably transfected cells were selected by culture in medium supplemented with puromycin (200 µg/ml). COS7 cells were cultivated with DMEM supplemented with 6% fetal calf serum, 1% penicillin/streptomycin, 1% glutamin.

Antibodies production

The anti-Ens-1 antibody used for immunofluorescence experiments was obtained by immunization of mice with the recombinant ENS-1 protein produced in *Escherichia coli* as follow. The full-length coding sequence of the ENS-1 gene was fused in C-terminal with a 6x-histidine Epitope tag in a pET-22 expression system and transformed in Rosetta (DE3)pLys Competent Cells (Novagen). Protein production was induced with 1 mM IPTG for 3 hours at 37°C and immediately checked by direct loading of cells on a SDS-polyacrylamide gel followed by Coomassie blue staining. Cells were then lysed by sonication and the different fractions (soluble and insoluble cell extracts) were kept for further analysis. As ENS-1 is predominantly detected in inclusion bodies, the insoluble fraction was isolated by centrifugation and solubilized in 8 M urea, 100 mM NaH$_2$PO$_4$, 10 mM Tris HCl (pH 8.0). ENS-1 His tag protein was purified on Ni-NTA columns (Qiagen) and refolded by overnight dialysis in 50 mM NaH$_2$PO$_4$, 300 mM NaCl, 10 mM Tris HCl (pH 8.0). Purification efficiency was checked by Coomassie blue staining of a SDS-polyacrylamide gel loaded with the purified ENS-1 protein before its use as immunogen for antibody production. Production of the monoclonal antibody was carried out by Covalab (France) and immunizations were performed using complete Freund adjuvant according to the relevant legislation and following protocols approved by the ethic committee of the University of Bourgogne (France). Antibody producing hybridomas were first selected by Covalab for their reactivity with ENS-1 coated wells in ELISA tests. Positive hybridomas were next screened for use in western-blotting and in immunostaining. Their reactivity was tested in CES cells and in cells devoid of the gene Ens-1 as negative control. The selected IgG (from clone 16h4g6e4) was purified on protein A-coupled to Sepharose and called 16h4 in the results part. This unique clone worked in immunostaining and its specificity is demonstrated in the results section.

As none of the monoclonal antibodies produced worked in western-blot, we developed polyclonal antibodies in rabbits using peptides of 16–20 amino acids isolated from the ENS-1 sequence as immunogens. Design and purification of the peptides, rabbit immunizations and antibody purification from the serum were carried out by Eurogentec using its proprietary adjuvant combination and following protocols approved by the ethic committee of the CER groupe (Belgium) according to the European legislation. One of the three tested peptides (named 3807: C-DRIRVLQNEARTRAGK-CONH$_2$) gave a specific polyclonal antibody working in western blot as described in the results section. This antibody did not work in immuno-staining. The 3807 peptide (DRIRVLQNEARTRAGK) is localized upstream the coiled-coil domain in the N-terminal part of the protein and spans amino-acids in position 47 to 62 of the protein (GenBank: AAK06824.1).

DNA constructs

For transient overexpression experiments, chick HP1γ and Ens-1 (Genbank: NM_001080873) were cloned in a pCi expression vector (Promega) modified to introduce a FLAG tag at the N-terminal part of the protein. HP1γ (Genbank: NM_204643) was amplified from chicken ES cDNA. For *in vitro* translation, Ens-1 was cloned in a pGBKT7 expression vector (Clontech). Genes expressed in fusion with a GFP protein at their N-terminal part were cloned in pEGFP-C1 vector (Clontech).

For stable overexpression, Ens-1 was cloned in a bicistronic construct (pCX Ens-1-HA-ires-puro) to express ENS-1 in fusion with two HA-tags at the C-terminal part of the protein. In this construct Ens-1 was separated from the puromycin resistance gene by an IRES, allowing translation of both proteins from a common mRNA transcribed under the control of a CAG promoter (a chicken β-actin promoter combined with a CMV enhancer) to ensure a strong expression [16].

To test the transcriptional activity of proteins using the CAT reporter assay, the genes were cloned in the pM vector (Clontech) to express HP1γ or ENS-1 in fusion with a Gal4 DNA binding domain at their N-terminal part. The reporter plasmid pG4-TK-CAT (Chloramphenicol Acetyl Transferase) contains six copies of the Gal4 binding sites upstream of a TK promoter to control CAT expression. pG4-TK-CAT was a gift from Dr E. Manet and has been described elsewhere [17]. The renilla luciferase reporter vector pRL-CMV, used as control, was from Promega.

Targeted mutations of ENS-1 were performed using the Quick Change Mutagenesis Kit (Stratagene) following the manufacturer's instructions. Mutation of the HP1 box (PxVxL) was performed using the following primers:

Sense:TTGATGAATGGATTAGCCACAGCCAGAGCCGA-GAAATTAGTTAAC; antisense:GTTAACTAATTTCTCGGC-TCTGGCTGTGGCTAATCCATTCATCAA. The control mutation was performed on a PxVxL motif that was not involved in the interaction of ENS-1 with HP1γ as assessed using a two-hybrid assay in yeast described previously [7]. The primers used were:

sense:TTGATGAATGGATTAGCCACAGCCAGAGCCGA-GAAATTAGTTAAC,

antisense: GTTAACTAATTTCTCGGCTCTGGCTGTGG-CTAATCCATTCATCAA.

In both PxVxL mutants the proline, valine and leucine amino acids were replaced by an alanine.

Deletion of the coiled-coil domain from position 238 to position 411 of the Ens-1 coding sequence was obtained by PCR amplification in two fragments of the sequence flanking the coiled-coil domain. Primers were designed to introduce an EcoRI motif that was used to ligate both amplified fragments. The fragment upstream of the coiled-coil domain was amplified using the following primers:

sense:ACTCGAGCCACCATGAGCAACAGTATGGCC;

antisense:ACCGGTGTCCTTACACGTGAACTCCACAGC-TGCTTT. To amplify the fragment downstream of the coiled-coil domain, the primers were: sense: ACGTGTAAGGACACCGGT;

antisense: TACGCGTTCAGCTCCCCTTGAGCTT. The deletion of the coiled-coil domain lowers the molecular weight of the ENS-1 protein by 5.5 kDa.

To measure the activity of both start codons in the ENS1 5' UTR, fragments were cloned downstream to the T7 promoter in the p0renilla vector previously described [18] as illustrated in the results. The human β-globin 5'-UTR with the authentic initiation codon was obtained by hybridizing two synthetic oligodeoxyribonucleotides (Eurogentec) downstream to the T7 promoter to generate the pGlobin-renilla vector used as control (CTR) [18].

Pull-down experiments

Chicken HP1γ cloned in the pCi-flag plasmid was produced in COS7 cells upon transfection with Exgene 500 (Fermentas) following the manufacturer's instructions. The next day, lysis of COS7 cells was performed (lysis buffer: 50 mM Tris-HCl pH 7.4; 150 mM NaCl; 1 mM EDTA; 1%Triton X-100; 1X Complete protease inhibitor cocktail (Roche); 1X Halt phosphatase inhibitor cocktail (Thermo Scientific)), the lysate was centrifuged and the supernatant was incubated with [^{35}S]-Met (Amersham) to label ENS-1 synthesized in vitro using a TnT Coupled Reticulocyte Lysate Systems kit (Promega) and pGBKT7-Ens1 plasmid. The mixture was incubated for 5 hours at 4°C on a rotating shaker. The flag-HP1γ protein was precipitated using Anti-flag M2 affinity gel (Sigma) incubated with the protein mixture overnight at 4°C on a rotating shaker. After centrifugation the agarose beads were washed four times in washing buffer (50 mM Tris-HCl pH 7.4; 150 mM NaCl). Bound proteins were eluted with loading buffer at 100°C and analyzed by SDS-PAGE and fluorography using a STORM Phosphorimager.

Immunofluorescence labeling

Cells were seeded on glass cover slides previously coated with gelatin 1% in 24 wells plates and cultivated to reach 80% confluency. CES cells were seeded with irradiated feeder cells and cultivated on the cover slides 48 h before analysis except for differentiated CES. In this case differentiation was induced 24 h after seeding without feeder cells, and using 10^{-6}M retinoic acid as previously described [8] for the indicated period of time. Culture medium was removed and fixation was performed with 2.5% paraformaldehyde in PBS for 20 min at room temperature. After washing tree times with PBS, the cells were saturated and permeabilized with saturation buffer (PBS, 0.1% Triton X-100, 10% fetal calf serum). Primary antibodies diluted in saturation buffer were incubated 1 h at room temperature. After washing three times with PBS, the secondary antibody diluted in saturation buffer was added for 1 h at room temperature. Following washing three times with PBS, cover slides were mounted on a slide using Gelmount (Biomeda) supplemented with Draq5 (Biostatus) diluted 1/1000. For co-localization studies, cells were incubated with a mixture of the anti-ENS1 monoclonal antibody 16h4 (5 µg/ml) produced in mice and the rabbit anti-human HP1γ antibody from Abcam (Ab10480, 1/200). After washing, the cells were incubated with both Alexa secondary antibodies (1/1000) simultaneously. F(ab')2 goat anti-mouse IgG conjugated with Alexa 488 or Alexa 555 or F(ab')2 goat anti-rabbit IgG conjugated with Alexa 555 were from Molecular Probes. In Fig. S1 immunostaining of CES was performed using the 42s2 mouse monoclonal antibody (Upstate) directed against HP1γ in addition to the Ab10480 (Abcam).

Acquisition of the images was performed with a Zeiss LSM510 Confocal microscope using a 63× (NA 1.4) Plan NeoFluor objective (PLATIM, UMS 3444 Biosciences Gerland-Lyon Sud). Each channel was imaged sequentially using the multi-track recording module before merging. Experiments were performed at least three times and gave similar results.

Electron microscopy

CES cells dissociated with trypsin-EDTA (Gibco) were pelleted by centrifugation and fixed with PFA 4% in PBS for 30 min in ice. These fixative conditions decreased the detection of the ENS-1 protein observed by immunofluorescence but were required to preserve most of the cell structures upon the following treatment.

Cells were then washed in PBS, mixed with 1% low melting point agarose and aspirated into a syringe. After agarose solidification, cells embedded into agarose were extruded from the syringe and the resulting extruded rods were cut into 3–5 mm^3 cylinders. Samples were washed in PBS at room temperature for 30 min, dehydrated in 70% ethanol and embedded in LR White resin (Electron Microscopy Sciences). Ultrathin sections were mounted on nickel grids and immunolabelling performed by floating the grids onto drops of the different solutions. We used PBS-1% BSA as saturation buffer. The anti-ENS-1 antibody (clone 16h4) and the anti-human androgen receptor antibody (441, Santa Cruz) used as IgG1 negative control were diluted in saturation buffer and incubated 1 h with the cells. After washing in PBS the cells were incubated for 1 h at room temperature with donkey anti-mouse IgG antibody conjugated with 6 nm diameter gold particles (EM grade, Jackson Immunoresearch) diluted in saturation buffer. After washing in PBS, the labeling was fixed with PBS-2% glutaraldehyde for 10 min, washed three times with PBS and once with water. Ultrathin sections were then contrasted with uranyl acetate and lead citrate, and examined with a Philips CM 120 electron microscope equipped with a Gatan Orius 200 2Kx2K digital camera (Centre Technique des Microstructures, Université Lyon 1, Villeurbanne, France).

Protein extractions and western blot

Whole cell lysates were prepared using RIPA lysis buffer (1% Triton X-100; 0.5% Sodium deoxycholate; 0.1% SDS; 150 mM NaCl; 0.1 mM DTT; 1 mM EDTA; 1 mM N-ethylmaleimide, Complete proteases inhibitor (Roche) in Tris-HCl 50 mM, pH 7.4). Following dissociation using trypsin-EDTA and washing in PBS, cells were suspended in cold RIPA lysis buffer (40×10^6 cells/ml) and incubated 30 min in ice. The lysate was centrifuged at 10000 g for 30 min at 4°C. The supernatant was harvested and protein concentration was measured using Bradford reagent (Sigma) and BSA in the standard curve. The indicated protein quantities were mixed with loading buffer (10% glycerol, 2% SDS, 0.7 M β-mercaptoethanol, 0.01% bromophenol blue in 62.5 mM Tris-HCl pH 6.8) and denaturation was performed 3 min at 100°C.

Fractionation experiments were performed using two protocols. In the first one, cells were disrupted using detergent before the cytoplasm and the nucleus were separated. Dissociated cells were suspended (60×10^6/ml) in cold buffer A (340 mM sucrose, 60 mM KCl, 15 mM NaCl, 2 mM EDTA, 0.5 mM EGTA, 0.1 mM DTT, 0.65 mM spermidine, 0.1% Triton X-100, Complete EDTA (Roche), 10 mM N-ethylmaleimide in 15 mM Tris-HCl pH7.4) and incubated 1 min in ice. The cytoplasm and the nuclei were separated by centrifugation 1500 g for 5 min at 4°C. The cytoplasm corresponding to the supernatant was clarified from residual nuclei by centrifugation 13000 g at 4°C for 5 min and the supernatant was harvested (cytoplasmic fraction, C). The nuclei containing pellets were suspended in buffer A and washed by deposition on a buffer A cushion in a 14 ml falcon tube followed by centrifugation at 9400 g for 10 min at 4°C. The nuclei were disrupted using the hypotonic buffer B (2 mM EDTA,

0.5 mM EGTA, 0.1 mM DTT, Complete EDTA (Roche), 10 mM N-ethylmaleimide in water). After 30 min incubation in ice, nuclei disruption was checked using a microscope. Next the soluble and insoluble nuclei components were separated by centrifugation 18000 g for 5 min at 4°C. Protein concentration was measured using Bradford reagent (Sigma) in the cytoplasmic fraction and in the soluble fraction of the nucleus containing fractions and the indicated quantities were mixed with protein loading buffer. The insoluble nucleus fraction was suspended in buffer A and mixed with the loading buffer (N) without measurement of the concentration but the volume corresponding to 6×10^6 cells was loaded to the gel.

In the second fractionation protocol the cells were disrupted without detergent according to the protocol provided by Abcam with some modifications. Briefly, cells were dissociated in PBS using a scraper, rapidly pelleted by centrifugation, re-suspended in cold fractionation buffer (250 mM sucrose, 20 mM Hepes, 10 mM KCl, 1.5 mM MgCl2, 1 mM EDTA, 1 mM EGTA, 1 mM DTT, 1 mM N-ethyl maleimide, Complete protease inhibitor (Roche)) and passed through a 26G needle before 30 min incubation on ice. Cell disruption was checked using a microscope. The nuclei were separated from the cytoplasm by centrifugation 1500 g for 5 min at 4°C. The supernatant contained the cytosol, the cytoplasmic membrane and mitochondria, the pellet contained nuclei. The supernatant was clarified by centrifugation 18000 g, 5 min at 4°C to separate the mitochondria in the pellet from the cytosol and the cytoplasmic membrane in the supernatant. The cytosol and the cytoplasmic membrane were separated by ultracentrifugation 100000 g for 1 h at 4°C. The supernatant corresponding to the cytosol (Cs) was harvested and the pellet was washed once in the fractionation buffer before suspension in RIPA lysis buffer (Cp). The nuclei were washed as indicated in the first protocol on a cushion of fractionation buffer, suspended in RIPA lysis buffer and sonicated (N).

For protein dosage and loading, all the fractions were treated as indicated for whole cell lysates.

Proteins separated on 10% SDS-polyacrylamide gels were transferred on Hybond ECL membranes (GE Healthcare). Transfer efficiency of the proteins was checked by transient staining of the membrane using Ponceau's red (Sigma). Blots were submitted to Western analysis using the following antibodies: anti-HP1γ (clone 2MOD-1G6AS, Euromedex), anti-HA (HA.7, Sigma), anti-β tubulin (Sigma), anti-HSP90 (AC88, Enzo), anti-LaminB1 (Abcam), 2MeH3K4 (Upstate), 3807 anti-ENS-1 polyclonal antibody. Anti-mouse and anti-rabbit secondary antibodies coupled with peroxidase were from Sigma. Saturation of the membrane was performed in saturation buffer (0.1% Tween, 5% low fat milk in PBS) 1 h at room temperature. Antibodies were diluted in saturation buffer and incubated with the membrane at room temperature for 1 h. After washing 3 times for 10 min in PBS-0.1% Tween, the membrane was incubated 1 h at room temperature with the HRP-conjugated secondary antibodies diluted in saturation buffer. Anti-mouse and anti-rabbit HRP conjugates were from Promega. After washing, peroxydase activity was revealed using the Amersham ECL western blotting detection reagent (GE Healthcare) as substrate and X-ray films (Fuji).

CAT assay

CES cells seeded in 12-well plates were co-transfected at 70% confluency using Lipofectamine 2000 (Invitrogen) as described by the manufacturer with a mixture of the following plasmids: pM plasmid (20 ng) encoding for proteins in fusion with Gal4 DNA binding domain (GBD-proteins), pG4-TK-CAT reporter plasmid (300 ng) allowing the tethering of the GBD-proteins upstream the TK promoter that controls the expression of the CAT gene, pCi plasmid (200 ng) allowing the expression of additional proteins and pRL-CMV plasmid (20 ng) as control to normalize transfection efficiency. After 24 h culture, the medium was removed and the cells were washed twice with PBS. Cell lysates were prepared and CAT expression was measured using the CAT ELISA kit from Roche according to the manufacturer's instructions. Renilla luciferase luminescence was measured on templates from the same lysates using the Renilla Luciferase Assay System (Promega) according to the manufacturer's protocol and using a Mithras LB940 microplate reader (Berthold technologies).

In silico analysis of the Ens-1 gene copies in the chicken genome

We used the sequence of Soprano [19] as a query to search the chicken genome (version WASHUC 2, retrieved from Ensembl http://www.ensembl.org/index.html) using BLASTN [20]. We recovered 78 copies among which 25 were very conserved compared to the reference sequence. We used ORF Finder (http://www.ncbi.nlm.nih.gov/gorf/gorf.html) on the 25 conserved sequences to determine the presence of ORFs and to retrieve the subsequent protein sequences. The coiled coil domains were predicted on the protein sequences using the MultiCoil program [21].

In vitro transcription

Capped RNAs were obtained by using 2 µg of linear DNA template mixed with 20 U of T7 RNA polymerase (Promega Co., Madison, WI, USA), 40 U of RNAsin (Promega Co, Madison, WI, USA), 10 mM rATP, rUTP, rCTP, 0.48 mM rGTP, 30 mM DTT in transcription buffer [40 mM Tris–HCl (pH 7.9), 6 mM MgCl2, 2 mM spermidine and 10 mM NaCl], the m7GpppG cap analogue (Invitrogen, Co) was added to a final concentration of 1.92 mM as previously described [22]. The transcription reaction was carried out at 37°C for 1 h and the RNAs were precipitated with LiCl at 2.5 M final concentration. The integrity of the RNAs was checked by electrophoresis on non-denaturating agarose gels and their concentration was quantified by spectrophotometry at 260 nm using Nanodrop (NanoDrop Technologies, Wilmington, Delaware, USA).

In vitro translation

In vitro transcribed RNAs were translated in 10 µl of the supplemented untreated RRL 50% (v/v) (Promega Co., Madison, WI, USA) in the presence of KCl (75 mM), MgCl$_2$ (0.5 mM), 20 µM of amino acids mix minus methionine and 0.6 µCi of [^{35}S]-methionine (GE Healthcare Life Sciences) for 30 min at 30°C. Reactions were stopped with 2× SDS-loading buffer and the products were resolved by 15% SDS–PAGE. Gels were dried and subjected to autoradiography using Biomax films (Eastman Kodak Co.).

Results

ENS-1 can control HP1γ activity in ES cells

The control exerted by ENS-1 on the transcriptional regulation mediated by HP1γ has been demonstrated in the neural plate [7]. Here we investigated whether this activity might also occur in chicken ES cells (CES). To this end we used a CAT reporter gene placed under the control of a minimal VP16 promoter downstream of the DNA motif recognized by the Gal4 protein. Cells were co-transfected with a pM vector encoding for HP1γ in fusion with the DNA binding domain of Gal4 (GBD) to target its interaction with the CAT reporter plasmid. In line with results from another study [23], Figure 1A shows that transcription was repressed when tethering HP1γ to the promoter. Transfection of CES cells with the previous cocktail supplemented with an ENS-1 expression vector restored the transcriptional activity to levels obtained in the absence of HP1γ while ENS-1 mutated in the HP1 box had no effect (Fig. 1A). To test whether ENS-1 acts directly on the promoter as a transcriptional activator, ENS-1 in fusion with the GBD was co-transfected with the CAT reporter plasmid. Transcription was not affected by the recruitment of ENS-1 to the promoter (Fig. 1B) confirming that the effect of ENS-1 on transcription was dependent on its interaction with HP1γ.

Therefore ENS-1 can directly interact with HP1γ and modulate its function as a transcriptional regulator in CES.

ENS-1 is both a cytoplasmic and nuclear protein

Since HP1γ is a nuclear protein, we looked for the distribution of the endogenous protein ENS-1 in these cells. To this end a monoclonal antibody was produced against the recombinant protein ENS-1 (as described in Material and Methods) and tested for use in confocal microscopy. Among a dozen of clones reacting with the immunogen, only one was suitable for immunostaining. Specificity for ENS-1 was first demonstrated by transfection of CES with the ENS-1-GFP fusion protein. As shown in Figure 2A, the antibody co-localized with ENS-1-GFP but not with GFP alone used as control. Secondly, as Ens-1 is restricted to the galliform species [5], we used STO murine embryonic fibroblasts as negative control. No signal was observed with the antibody in these cells (Fig. 2B). Therefore the antibody is specific for ENS-1 and represents a relevant tool to study expression and distribution of this protein in CES. As shown in Figure 2C almost all the CES cells expressed ENS-1 but inside the cells the labeling was not uniform, notably in the cytoplasm when compared with the nucleus (see Fig. 2D). In the cytoplasm the distribution of ENS-1 appeared as organized, forming a scaffold of spots and stretches that sometimes crossed-over the nuclear membrane. In the nucleus that was labelled with Draq5, a stoechiometric DNA-intercaling agent, the distribution of ENS-1 was more uniform. Nucleolus that were not labeled with Draq5 were poorly stained with the anti-ENS-1 antibody (Fig. 2D).

The distribution of ENS-1 was further examined by electron microscopy with the same antibody (Fig. 3). To preserve the intracellular structures this approach required fixative conditions that were stronger than those used for confocal analysis and thus altered the ENS-1 epitope recognized by the 16h4 antibody. A compromise was found using 4% PFA only. Most of the cell structure was preserved as well as the ENS-1 epitope but some cytoplasmic structures were lost, leaving empty areas in the cytoplasm that likely correspond to the endoplasmic reticulum (Fig. 3). ENS-1 was abundantly detected at the periphery of these areas (Fig. 3A) and in the nucleus (Fig. 3B). Examination of the cytoplasmic compartment shows that ENS-1 accumulates at the cytoplasmic membrane (Fig. 3A). In the nucleus, the protein was found in regions that were more or less dense to electrons, and corresponding to chromatin and to non-chromatin spaces respectively, suggesting that ENS-1 location in the nucleus is not restricted to protein complexes formed on DNA (Fig. 3B). No particles were observed in controls where anti-ENS-1 antibody was replaced by an irrelevant antibody (Fig. 3D) or in the absence of primary antibody (Fig. 3C).

Therefore, we conclude that the protein ENS-1 is expressed in all CES cells and is largely present in the cytoplasm and in the nucleus.

ENS-1 sparsely co-localizes with HP1γ in the nucleus and in the cytoplasm

To explore the context in which the interaction between the endogenous proteins ENS-1 and HP1γ may occur, co-localization experiments were performed in CES. Consistent with a previous study in mouse ES cells [24], immunostaining of HP1γ heterogeneously formed a speckled and diffuse pattern in the nucleus of CES (Fig. 4A). Yellow spots in the orthogonal sections of the cells (Fig. 4B, panel a) indicated that both proteins co-localized in interphasic nuclei while other ENS-1 and HP1γ molecules were visualized as neighboring green and red spots respectively. Merging with the DNA staining, in blue, produced white spots reflecting co-localizations of ENS-1 with HP1γ in chromatin (Fig. 4B, panel b) and a majority of purple spots corresponding to HP1γ on chromatin. These results are in agreement with the ability of ENS-1/HP1γ heterodimers in the

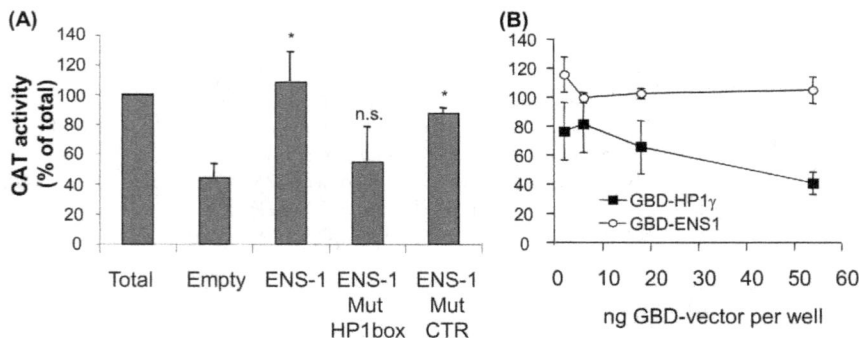

Figure 1. Transcriptional repression by HP1γ is modified by its interaction with ENS-1 in CES. CES cells were co-transfected with the pG4-TK-CAT reporter plasmid and the plasmid encoding for the indicated protein in fusion with a Gal4-DNA binding domain (GBD). An equal amount of pRL-CMV plasmid encoding for Renilla Luciferase was added to each well for normalization. (A) Transcription inhibition of the CAT reporter gene by GBD-HP1γ in the presence of pCi expression vectors encoding for ENS-1, for ENS-1 mutated in the HP1box. Mutation in an irrelevant PxVxL sequence was used as control. Maximal inhibition by HP1γ was given by co-transfection with the empty pCi vector (Empty). (B) Promoter activity was compared in the presence of GBD-ENS-1 and GBD-HP1γ. In (A) and (B) the promoter activity was assayed by quantifying the amount of CAT protein expressed by CAT-ELISA. Data were normalized on the basis of the luciferase activity in each well. Results represent CAT expressions in the presence of GBD-fusion proteins as percent of the maximal expression obtained with GBD alone in the same conditions. They are mean of four (A) and two (B) independent experiments +/− SD. Statistics are results of t test relative to the value obtained in the Empty condition: * p<0.005; n.s. not significant (p>0.1). In (B), CAT activity difference between 2 and 6 ng of GBD-ENS-1 vector is not significant.

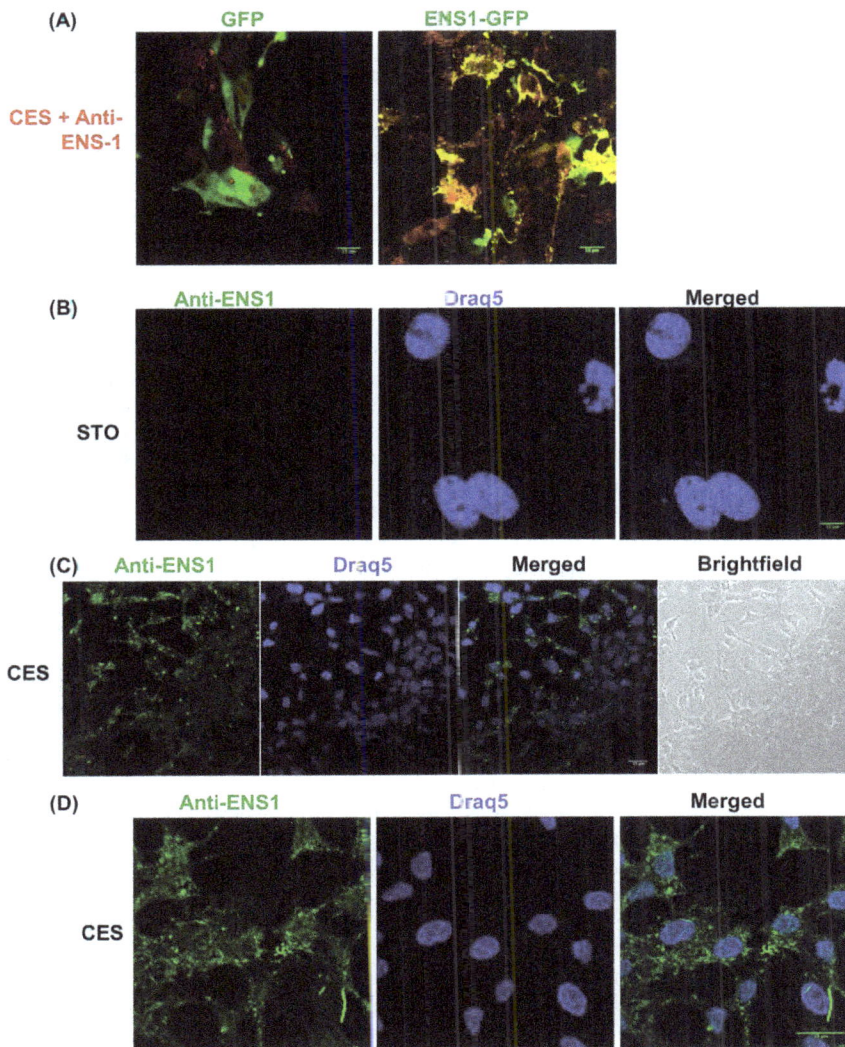

Figure 2. Localization of ENS-1 in the cytoplasm and in the nucleus of CES. (**A**) CES transfected with an expression vector for GFP alone or in fusion with the ENS1 protein were incubated with the 16h4antibody produced against ENS1. An anti-mouse antibody coupled with Alexa555 was used as secondary antibody (red) while GFP proteins were in green. Overlapping signals gave a yellow staining. In (**B**), (**C**) and (**D**) immunostaining with the 16h4 antibody was detected with an Alexa488 anti-mouse antibody (green) and the nucleus was localized by DNA staining with Draq5 (blue). (**B**) The 16h4 anti-ENS-1 antibody does not label STO murine embryonic fibroblasts (used as negative control). (**C**) ENS-1 detected in CES cells cultivated in proliferating conditions. (**D**) ENS-1 in CES cells at a higher magnification. All observations were performed by confocal microscopy. Is shown: the signal over the background obtained with the secondary antibodies in the absence of primary antibody. Scale bar 15 μm. The images are representative of three independent experiments.

embryo to specifically regulate gene transcription [7]. Panels c, d and e, showing each of the labeling that were merged in panels a and b (Fig. 4B), revealed that ENS-1 was present in all the thickness of the nucleus (panel c) while HP1γ and the chromatin had a more restricted distribution. Notably HP1γ proteins were concentrated in the center of the cells (panel d) while chromatin was mostly observed in the upper half (panel e). Measurement of the fluorescence intensity along the longitudinal axis of two cells confirmed that most of the HP1γ dots did not overlap with those of ENS-1 (Fig. 4C, upper panel) and that co-localization between both proteins was a restricted event that also occurred in regions with low DNA density (Fig. 4C, lower panel). These results suggest that ENS-1/HP1γ heterodimers can exist independently on the interaction with chromatin. In agreement with this hypothesis, pull-down experiments performed between in vitro translated ^{35}S-

ENS-1 and cell lysates from COS7 cells overexpressing chicken HP1γ demonstrate that the interaction between both proteins is direct, involving the HP1 box of the ENS-1 protein but not its coiled-coil domain as illustrated by the use of mutated ENS-1 proteins (Fig.4D). During mitosis, HP1γ is released from the chromatin and is distributed diffusely throughout the cell [25]. Accordingly, in prometaphasic cells when the nuclear membrane was dissolved, HPγ and ENS-1 were progressively released from the condensing chromatin and accumulated in the cytoplasm (Fig. 4E). Co-localizations were observed (Fig. 4E panel b) but most of the proteins were maintained separated as illustrated by the few number of yellow spots in comparison with the red and green ones (see panels a and c, Fig. 4E). In interphasic cells some HP1γ molecules were also detected in the cytoplasm as confirmed

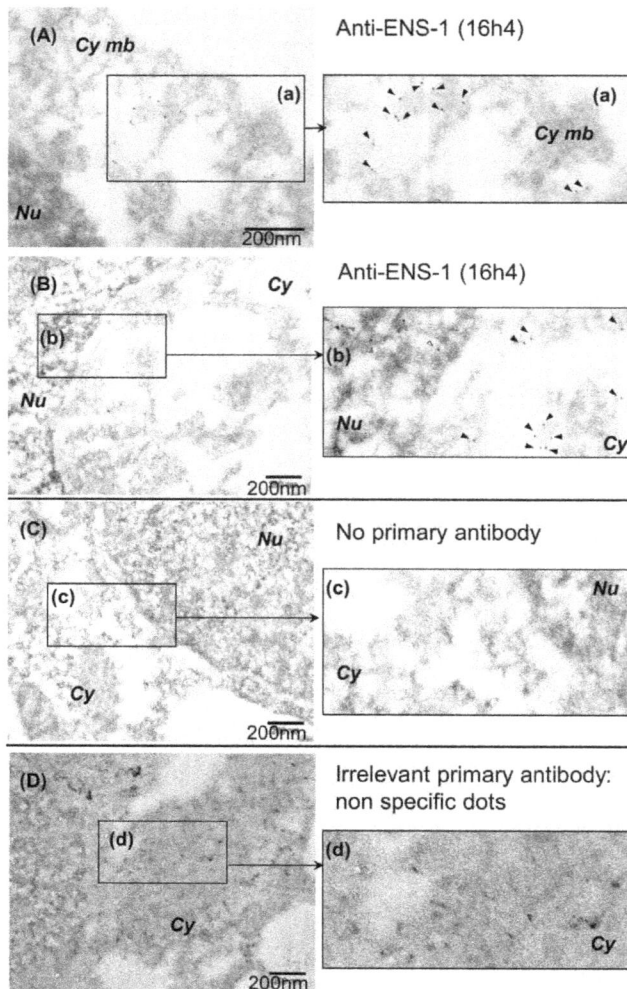

Figure 3. Subcellular localization of ENS-1 by electron micros-copy. CES were labeled with the 16h4 anti-ENS1 antibody. The secondary antibody was a gold-conjugated anti-mouse antibody (gold particles are seen as black dots). To preserve the affinity of the antibody for ENS-1 a moderate fixative procedure (4% PFA) was performed. Some sub-cellular structures were not well preserved and showed an empty appearance in the cytoplasm. (A) Observation of ENS-1 close to the cytoplasmic membrane *Cy mb*. (a) Magnification of the rectangle designed in (A).Black arrowheads show the 16h4 dots. (B) Observation at the junction between the cytoplasm *Cy* and the nucleus *Nu*. (b) Magnification of the rectangle designed in (B) shows ENS-1 in the chromatin (red arrowheads), in interchromatin spaces (white arrowheads) and in the cytoplasm (black arrowheads). (C) No labeling was obtained in the absence of the 16h4 antibody nor (D) with the anti-human androgen receptor 441 used as negative control in the same conditions. (c) and (d) show magnification of the rectangles designed in (C) and (D) respectively. Non specific dots with the irrelevant primary antibody in (D) and (d) were larger and have irregular outlines when compared with the 16h4 antibody (C) and (c).

by the use of two distinct antibodies (Fig. S1) and a partial co-localization with ENS-1 was found in this compartment (Fig. S2 a).

Therefore the interaction between ENS-1 and HP1γ does not require any CES specific co-factor nor chromatin binding and might occur in the cytoplasm and in the nucleus. However despite a close contact between both proteins outside the nucleus during mitosis, the majority of them co-localize in the nucleus of interphasic cells.

ENS-1 protein is differently regulated in the cytoplasm and in the nucleus

To better characterize the protein ENS-1 in the cytoplasm and in the nucleus, we performed western blot analysis of whole cell lysates. To this end we developed another ENS-1 specific antibody called 3807 (see Material and Methods). As shown in Figure 5A, this antibody gave a unique band of about 55 kDa in CES, but not in mouse ES cells, indicating its specificity for chicken cells. This molecular weight is in accordance with the 54 kDa prediction made from the protein sequence (490aa, Genebank: AAK06824). To confirm that the protein recognized by the 3807 antibody is ENS-1, CES cells were transfected with an expression vector (pCx ENS-1-HA-ires-puro) encoding for an HA-tagged ENS-1 protein. To be sure that antibiotic resistant cells will also express Ens-1 transcripts, both genes were placed under the control of the same promoter and were separated by an IRES sequence to generate a unique transcript for both genes that are next translated independently. Indeed, culture with puromycin of CES cells transfected with pCx ENS-1 but not with the empty vector, induced the expression of a unique HA-tagged protein as observed with the anti-HA antibody in Figure 5B. This protein is also recognized by the 3807 antibody in addition to the endogenous ENS-1 protein detected below and in control cells. These results confirm the specificity of the 3807 antibody for ENS-1.

In a previous report we have shown that the promoter of *Ens-1* is repressed while the expression of Sox2 is induced when CES differentiate [8]. These results are in agreement with the idea that differentiation of CES *in vitro* may mimic the release of the repression mediated by the dimer ENS-1/HP1γ on the promoter of Sox2 before the emergence of the neural plate. To explore the relevance of this hypothesis, variations of ENS-1 expression during differentiation were examined at the protein level. Western-blot results in Figure 5C show that the protein content decreased to become quite undetectable 24 h after retinoic acid addition and was maintained at this residual level after 48 h. In the same samples the expression level of HP1γ was not affected by differentiation (Fig. 5C) in agreement with previous observations in mouse ES cells [12]. These results are in accordance with the known regulation of the promoter of Ens-1 [8]. To confirm with another approach, ENS-1 expression was analyzed during CES cells differentiation by immunofluorescence experiments using the 16h4 antibody. This approach confirmed the decrease of the cytoplasmic ENS-1 during differentiation but with changes in the distribution of ENS-1 (Fig. 6A). In contrast to undifferentiated cells where the protein was mostly cytoplasmic, the localization was mostly nuclear in differentiated cells (Fig. 6A, Fig. S2). Notably, the organized distribution of ENS-1 in the cytoplasm was lost, giving a diffuse signal after 24 h differentiation that became undistinguishable from the back-ground after 48 h while the signal in the nucleus was still over the background (Fig. 6A). These results suggest that variations in ENS-1 contents are not similar in the cytoplasm and in the nucleus.

To further explore these changes, enrichments in the nuclear and in the cytoplasmic proteins were obtained by fractionation of cell lysates before analysis by western blot using the 3807 antibody.

Results in Figure 6B show that the 55 kDa ENS-1 protein is present in the cytoplasm and in the nucleus of undifferentiated cells. However in the nucleus only, this band is flanked by two other ones of 58 and 47 kDa that were not detected when using whole cell lysates. During differentiation, the 55 kDa protein was decreased in the cytoplasm but maintained in the nucleus. This was also generally true of the two other proteins. More precisely, the relative proportions of the three ENS-1 related bands in the nucleus were modified during differentiation. In undifferentiated

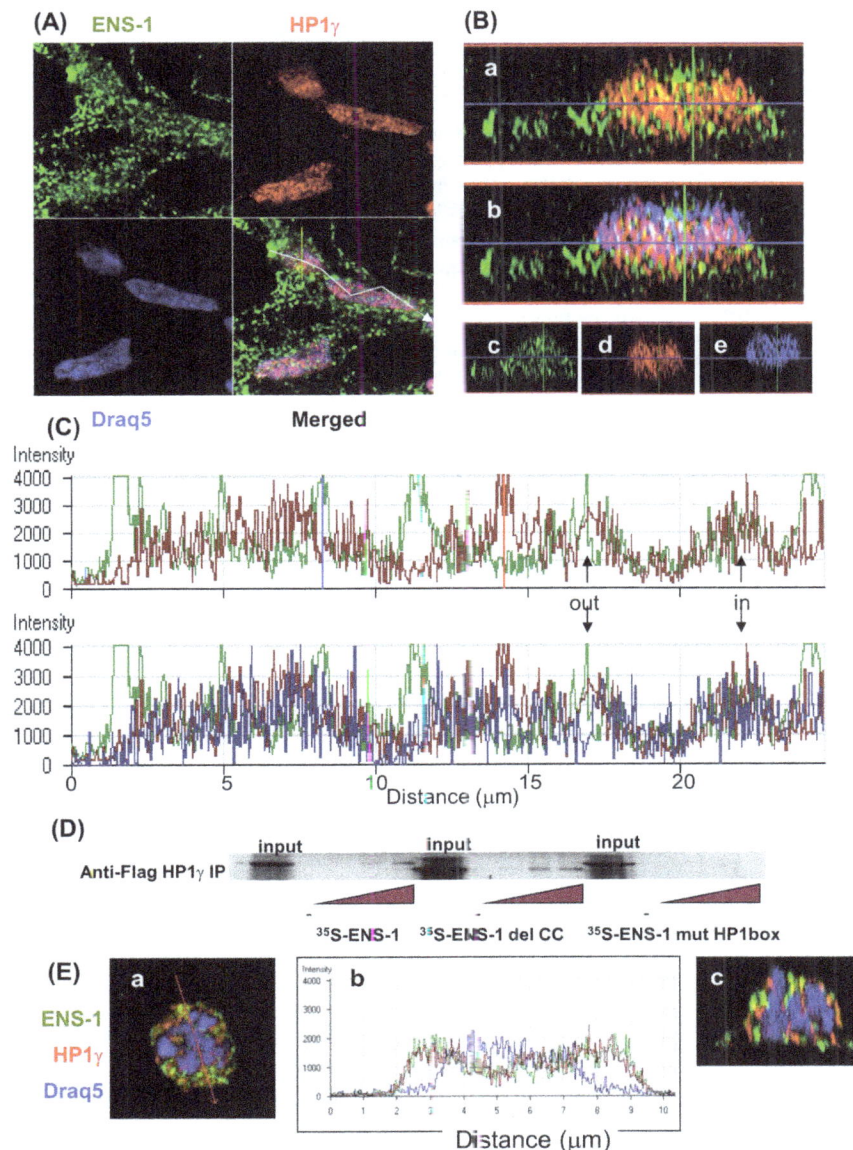

Figure 4. Sparse co-localization of ENS-1 and HP1γ in CES nuclei. (**A**) Immunostaining of CES cells was performed simultaneously with the rabbit anti- HP1γ (red) and the mouse anti-ENS-1 antibodies (green) detected using Alexa 555 and Alexa 488 conjugated secondary antibodies respectively. DNA was stained with Draq5 (blue). Observations were performed by confocal microscopy. (**B**) Distribution of ENS-1 and HP1γ in the thickness of CES. The z-cut axis is reported in (A) with a yellow bar on the merge panel. Distribution of HP1γ and ENS-1 is shown in panel a, co-localization of both proteins gives a yellow spot. In panel b the DNA staining was shown and white spots indicate ENS-1 and HP1γ proteins co-localizing on DNA. Individual distributions of ENS-1, HP1γ and DNA merged in a and b are represented in panes c, d and e respectively. (**C**) Fluorescence intensity for the three labels along the line represented in (A) by the white arrow. The upper panel represents fluorescence intensities obtained for HP1γ (red) and ENS-1 (green), the intensities obtained for chromatin (blue) have been added to the lower panel. Arrows point out co-localizations of HP1γ and ENS-1 in or out the chromatin. Intensity axis is in arbitrary units. (**D**) Pull-down experiments of ^{35}S-ENS-1 with Flag-HP1γ. Lysates from COS cells transfected with pCi-flag- HP1γ were used as a source for the flag tagged chicken HP1γ protein. ^{35}S labeled ENS-1 proteins were obtained by *in vitro* translation from pGBKT7 constructs coding for intact or mutated ENS-1 proteins: del CC indicates deletion of the coiled-coil domain (-50AA, 5.56 kDa), mutHP1 box indicates mutation in the HP1 box. After incubation of COS7 cell lysates with different dilutions of the ENS-1 proteins, HP1γ was precipitated with an anti-flag antibody and the association with ^{35}S-ENS-1 was revealed by fluorography. Representative results of at least three different experiments. (**E**) Confocal analysis of a CES cell entering in mitosis (prometaphase) from the same experiment as in A. In panel a, merge representation of the labeling for ENS-1, HP1γ and DNA as in A. In panel (b), fluorescence intensity along the line represented in panel (a) by the red arrow for the three markers labeled. The panel (c) is a z-cut representation of the same cell.

cells, the 58 kDa nuclear protein was more abundant than the two others. After 24 h differentiation, the proportion of the 55 and the 47 kDa bands increased relatively to the 58 kDa band and became more abundant. Later, in 48h-differentiated cells, all three proteins were equally and strongly detected in the nucleus (Fig. 6B).

Altogether these results confirm the observations made by immunostaining with the 16h4 antibody and indicate that the 55 kDa protein ENS-1 is maintained in the nucleus but not in the cytoplasm upon differentiation. They also reveal that ENS-1-like

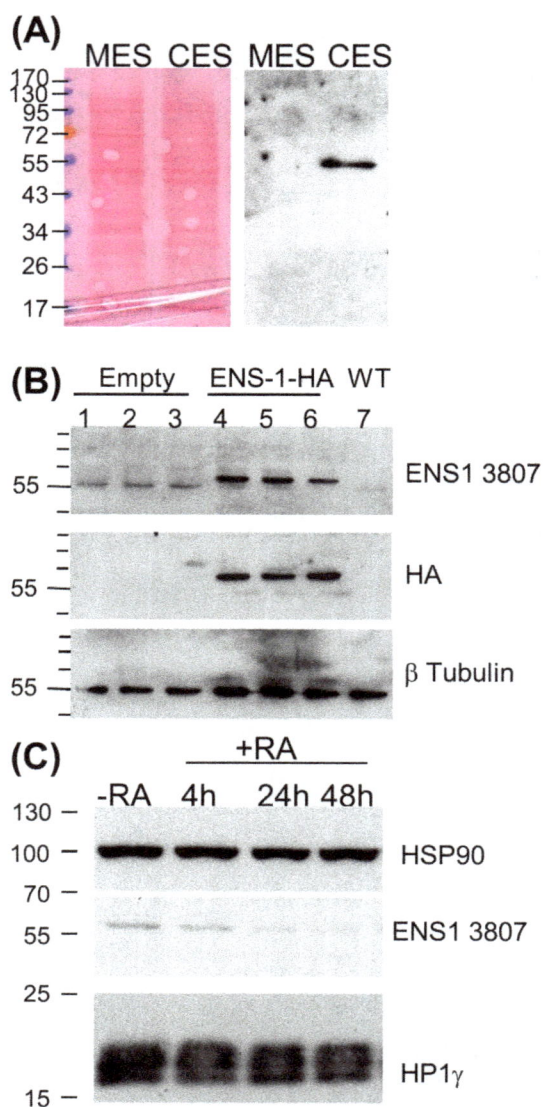

Figure 5. Western blot analysis of the protein ENS-1. (**A**) Proteins (15 µg) of whole cell lysates from murine (MES) and chicken (CES) cells were analyzed by western blot using the anti-ENS-1 3807 antibody. Protein loading was equivalent in both conditions as illustrated by the Ponceau's red staining of the blot. (**B**) Proteins lysates (20 µg) from CES cells transfected with an HA-tagged ENS-1 protein (lanes 4,5,6) or with an empty vector (lanes 1,2,3) were compared with untransfected cells (WT, lane 7). Triplicates were from three independent transfection experiments. The HA-tagged protein was detected by the 3807 antibody and by the anti-HA antibody at a molecular weight of 60 kDa. The anti-ENS1 antibody also detected the endogenous protein at 55 kDa. Both proteins differed in size by 3 kDa corresponding to the 33 additional amino acids added at the C-terminal part of ENS-1 in the transgenic protein in addition to the two HA tags (2.4 kDa). (**C**) Western blot analysis of ENS-1 and HP1γ (Chemicon antibody) in whole cell lysates (12 µg) from undifferentiated CES (-RA) or CES induced to differentiate with retinoic acid (RA, 10^{-6}M) for 4 to 48 h as indicated. HSP90 was used as protein loading control.

proteins recognized by the 3807 antibody are expressed and are restricted to the nucleus.

It is of note that all these proteins are associated with the insoluble components of the nucleus (Fig. 6B) and were not found in the soluble fraction (Fig. S3). The decrease observed in

Figure 5C with whole cell lysates reflects the cytoplasmic ENS-1 protein only since insoluble components were discarded by centrifugation before analysis.

To confirm that the cytoplasm does not contain the additional 47 and 58 kDa proteins detected in the nucleus, a fractionation protocol disrupting the cells without detergent was managed to preserve the soluble and the insoluble parts of the cytoplasm that were next separated by centrifugation. Results in Figure 6C show that ENS-1 was restricted to the pellet confirming that ENS-1 is not a soluble protein. Contamination with nuclear components was excluded since a unique band of 55 kDa was recognized with the anti-ENS-1 antibody and only traces of histone (2MeH3K4) and laminB1 coming from the nucleus were found. In parallel, analysis of the nuclear content showed the additional bands already observed with the previous protocol. The homogenous protein loading rendered possible the quantitative comparisons between the distinct sub-cellular fractions. The 55 kDa protein ENS-1 was equivalently distributed between the cytoplasm and the nucleus in CES cells. In the nucleus only, the 47 kDa was present but poorly detectable and the abundance of the 58 kDa was confirmed. Interestingly, HP1γ was found in all fractions, in agreement with immunofluorescence data. Therefore, it seems that differently from HP1γ, ENS-1 is not a free protein in the cytoplasm nor in the nucleus but is rather associated with insoluble structures.

Altogether the results obtained by a combination of immuno-fluorescence and western blot approaches using two distinct ENS-1 specific antibodies support the conclusion that the endogenous protein ENS-1 is differently regulated in the cytoplasm and in the nucleus and that ENS-1-like proteins also contribute to this difference between both compartments.

In silico analysis of Ens-1 gene copies

Due to its retroviral origin, distinct copies of the gene Ens-1 exist [5] that may generate distinct ENS-1 like proteins depending on the conservation of their ORF sequence. Those with the most conserved ORF in an updated sequence of the chicken genome are listed in Table 1. Seven copies encode for a 54 kDa ENS-1 protein and one copy for an ENS-1 like protein of 47 kDa but none for a 58 kDa protein. Based on a full sequence homology with the 3807 peptide and on conservation of the promoter active domains [8], the 47 kDa and five of the 54 kDa proteins encoding copies may account for the protein pattern detected with the 3807 antibody. Copies encoding for lower or for higher molecular weight proteins ranking from 34 to 69 kDa fulfill the same criteria but were not detected by this antibody (see Fig.5A, 6C and S3 to compare distinct cell lysis protocols). Of note, all the ENS-1 like proteins with an intact 3807 sequence have a conserved coiled-coil domain but only the 54 kDa proteins have an intact HP1box. Thus, it can be assumed that ENS-1-like proteins would be inactive towards HP1γ if generated from distinct gene copies.

Alternatively, the 58 kDa ENS-1 like protein may result from translation initiation starting from a start codon distinct from the ATG that accounts for the 54 kDa protein. This hypothesis is supported by the existence of a GTG sequence in the 5′UTR of the gene Ens-1 that is located 96 nucleotides upstream the ATG (Fig. 7A) and may extend the Ens-1 coding sequence, as already reported for human genes [26]. Both codons are in frame and the N-terminal extension generated from the GUG would overweight the ENS-1 protein of 3 kDa (Fig. 7B), corresponding to the difference between the 55 and the 58 kDa proteins detected by western blot. We checked the translation efficiency from the GUG codon alone or in the presence of the ATG. Using the constructs containing the 5′UTR of ENS-1 upstream of a luciferase reporter

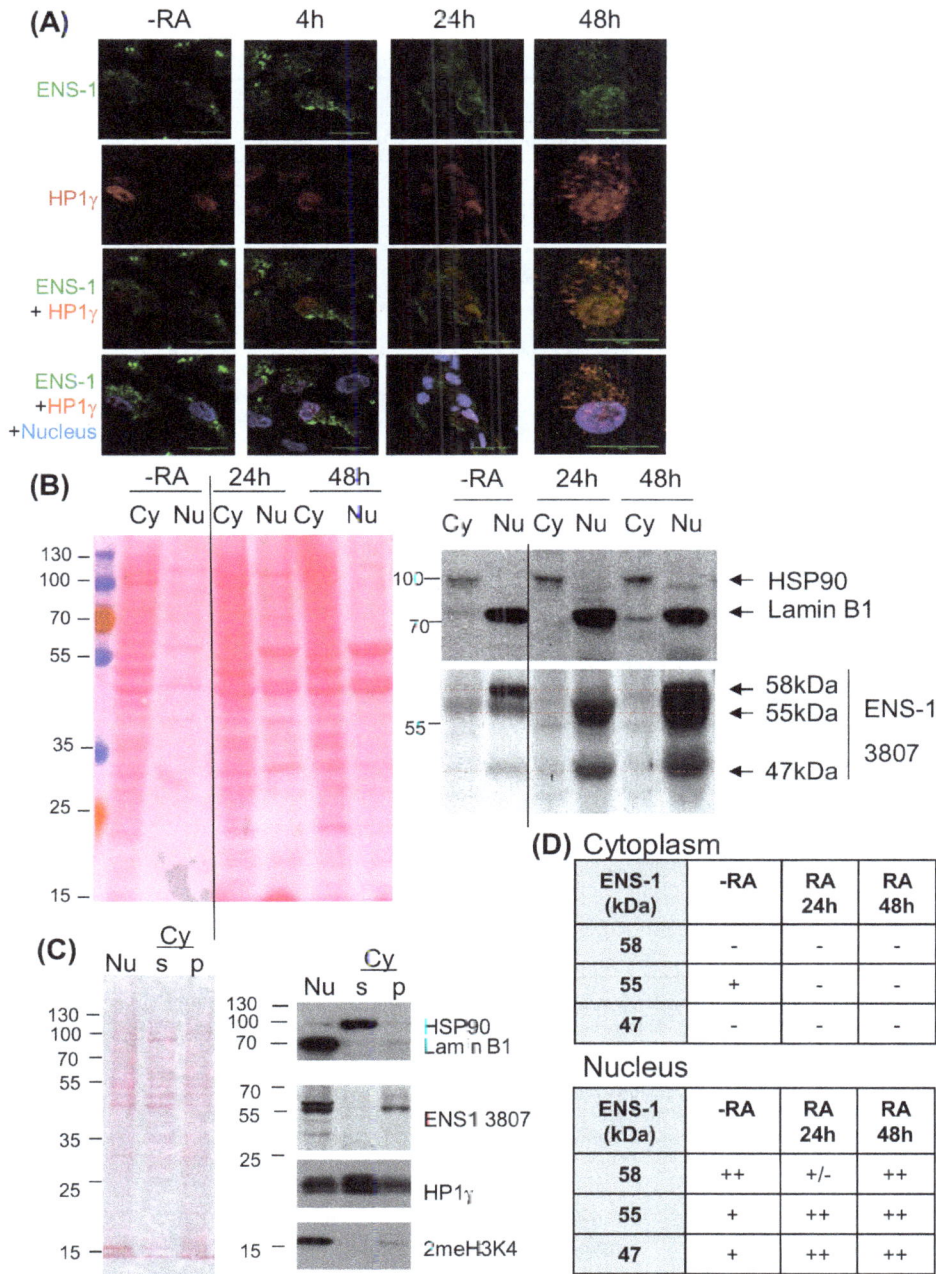

Figure 6. Subcellular distribution of ENS-1 in CES and during differentiation. (**A**) Immunostaining of ENS-1 (16h4 antibody) and HP1γ (Abcam antibody) in CES (-RA) and CES differentiated with retinoic acid for the indicated period. Nuclei are labeled with Draq5 and are in blue in merging panels. Image acquisition was optimized to observe the distribution of the proteins and does not reflect the real expression level. Scale bar 15 μm. (**B**) Western blot analysis of ENS-1 in the cytoplasmic (Cy) and in the nuclear (Nu) fractions of undifferentiated CES (-RA) or CES induced to differentiate with retinoic acid for 24h or 48h. Ponceau's red staining serves as a protein loading control between both fractions. In the nucleus soluble and insoluble components were separated and only the precipitating fraction that contains ENS-1 is shown. The volume corresponding to 6×10⁶ cells was loaded for the N fraction and Lamin B1 was used as loading control. In the cytoplasm (Cy) ENS-1 was in the supernatant, 15 μg of proteins were loaded and HSP90 was used as loading control. The vertical line indicates missing lanes between the two presented parts of the same gel. Dotted lines in red indicate the position of three ENS-1 related proteins. (**C**) Separation of soluble and insoluble components from the cytoplasm. The cytoplasmic fraction separated from the nucleus was subjected to extended centrifugation. The pellet (Cy p) and the supernatant (Cy s) were analyzed for ENS-1 protein and HP1γ. The nuclear fraction corresponding to the whole nuclear proteins is represented (Nu). Protein loading was 10 μg for each fraction. HSP90 identifies the cytoplasmic soluble fraction, Lamin B1 identifies fractions containing membrane proteins and 2MeH3K4 identifies fractions containing chromatin. Results in (B) and (C) were obtained from two independent experiments using distinct fractionation protocols. (D) Summary tables of the ENS-1 forms found in the cytoplasm (upper table) and in the nucleus (lower table) from CES cells (-RA) and from CES induced to differentiate with retinoic for 24 h or 48 h.

Table 1. List of the ENS-1 like sequences and homology with the 3807 peptide.

Sequence	Position of the gene			Protein Size		AA	Promoter
Name	Chrom.	Start	End	Amino Acids (AA)	kDa (1)	Identity (2)	activity (3)
Seq.12	chr1	153080201	153086086	490	54	16/16	Y
Seq.7	chr1	105099609	105104886	490	54	16/16	Y
Seq.32	chr2	95636316	95642444	488	54	16/16	Y
Seq.62	chrUn_ran.	23794802	23799438	487	54	16/16	Y
Seq.10	chr1	146896807	146900813	490	54	16/16	No
Seq.31	chr2	95607871	95610941	427	47	16/16	Y
Seq.38	chr4	29062470	29069620	392	43	16/16	Y
Seq.41	chr5	56151532	56156408	475	52	16/16	Y
Seq.46	chr9	3145510	3148390	627	69	16/16	Y
Seq.52	chrUn_ran.	14717478	14720393	307	34	16/16	Y
Seq.30 (ENS-3)	chr2	72991670	73001955	698	77	16/16	Y
Seq.18	chr1	166123942	166128607	490	54	15/16	Y
Seq.2	chr1	49392030	49395290	490	54	15/16	No
Seq.69	chrUn_ran.	48116551	48120222	265	29	15/16	Y
Seq.54	chrUn_ran.	18990010	18992568	160	18	15/16	Y
Seq.15	chr1	164856462	164859682	213	23	N/A	Y
Seq.34	chr2	138621760	138623256	210	23	N/A	Y
Seq.26	chr2	36219087	36223257	378	42	N/A	Y
Seq.24	chr2	11703874	11708540	293	32	N/A	Y
Seq.40	chr5	3991240	3997526	182	20	N/A	Y
Seq.8	chr1	134745214	134749487	182	20	N/A	Y
Seq.22	chr12	18117921	18121939	182	20	N/A	Y
Seq.35	chr3	92605110	92611133	182	20	N/A	Y
Seq.20	chr1	167194063	167197535	182	20	N/A	No
Seq.3	chr1	49395084	49396779	<10	N/A	N/A	Y

Among 78 copies detected in the chicken genome, 25 were very conserved compared to the reference sequence and are listed below. ORF Finder was used to determine the presence of ORF and to retrieve the subsequent protein sequence. Identity with the 3807 peptide (16 amino acids) defined three categories: total identity (16/16 AA), partial identity (15/16 AA) or no identity (N/A).
(1) The molecular weight of the protein was calculated with the formula: AA number X 110/1000.
In bold are indicated the protein sizes detected in CES cell lysates with the 3807 antibody in western-blotting.
(2) Identity between the ENS-1 like protein sequences and the 3807 peptide (DRIRVLQNEARTRAGK)
(3) Potential activity based on the presence (Y) or the absence (No) of the Nanog, Gata and Ets transcription factors binding sites controling promoter activity.

gene under the control of a T7 promoter (Fig. 7C), we generated RNAs that were used for *in vitro* translation. Results in Figure 7D show that the GUG start codon efficiently initiated the translation of the luciferase gene (as indicated by the arrow (LUC) in lanes II and I). When the ATG start codon was also present, translation was initiated from both codons to yield two proteins that differ in size by 3 kDa. Luciferase was produced from the ATG while the upstream GTG was responsible for the synthesis of the extended protein by translation of the 99 nucleotides segment from GTG to ATG (in lanes IV and III). It is noteworthy that protein expression from the GTG was less efficient when the latter was placed upstream of the ATG. Interestingly, when the Ens-1 5′UTR was driving protein synthesis (constructs I and III), expression was severely reduced from both start codons and translation from GUG became residual. These data indicate that an N-terminal extension of ENS-1 during RNA translation may account for the 58 kDa ENS-1 like protein but the low activity of the GUG start in the context of the Ens-1 5′UTR suggests that an additional regulation of translation would be involved in CES for a real

contribution to the protein pattern revealed with the 3807 antibody.

Finally, it cannot be ruled out that ENS-1 like proteins result from post-translational modifications of the ENS-1 protein itself. In yeast two-hybrid assay ENS-1 interacts with the conjugating enzyme Ubc9 (unpublished data) that transfers a SUMO protein to its protein substrate [27]. SUMO are about 10 kDa proteins in size, that is compatible with a modification of the 47 kDa ENS1-like protein. We found in the sequence of ENS-1 two sumoylation motifs that are conserved in all the copies with an intact 3807 peptide sequence (Table 2), thus supporting the potential involvement of post-translational modifications in the production of the 5 kDa protein. It is of note that one of the SUMO binding sites is located inside the coiled-coil domain involved in dimerization of the protein and in its recruitment to Sox2 promoter [7]. In addition, as SUMOylation has been reported to regulate intracellular distribution of proteins [28], this may explain the nuclear localization of ENS-1 in spite of lack of nuclear localization signal.

(A) ttctttagtt **gtg**gtgtgcc tccgtgggag aggcgataag gagttatttg tacttttgaa

taggagtacc tcctctctca gtgtatatct ttctgtgtat ttggga atgagcaaca...

(B)

```
VVCLRGRGDKELFVLLNRST SSLSVYLSVYLGMSNSMASM KSEDVLFDLLEKHGARPSVS
GVDWARQNWYNLQSVSDRIR VLQNEARTRAGKGKSFICAV LGAALKAAVEFREEKNSTET
QTIQALQESVKVTQELVKSL QSQIRSLEDQLEREKHNSVL LQTAFKELITCKDTGDTVIH
SAPQEKVYPQGKLQEVKERL DKLEASPAHIRPLIKTEYTF DNSENLDPQMNVKEIPFSAT
ELAKLKKDFSRSPKESETEY VWRVSLTGGDQILLTEKEAE GYWGPGVFLTTGNNRAPWSL
TQRAAYWAGGLNPLERGDPL AITGTIDQLVENVQKAACLQ MMYDRKLQPHNESPMMLPVN
PERLTPLIRGLPESLKPIGI QLQGKIQAMSQGERTWAALE GSVAPNHQSGPKVWTWGEVA
QELINYGRKYGPVVSTCSKF EPRGVRLAVASLASRPPSPR LIGTKKVSSPVKTGTRCIDH
KRNGLWTLGWTKGIPRDLMN GLPTVRLEKLVNCWPEQKLK GS
```

(C)

(D)

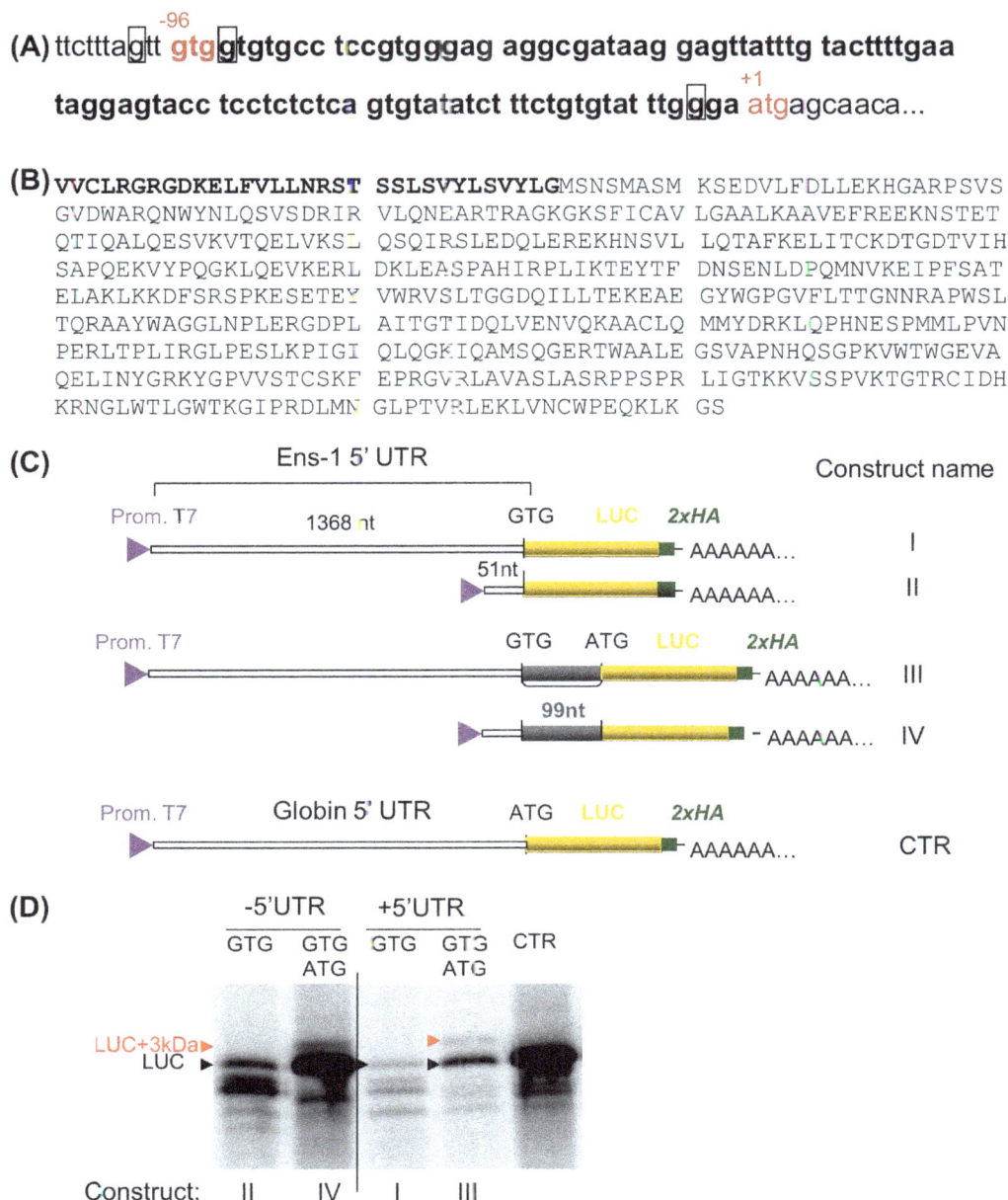

Figure 7. Presence of two start codons in the Ens-1 5'UTR. (A) The region containing the two putative start codons in the sequence encoding the whole mRNA (NM_001080873) is represented. The ATG start codon (in red) is the authentic initiation site and generates the 54 kDa protein that has been published previously (NM_001080873.1). The GTG codon (in red) is positioned 96 nucleotides. upstream of the AUG and generates a 57 kDa protein. The purine in positions -3 or +4 from the first base of each start codon is squared (Kozak consensus sequence). **(B)** Putative sequence of both proteins that only differ by the 3 kDa peptide in N-terminal position represented in bold. **(C)** Schematic representation of the constructs used to validate the translation initiation from the GTG and the ATG start codons in the 5' UTR of Ens-1. **(D)** RNA generated from the constructs depicted in (C) were used for in vitro translation of the luciferase protein in the presence of radioactive methionine. [35S]methionine-labelled proteins were separated on SDS-PAGE and revealed by autoradiography. The position of the luciferase protein (LUC) generated by constructs with only one start codon is indicated (GTG in Ens-1 5'UTR or ATG in CTR). Initiation from the two start codons generates LUC with ATG and a larger protein (LUC+3 kDa) resulting from initiation at the upstream GTG.

Discussion

This paper describes the endogenous form of the protein ENS-1 that controls the timing of the neural plate emergence in chicken by interacting with the epigenetic regulator HP1γ. Using different approaches, based on new antibodies raised against ENS-1, we show that this protein is strongly expressed in pluripotent CES cells isolated from the epiblast, in agreement with its transcrip-

tional pattern of expression [8]. These antibodies are specific for ENS-1 since they react with a tagged-ENS-1 transgene but not with cells devoid of the Ens-1 gene. Each antibody was screened for immunofluorescence and for western-blot applications but none worked in both approaches. Therefore both antibodies were used in parallel in western blot or in immunofluorescence to characterize the protein ENS-1. They gave converging results about localization of the protein both in the nucleus where HP1γ

Table 2. Conservation of functional domains in ENS-1 like copies.

	Protein size	Domains conservation and their position (AA)			
	kDa	HP1box (1)	cc-domain (2)	Sumo site 1 (3)	Sumo site 2 (4)
Sop.12	**54**	469	92-126	180	96
Sop.7	**54**	469	92-126	180	96
Sop.32	**54**	467	92-124	178	96
Sop.62	**54**	466	92-123	177	93
Sop.10	54	469	92-126	180	96
Sop.31	**47**	/	109-144	198	114
Sop.38	**43**	/	96-123	180	96
Sop.41	**52**	/	92-126	180	96
Sop.46	**69**	/	92-126	180	96
Sop.52	**34**	/	95-123	180	96
Sop.30 (= ENS-3)	**77**	/	422-450	507	423
Sop.18	54	469	92-126	180	96
Sop.2	54	469	92-126	180	96
Sop.69	29	/	92-124	180	96
Sop.54	18	/	85-124	/	92
Sop.15	23	/	/	36	/
Sop.34	23	/	/	77	/
Sop.26	42	357	/	68	/
Sop.24	32	272	/	/	/
Sop.40	20	161	/	/	/
Sop.8	20	161	/	/	/
Sop.22	20	161	/	/	/
Sop.35	20	161	/	/	/
Sop.20	20	161	/	/	/
Sop.3	N/A	113	/	/	/

In red are indicated the copies with an intact 3807 sequence and with an intact promoter.
(1) Based on the HP1 box motif involved in the interaction of ENS-1 with HP1γ: GLPTVRLE.
(2) Coiled-coil(cc) domain defined using the MultiCoil program.
(3) Sequence motif PLIKTEY.
(4) Sequence motif ESVKVTQ, inside the coiled-coil domain.

is concentrated and in the cytoplasm, but with distinct features as revealed by western blot. In the nucleus, but not in the cytoplasm, three anti-ENS-1 reactive bands were found. Our results support the idea that the relative proportion of these ENS-1 like proteins is part of the regulation of ENS-1.

Indeed, differentiation of ES cells using retinoic acid promotes their commitment preferentially toward the neurectoderm [12]. Under these conditions, the expression of ENS-1 at the transcriptional level is only partially repressed [8]. With both antibodies we show in parallel, that the ENS-1 protein content is decreased in the cytoplasm but not in the nucleus where it is maintained but with deep changes in the relative proportions of all three proteins.

In the prospective neural plate the recruitment of ENS-1 to the promoter of Sox2 represses the gene, and activation [29] [30] is induced by competition with the neo-synthesized protein BERT that releases ENS-1 [7]. In mice, Sox2 [31] is known to be, along with Oct4 [32] and Nanog [33], one of the key players of pluripotency that are repressed during differentiation. Differently, in CES cells we reported that Sox2 is induced upon differentiation

[8], suggesting that ENS-1 is fully active to repress Sox2 in CES. In agreement, we show here that ENS-1 can control the function of HP1γ in a reporter system and that both proteins locally co-localize on chromatin in the nucleus of CES cells.

Identification of genes targeted by the ENS-1/HP1γ heterodimer in CES represents a future challenge to understand integration of these proteins in the pluripotency network of CES cells, and the data presented here strongly support the development of approaches directly addressing this question at the protein level. The development of additional antibodies suitable for immune-precipitation will be required, this cannot be achieved with the antibodies presented here. Targeting other epitopes of the protein for antibody preparation will also be essential to better characterize the nature of the ENS-1 like proteins. The analysis of the different Ens-1 gene sequences suggests that distinct copies of the gene Ens-1 or another translation initiation start could at least partially explain the ENS-1 protein pattern observed, but the contribution of post-translational modifications cannot be excluded.

In conclusion our results demonstrate that understanding the regulation of the protein ENS-1 is more complex than could be anticipated from analysis of its transcripts and this study provides new tools to track this important protein during chick development.

Supporting Information

Figure S1   Immunostaining of the chicken HP1γ with the commercial antibody. (A) CES cells transiently transfected with HP1γ in fusion with GFP (green) were labeled with the Ab1080 (Abcam) antibody used in Fig. 4. (B) CES cells were labeled with a mixture of the rabbit Ab1080 (red) and the mouse 42s2 (green, Upstate) anti- HP1γ antibody. The secondary antibodies used were an anti-rabbit conjugated with Alexa 555 (red) and an anti-mouse antibody conjugated with Alexa 488 (green). Overlapping signals gave a yellow color. Results from two independent experiments are presented. In the nucleus both antibodies gave similar staining and both detected traces of HP1γ in the cytoplasm but with more or less intensity depending on the experiment.

Figure S2   Distribution and colocalization of ENS-1 with HP1γ during differentiation of CES cells. Images are those presented in Figure 5C complemented with fluorescence intensity for the three labeling along the line represented in the merged image by the red arrow. CES cells (a) or cells differentiated for 4 h (b), 24 h (b) or 48 h (d) with retinoic acid were stained as in

Figure 4 with anti-ENS-1 (green) and anti-HP1γ antibody (red). Signal intensities are presented as arbitrary units.

Figure S3   Western blot of soluble and insoluble protein fractions in the nucleus of CES cells. The whole blot corresponding to Fig. 5D is represented. N1 to N3 and S1 to S3 are loading replicates of respectively the insoluble and the soluble fractions of the nucleus. The S fractions had proteins concentrations lower (5 μg) than the N fractions (15 μg) but even in the S3 fraction that was slightly colored by Ponceau's red, no ENS-1 protein was detected.

Acknowledgments

We thank Dr Maïa Caillier for technical assistance. We acknowledge Isabelle Grosjean and the contribution of the CelluloNet and of the PAP facilities of UMS3444/US8 Biosciences Gerland-Lyon Sud respectively for amplification of the 16h4g6e4 hybridoma and for antibody purification. We are very grateful to Dr Luciano Pirola for the critical reading of this manuscript.

Author Contributions

Conceived and designed the experiments: AM FR AMB HA TO DD EL. Performed the experiments: SB FR AM AMB HA DD EL. Analyzed the data: AM FR JS EL DD TO. Contributed reagents/materials/analysis tools: HA. Wrote the paper: AM FR JS TO HA EL.

References

1. Macfarlan TS, Gifford WD, Driscoll S, Lettieri K, Rowe HM, et al. (2012) Embryonic stem cell potency fluctuates with endogenous retrovirus activity. Nature 487: 57–63.
2. Kunarso G, Chia N-Y, Jeyakani J, Hwang C, Lu X, et al. (2010) Transposable elements have rewired the core regulatory network of human embryonic stem cells. Nat Genet 42: 631–634.
3. Pi W, Yang Z, Wang J, Ruan L, Yu X, et al. (2004) The LTR enhancer of ERV-9 human endogenous retrovirus is active in oocytes and progenitor cells in transgenic zebrafish and humans. Proceedings of the National Academy of Sciences of the United States of America 101: 805–810.
4. Blond J-L, Lavillette D, Cheynet Vr, Bouton O, Oriol G, et al. (2000) An Envelope Glycoprotein of the Human Endogenous Retrovirus HERV-W Is Expressed in the Human Placenta and Fuses Cells Expressing the Type D Mammalian Retrovirus Receptor. Journal of Virology 74: 3321–3329.
5. Lerat E, Birot AM, Samarut J, Mey A (2007) Maintenance in the chicken genome of the retroviral-like cENS gene family specifically expressed in early embryos. J Mol Evol 65: 215–227.
6. Streit A, Berliner AJ, Papanayotou C, Sirulnik A, Stern CD (2000) Initiation of neural induction by FGF signalling before gastrulation. Nature 406: 74–78.
7. Papanayotou C, Mey A, Birot AM, Saka Y, Boast S, et al. (2008) A mechanism regulating the onset of Sox2 expression in the embryonic neural plate. PLoS Biol 6: e2.
8. Mey A, Acloque H, Lerat E, Gounel S, Tribollet V, et al. (2012) The endogenous retrovirus ENS-1 provides active binding sites for transcription factors in embryonic stem cells that specify extra embryonic tissue. Retrovirology 9: 21.
9. Acloque H, Risson V, Birot AM, Kunita R, Pain B, et al. (2001) Identification of a new gene family specifically expressed in chicken embryonic stem cells and early embryo. Mech Dev 103: 79–91.
10. Niwa H, Ogawa K, Shimosato D, Adachi K (2009) A parallel circuit of LIF signalling pathways maintains pluripotency of mouse ES cells. Nature 460: 118–122.
11. Fong H, Hohenstein KA, Donovan PJ (2008) Regulation of Self-Renewal and Pluripotency by Sox2 in Human Embryonic Stem Cells. Stem Cells 26: 1931–1938.
12. Caillier M, Thenot S, Tribollet V, Birot AM, Samarut J, et al. (2010) Role of the Epigenetic Regulator HP1gamma in the Control of Embryonic Stem Cell Properties. PLoS One 5: e15507.
13. Pain B, Clark ME, Shen M, Nakazawa H, Sakurai M, et al. (1996) Long-term in vitro culture and characterisation of avian embryonic stem cells with multiple morphogenetic potentialities. Development 122: 2339–2348.
14. Lavial F, Acloque H, Bertocchini F, Macleod DJ, Boast S, et al. (2007) The Oct4 homologue PouV and Nanog regulate pluripotency in chicken embryonic stem cells. Development 134: 3549–3563.
15. Acloque H, Mey A, Birot AM, Gruffat H, Pair B, et al. (2004) Transcription factor cCP2 controls gene expression in chicken embryonic stem cells. Nucleic Acids Res 32: 2259–2271.
16. Niwa H, Yamamura K, Miyazaki J (1991) Efficient Selection for High-Expression Transfectants with a Novel Eukaryotic Vector. Gene 108: 193–199.
17. Waltzer L, Perricaudet M, Sergeant A, Manet E (1996) Epstein-Barr virus EBNA3A and EBNA3C proteins both repress RBP-J kappa-EBNA2-activated transcription by inhibiting the binding of RBP-J kappa to DNA. Journal of Virology 70: 5909–5915.
18. Rifo RS, Ricci EP, Décimo D, Moncorgé O, Ohlmann T (2007) Back to basics: the untreated rabbit reticulocyte lysate as a competitive system to recapitulate cap/poly(A) synergy and the selective advantage of IRES-driven translation. Nucleic Acids Research 35: e121.
19. Wicker T, Robertson JS, Schulze SR, Feltus FA, Magrini V, et al. (2005) The repetitive landscape of the chicken genome. Genome Res 15: 126–136.
20. Altschul SF, Madden TL, Schäffer AA, Zhang J, Zhang Z, et al. (1997) Gapped BLAST and PSI-BLAST: a new generation of protein database search programs. Nucleic Acids Research 25: 3389–3402.
21. Kim PS, Berger B, Wolf E (1997) MultiCoil: A program for predicting two-and three-stranded coiled coils. Protein Science 6: 1179–1189.
22. Prévôt D, Décimo D, Herbreteau CH, Roux F, Garin J, et al. (2003) Characterization of a novel RNA-binding region of eIF4GI critical for ribosomal scanning. The EMBO Journal 22: 1909–1921.
23. Smallwood A, Black JC, Tanese N, Pradhan S, Carey M (2008) HP1-mediated silencing targets Pol II coactivator complexes. Nat Struct Mol Biol 15: 318–320.
24. Dialynas GK, Terjung S, Brown JP, Aucott RL, Baron-Luhr B, et al. (2007) Plasticity of HP1 proteins in mammalian cells. J Cell Sci 120: 3415–3424.
25. Fischle W, Tseng BS, Dormann HL, Ueberheide BM, Garcia BA, et al. (2005) Regulation of HP1-chromatin binding by histone H3 methylation and phosphorylation. Nature 438: 1116–1122.
26. Ivanov IP, Firth AE, Michel AM, Atkins JF, Baranov PV (2011) Identification of evolutionarily conserved non-AUG-initiated N-terminal extensions in human coding sequences. Nucleic Acids Research 39: 4220–4234.
27. Hochstrasser M (2001) SP-RING for SUMO: New Functions Bloom for a Ubiquitin-like Protein. Cell 107: 5–8.
28. Geiss-Friedlander R, Melchior F (2007) Concepts in sumoylation: a decade on. Nat Rev Mol Cell Biol 8: 947–956.
29. Rex M, Orme A, Uwanogho D, Tointon K, Wigmore PM, et al. (1997) Dynamic expression of chicken Sox2 and Sox3 genes in ectoderm induced to form neural tissue. Developmental Dynamics 209: 323–332.
30. Acloque H, Ocaña Oscar H, Matheu A, Rizzoti K, Wise C, et al. (2011) Reciprocal Repression between Sox3 and Snail Transcription Factors Defines Embryonic Territories at Gastrulation. Developmental Cell 21: 546–558.

31. Avilion AA, Nicolis SK, Pevny LH, Perez L, Vivian N, et al. (2003) Multipotent cell lineages in early mouse development depend on SOX2 function. Genes Dev 17: 126–140.

32. Nichols J, Zevnik B, Anastassiadis K, Niwa H, Klewe-Nebenius D, et al. (1998) Formation of pluripotent stem cells in the mammalian embryo depends on the POU transcription factor Oct4. Cell 95: 379–391.

33. Chambers I, Colby D, Robertson M, Nichols J, Lee S, et al. (2003) Functional expression cloning of Nanog, a pluripotency sustaining factor in embryonic stem cells. Cell 113: 643–655.

18

Subunits of the *Drosophila* Actin-Capping Protein Heterodimer Regulate Each Other at Multiple Levels

Ana Rita Amândio, Pedro Gaspar, Jessica L. Whited[¤], Florence Janody*

Instituto Gulbenkian de Ciência, Oeiras, Portugal

Abstract

The actin-Capping Protein heterodimer, composed of the α and β subunits, is a master F-actin regulator. In addition to its role in many cellular processes, Capping Protein acts as a main tumor suppressor module in *Drosophila* and in humans, in part, by restricting the activity of Yorkie-YAP/TAZ oncogenes. We aimed in this report to understand how both subunits regulate each other *in vivo*. We show that the levels and capping activities of both subunits must be tightly regulated to control F-actin levels and consequently growth of the *Drosophila* wing. Overexpressing *capping protein* α and β decreases both F-actin levels and tissue growth, while expressing forms of Capping Protein that have dominant negative effects on F-actin promote tissue growth. Both subunits regulate each other's protein levels. In addition, overexpressing one of the subunit in tissues knocked-down for the other increases the mRNA and protein levels of the subunit knocked-down and compensates for its loss. We propose that the ability of the α and β subunits to control each other's levels assures that a pool of functional heterodimer is produced in sufficient quantities to restrict the development of tumor but not in excess to sustain normal tissue growth.

Editor: Franck Pichaud, MRC, University College of London, United Kingdom

Funding: This work was supported by grants from Fundação para a Ciência e Tecnologia (FCT) (Grants PTDC/BIA-BCM/121455/2010). Pedro Gaspar was the recipients of fellowships from FCT SFRH/BD/47261/2008. The funders had no role in study design, data collection and analysis, decision to publish, or preparation of the manuscript.

Competing Interests: The authors have declared that no competing interests exist.

* Email: fjanody@igc.gulbenkian.pt

¤ Current address: Regenerative Medicine Center and Department of Orthopedics, Brigham & Women's Hospital, Harvard Medical School, Cambridge, Massachusetts, United States of America

Introduction

The actin cytoskeleton controls numerous processes, including cell shape, mobility, division and intracellular transport. In normal cells, the actin cytoskeleton is tightly controlled to regulate these essential functions; however, it can be subverted by cancer cells and contributes to changes in cell growth, proliferation, stiffness, movement and invasiveness [1,2]. Moreover, alterations in the activity or expression of actin-binding proteins (ABPs) *per se*, have been linked to cancer initiation and progression [2,3,4,5,6].

Among these actin regulators, the actin Capping Protein (CP) heterodimer, composed of an α and a β subunit, appears to act as a main tumor suppressor module [7,8,9,10]. CP was named based on its ability to bind and cap actin filament barbed ends, inhibiting the addition and loss of actin monomers [11,12,13]. CP has homologs in nearly all eukaryotic cells, including vertebrates, invertebrates, plants, fungi, insects and protozoa [14]. *Drosophila* and organisms other than vertebrates have single genes encoding *capping protein* α *(cpa)* or β *(cpb)*. In contrast, vertebrates contain two genes expressed somatically that encode two α subunits (α1 and α2), and one single gene that produce two β isoforms (β1 and β2) through alternative splicing [15,16,17]. Although the amino acid sequences of the α and β subunits are not more similar to each other than they are to other ABPs, nor they share common sequences with other proteins, they have extremely similar secondary and tertiary structures [18]. When in complex, the heterodimer resembles a mushroom with the C-terminus of each

subunit forming tentacles located on the top surface of the heterodimer [19,20]. *In vitro* analyses of chicken and budding yeast CP revealed that deletions or point mutations in either the α or β tentacles do not affect protein stability but reduce the capping affinity, while a complete removal of both tentacles fully abrogates the actin-binding activity [12,20]. Thus, CP appears to cap F-actin barbed ends via the independent interaction of both tentacles with actin. *In vivo*, a truncated form of *Drosophila cpa* deleted of the C-terminal 28 amino acids has no effect on F-actin when expressed alone but promotes F-actin accumulation when co-expressed with full length *cpb* [21]. Similarly, a chicken β subunit containing a point mutation changing a conserved leucine to arginine at position 262, which caps actin poorly, disrupts the early steps in myofibrillogenesis of cultured myotubes and the sarcomere of mouse heart [22,23,24].

In yeast and *Drosophila*, removing either *cpa* or *cpb* induces F-actin accumulation and identical phenotypes [25,26,27]. In the fly, CP is required for proper differentiation of adult bristles, survival of the adult retina, determination of the oocyte and cortical integrity of nurse cells in the egg chamber [27,28,29,30]. In addition, CP has a key role in restricting tissue growth. In the whole wing disc epithelium, CP-dependent F-actin regulation suppresses inappropriate tissue growth by inhibiting the activity of the Yorkie (Yki) oncogene, which mediates Hippo signalling activity [7,9]. This function is conserved, as the α1 subunit is also required to limit the activity of the Yki orthologs YAP and TAZ in mammary epithelial cells [31]. In addition, in the distal *Drosophila* wing disc epithelium, CP prevents JNK-mediated apoptosis or proliferation and counteracts the oncogenic ability of Src [8,21,32]. Furthermore, underexpression of the human α1 subunit correlates with cancer-related death and causes a significant increase in gastric cancer cell migration and invasion *in vitro*, whereas its overexpression has the opposite effect [10].

We aimed in this report to understand how both subunits regulate each other *in vivo* to control F-actin levels and tissue growth. We show that Cpa and Cpb stabilize each other's protein levels and can stimulate the production of each other's mRNA when the level of one of the subunit is reduced. Because overexpressing *CP* decreases F-actin levels and tissue growth, while expressing forms of CP mutated in their actin-binding domains has opposite effects, we propose that by regulating each other, Cpa and Cpb assure that a pool of functional CP heterodimer is produced in sufficient quantities to restrict tissue growth and therein prevent tumor development but not in excess to sustain proper tissue growth.

Materials and Methods

Molecular Biology

To generate $UAS\text{-}cpb^{L262R}$, site-directed mutagenesis was performed on the plasmid UAS-*cpb*, using the QuikChange kit (Stratagene, # 200519). The mutated plasmid was confirmed by sequencing and transgenic flies were generated by standard methods.

Fly strains and genetics

Fly stocks used were *sd*-Gal4 [33]; *nub*-Gal4 [34]; *hh*-Gal4 (a gift from T. Tabata); *da*-Gal4 [35]; UAS-*cpa*-IR^{C10}, UAS-*cpa*-IR^{B4} [7]; UAS-*HA*-*cpa*^{89E}, UAS-*HA*-*cpa*^{AABD} [21]; UAS-cpb7 [36]; UAS-*cpb*-RI^{45668} (Vienna *Drosophila* Research Center, VDRC); *cpa*^{107E} [25], *cpb*^{M143} (FlyBase). To generate *cpa* mutant clones marked by the absence of GFP and expressing or not UAS-*HA*-*cpa*^{89E} or UAS-*HA*-*cpa*^{AABD} or UAS-*cpb*^{7}, *w*; FRT42D, *cpa*^{69E}/CyO or *w*; FRT42D, *cpa*^{107E}/CyO; UAS-*HA*-*cpa*^{89E}/Tm6β or *w*; FRT42D, *cpa*^{107E}/CyO; UAS-*HA*-*cpa*^{AABD}/Tm6β or *w*; FRT42D, *cpa*^{107E}, UAS-*cpb*^{7}/CyO males were crossed to *y*, *w*, FRT42D, *ubi*-GFP; *T155*-Gal4, UAS-*flp*/ST females. To generate *cpb* mutant clones marked by the absence of GFP, *w*, *y*; FRT40A, *cpb*^{M143}/CyOy^{+} males were crossed to *y*, *w*, *hs*FLP122; FRT40A, *ubi*-GFP females and the progeny was heat-shocked at first and second instar larvae. All crosses were maintained at 25°C and the progeny was dissected at end of third instar larvae.

Antibody Generation

The rabbit anti-Cpa and rabbit anti-Cpb polyclonal antibodies were generated by Metabion International AG using full length Cpa or Cpb tagged with Histidine.

Immunohistochemistry and quantification

We performed immunocytochemistry using the procedure described in Lee and Treisman [37]. Primary antibodies used were mouse anti-Arm (N2 7A1, Developmental Studies Hybridoma Bank (DSHB); 1:10), rat anti-DE-Cad (1:50, CAD2, DSHB), rabbit anti-Cpa (1:200); rabbit anti-Cpb (1:200); mouse anti-HA (Covance 11 MMS101P; 1:1000) and rabbit anti-Caspase 3 (Cell Signalling #9661; 1:50). Rhodamine conjugated phalloidin (Sigma) was used at a concentration of 0.3 μM. Secondary antibodies were from Jackson Immunoresearch, used at 1:200. Wing discs were mounted in VECTASHIELD Mounting Media (Vector Laboratories, Inc. #H-1000). Fluorescence images were obtained on a Leica SP5 confocal microscope or on a LSM 510

Zeiss confocal microscope. The NIH Image J program was used to perform measurements. Quantifications of the intensity of Caspase 3 signals were performed as described in [21]. Quantifications of the ratio of Phalloidin signal between posterior and anterior wing compartments were performed as described in [7]. To quantify the ratio of Cpa or Cpb signals between the anterior and posterior wing disc compartments, a region of interest (ROI) of 100 per 50 pixels was selected. The sum of the gray values was measured for each ROI, applied to each compartments for each disc on optical cross sections through distal wing disc epithelium comprising the apical surface. To measure wing size, wing were dissected one to two days after eclosion and imaged using the Hamamatsu Orca-ER camera attached to a Zeiss' Stereo Lumar V12 stereoscope. The total area of each wing was outlined and measured using the *area measurement* function. Statistical significance was calculated using a two-tailed *t*-test.

Western Blotting

For each genetic background, proteins were extracted from either four wing imaginal discs or four dechorionated embryos using a 2x SDS sample buffer (Sigma #S3401). Samples were frozen in liquid nitrogen, boiled for 5 minutes in 5 μl Sample Buffer 2x, spun at 13,000 g for 1 minute, loaded on a 10% SDS-PAGE gel and transferred to a PVDF membrane (Amersham Hybond-P, GE Healthcare). Proteins were visualized by immunoblotting using rabbit anti-Cpa (1:2500) or rabbit anti-Cpb (1:2500) or mouse anti-HA (Covance 11 MMS101P; 1:1000) or rabbit anti-Histone H3 (Cell Signalling #9715; 1:3000). HRP-conjugated donkey anti-mouse or donkey anti-rabbit secondary antibodies were used at 1:5000 (Jackson ImmunoResearch Laboratories, Inc.). Blots were developed using Amersha ECL Plus Western Blotting Detection System (GE Healthcare). Densitometric analysis of signal intensity was performed using the GelQuant.NET software (biochemlabsolutions.com) and normalized with the loading control. Statistical significance was calculated using a *Paired t-test*.

Isolation of RNA and Real-Time qRT-PCR

Total RNAs were extracted from either 10 first instar larvae or 50 wing imaginal discs for each genetic background. Samples were homogenized in RLT buffer treated with DNase (Qiagen) at 4 degree C and total RNAs were isolated using the RNeasy mini kit (Qiagen) following manufacturer instructions. First Strand cDNA Synthesis Kit for RT-PCR (Roche) was used to produce cDNAs from 1 μg of total RNA. To quantify mRNA levels, qPCRs were carried out on reverse-transcribed total mRNA using intron-exon-specific primers (Table S1), designed using the Primer3 software [38,39], and ensuring that efficiency is at least 90% and restricting primer dimmer formation. Real-time qPCR was performed using PerfeCTa SYBR Green FastMix (Quanta Biosciences) in 384 well skirted PCR microplates (Axygen) sealed with optically clear sealing tape (STARSTEDT) in the Applied Biosystems 7900HT Fast Real-Time PCR System. The relative amount of mRNA for each condition was calculated after normalization to the *RpL32* transcript. Statistical significance was calculated using a *Paired t-test* with significance at $P<0.05$.

Results

Cpa and Cpb stabilize each other's protein levels and accumulate at Adherens Junctions

To understand how Cpa and Cpb are regulated to restrict growth of *Drosophila* epithelia, we generated polyclonal antibodies to each CP subunit. In lysates from embryos expressing UAS-

mCD8-GFP under the control of the ubiquitous *daughterless*-Gal4 (*da*-Gal4) driver, the Cpa (Fig. 1A) and Cpb (Fig. 1B) antibodies revealed a band at around 32 and 31 kDa respectively by Western Blot. These signals were lost in embryonic extracts from homozygous *cpa* (Fig. 1A) or *cpb* mutants (Fig. 1B) respectively. Conversely, overexpressing full length *cpa*, tagged with HA (UAS-*HA-cpa*⁺; Fig. 1A) or *cpb* (UAS-*cpb*⁺; Fig. 1B) with *da*-Gal4, enhanced the anti-Cpa or anti-Cpb signals respectively. Similarly, Cpa levels were increased in wing disc lysates overexpressing *HA-cpa* under *scalloped*-Gal4 control (*sd>HA-cpa*⁺; Fig. 1C), while endogenous Cpb levels were similar to control *sd>GFP* lysates (Fig. 1D). Forcing *cpb* expression in this tissue also induced a significant increase in Cpb levels by Western Blot (Fig. 1D) but did not significantly affect endogenous Cpa levels (Fig. 1C). Cross-sections through wing disc epithelia expressing UAS-*mCD8-GFP* in the posterior compartment using the *hedgehog*-Gal4 (*hh*-Gal4) driver showed that Cpa (Fig. 1E–E''') and Cpb (Fig. 1F–F''') accumulated at the apical cell membrane and co-localized with components of Adherens Junctions, including the β-Catenin homolog Armadillo (Arm). Co-expressing *cpb* and *mCD8-GFP* in this domain strongly enhanced the anti-Cpb signals but did not affect Cpa levels (Fig. 2D–D''). Conversely, *hh>HA-cpa*⁺ wing disc epithelia displayed an apical localization of HA-Cpa, like endogenous Cpa (Fig. 1E–E''), but no change in Cpb levels (Fig. 2E–E''). Thus, the anti-Cpa and Cpb antibodies recognize specifically Cpa and Cpb respectively.

Strikingly, Cpa levels were strongly reduced not only in wing disc extracts expressing double-stranded RNAs (dsRNA) for *cpa* under *sd*-Gal4 control (*sd>cpa-IR*) but also in discs knocked-down for *cpb* (*sd>cpb-IR*; Fig. 1C). In the converse experiment, the amount of Cpb was also strongly reduced in both *sd>cpb-IR* and *sd>cpa-IR* wing disc extracts (Fig. 1D). Similarly, knocking down *cpa* (Fig. 1G–G' and H–H') or *cpb* (Fig. 1I–I' and J–J') in the posterior wing disc compartment with *hh*-Gal4 significantly reduced the apical accumulation of both Cpa and Cpb when compared to anterior compartments used as internal controls. Moreover, both Cpa and Cpb levels were also strongly reduced in lysates from first instar larvae homozygote mutant for *cpa* or *cpb* (Fig. S1A) and in clones mutant for *cpa* or *cpb* (Fig. S1B–B'' to E–E''). To verify that the *cpa* dsRNA did not affect *cpb* mRNA and *vice versa*, we performed quantitative RT-PCR (qRT-PCR) experiments on wing imaginal discs knocked down for *cpa* or *cpb*. As expected, *sd>cpa-IR* or *sd>cpb-IR* wing discs showed a significant reduction of *cpa* (Fig. 1K, 2.5±0.43 folds) or *cpb* mRNA (Fig. 1L, 2.6±0.41 folds) levels respectively, relative to control *sd>GFP*. However, *cpa* mRNA levels were not significantly affected by a reduction in *cpb* (Fig. 1K), nor were *cpb* mRNA levels reduced in wing discs knocked-down for *cpa* (Fig. 1L). Similarly, a reduction in *cpa* or *cpb* levels had no effect on *cpb* or *cpa* mRNA levels, respectively, in first instar larvae expressing *cpa-IR* or *cpb-IR* under *da*-Gal4 control (Fig. S1F and G). Taken together, we conclude that Cpa and Cpb accumulate at apical cell membrane and enhance each other's protein levels.

Cpa and Cpb levels are rate limited to form a functional heterodimer

The Capping Protein α and β subunits form a functional heterodimer, which caps F-actin barbed ends via the interaction of the α and β tentacles with actin (Fig. 1A and [11,12,13,20]). To confirm that the stabilization of Cpa and Cpb's protein levels by each other promotes the formation of a functional heterodimer, we first tested if co-expressing *cpb* and *HA-cpa* would enhance the levels of both subunits by comparing the levels of HA-Cpa and Cpb when overexpressed alone or together, ensuring that each

genetic combination contained the same number of UAS transgenes. Indeed, by Western Blot (Fig. 2B, *P<0.0092*) and in wing disc epithelia (Fig. 2 compare F–F'' with E–E''), HA levels were strongly enhanced when *HA-cpa* was co-expressed with *cpb*. Similarly, the co-expression of *HA-cpa* and *cpb* strongly increased Cpb levels compared to wing disc lysates overexpressing *cpb* alone (Fig. 2C). Overexpressed *HA-cpa* and *cpb* appeared to form a functional heterodimer as their co-expression in the posterior wing disc compartment with *hh*-Gal4 decreased the apical F-actin ratio between both compartments compared to *hh>GFP* control (Fig. 3F, *P<0.0001*). In contrast, overexpressing either *HA-cpa* or *cpb* alone has no effect on F-actin levels [21]. We conclude that the levels of endogenous Cpa and Cpb available are rate limited to form a functional heterodimer.

Forms of CP mutated in α or β tentacle counteract the ability of wild type CP to restrict F-actin accumulation

Surprisingly, expressing an HA-tagged form of Cpa deleted of the α tentacle (UAS-*HA-cpa^ΔABD*) has no significant effect on F-actin when expressed alone [21] but triggered apical F-actin accumulation when co-expressed with *cpb* (Fig. 3F, *P<0.0001* and [21]), indicating that HA-Cpa^ΔABD affects F-actin only in the presence of overexpressed *cpb*. We therefore tested if the co-expression of *cpb* would also enhance the levels of HA-Cpa^ΔABD. In contrast to full length HA-Cpa, which accumulated apically (Fig. 2E–E''), HA-Cpa^ΔABD localized uniformly along the apical-basal axis in the posterior compartment of *hh>HA-cpa^ΔABD* wing discs (Fig. 2G–G''). Strikingly, co-expressing *cpb* not only enhanced strongly HA-Cpa^ΔABD levels as assessed by Western Blot (Fig. 2B, *P<0.0002*), but also relocalized HA-Cpa^ΔABD at the apical cell membrane (Fig. 2H–H''). Thus, forcing Cpb levels enhances the levels of HA-Cpa^ΔABD and promotes its apical localization.

The heterodimer formed between HA-Cpa^ΔABD and Cpb appears to have reduced capping activity and may be recruited to F-actin barbed ends, preventing the binding of wild type CP. If so, we would expect that a form of Cpb truncated of its β tentacle would also promote F-actin accumulation in the presence of endogenous CP. To test this possibility, we expressed a form of *cpb* mutated in the highly conserved Leucine 262 (UAS-*cpb^L262R*), which has been proposed to directly interact with actin [12]. While overexpressing full length *cpb* had no significant effect on F-actin (Fig. 3 compare B–B'' with A–A' and F), *hh>cpb^L262R* wing discs accumulated apical F-actin in the posterior compartment (Fig. 3C–C'' and F, *P<0.0001*). However, co-expressing full length *HA-cpa* in these tissues suppressed the apical F-actin accumulation due to the presence Cpb^L262R (Fig. 3D–D'' and F, *P<0.0001*). Thus, forcing Cpa levels tethers the effects of Cpb^L262R on F-actin. In contrast, F-actin accumulation was strongly enhanced when *cpb^L262R* was co-expressed with *HA-cpa^ΔABD* (Fig. 3E–E'' and F, *P<0.0001*). Moreover, Cpb^L262R, like full length Cpb, enhances HA-Cpa^ΔABD levels and triggered its relocalization to the apical cell membrane (Fig. 2I–I''). We conclude that forms of CP with reduced capping activity inhibit wild type CP to restrict F-actin accumulation, most likely by tethering barbed ends, preventing the recruitment of wild type CP.

CP and forms of CP with dominant negative effects on F-actin have opposite effects on tissue growth

Decreasing or increasing CP levels has opposite effects on F-actin levels (Fig. 3F and [25]). Because loss of *CP* induces overgrowth of the wing disc epithelium by promoting Yki activity [7,9], we asked of overexpressing *cpa* and *cpb* has an opposite effect on tissue growth. Indeed, overexpressing full length *HA-cpa* and *cpb*

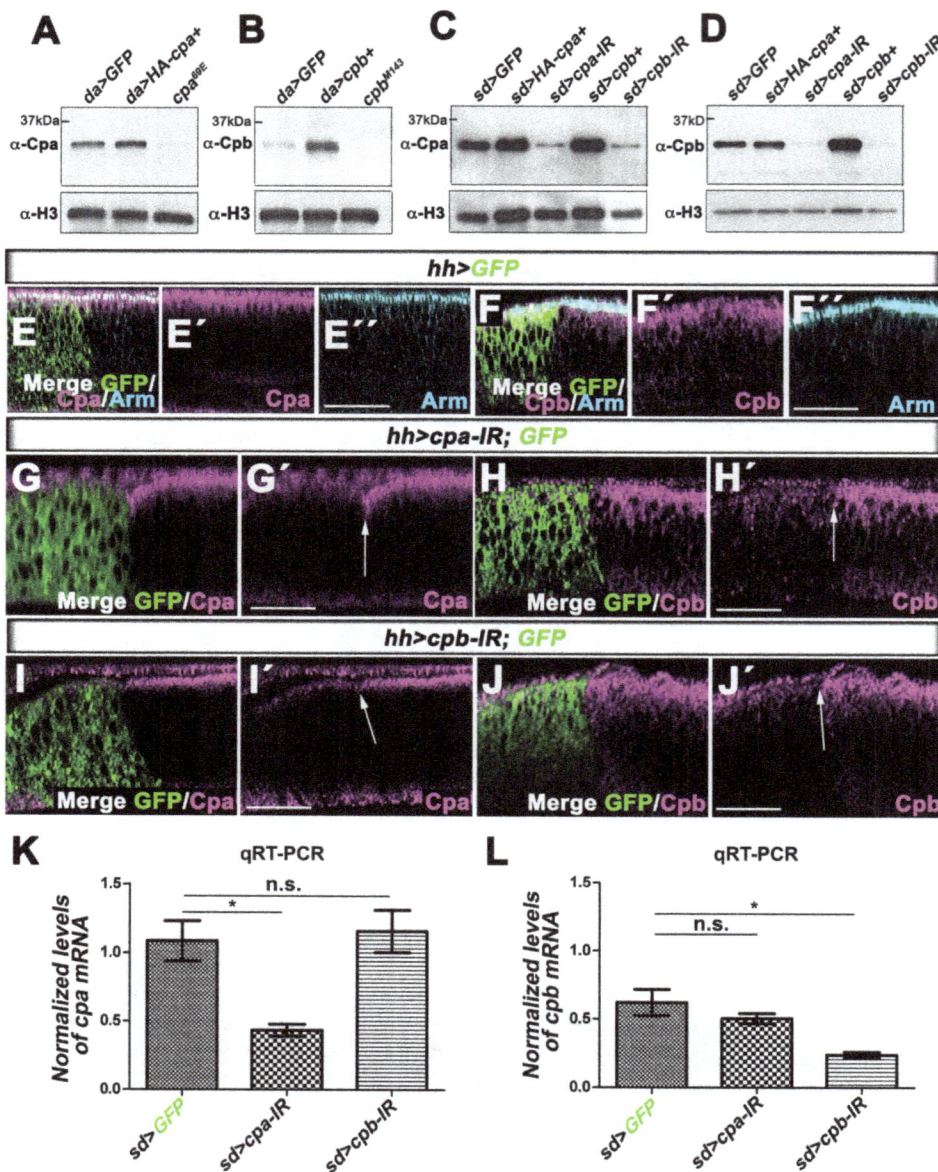

Figure 1. Loss of *cpa* or *cpb* reduces both Cpa and Cpb protein levels. (A) western blot on protein extracts from embryos expressing UAS-*mCD8GFP* (lane 1) or *UAS-HA-cpa^{89E}* (lane 2) under *da*-Gal4 control or homozygote mutant for the *cpa^{69E}* allele (lane 3), blotted with anti-Cpa (upper panel) and anti-H3 (lower panel). (B) western blot on protein extracts from embryos expressing UAS-*mCD8GFP* (lane 1) or *UAS-cpb^7* (lane 2) under *da*-Gal4 control or homozygote mutant for the *cpb^{M143}* allele (lane 3), blotted with anti-Cpb (upper panel) and anti-H3 (lower panel). (C and D) western blots on protein extracts from wing imaginal discs expressing UAS-*mCD8GFP* (lane 1) or *UAS-HA-cpa^{89E}* (lane 2) or *UAS-cpa-IR^{C10}* (lane 3) or *UAS-cpb^7* (lane 4) or *UAS-cpb-IR^{45668}* (lane 5) under *sd*-Gal4 control, blotted with (C) anti-Cpa (upper panel) and anti-H3 (lower panel) or (D) anti-Cpb (upper panel) and anti-H3 (lower panel). (E–E" to J–J') optical cross sections through distal wing disc epithelium of third instar larvae with apical side up in which *hh*-Gal4 drives (E–E" and F–F") UAS-*mCD8-GFP* (green in E and F) and (G–G' and H–H') *UAS-cpa-IR^{C10}* or (I–I' and J–J') *UAS-cpb-IR^{45668}*. Discs are stained with (E–E", G–G' and I–I') anti-Cpa (magenta) or (F–F", H–H' and J–J') anti-Cpb (magenta) and (E–E" and F–F") anti-Arm. The arrows in G', H', I' and J' mark the limits of the posterior compartment boundary. The scale bars represent 15 μm. (K and L) graphs of (K) *cpa* or (L) *cpb* mRNA levels measured by five independent qRT-PCR in wing imaginal discs expressing UAS-*mCD8GFP* (lane 1) or *UAS-cpa-IR^{C10}* (lane 2) or *UAS-cpb-IR^{45668}* (lane 3) under *sd*-Gal4 control. (K) the mean for *sd>GFP* is 1.084; for *sd>cpa-IR^{C10}* is 0.4328; for *sd>cpb-IR^{45668}* is 1.155. $P<0.0027$ for comparison of lane 1 and 2. (L) the mean for *sd>GFP* is 0.6210; for *sd>cpa-IR^{C10}* is 0.5037; for *sd>cpb-IR^{45668}* is 0.2375. $P<0.0049$ for comparison of lane 1 and 3. n.s. indicates a non-significant P. Error bars indicate s.e.m.

in the wing primordium using the *nubbin*-Gal4 (*nub*-Gal4) driver significantly reduced the size of the adult wing (Fig. 4A, compare *nub>GFP* control wing in green to *nub>cpa$^+$*, *cpb$^+$* wing in magenta and F; $P<0.0151$), but does not affect cell survival [21]. Thus, tight CP levels are critical to control tissue growth.

To determine if CP controls tissue growth via F-actin regulation, we analyzed the effect of expressing forms of *cpa* and *cpb* that have dominant negative effects on F-actin on wing growth. Expressing *HA-cpa$^{\Delta ABD}$* and *cpb* (Fig. 4B and F, $P<0.0001$) or *cpb^{L262R}* alone (Fig. 4C and F, $P<0.0001$) or combined with *HA-cpa$^{\Delta ABD}$* (Fig. 4E and F, $P<0.0001$) under *nub*-Gal4 control, not

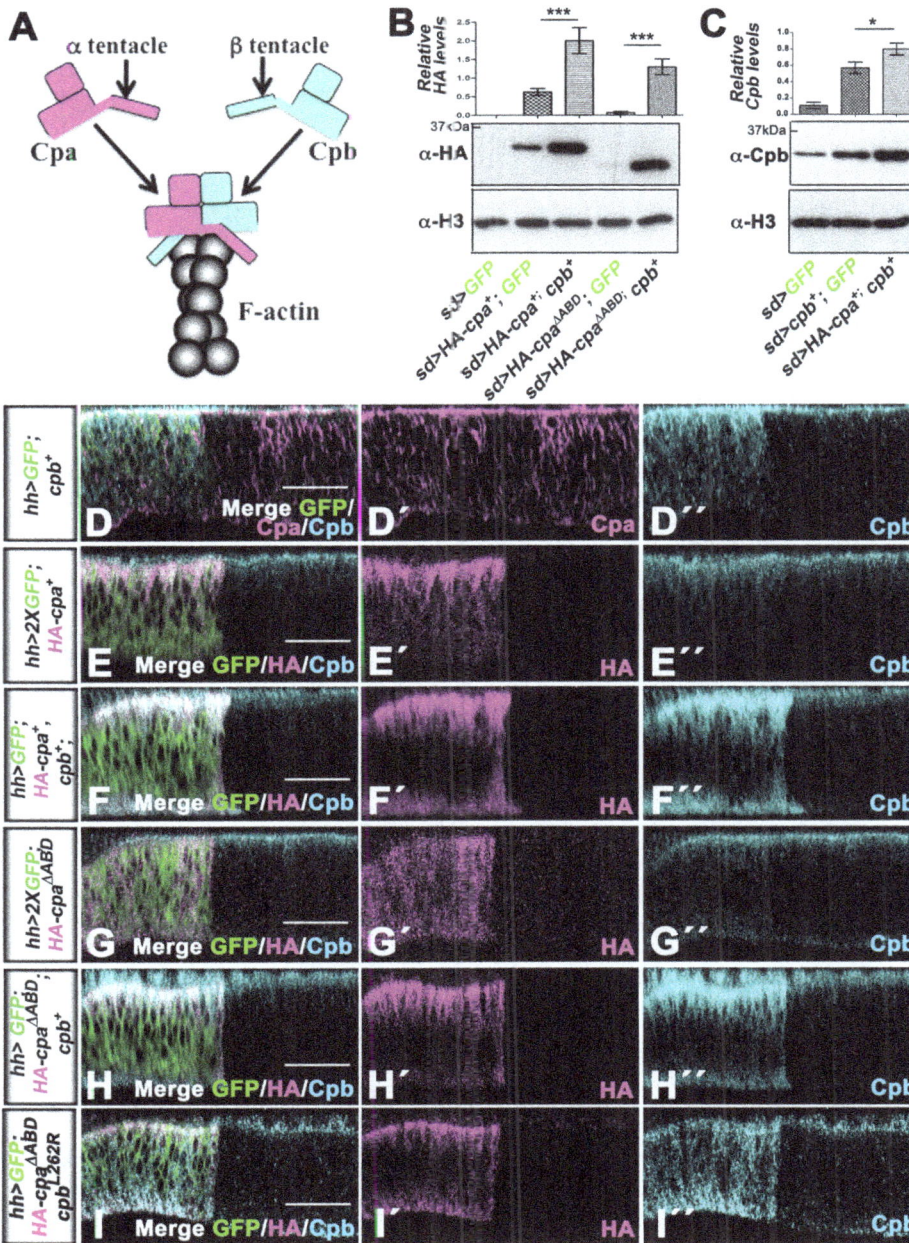

Figure 2. Increasing the levels of individual CP subunits alone has no effect on the endogenous levels of the other subunit, while co-expressing *HA-cpa* or *HA-cpa^ΔABD* and *cpb* enhance synergistically the levels of both subunits. (A) model by which Cpa and Cpb cap F-actin barbed ends via the interaction of α and β tentacles with actin. (B) western blot on protein extracts from wing discs expressing UAS-*mCD8-GFP* (lane 1) or UAS-*mCD8-GFP* and UAS-*HA-cpa^89E* (lane 2) or UAS-*HA-cpa^89E* and UAS-*cpb^7* (lane 3) or UAS-*mCD8-GFP* and UAS-*HA-cpa^ΔABD* (lane 4) or UAS-*HA-cpa^ΔABD* and UAS-*cpb^7* (lane 5) under *sd*-Gal4 control, blotted with anti-HA (middle panel) and anti-H3 (lower panel). The means for lane 1 is 0, for lane 2 is 0.6250, for lane 3 is 2, for lane 4 is 0.0667, for lane 5 is 1.300. Error bars indicate s.e.m.. $P<0.0092$ for comparison of lanes 2 and 3 and of lanes 4 and 5. (B) western blot on protein extracts from wing discs expressing UAS-*mCD8-GFP* (lane 1) or UAS-*mCD8-GFP* and UAS-*cpb^7* (lane 2) or UAS-*cpb^7* and UAS-*HA-cpa^89E* (lane 3) under *sd*-Gal4 control, blotted with anti-Cpb (middle panel) and anti-H3 (lower panel). The upper panels in B and C represent a quantification of relative (B) HA or (C) Cpb intensity signals for each genetic combination, measured by 4 independent blots. The means for lane 1 is 0.1088, for lane 2 is 0.5699, for lane 3 is 0.7982. Error bars indicate s.e.m. $P<0.0182$ for comparison of lanes 2 and 3. (D–D″ to I–I″) optical cross sections through distal epithelia of third instar wing imaginal discs with apical sides up and posterior sides to the left in which *hh*-Gal4 drives (D–D″) UAS-*cpb^7* and one copy of UAS-*mCD8-GFP* (green in D) or (E–E″) UAS-*HA-cpa^89E* and two copies of UAS-*mCD8-GFP* (green in E) or (F–F″) UAS-*HA-cpa^89E*, UAS-*cpb^7* and one copy of UAS-*mCD8-GFP* (green in F) or (G–G″) UAS-*HA-cpa^ΔABD* and two copies of UAS-*mCD8-GFP* (green in G) or (H–H″) UAS-*HA-cpa^ΔABD*, UAS-*cpb^7* and one copy of UAS-*mCD8-GFP* (green in H) or (I–I″) UAS-*cpb^L262R*, UAS-*HA-cpa^ΔABD* and one copy of UAS-*mCD8-GFP* (green in I). Discs are stained with anti-Cpb (Cyan blue) and (D–D″) anti-Cpa (magenta) or (E–E″ to I–I″) anti-HA (magenta), which reveals (E–E″ and F–F″) *HA-cpa^89E* or (G–G″ to I–I″) *HA-cpa^ΔABD* expression. The scale bars represent 15 μm.

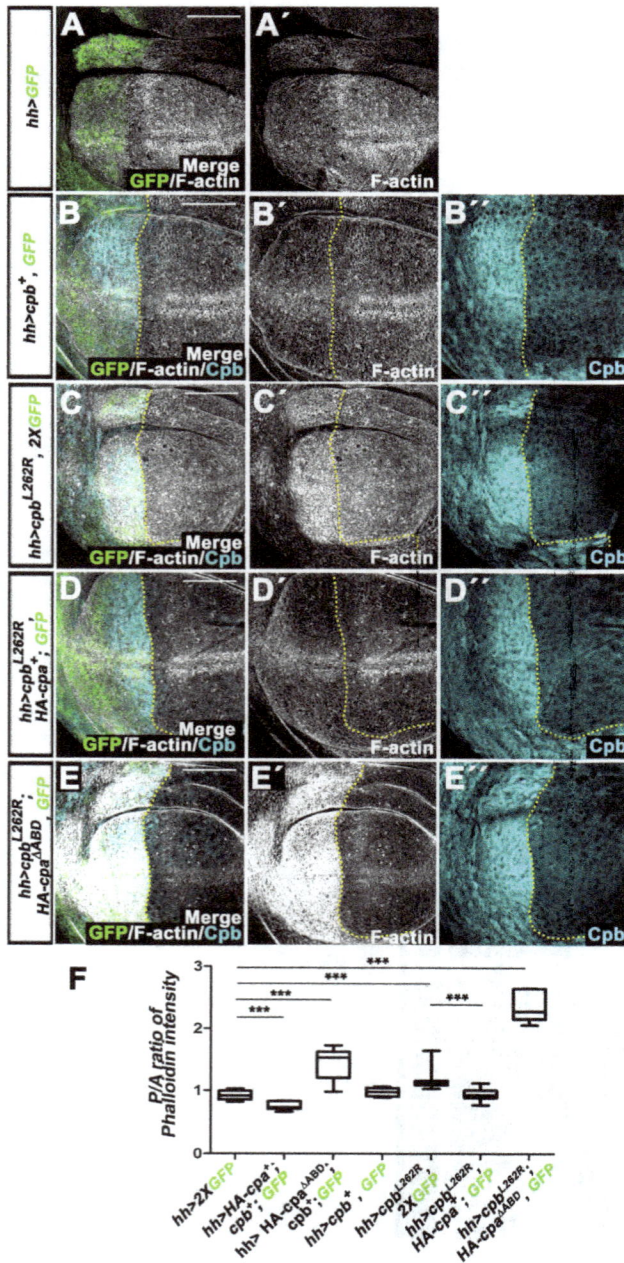

Figure 3. Overexpressing *HA-cpa* suppresses the apical F-actin accumulation of *cpb^{L262R}*-expressing wing discs, whereas *HA-cpa^{ΔABD}* expression has the opposite effect. (A–A″ to E–E″) standard confocal sections of the apical cell membrane of third instar wing imaginal discs with dorsal sides up and posterior sides to the left, expressing (A–A′) one copy of UAS-*mCD8-GFP* (green in A) or (B–B″) UAS-*cpb^7* and one copy of UAS-*mCD8-GFP* (green in B) or (C–C″) UAS-*cpb^{L262R}* and two copies of UAS-*mCD8-GFP* (green in C) or (D–D″) UAS-*cpb^{L262R}*, UAS-*HA-cpa^{89E}* and one copy of UAS-*mCD8-GFP* (green in D) or (E–E″) UAS-*cpb^{L262R}*, UAS-*HA-cpa^{ΔABD}* and one copy of UAS-*mCD8-GFP* (green in E) under *hh*-Gal4 control. Discs are stained with Phalloidin (white) to mark F-actin and (B–B″ to E–E″) anti-Cpb (cyan blue). The yellow lines outline the anterior-posterior compartment boundary. The scale bars represent 30 μm. (F) Mean intensity of the ratio of Phalloidin signal between posterior and anterior wing compartments of *hh*-Gal4 driving two copies of UAS-*mCD8-GFP* (lane 1) or UAS-*HA-cpa^{89E}*, UAS-*cpb^7* and one copy of UAS-*mCD8-GFP* (lane 2) or UAS-*HA-cpa^{ΔABD}*, UAS-*cpb^7* and one copy of UAS-*mCD8-GFP* (lane 3) or UAS-*cpb^7* and one copy of UAS-*mCD8-GFP* (lane 4) or UAS-*cpb^{L262R}* and two copies of UAS-

mCD8-GFP (lane 5) or UAS-*cpb^{L262R}*, UAS-*HA-cpa^{89E}* and one copy of UAS-*mCD8-GFP* (lane 6) or UAS-*cpb^{L262R}*, UAS-*HA-cpa^{ΔABD}* and one copy of UAS-*mCD8-GFP* (lane 7). The mean for lane 1 is 0.922 (n = 12) for lane 2 is 0.775 (n = 8), for lane 3 is 1.435 (n = 10), for lane 4 is 0.977 (n = 10), for lane 5 is 1.175 (n = 16), for lane 6 is.0.937 (n = 14), for 7 is 2.348 (n = 6). Error bars indicate s.e.m.. *** indicate P<0.0001.

only promoted apical F-actin accumulation (Fig. 3), but also enhanced significantly the growth of adult wings. Strikingly, expressing *HA-cpa* suppressed the overgrowth of *nub>cpb^{L262R}* wings (Fig. 4D and F, P<0.0001), indicating that the effect of Cpb^{L262R} on F-actin and tissue growth is dependent on the levels of full length Cpa. Because altering the levels or activity of CP did not affect the density of wing hairs (Fig. 4A′, B′, C′ D′ and E′), which develop from one single cell, the CP-dependent growth defects most likely result from changes in proliferation rate rather than alteration of cell size. We conclude that a CP-dependent reduction of F-actin levels correlates with tissue undergrowth, while a CP-dependent increase in F-actin levels is associated with tissue overgrowth.

The α tentacle is not absolutely required to form a functional heterodimer

Because the heterodimer formed between HA-Cpa^{ΔABD} and Cpb appears to be recruited at F-actin barbed ends, we tested if HA-Cpa^{ΔABD} can partially compensate for the loss of endogenous Cpa. Expressing *cpa-IR* under *sd*-Gal4 control induced the activation of Caspase 3 in numerous cells in the distal wing disc epithelium (Fig. 5A–A′). Apoptosis was almost fully suppressed by overexpressing full length *HA-cpa* (Fig. 5B–B′ and G; P<0.0001). Expressing *HA-cpa^{ΔABD}* also significantly prevented apoptosis of *sd>cpa-IR* wing discs, although to a much weaker extent than *HA-cpa* (Fig. 5C–C′ and G; P<0.0005). These effects were not only due to titration of the *cpa* dsRNAs by the overexpressed *cpa* constructs as *HA-cpa* (Fig.5E–E″ and H) or *HA-cpa^{ΔABD}* (Fig. 5F–F″ and H; P<0.0048) also rescued apoptosis of clones mutant for a *cpa* allele. Expressing *HA-cpa* or *HA-cpa^{ΔABD}* in *sd>cpa-IR* wing discs also partially restored Cpa (Fig. 5I) and Cpb (Fig. 5J) levels, as assessed by Western blot. Quantification of the ratio of Cpb signals between the posterior and anterior compartments of wing discs expressing *cpa-IR* under *hh*-Gal4 control showed that knocking-down *cpa* reduced Cpb levels in the posterior compartment compared to *hh>GFP* control (Fig. 5K). This decrease in Cpb levels was significantly alleviated by the presence of HA-Cpa^{ΔABD} (Fig. 5K P<0.0085). We conclude that in the absence of wild type Cpa, Cpa^{ΔABD} s capable of forming a functional heterodimer with Cpb, which prevents apoptosis.

Cpb compensates for a reduction in *cpa* by enhancing *cpa* mRNA levels and *vice versa*

Interestingly, co-expressing *cpb* with *HA-cpa^{ΔABD}* almost fully suppressed apoptosis of wing discs knocked-down for *cpa* (Fig. 6 compare B–B′ with A–A′ and D; P<0.0001). This effect could be due to the stabilization and apical relocalization of HA-Cpa^{ΔABD} when co-expressed with *cpb* (Fig. 2H–H″). However, apoptosis of *sd>cpa-IR* wing discs was also significantly suppressed by overexpressing *cpb* alone (Fig. 6C–C′ and D; P<0.0001). Conversely, expressing *HA-cpa* in tissues knocked-down for *cpb* (*sd>cpb-IR*) also prevented apoptosis (Fig. 7 compare B–B′ with A–A′ and C; P<0.0001).

To understand the mechanisms by which Cpa and Cpb compensate for each other's function, we tested the effect of overexpressing *cpb* on Cpa levels in *cpa*-depleted tissues. As

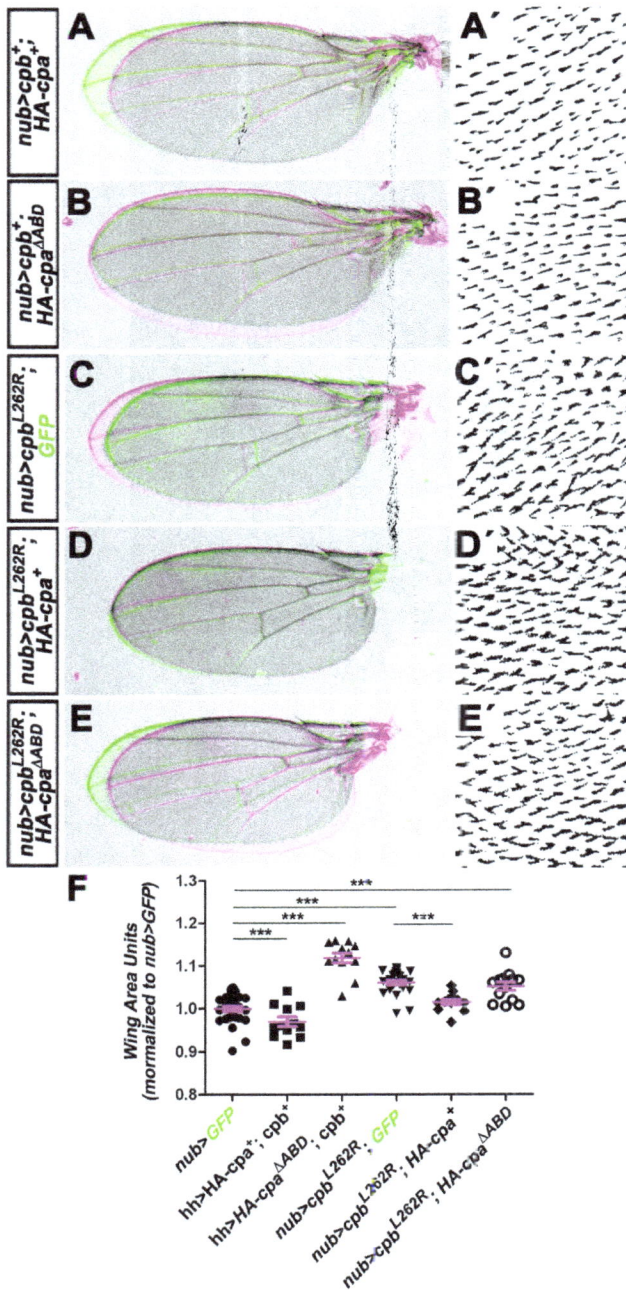

Figure 4. Overexpressing full length *HA-cpa* **and** *cpb* **prevents wing growth, while ectopic expression of** *HA-cpa^ΔABD* **and/or** *cpb^L262R* **has the opposite effect.** (A, B, C, D and E) merge between adult wings expressing in green UAS-mCD8GFP under *nub*-Gal4 control and in magenta (A) UAS-HA-cpa^89E and UAS-HA-cpa^ΔABD or (B) UAS-HA-cpa^ΔABD and UAS-cpb^7 or (C) UAS-cpb^L262R and one copy of UAS-mCD8-GFP or (D) UAS-cpb^L262R and UAS-HA-cpa^89E or (E) UAS-cpb^L262R and UAS-HA-cpa^ΔABD under *nub*-Gal4 control. (A', B', C' D' and E') magnification of hairs on adult wings for the genotypes shown in A, B, C, D and E. (F) quantification of relative wing size normalized to *nub*>GFP control for *nub*-Gal4 driving UAS-mCD8-GFP (lane 1) or UAS-HA-cpa^89E cpb^7 (lane 2) or UAS-HA-cpa^ΔABD and UAS-cpb^7 (lane 3) or UAS-cpb^L262R and one copy of UAS-mCD8-GFP (lane 4) or UAS-cpb^L262R and UAS-HA-cpa^89E (lane 5) or UAS-cpb^L262R and UAS-HA-cpa^ΔABD (lane 6). The mean for lane 1 is 1(n = 32), for lane 2 is 0.9702 (n = 12), for lane 3 is 1.119 (n = 13), for lane 4 is 1.061 (n = 24), for lane 5 is 1.015 (n = 13), for lane 6 is 1.051 (n = 13). Error bars indicate s.e.m.. P<0.015 for comparison of lanes 1 and 2. P<0.0001 for comparison of lanes 1 and 3 or 4 or 6 and for comparison of lane 4 and 5.

previously observed, by Western Blots, Cpa (Fig. 6F) and Cpb (Fig. 6G) levels were strongly reduced in wing disc extracts knocked-down for *cpa*. Forcing *cpb* levels in these tissues enhanced the levels of both Cpa (Fig. 6F and Fig. S2) and Cpb (Fig. 6G and Fig. S2). We quantified this effect by measuring the ratio of Cpa signals between the posterior and anterior compartments of *hh*>*cpa-IR*-expressing wing discs, in the presence or absence of overexpressed *cpb*. While in control *hh*>*GFP* tissues this ratio was 0.95, knocking down *cpa* reduced this ratio to 1,34 folds (Fig. 6H; *P<0.0001*). This effect was significantly alleviated by the overexpression of *cpb* (Fig. 6H; *P<0.01*). In contrast, overexpressing *cpb* in control *hh*>*GFP* wing discs did not affect Cpa levels (Fig. 6H), indicating that Cpb enhances Cpa levels only when cells contain reduced Cpa levels. By Western Blots, *HA-cpa* also enhanced both Cpa (Fig. 7D) and Cpb (Fig. 7E) levels when expressed in tissues knocked-down for *cpb*. Thus, Cpa compensates for a reduction in *cpb* by stimulating the production of Cpb, and *vice versa*.

Using qRT-PCR, we next analyzed if overexpressing either subunits affects the mRNA levels of the other. After normalization to the *RpL32* transcript used as an internal control, we observed that whereas *cpa* (Fig. 6I, *P<0.0027*) but not *cpb* (Fig. 6K) mRNA levels were strongly reduced in wing discs knocked-down for *cpa* (*sd*>*cpa-IR*), forcing *cpb* levels in these tissues fully restored *cpa* mRNA to wild type levels (Fig. 6I; *P<0.0003*). In contrast, in wing discs that contained endogenous *cpa* and *cpb*, overexpressing *cpb*, which strongly enhanced *cpb* mRNA levels (Fig. 6L), had no significant effect on *cpa* mRNA levels (Fig. 6J). Thus, Cpb stimulates the production or stabilization of *cpa* mRNA only when Cpa levels are reduced. In the converse experiment, overexpressing *HA-cpa* in *sd*>*cpb*-depleted wing discs enhanced the levels of both *cpa* (Fig. 7F) and *cpb* (Fig. 7H; *P<0.0918*) mRNA. However, in wing discs that contained endogenous *cpa* and *cpb*, only *cpa* mRNA levels were strongly increased (Fig. 7G and I). The ability of Cpb to suppress apoptosis of *cpa*-depleted wing discs was due to the increase in *cpa* mRNA and protein levels as clones mutant for a *cpa* allele showed similar apoptotic levels in the absence or presence of overexpressing *cpb* (Fig. 6E). We conclude that Cpa compensates for a reduction in *cpb* by increasing *cpb* mRNA levels and *vice versa*.

Discussion

Cpa and Cpb regulate each other at multiple levels

Our data argue that in *Drosophila*, different pools of Cpa and/or Cpb co-exist, and they regulate each other at various levels. One level of regulation involves their reciprocal stabilization of their protein levels. First, in *Drosophila*, like in yeast, the loss of one CP subunit reduces the protein levels of the other subunit ([26] and Fig. 1) but does not affect its mRNA levels (Fig. 1 and Fig. S1). Second, co-expressing *cpa* and *cpb* in *Drosophila* tissues enhances synergistically the levels of both subunits relative to the levels of each subunit overexpressed alone (Fig 2). Third, large quantities of soluble active chicken CP can be produced in bacteria only when both subunits are co-expressed [40]. Cpa and Cpb may stabilize each other's protein levels via direct protein-protein interactions [19]. The tight interaction between both subunits may prevent the recruitment of E3 ubiquitin ligases that would otherwise target individual CP subunits for degradation by the 26S proteasome. As an heterodimer, CP has been shown to bind

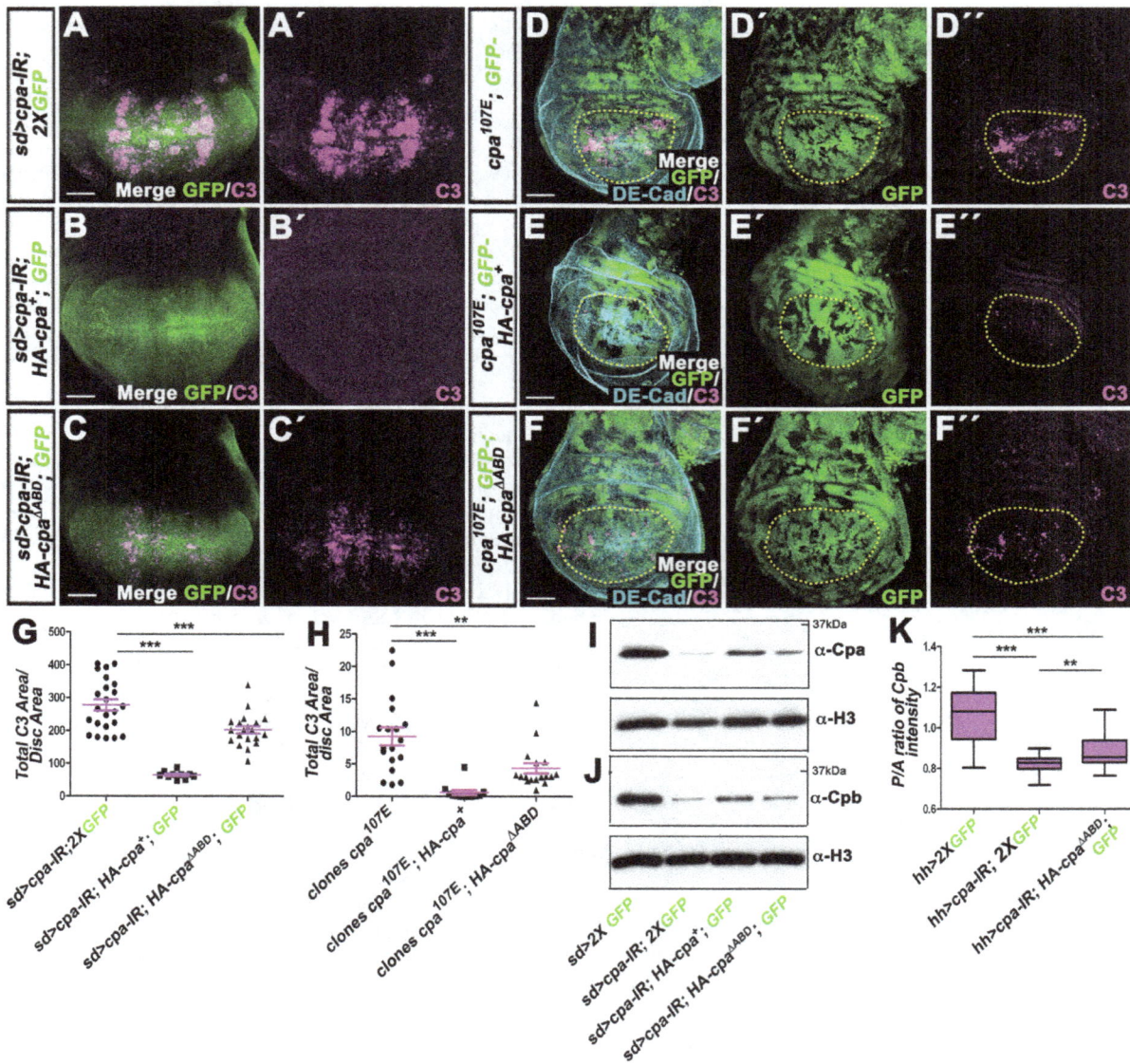

Figure 5. Expressing *HA-cpa* or *HA-cpa^{ΔABD}* suppresses apoptosis and restores Cpb levels of wing discs knocked-down for *cpa*. (A–A' to F–F') standard confocal sections of third instar wing imaginal discs with dorsal sides up. (A–A' to C–C') *sd*-Gal4 driving (A–A') UAS-*cpa-IR^{C10}* and two copies of UAS-*mCD8-GFP* (green in A) or (B–B') UAS-*cpa-IR^{C10}*, UAS-*HA-cpa^{89E}* and one copy of UAS-*mCD8-GFP* (green in B) or (C–C') UAS-*cpa-IR^{C10}*, UAS-*HA-cpa^{ΔABD}* and one copy of UAS-*mCD8-GFP* (green in C). (D–D'' to F–F'') *T155*-Gal4; UAS-*flp* induced *cpa^{107E}* mutant clones marked by the absence of GFP (green) and expressing (E–E'') UAS-*HA-cpa^{89E}* or (F–F'') UAS-*HA-cpa^{ΔABD}* in the whole wing disc epithelium. Discs are stained with anti-activated-Caspase 3 (magenta), which monitors DRONC activation and (D–D'' to F–F'') anti-DE-Cad (cyan blue). The scale bars represent 30 μm. (G) quantification of total C3 area per disc area for the three genotypes shown in A–A' to C–C'. The mean for *sd*>*cpa-IR^{C10}*, 2XGFP is 92.4 (n = 23); for *sd*>*cpa-IR^{C10}*, HA-cpa^{89E}, 1XGFP is 0.7 (n = 10); for *sd*>*cpa-IR^{C10}*, HA-cpa^{ΔABD}, 1XGFP is 51.4 (n = 20). Error bars indicate s.e.m. P<0.0001 for comparison of lane 1 and 2. P<0.0005 for comparison of lane 1 and 3. (H) quantification of total C3 area per disc area for the three genotypes shown in D–D'' to F–F''. The means for *T155*>*flp*; *cpa^{107E}* is 9.228 (n = 18); for *T155*>*flp*; *cpa^{107E}*; UAS-*HA-cpa^{89E}* is 0.608 (n = 12); for *T155*>*flp*; *cpa^{107E}*; UAS-*HA-cpa^{ΔABD}* is 4.329 (n = 17). Error bars indicate s.e.m. P<0.0001 for comparison of *T155*>*flp*; *cpa^{107E}* and *T155*>*flp*; *cpa^{107E}*; UAS-*HA-cpa^{89E}* and P<0.0048 for comparison of *T155*>*flp*; *cpa^{107E}* and *T155*>*flp*; *cpa^{107E}*; UAS-*HA-cpa^{ΔABD}*. (I and J) western blots on protein extracts from wing discs expressing two copies of UAS-*mCD8-GFP* (lane 1) or UAS-*cpa-IR^{C10}* and two copies of UAS-*mCD8-GFP* (lane 2) or UAS-*cpa-IR^{C10}* and UAS-*HA-cpa^{89E}* and one copy of UAS-*mCD8-GFP* (lane 3) or UAS-*cpa-IR^{C10}* and UAS-*HA-cpa^{ΔABD}* and one copy of UAS-*mCD8-GFP* (lane 4) under *sd*-Gal4 control, blotted with (I) anti-Cpa (upper panel) and anti-H3 (lower panel) or (J) anti-Cpb (upper panel) and anti-H3 (lower panel). (K) mean intensity of the ratio of Cpb intensity signals between posterior and anterior wing compartments of *hh*-Gal4 driving two copies of UAS-*mCD8-GFP* (lane 1) or UAS-*cpa-IR^{C10}* and two copies of UAS-*mCD8-GFP* (lane 2) or UAS-*cpa-IR^{C10}* and UAS-*HA-cpa^{ΔABD}* and one copy of UAS-*mCD8-GFP* (lane 3). The mean for lane 1 is 1.064 (n = 20), for lane 2 is 0.822 (n = 17), for lane 3 is 0.883 (n = 24). Error bars indicate s.e.m.. P<0.0001 for comparison of lanes 1 and 2 or 3 or P<0.0085 for comparison of lanes 2 and 3.

to the fast polymerizing ends of actin filaments, preventing further addition of actin monomers [41,42] and to restrict F-actin accumulation in *Drosophila* tissues [25,27]. In addition, Cpa and Cpb appear to show some function on their own as overexpressing *cpb* rescues apoptosis of wing discs knocked-down for *cpa* and *vice versa* (Fig. 6 and 7). Overexpression of *cpb* alone is also sufficient to

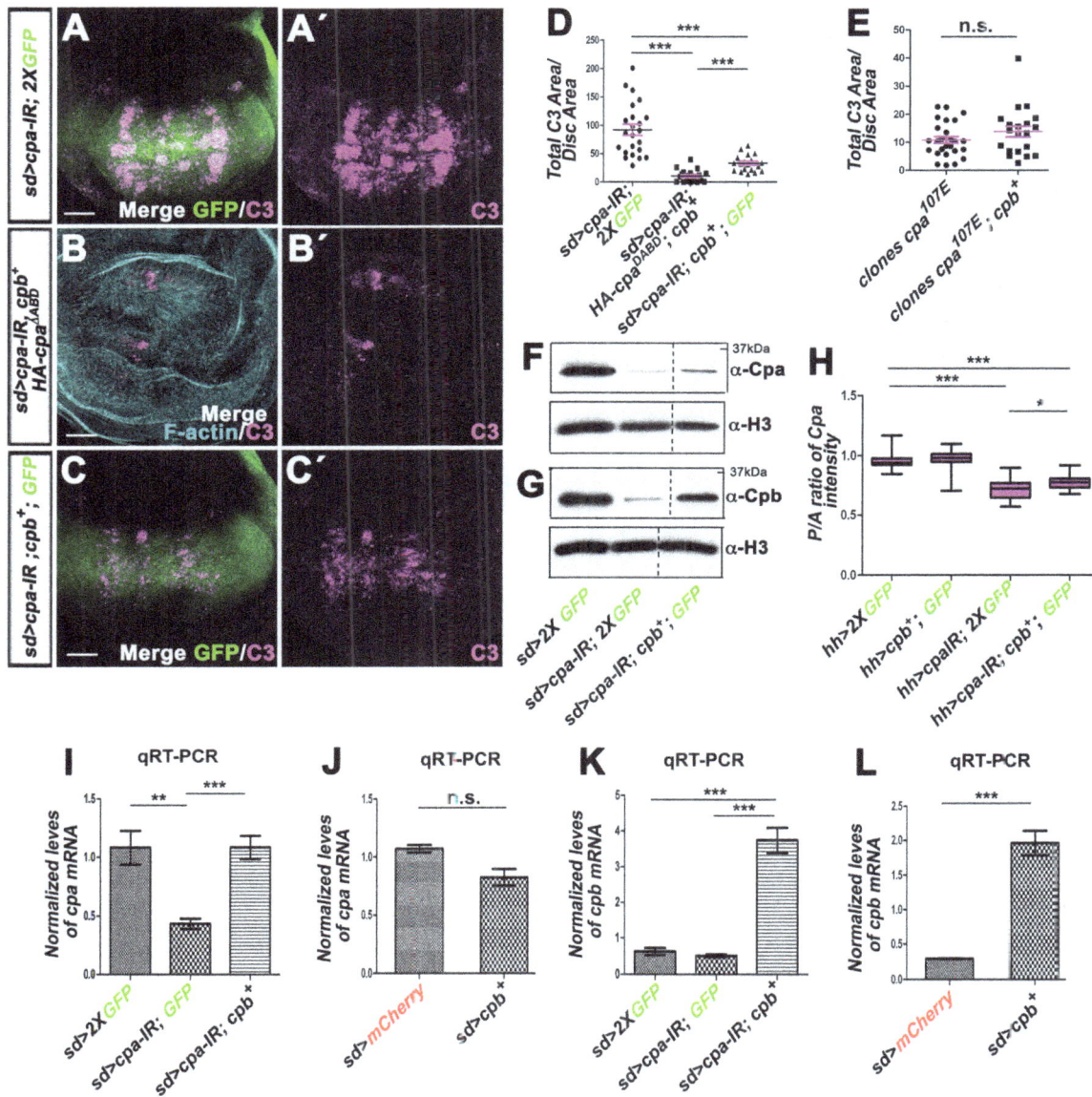

Figure 6. Overexpressing *cpb* in wing discs knocked-down for *cpa*, restores *cpa* mRNA and protein levels and suppresses apoptosis.
(A–A′ to C–C′) standard confocal sections of third instar wing imaginal discs with dorsal sides up, expressing (A–A″) UAS-*cpa-IR^{C10}* and two copies of UAS-*mCD8-GFP* (green in A) or (B–B′) UAS-*cpa-IR^{C10}*, UAS-HA-*cpa^{ΔABD}* and UAS-*cpb^7* or (C–C′) UAS-*cpa-IR^{C10}*, UAS-*cpb^7* and one copy of UAS-*mCD8-GFP* (green in C) under *sd*-Gal4 control. Discs are stained with anti-activated-Caspase 3 (magenta), which monitors DRONC activation and (B–B′) Phalloidin (cyan blue in B) to underline wing disc shape. The scale bars represent 30 μm. (D) quantification of total C3 area per disc area for the genotypes *sd>cpa-IR^{C10}*, *2XGFP* (lane 1); *sd>cpa-IR^{C10}*, HA-*cpa^{ΔABD}*, *cpb^7* (lane 2) and *sd>cpa-IR^{C10}*, *cpb^7*, *1XGFP* (lane 3). The means for lane 1 is 92.4 (n = 23); for lane 2 is 10.61 (n = 19); for lane 3 is 32.9 (n = 20). Error bars indicate s.e.m. $P<0.0001$ for comparison of lane 1 and 2 or 3 or lane 2 and 3. (E) quantification of total C3 area per disc area for wing discs containing *T155>flp; cpa^{107E}* mutant clones (lane 1) or *T155>flp, cpa^{107E}* mutant clones expressing UAS-*cpb^7* (lane 2). The means for lane 1 is 10.80 (n = 26); for lane 2 is 13.77 (n = 20). n.s. indicates non-significant P value. (F and G) western blots on protein extracts from wing discs expressing two copies of UAS-*mCD8-GFP* (lane 1) or UAS-*cpa-IR^{C10}* and two copies of UAS-*mCD8-GFP* (lane 2) or UAS-*cpa-IR^{C10}*, UAS-*cpb^7* and one copy of UAS-*mCD8-GFP* (lane 3) under *sd*-Gal4 control, blotted with (F) anti-Cpa (upper panel) and anti-H3 (lower panel) or (G) anti-Cpb (upper panel) and anti-H3 (lower panel). Panels derive from the same experiment shown in Figure 5E and F and blots were processed in parallel (see Figure S2 showing the whole experiment). (H) mean intensity of the ratio of Cpa signals between posterior and anterior wing compartments of *hh*-Gal4 driving two copies of UAS-*mCD8-GFP* (lane 1) or UAS-*cpb^7* and one copy of UAS-*mCD8-GFP* (lane 2) or UAS-*cpa-IR^{C10}* and two copies of UAS-*mCD8-GFP* (lane 3) or UAS-*cpa-IR^{C10}* and UAS-*cpb^7* and one copy of UAS-*mCD8-GFP* (lane 4). The mean for lane 1 is 0.959 (n = 15), for lane 2 is 0.970 (n = 20), for lane 3 is 0.716 (n = 21), for lane 4 is 0.776 (n = 20). Error bars indicate s.e.m.. $P<0.0001$ for comparison of lanes 1 and 3 or 4 or $P<0.01$ for comparison of lanes 3 and 4. (I to L) graph of (I and J) *cpa* or (K and L) *cpb* mRNA levels measured by five independent qRT-PCR in wing imaginal discs expressing (I and K) two copies of UAS-*mCD8GFP* (lane 1) or UAS-*cpa-IR^{C10}* and UAS-*mCD8GFP* (lane 2) or UAS-*cpa-IR^{C10}* and UAS-*cpb^7* (lane 3) or (J and L) UAS-*mCherry* (lane 1) or UAS-*cpb^7* (lane 2) under *sd*-Gal4 control. (I) the means for lane 1 is 1.084; for lane 2 is 0.4328; for lane 3 is 1.086. $P<0.0027$ for comparison of lane 1 with 2 or $P<0.0003$ for comparison of lane 2 with 3. (J) the means for lane 1 is 1.07; for lane 2 is 0.824. n.s. indicates non-significant P value. (K) the means for lane 1 is 0.621; for lane 2 is 0.5031; for lane 3 is 3.735. $P<0.0001$ for comparison of lane 3 with 1 or 2. (L) the means for lane 1 is 0.292; for lane 2 is 1.961. $P<0.0001$ for comparison of lane 1 and 2. Error bars indicate s.e.m.

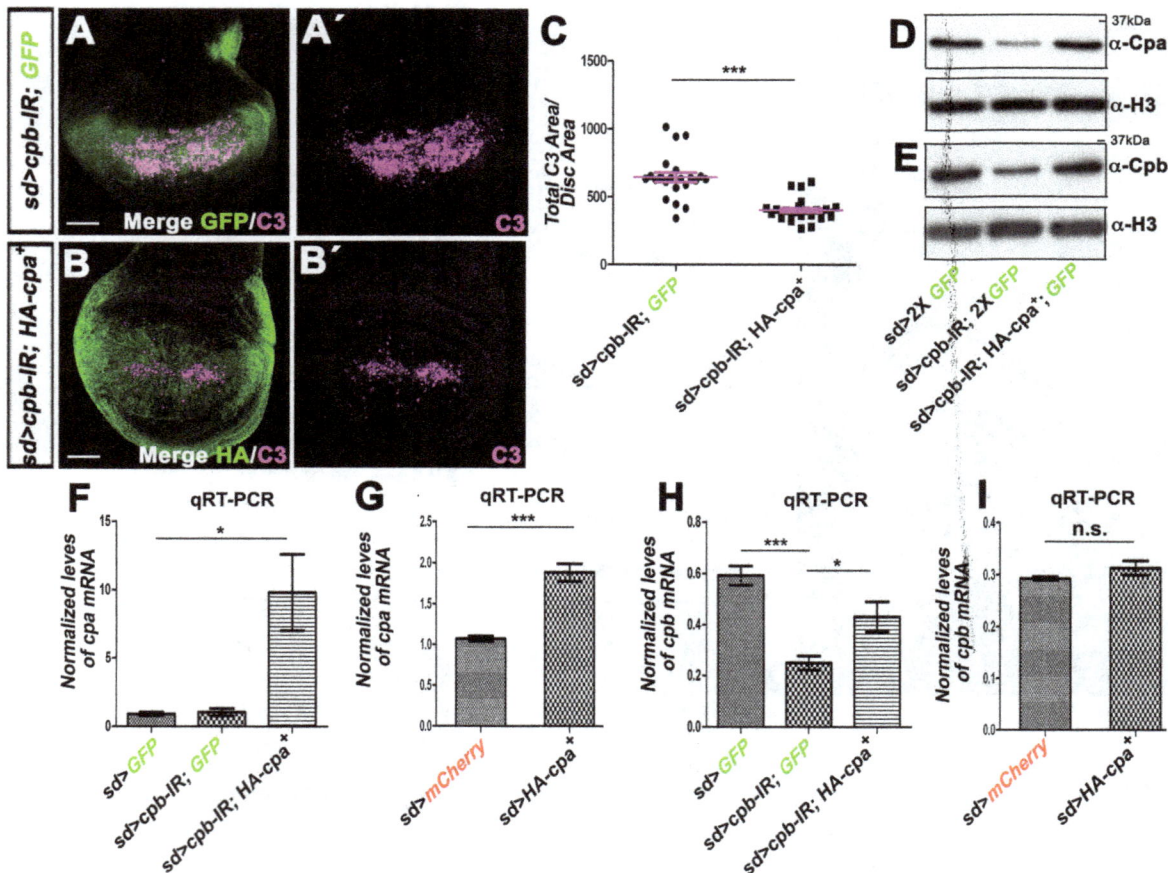

Figure 7. Overexpressing *HA-cpa* **in wing discs knocked-down for** *cpb* **restores** *cpb* **mRNA and protein levels and suppresses apoptosis.** (A–A′ and B–B′) standard confocal sections of third instar wing imaginal discs with dorsal sides up, expressing (A–A″) UAS-*cpb-IR*[45668] and one copy of UAS-*mCD8-GFP* (green in A) or (B–B′) UAS-*cpb-IR*[45668] and UAS-*HA-cpa*[89E] under *sd*-Gal4 control. Discs are stained with anti-activated-Caspase 3 (magenta), which monitors DRONC activation and (B–B′) anti-HA (green in B), reflecting *HA-cpa*[89E] expression. The scale bars represent 30 μm. (C) quantification of total C3 area per disc area for the two genotypes shown in A–A′ and B–B′. The means for *sd*>*cpb-IR*[45668], GFP is 62.19 (n = 22); for *sd*>*cpb-IR*[4566], *HA-cpa*[89E] is 26.67 (n = 32). Error bars indicate s.e.m. *P*<0.0001 for comparison of between both genotypes. (D and E) western blots on protein extracts from wing discs expressing two copies of UAS-*mCD8-GFP* (lane 1) or UAS-*cpb-IR*[45668] and two copies of UAS-*mCD8-GFP* (lane 2) or UAS-*cpb-IR*[45668] and UAS-*HA-cpa*[89E] and one copy of UAS-*mCD8-GFP* (lane 3) under *sd*-Gal4 control, blotted with (D) anti-Cpa (upper panel) and anti-H3 (lower panel) or (E) anti-Cpb (upper panel) and anti-H3 (lower panel). (F to I) graphs of (F and G) *cpa* or (H and I) *cpb* mRNA levels measured by (F and H) three or (G and I) five independent qRT-PCR in wing imaginal discs expressing (F and H) two copies of UAS-*mCD8GFP* (lane 1) or UAS-*cpb-IR*[45668] and UAS-*mCD8GFP* (lane 2) or UAS-*cpb-IR*[45668] and UAS-*HA-cpa*[89E] (lane 3) or (G and I) UAS-*mCherry* (lane 1) or UAS-*HA-cpa*[89E] under *sd*-Gal4 control. (F) the means for lane 1 is 0.90; for lane 2 is 1.04; for lane 3 is 9.8. *P*<0.033 for comparison of lane 1 with 3. (G) the means for lane 1 is 1.07; for lane 2 is 1.88. *P*<0.0001 for comparison of lane 1 and 2. (H) the means for lane 1 is 0.59; for lane 2 is 0.25; for lane 3 is 0.4319. *P*<0.0018 for comparison of lane 1 and 2 or *P*<0.048 for comparison of lane 2 and 3. (I) the means for lane 1 is 0.29; for lane 2 is 0.31. n.s. indicates non-significant *P* value. Error bars indicate s.e.m.

enhance the retinal defects of flies knocked down for the Cbl-interacting protein *cindr* [43] and to rescue the migration and F-actin polarization defects of *Drosophila* border cells mutant for *warts* [44]. Because individual chicken CP subunits expressed in bacteria are mainly deposited into insoluble cytoplasmic inclusion bodies but can be renaturated as active heterodimers [45], individual subunit may exist in the cell as pools of insoluble monomers. The molecular mechanism by which individual CP subunit compensates for each other's function remains to be determined. Several observations argue that this mechanism involves the production of the subunit knocked-down by the other subunit via an increase of its mRNA levels (Fig. 6 and 7). CP has been observed in the nuclei of chicken retinal and kidney epithelial cells in culture, in Madin-Darby canine kidney (MDCK) cells, in *Xenopus laevis* oocytes and bovine lens epithelial cells in culture [46,47]. Whether Cpa and Cpb influence each other's transcription in the nucleus is an interesting possibility to be tested. The protein-mRNA feedbacks between Cpa and Cpb may guarantee that a pool of functional heterodimer is present to limit F-actin polymerization. However, a CP-dependent negative feedback mechanism must exist that restricts the production of CP in excess, as forcing the expression of one of the subunit in tissues that contain endogenous CP does not enhance the mRNA and protein levels of the other subunit (Fig. 6 and 7). Because the loss of one subunit has no effect on the mRNA levels of the other subunit (Fig. 1 and Fig. S1), the CP-dependent negative feedback may act by limiting the ability of individual subunits to stimulate the production of each other's mRNAs. Thus, in addition to regulate each other's protein levels, individual CP subunit stimulates each other's mRNA production up to an optimal physiological threshold of functional heterodimers. Further experiments are necessary to elucidate the protein-mRNA feedback loop mechanisms, which operate between both subunits.

Capping activity of the CP heterodimer at actin filament barbed ends

Our observations argue that *in vivo* the actin-binding domain of Cpa is not absolutely required to form a functional CP heterodimer, as HA-Cpa$^{\Delta ABD}$ partially compensates for the loss of endogenous Cpa (Fig. 5). Consistent with our observations, in actin assembly assays, a mutant form of the chicken α subunit that lacks the α tentacle is able to cap F-actin [12]. Nevertheless, the α tentacle may favor the interaction and therefore stabilization of the α subunit by the β. This possibility is consistent with the observation that HA-Cpa$^{\Delta ABD}$ is found in the cell at much lower levels than full length HA-Cpa (Fig. 2) despite both transgenes being inserted at the same locus in the fly genome and therefore likely expressed at similar levels [21]. Consistent with this hypothesis, Arginine 259 of the chicken α1 tentacle forms side-chain hydrogen bonds with three residues of the β subunit, all residues being conserved across isoforms and species [19]. Moreover, *in vitro*, a truncated form of the chicken α1 subunit, consisting only of the C-terminal domain, retains the ability to form a heterodimer [48]. The reduced ability of HA-Cpa$^{\Delta ABD}$ to interact with Cpb may explain its inability to fully suppress apoptosis of Cpa-depleted tissues (Fig. 5) and to affect F-actin levels when overexpressed alone [21]. However, several observations indicate that the α and β tentacles also enable full capping activity *in vivo*. First, in actin assembly assays, the C-terminus of the chicken α1 and β1 subunits are required for high-affinity capping [12]. Second, in the presence of endogenous CP, stabilizing HA-Cpa$^{\Delta ABD}$ levels by forcing *cpb* expression does not reduce F-actin levels, as does overexpressed *HA-cpa/cpb*, but instead, promotes F-actin accumulation (Fig. 3 and [21]). Third, replacing leucine 262 of the chicken β subunit has no effect on protein stability and global structure but decreases the capping affinity significantly [12,20]. Fourth, identical mutations in the β orthologs induces F-actin accumulation in *Drosophila* tissues (Fig. 3) and disrupts the sarcomere of mouse heart [24]. Thus, we propose that the heterodimers formed between HA-Cpa$^{\Delta ABD}$ and Cpb or between Cpb^{L262R} and Cpa are recruited to F-actin barbed ends and cap actin filaments less efficiently than wild type CP. The low capping activity of the HA-Cpa$^{\Delta ABD}$/Cpb heterodimer is sufficient to partially compensate for the loss of Cpa. However, in the presence of endogenous CP, the HA-Cpa$^{\Delta ABD}$/Cpb heterodimers compete with wild type Cpa/Cpb heterodimers for binding the barbed ends of F-actin, which can lead to defects in F-actin.

Tight regulation of CP levels is critical to control tissue growth

CP appears to act as a gatekeeper, which limits the development of cancer-related processes. Loss of the α subunit promotes Yki/YAP/TAZ-dependent proliferation in *Drosophila* epithelia and in human cells [9,31], causes a significantly increase in gastric cancer cell migration and is associated with cancer-related death [10]. In contrast, increasing CP levels has opposite effects: it reduces tissue growth (Fig. 4) and prevents Src-mediated tumour development in *Drosophila* [21], and significantly restricts gastric cancer cell migration [10]. Several of our observations argue that the function of CP on tissue growth involves its F-actin capping activity. First expressing *cpb*^{L262R}, which contains a single point mutation affecting the capping activity [23], induces F-actin accumulation (Fig. 3) and wing overgrowth (Fig. 4). Moreover, CP-dependent F-actin accumulation correlates with tissue overgrowth, whereas

tissue undergrowth is associated with a CP-dependent reduction in F-actin (Fig. 3 and 4). Consistent with these observations, other actin regulators have been shown to control Yki/YAP/TAZ dependent tissue growth [7,9,31]. Thus, a reduction or an increase of CP levels has deleterious consequences on tissue growth, implying that it must be tightly regulated. This may be achieved in part by the ability of Cpa and Cpb to stimulate or limit the production of each other in conditions of lower or higher CP levels respectively, assuring that a pool of functional CP heterodimer is produced in sufficient quantities in the cell to prevent cancer development but not in excess to sustain proper tissue growth.

Supporting Information

Figure S1 Reducing *cpa* or *cpb* levels reduces both Cpa and Cpb protein levels. (A) western blot on protein extracts from first instar larvae, either *white minus* (lane 1) or homozygote mutant for *cpa*^{69E} (lane 2) or homozygote mutant for *cpb*^{M143} (lane 3), blotted with (upper panel) anti-Cpa (upper bands) and anti-Cpb (lower band) and (lower panel) anti-H3. (B–B″ to E–E″) standard confocal sections of third instar wing imaginal discs, containing (B–B″ and C–C″) *T155*-Gal4; UAS-*flp* induced *cpa*^{69E} mutant clones marked by the absence of GFP (green) or (D–D″ and E–E″) heat shocked-induced *cpb*^{M143} mutant clones marked by the absence of GFP (green). Discs are stained with (B–B″ and E–E″) anti-Cpa (magenta) or (C–C″ and E–E″) anti-Cpb (magenta). The scale bars represent 15 μm. (F and G) graphs of (F) *cpa* or (G) *cpb* mRNA levels measured by three independent qRT-PCR in first instar larvae expressing UAS-mCD8-GFP (lane 1) or UAS-*cpa*-IR^{C10} (lane 2) or UAS-*cpb*-IR^{45668} (lane 3) under *da*-Gal4 control. (F) The means for lane 1 is 7.04; for lane 2 is 1.13; for lane 3 is 5.91. Error bars indicate s.e.m.. *P<0.015* for comparison of lane 1 and 2. (F) The means for lane 1 is 1.97; for lane 2 is 1.96; for lane 3 is 0.46. Error bars indicate s.e.m.. *P<0.021* for comparison of lane 1 and 3. n.s. indicates non-significant *P* values.

Figure S2 Expressing *HA-cpa* or *HA-cpa*$^{\Delta ABD}$ or *cpb* in wing discs knocked down for *cpa* restores Cpa and Cpb levels. Western blots on protein extracts from wing discs expressing two copies of UAS-*mCD8-GFP* (lane 1) or UAS-*cpa-IR*^{C10} and two copies of UAS-*mCD8-GFP* (lane 2) or UAS-*cpa-IR*^{C10} and UAS-*HA-cpa*^{89E} and one copy of UAS-*mCD8-GFP* (lane 3) or UAS-*cpa-IR*^{C10} and UAS-*HA-cpa*$^{\Delta ABD}$ and one copy of UAS-*mCD8-GFP* (lane 4) or UAS-*cpa-IR*^{C10} and UAS-*HA-cpa*^{ABD}, which contains the last 28 amino acids of the Cpa C-terminus and one copy of UAS-*mCD8-GFP* (lane 5) or UAS-*cpa-IR*^{C10}, UAS-*cpb*^{7} and one copy of UAS-*mCD8-GFP* (lane 6) under *sd*-Gal4 control, blotted with (A) anti-Cpa (upper panel) and anti-H3 (lower panel) or (B) anti-Cpb (upper panel) and anti-H3 (lower panel).

Table S1 Intron-exon-specific primers used to quantify *cpa*, *cpb* and *RpL32* mRNA levels by qRT-PCR.

Acknowledgments

We thank the Bloomington *Drosophila* Stock Center, the National Institute of Genetics and the Developmental Studies Hybricoma Bank for fly stocks. We are grateful to Barbara Jezowska, Beatriz García Fernández, Francesca Vasconcelos, Sonia Rosa, Vania Neves and Claire Brombey for technical support and advice on qRT-PCR. The manuscript was improved by the critical comments of Claudine Chaouiya and Alekos Athanasiadis.

Author Contributions

Conceived and designed the experiments: ARA PG JLW FJ. Performed the experiments: ARA PG JLW FJ. Analyzed the data: ARA PG FJ. Wrote the paper: FJ.

References

1. Lambrechts A, Van Troys M, Ampe C (2004) The actin cytoskeleton in normal and pathological cell motility. Int J Biochem Cell Biol 36: 1890–1909.
2. Stevenson RP, Veltman D, Machesky LM (2012) Actin-bundling proteins in cancer progression at a glance. J Cell Sci 125: 1073–1079.
3. Shah V, Braverman R, Prasad GL (1998) Suppression of neoplastic transformation and regulation of cytoskeleton by tropomyosins. Somat Cell Mol Genet 24: 273–280.
4. Mahadev K, Raval G, Bharadwaj S, Willingham MC, Lange EM, et al. (2002) Suppression of the transformed phenotype of breast cancer by tropomyosin-1. Exp Cell Res 279: 40–51.
5. Mammoto A, Ingber DE (2009) Cytoskeletal control of growth and cell fate switching. Curr Opin Cell Biol 21: 864–870.
6. Yamamoto N, Okano T, Ma X, Adelstein RS, Kelley MW (2009) Myosin II regulates extension, growth and patterning in the mammalian cochlear duct. Development 136: 1977–1986.
7. Fernandez BG, Gaspar P, Bras-Pereira C, Jezowska B, Rebelo SR, et al. (2011) Actin-Capping Protein and the Hippo pathway regulate F-actin and tissue growth in Drosophila. Development 138: 2337–2346.
8. Jezowska B, Fernandez BG, Amandio AR, Duarte P, Mendes C, et al. (2011) A dual function of Drosophila capping protein on DE-cadherin maintains epithelial integrity and prevents JNK-mediated apoptosis. Dev Biol 360: 143–159.
9. Sansores-Garcia L, Bossuyt W, Wada K, Yonemura S, Tao C, et al. (2011) Modulating F-actin organization induces organ growth by affecting the Hippo pathway. EMBO J 30: 2325–2335.
10. Lee YJ, Jeong SH, Hong SC, Cho BI, Ha WS, et al. (2013) Prognostic value of CAPZA1 overexpression in gastric cancer. Int J Oncol 42: 1569–1577.
11. Isenberg G, Aebi U, Pollard TD (1980) An actin-binding protein from Acanthamoeba regulates actin filament polymerization and interactions. Nature 288: 455–459.
12. Wear MA, Yamashita A, Kim K, Maeda Y, Cooper JA (2003) How capping protein binds the barbed end of the actin filament. Curr Biol 13: 1531–1537.
13. Kim K, McCully ME, Bhattacharya N, Butler B, Sept D, et al. (2007) Structure/function analysis of the interaction of phosphatidylinositol 4,5-bisphosphate with actin-capping protein: implications for how capping protein binds the actin filament. J Biol Chem 282: 5871–5879.
14. Wear MA, Cooper JA (2004) Capping protein: new insights into mechanism and regulation. Trends Biochem Sci 29: 418–428.
15. Hart MC, Korshunova YO, Cooper JA (1997) Vertebrates have conserved capping protein alpha isoforms with specific expression patterns. Cell Motil Cytoskeleton 38: 120–132.
16. Schafer DA, Korshunova YO, Schroer TA, Cooper JA (1994) Differential localization and sequence analysis of capping protein beta-subunit isoforms of vertebrates. J Cell Biol 127: 453–465.
17. Hurst S, Howes EA, Coadwell J, Jones R (1998) Expression of a testis-specific putative actin-capping protein associated with the developing acrosome during rat spermiogenesis. Mol Reprod Dev 49: 81–91.
18. Cooper JA, Sept D (2008) New insights into mechanism and regulation of actin capping protein. Int Rev Cell Mol Biol 267: 183–206.
19. Yamashita A, Maeda K, Maeda Y (2003) Crystal structure of CapZ: structural basis for actin filament barbed end capping. Embo J 22: 1529–1538.
20. Kim K, Yamashita A, Wear MA, Maeda Y, Cooper JA (2004) Capping protein binding to actin in yeast: biochemical mechanism and physiological relevance. J Cell Biol 164: 567–580.
21. Fernandez BG, Jezowska B, Janody F (2014) Drosophila actin-Capping Protein limits JNK activation by the Src proto-oncogene. Oncogene.
22. Barron-Casella EA, Torres MA, Scherer SW, Heng HH, Tsui LC, et al. (1995) Sequence analysis and chromosomal localization of human Cap Z. Conserved residues within the actin-binding domain may link Cap Z to gelsolin/severin and profilin protein families. J Biol Chem 270: 21472–21479.
23. Schafer DA, Hug C, Cooper JA (1995) Inhibition of CapZ during myofibrillogenesis alters assembly of actin filaments. J Cell Biol 128: 61–70.
24. Hart MC, Cooper JA (1999) Vertebrate isoforms of actin capping protein beta have distinct functions In vivo. J Cell Biol 147: 1287–1298.
25. Janody F, Treisman JE (2006) Actin capping protein {alpha} maintains vestigial-expressing cells within the Drosophila wing disc epithelium. Development 133: 3349–3357.
26. Amatruda JF, Gattermeir DJ, Karpova TS, Cooper JA (1992) Effects of null mutations and overexpression of capping protein on morphogenesis, actin distribution and polarized secretion in yeast. J Cell Biol 119: 1151–1162.
27. Delalle I, Pfleger CM, Buff E, Lueras P, Hariharan IK (2005) Mutations in the Drosophila orthologs of the F-actin capping protein alpha- and beta-subunits cause actin accumulation and subsequent retinal degeneration. Genetics 171: 1757–1765.
28. Hopmann R, Cooper JA, Miller KG (1996) Actin organization, bristle morphology, and viability are affected by actin capping protein mutations in Drosophila. J Cell Biol 133: 1293–1305.
29. Frank DJ, Hopmann R, Lenartowska M, Miller KG (2006) Capping Protein and the Arp2/3 Complex Regulate Non-Bundle Actin Filament Assembly to Indirectly Control Actin Bundle Positioning during Drosophila melanogaster Bristle Development. Mol Biol Cell.
30. Gates J, Nowotarski SH, Yin H, Mahaffey JP, Bridges T, et al. (2009) Enabled and Capping protein play important roles in shaping cell behavior during Drosophila oogenesis. Dev Biol 333: 90–107.
31. Aragona M, Panciera T, Manfrin A, Giulitti S, Michielin F, et al. (2013) A mechanical checkpoint controls multicellular growth through YAP/TAZ regulation by actin-processing factors. Cell 154: 1047–1059.
32. Rudrapatna VA, Bangi E, Cagan RL (2013) A Jnk-Rho-Actin remodeling positive feedback network directs Src-driven invasion. Oncogene.
33. Klein T, Arias AM (1998) Different spatial and temporal interactions between Notch, wingless, and vestigial specify proximal and distal pattern elements of the wing in Drosophila. Dev Biol 194: 196–212.
34. Calleja M, Moreno E, Pelaz S, Morata G (1996) Visualization of gene expression in living adult Drosophila. Science 274: 253–255.
35. Wodarz A, Hinz U, Engelbert M, Knust E (1995) Expression of crumbs confers apical character on plasma membrane domains of ectodermal epithelia of Drosophila. Cell 82: 67–76.
36. Whited JL, Cassell A, Brouillette M, Garrity PA (2004) Dynactin is required to maintain nuclear position within postmitotic Drosophila photoreceptor neurons. Development 131: 4677–4686.
37. Lee JD, Treisman JE (2001) Sightless has homology to transmembrane acyltransferases and is required to generate active Hedgehog protein. Curr Biol 11: 1147–1152.
38. Untergasser A, Cutcutache I, Koressaar T, Ye J, Faircloth BC, et al. (2012) Primer3—new capabilities and interfaces. Nucleic Acids Res 40: e115.
39. Koressaar T, Remm M (2007) Enhancements and modifications of primer design program Primer3. Bioinformatics 23: 1289–1291.
40. Soeno Y, Abe H, Kimura S, Maruyama K, Obinata T (1998) Generation of functional beta-actinin (CapZ) in an E. coli expression system. J Muscle Res Cell Motil 19: 639–646.
41. Schafer DA, Cooper JA (1995) Control of actin assembly at filament ends. Annu Rev Cell Dev Biol 11: 497–518.
42. Fowler VM (1996) Regulation of actin filament length in erythrocytes and striated muscle. Curr Opin Cell Biol 8: 86–96.
43. Johnson RI, Seppa MJ, Cagan RL (2008) The Drosophila CD2AP/CIN85 orthologue Cindr regulates junctions and cytoskeleton dynamics during tissue patterning. J Cell Biol 180: 1191–1204.
44. Lucas EP, Khanal I, Gaspar P, Fletcher GC, Polesello C, et al. (2013) The Hippo pathway polarizes the actin cytoskeleton during collective migration of Drosophila border cells. J Cell Biol 201: 875–885.
45. Remmert K, Vullhorst D, Hinssen H (2000) In vitro refolding of heterodimeric CapZ expressed in E. coli as inclusion body protein. Protein Expr Purif 18: 11–19.
46. Ankenbauer T, Kleinschmidt JA, Walsh MJ, Weiner OH, Franke WW (1989) Identification of a widespread nuclear actin binding protein. Nature 342: 822–825.
47. Schafer DA, Mooseker MS, Cooper JA (1992) Localization of capping protein in chicken epithelial cells by immunofluorescence and biochemical fractionation. J Cell Biol 118: 335–346.
48. Casella JF, Torres MA (1994) Interaction of Cap Z with actin. The NH2-terminal domains of the alpha 1 and beta subunits are not required for actin capping, and alpha 1 beta and alpha 2 beta heterodimers bind differentially to actin. J Biol Chem 269: 6992–6998.

19

Identification of Novel Regulatory Genes in Development of the Avian Reproductive Tracts

Whasun Lim, Gwonhwa Song*

Department of Biotechnology, College of Life Sciences and Biotechnology, Korea University, Seoul, Republic of Korea

Abstract

The chicken reproductive system is unique in maintaining its functions including production of eggs or sperm, fertilization of the egg by sperm maintained in sperm nests, production of hormones regulating its growth, development and function, and reproduction. Development of the reproductive organs is a highly regulated process that results in differentiation and proliferation of germ cells in response to predominant regulatory factors such as hormones and transcription factors. However, only a few genes are known to determine morphogenesis of the chicken reproductive tract and their mechanisms are unknown. Therefore, in the present study, we investigated the expression patterns of four genes including SNCA, TOM1L1, TTR and ZEB1 in the gonads at embryonic days 14 and 18, and in immature (12-week-old) and mature (50-week-old) chickens, as well as the reproductive tract including ovary, oviduct and testes of the respective sexes by qRT-PCR, in situ hybridization and immunofluorescence analyses. The expression of SNCA, TOM1L1 and ZEB1 genes was higher in immature and mature female reproductive tracts than expression of TTR. In addition, different temporal and spatial patterns of expression of the four genes were observed during maturation of testis in chickens. Specifically, SNCA, TOM1L1 and TTR were highly expressed in testes of 12-week-old chickens. Moreover, several chicken specific microRNAs (miRs) were demonstrated to affect expression of target gene mRNAs by directly binding to the 3'-UTR of their target genes through actions at the post-transcriptional level as follows: *miR-153* and *miR-1643* for SNCA; *miR-1680** for TTR; and *miR-200b* and *miR-1786* for ZEB1. These results suggest that four-selected genes play an important role in development of the male and female reproductive tract in chickens and expression of most candidate genes is regulated at the post-transcriptional level through specific microRNAs.

Editor: Helen White-Cooper, Cardiff University, United Kingdom

Funding: This research was funded by Basic Science Research Program (2013R1A1A2A10005948) through the National Research Foundation of Korea (NRF) funded by the Ministry of Education, Science, and Technology, and also by a grant from the Next-Generation BioGreen 21 Program (No. PJ008142), Rural Development Administration, Republic of Korea. The funders had no role in study design, data collection and analysis, decision to publish, or preparation of the manuscript.

Competing Interests: The authors have declared that no competing interests exist.

* E-mail: ghsong@korea.ac.kr

Introduction

The chicken is an invaluable animal model for research on embryology and reproductive developmental biology. In avian models, sexual differentiation of the reproductive system is initiated in the embryonic gonads from embryonic day 6.5 (E6.5, HH30) to be either ovaries or testes and this occurs asymmetrically in females and symmetrically in males. It depends on which component of the embryonic gonad, cortex or medullary, is colonized by germ cells that migrate there through bloodstream [1,2,3]. In the female (ZW), germ cells asymmetrically populate the left and right gonads. The thickened outer cortex only develops in the left gonad while medulla cords form lacunae. However, the right gonad regresses and fails to develop the cortex. Female germ cells enter meiosis that commences between E14 and E18 in the left gonad [4,5,6]. Then, by beginning folliculogenesis, granulosa and theca cells surrounding the oocyte are formed from cells in the cortex of the left functional gonad. There are several candidate genes, forkhead box L2 (FOXL2), wingless-related MMTV integration site 4 (WNT4), proprotein convertase subtilisin/kexin type 6 (PCSK6) and bone morphogenetic protein 6 (BMP6) known to regulate follicular development [3,7,8].

The chicken oviduct has a major role in that it is a reproductive tract that produces eggs. The oviduct derives from the Mullerian duct that develops only on the left side in female chicks whereas the right Mullerian duct degenerates in female chicks and both of the Mullerian ducts degenerate in male chicks. The immature oviduct develops rapidly after 16 weeks of age and egg laying begins when chickens are 16- to 20-weeks-old [9,10]. The oviduct of laying hens consists of four specific segments which are the infundibulum (fertilization), magnum (production of egg-white proteins), isthmus (formation of the soft shell membrane) and shell gland (formation of the outer egg shell) [11]. Development of the oviduct is stimulated in response to estrogen and previous studies have shown the regulatory genes, serpin peptidase inhibitor, clade B (ovalbumin), member 3 (SERPINB3) [12], SERPINB11 [13], adenosylhomocysteinase-like 1 (AHCYL1) [14] alpha 2 macro-globulin (A2M) [15] and pleiotrophin (PTN) [16] are highly expressed during development of the immature oviduct in chickens.

In contrast to female reproductive organs, gonadal morphogenesis leading to a mature testis is symmetrical between left and

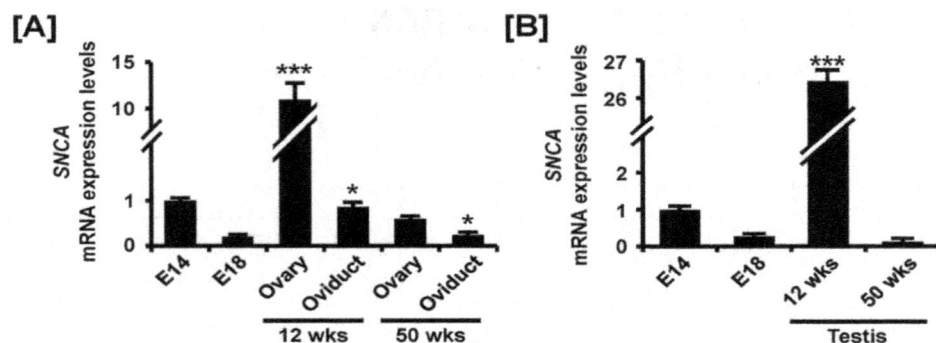

Figure 1. Quantitative analysis of *SNCA* mRNA expression in female and male reproductive tracts during their development. Quantitative RT-PCR was conducted using cDNA templates from female (A) and male (B) gonads at embryonic days 14 and 18, 12-week-old ovary and oviduct and 50-week-old ovary and oviduct. The asterisks denote statistically significant differences (*** $p < 0.001$ and * $p < 0.05$).

right gonads and originates in thickened medullary cords via proliferation of Sertoli cells within the cords that are anlage of the seminiferous tubules of male embryos (ZZ). Testes of 6-week-old chicken have seminiferous tubules that include a simple layer of spermatogonia, Sertoli cells, basal lamina and myoid cells. In testes of 50-week-old chickens all stage of spermatogenesis from spermatogonia to spermatozoa are found along with Sertoli cells surrounded by basal lamina and myoid cells [3,17]. In development of testis, testes-determining genes such as doublesex and mab-3 related transcription factor 1 (DMRT1) (Z-linked gene) and sex determining region Y-box 9 (SOX9) (sertoli cell differentiation factor) participate in testicular morphogenesis [18]. However, cell- and tissue-specific regulation for spermatogenesis remains unknown.

In a previous study, we identified several novel genes based on significant changes in their expression and functional categorization of genes changed between left and right gonads at embryonic days 6 and 9 through microarray analysis that may regulate gonadal morphogenesis in the both sexes of chicken embryos. We focused on four genes including synuclein alpha (SNCA), target of myb 1 (chicken) like 1 (TOM1L1), transthyretin (TTR) and zinc finger E-box binding homeobox 1 (ZEB1) that are associated with cellular proliferation and embryonic development and regulated by FSH and LH for E6 gonadal cells. However, these genes have not been investigated with respect to their influence on development of chicken reproductive tract. Therefore, we hypothesized that these selected genes effect changes in morphogenesis of reproductive organs in chickens. Accordingly, we determined differential patterns of mRNA expression and verified cell- and tissue-specific localization of mRNAs and proteins encoded by the four genes of interesting during development of female and male reproductive tracts in chickens. Moreover, we investigated post-transcriptional regulation of expression of three of the genes (SNCA, TTR and ZEB1) using a miRNA target validation assay. Results of present study provide novel insights into SNCA, TOM1L1, TTR and ZEB1 genes with respect to their tissue-specific expression during differentiation of germ cells into mature reproductive organs and post-transcriptional regulation of their expression by specific miRNAs in chickens.

Results

Comparative Expression of SNCA during Development of Reproductive Organs in both Sexes of Chickens

As illustrated in Figure 1A, the results from quantitative RT-PCR analyses indicated that expression of *SNCA* mRNA decreased 80% in left gonads at E18, 20% ($P < 0.05$) in oviducts of 12 week old chicks and 40% and 76% ($P < 0.05$) in the adult ovary and oviduct at 50 weeks, respectively as compared with *SNCA* expression in the gonads at E14. On the other hand, *SNCA* mRNA expression increased 11.2-fold ($P < 0.001$) in the ovaries of 12-week-old chickens. Moreover, expression of *SNCA* decreased 70% and 90% in the gonads at E18 and testis of 50-week-old chickens, respectively as compared with *SNCA* expression in the gonads at E14 (Figure 1B). However, *SNCA* increased 26.3-fold ($P < 0.001$) in the testis of 12-week-old chickens as compared with expression of *SNCA* in E14 gonads. *In situ* hybridization and immunofluorescence analyses detected *SNCA* mRNA and protein localized mainly in the cortex region of embryonic gonads and both were highly expressed in ovarian follicles of the immature chicken oviduct (Figure 2A and 2B). In males, in accordance with mRNA expression, SNCA protein was localized to the seminiferous cord of gonads at E14 and abundance decreased to E18. Interestingly, SNCA protein was abundant in the seminiferous tubules of 12-week-old testis and weakly expressed in Sertoli cells of 50-week-old testes (Figure 3A and 3B).

Comparative Expression of TOM1L1 during Development of Reproductive Organs in both Sexes of Chickens

Tissue specific expression of *TOM1L1* mRNA was found to increase 9.4- ($P < 0.001$) and 4.9- ($P < 0.01$) fold in oviducts at 12-weeks and ovaries at 50 weeks, respectivelyand decrease 70% in oviducts at 50 weeks as compared with *TOM1L1* expression in the female gonads at E14 by quantitative RT-PCR (Figure 4A). In the reproductive tract of male chickens, *TOM1L1* expression increased 5.2- fold ($P < 0.001$) in testis at 12weeks and decreased 70% ($P < 0.05$) in testes of at 50weeks in male chickens as compared with *TOM1L1* expression in the gonads at E14 (Figure 4B). In addition, TOMIL1 expression was strong in the oviduct of 12-week-old chickens and moderately expressed in ovarian follicles of 12- and 50-week-old female chickens (Figure 5A and 5B). In male chickens, *TOM1L1* mRNA and protein were most abundant in the seminiferous cord of embryonic gonads at E14 and E18,

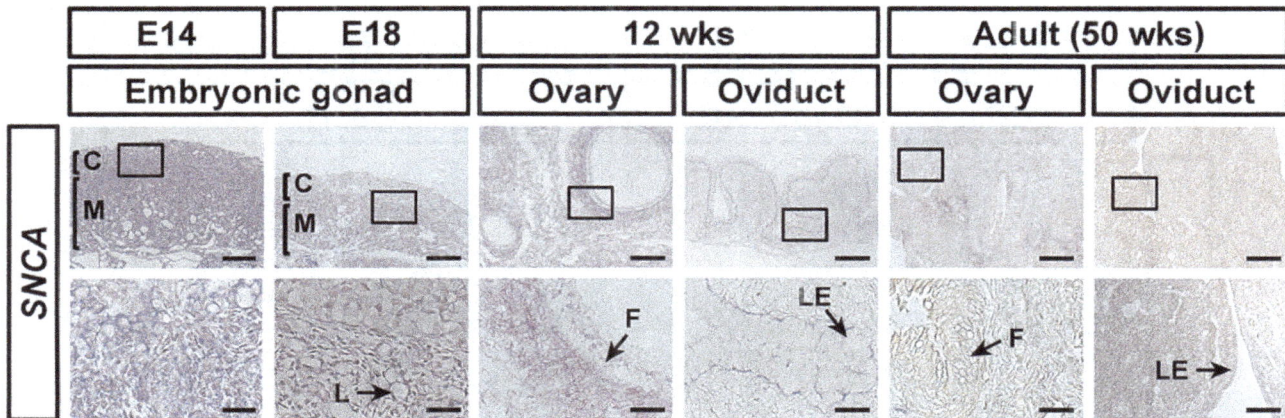

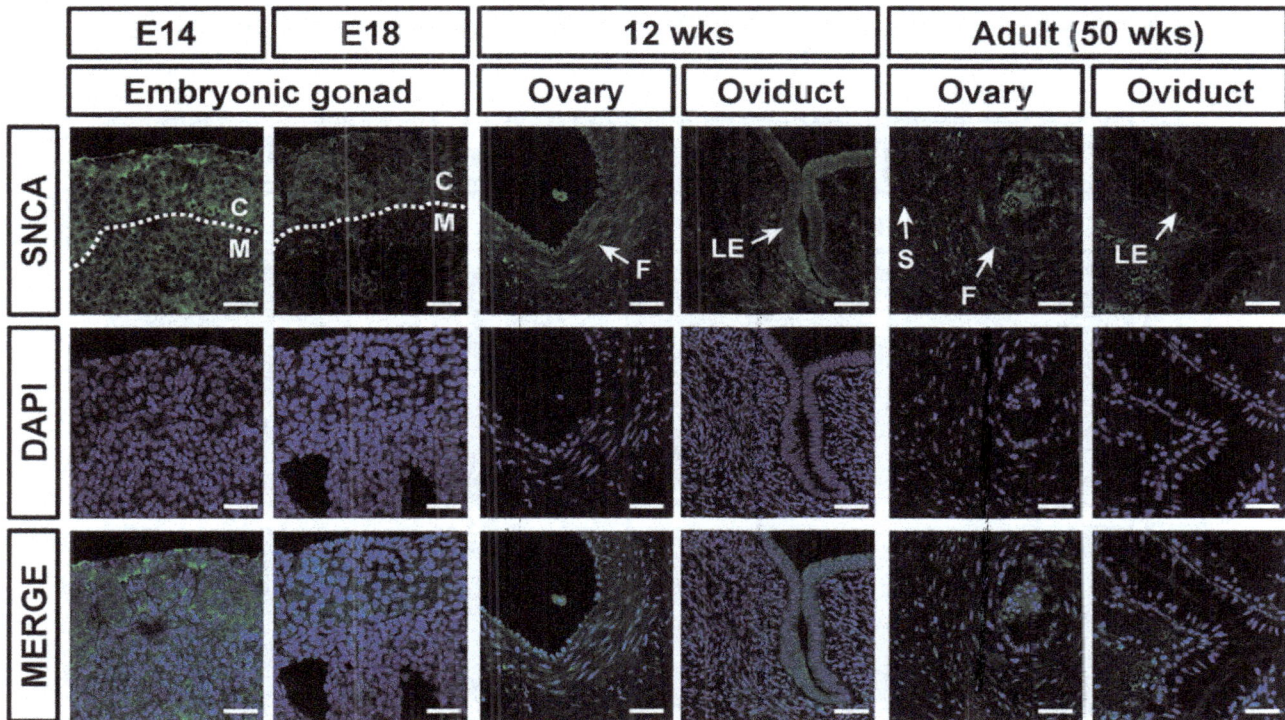

Figure 2. Cell-specific localization of mRNA and protein for SNCA in female reproductive tracts during their development. Cell-specific expression of *SNCA* mRNA and protein in development of the female reproductive tract was demonstrated by *in situ* hybridization (A) and immunofluorescence analyses (B). Cell nuclei were stained with DAPI (blue). Legend: C, cortex; F, follicle; L, lacunae; LE, luminal epithelium; M, medullar. Scale bar represents 100 μm and 20 μm for first and second horizontal panels of (A) and 50 μm for (B). See *Materials and Methods* for a complete description of the methods.

seminiferous tubules of immature testis (12 wks) and spermatogonia of testes at 50 weeks (Figures 6A and 6B).

Comparative Expression of TTR during Development of Reproductive Organs in both Sexes of Chickens

Chicken TTR gene expression was demonstrated in the reproductive tract during development of ovaries, oviduct and testes. In females, *TTR* was expressed weakly during development

of the ovary and oviduct. The expression levels indicated 0.01- (*P*<0.001), 0.08- (*P*<0.01), 0.02-, 0.3- (*P*<0.001) and 0.02- fold changes in *TTR* mRNA in the embryonic gonads at E18, 12-week-old ovaries and oviducts and 50-week-old ovaries and oviducts as compared with TTR expression in the embryonic gonads at E14 (Figure 7A). Next, *TTR* mRNA was evaluated during testis development in chickens. The results showed that TTR expression decreased 92% in the embryonic gonads at E18 and increased 5.1-

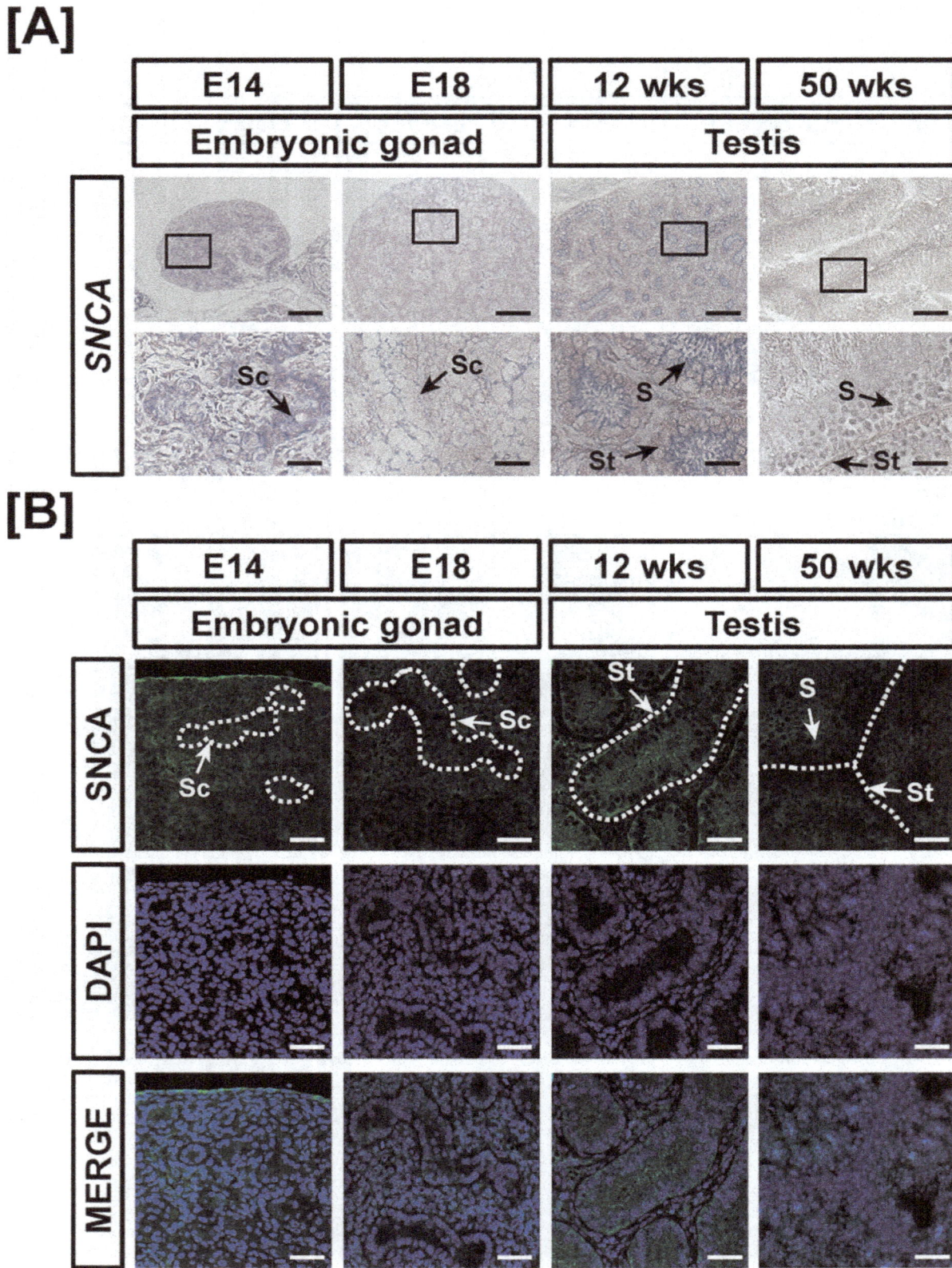

Figure 3. Cell-specific localization of mRNA and protein for SNCA in male reproductive tracts during their development. Localization of SNCA expression was analyzed in the male reproductive tract of chickens during their development by *in situ* hybridization (A) and immunofluorescence analyses (B). Cell nuclei were stained with DAPI (blue). Legend: S, Sertoli cell; Sc, seminiferous cord; St, seminiferous tubule. Scale bar represents 100 μm and 20 μm for first and second horizontal panels of (A) and 50 μm for (B). See *Materials and Methods* for a complete description of the methods.

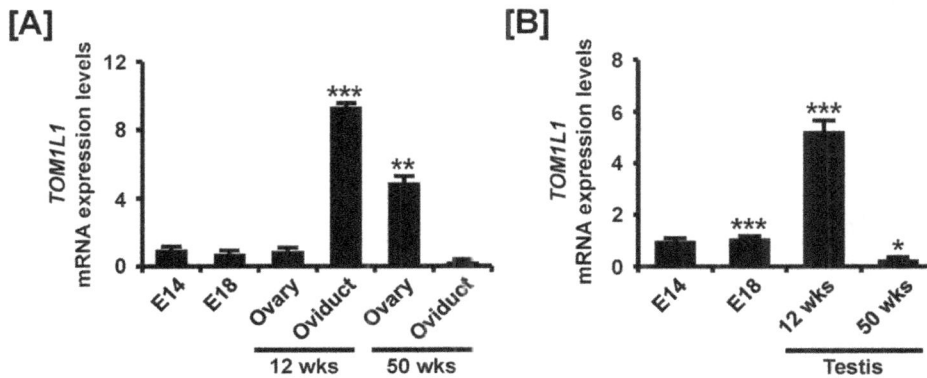

Figure 4. Quantitative analysis of *TOM1L1* mRNA expression in female and male reproductive tracts during their development. Quantitative RT-PCR was conducted using cDNA templates from female (A) and male (B) gonads at embryonic days 14 and 18, 12-week-old ovary and oviduct and 50-week-old ovary and oviduct. The asterisks denote statistically significant differences (*** $p < 0.001$, ** $p < 0.01$ and * $p < 0.05$).

($P < 0.001$) and 1.3- ($P < 0.01$) fold in testes of 12- and 50-week-old chickens, respectively as compared with expression at E14 (Figure 7B). In accordance with quantitative mRNA expression, cell-specific expression, based on results from *in situ* hybridization and immunofluorescence analyses, revealed that TTR is expressed mainly in the cortex of embryonic gonads, whereas its expression is rarely detected in other tissues of the female reproductive tract (Figure 8A and 8B). Furthermore, TTR was localized predominantly to the seminiferous cords of embryonic gonads (E14), seminiferous tubules of immature testis (12 wks) and Sertoli cells of adult testes (50 wks) as shown in Figures 9A and 9B.

Comparative Expression of ZEB1 during Development of Reproductive Organs in Female and Male Chickens

Tissue-specific expression of ZEB1 in the female and male reproductive tracts of chickens during development was demonstrated using quantitative RT-PCR, *in situ* hybridization and immunofluorescence analyses. As illustrated in Figure 10A, *ZEB1* mRNA expression increased in the reproductive tracts of 12- and 50-week-old female chickens as compared to embryonic gonads. Its expression increased 11.9- ($P < 0.01$), 13- ($P < 0.001$), 2.2- ($P < 0.05$) and 7.6- ($P < 0.001$) fold in 12-week-old ovaries and oviducts and 50-week-old ovaries and oviducts, respectively as compared with *ZEB1* expression in the gonads at E14. In male chickens, ZEB1 expression increased gradually from E18 gonads to adult testes. Expression of *ZEB1* mRNA increased 0.2- ($P < 0.001$), 0.6- ($P < 0.01$) and 1.3- ($P < 0.01$) fold in the gonads at E14 and E18 and testis of 12- and 50-week-old male chickens, respectively as compared with *ZEB1* expression in the gonads at E14 (Figure 10B). *ZEB1* mRNA and protein were highly expressed in luminal epithelium of the oviduct at 12- and 50-weeks of age and in ovarian follicles of ovaries from 12-week-old female chickens (Figure 11A and 11B). In addition, ZEB1 was weakly expressed in the cortex and medullary region of gonads at E14 and E18. In male reproductive organs, *ZEB1* mRNA and protein were localized predominantly to seminiferous cords of E14 gonads and Sertoli cells of testes from 50-week-old male chickens (Figure 12A and 12B). These results suggest that ZEB1 participates in development of both the oviduct and testis in chickens.

Post-transcriptional Regulation of Genital Ridge Development Regulatory Genes by Chicken microRNAs

We next investigated the possibility that expression of *SNCA*, *TTR* and *ZEB1* is regulated at the post-transcriptional level by microRNAs (miRNAs) using a miRNA target validation assay. In order to find target miRNAs and their binding sites within the 3'-UTR of *SNCA*, *TTR* and *ZEB1* genes, the miRNA target prediction database (miRDB: http://mirdb.org/miRDB/) was used. It revealed several putative binding sites for miRNAs including *miR-153* and *miR-1643* for SNCA, *miR-1680** for TTR and *miR-200b* and *miR-1786* for ZEB1 (Figures 13, 14 and 15). However, no specific target miRNA was detected for TOM1L1. Thus, we determined if these specific miRNAs influence expression of SNCA, TTR and ZEB1 via their 3'-UTR. A fragment of each 3'-UTR with binding sites for the miRNAs was cloned downstream of the green fluorescent protein (GFP) reading frame, thereby creating a fluorescent reporter for function of the 3'-UTR region (Figure 13B, 14B and 15B). After co-transfection of eGFP-3'-UTR and DsRed-miRNA, analyses for intensity of GFP expression and percentage of GFP-expressing cells were conducted using FACS and fluorescence microscopy. In the presence of *miR-153* and *miR-1643* decreased the intensity and percentage of GFP-SNCA-expressing cells 58% and 61% (Figure 13). In addition, *miR-1680** decreased the intensity and percentage of cells expressing TTR by 58% (Figure 14). Furthermore, *miR-200b* and *miR-1786* decreased the intensity and percentage of GFP-ZEB1-expressing-cells by 63% and 66%, respectively (Figure 15). These results indicate that specific miRNAs associated with target transcripts may be involved in development of reproductive organs in chickens and regulate their expression at the post-transcriptional level during morphogenesis of the ovary, oviduct and testis.

Discussion

Results of the current study revealed differential temporal and spatial expression patterns for key genes, SNCA, TOM1L1, TTR and ZEB1 that are important for development and differentiation of chicken reproductive tract in both sexes. In addition, the results indicate that among the four selected genes, expression of SNCA, TTR and ZEB1 is post-transcriptionally regulated via specific miRNAs binding directly the 3'-UTR of these target genes. These results support our hypothesis that molecular patterning of the

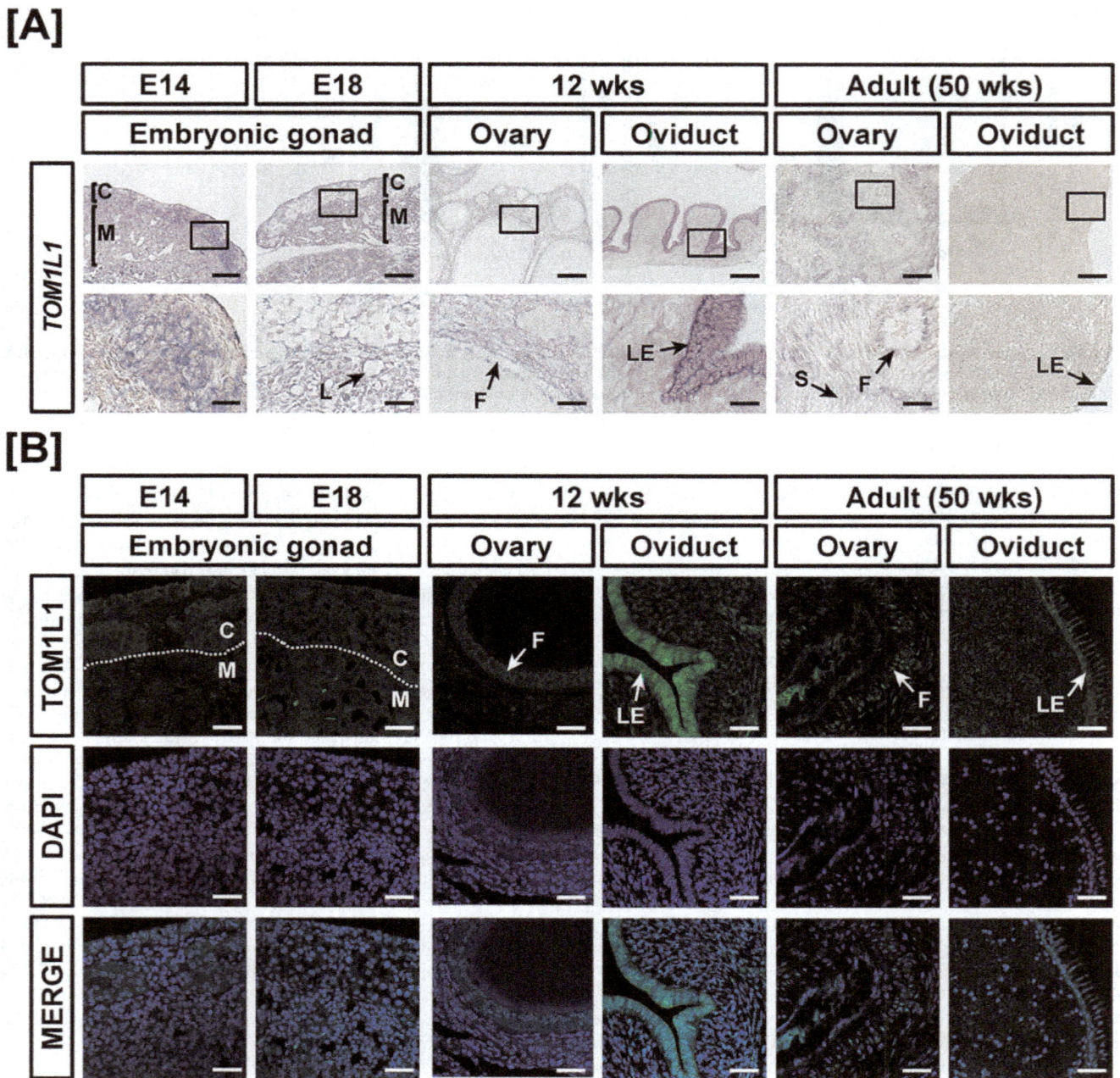

Figure 5. Cell-specific localization of mRNA and protein for TOM1L1 in female reproductive tracts during their development. Cell-specific expression of *TOM1L1* mRNA and protein in development of the female reproductive tract was demonstrated by *in situ* hybridization (A) and immunofluorescence analyses (B). Cell nuclei were stained with DAPI (blue). Legend: C, cortex; F, follicle; L, lacunae; LE, luminal epithelium; M, medullar. Scale bar represents 100 μm and 20 μm for first and second horizontal panels of (A) and 50 μm for (B). See *Materials and Methods* for a complete description of the methods.

reproductive system is affected by prominent transcripts crucial for development of female and male reproductive organs in chickens.

The chicken is a firmly entrenched animal model for research in embryology and reproductive developmental biology, but little is known about regulatory genes that control development of the reproductive tract in female and male chickens. In our previous study (Lim and Song, 2014, in submission), we reported novel genes and hormonal regulation of gonad morphogenesis in

chicken embryos. To demonstrate the differential patterns of expression of mRNAs and proteins of SNCA, TOM1L1, TTR and ZEB1 during development and differentiation of germ cells and primordial tissues to mature reproductive organs, we performed quantitative RT-PCR, *in situ* hybridization and immunofluorescence analyses.

SNCA belongs to a family of small and highly conserved proteins in vertebrates including alpha-, beta- and gamma-

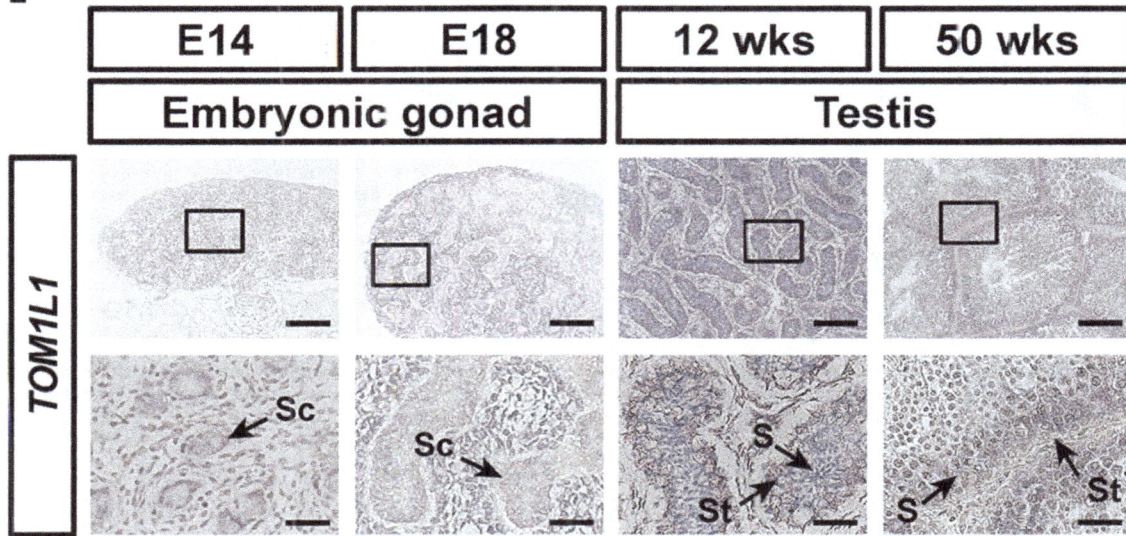

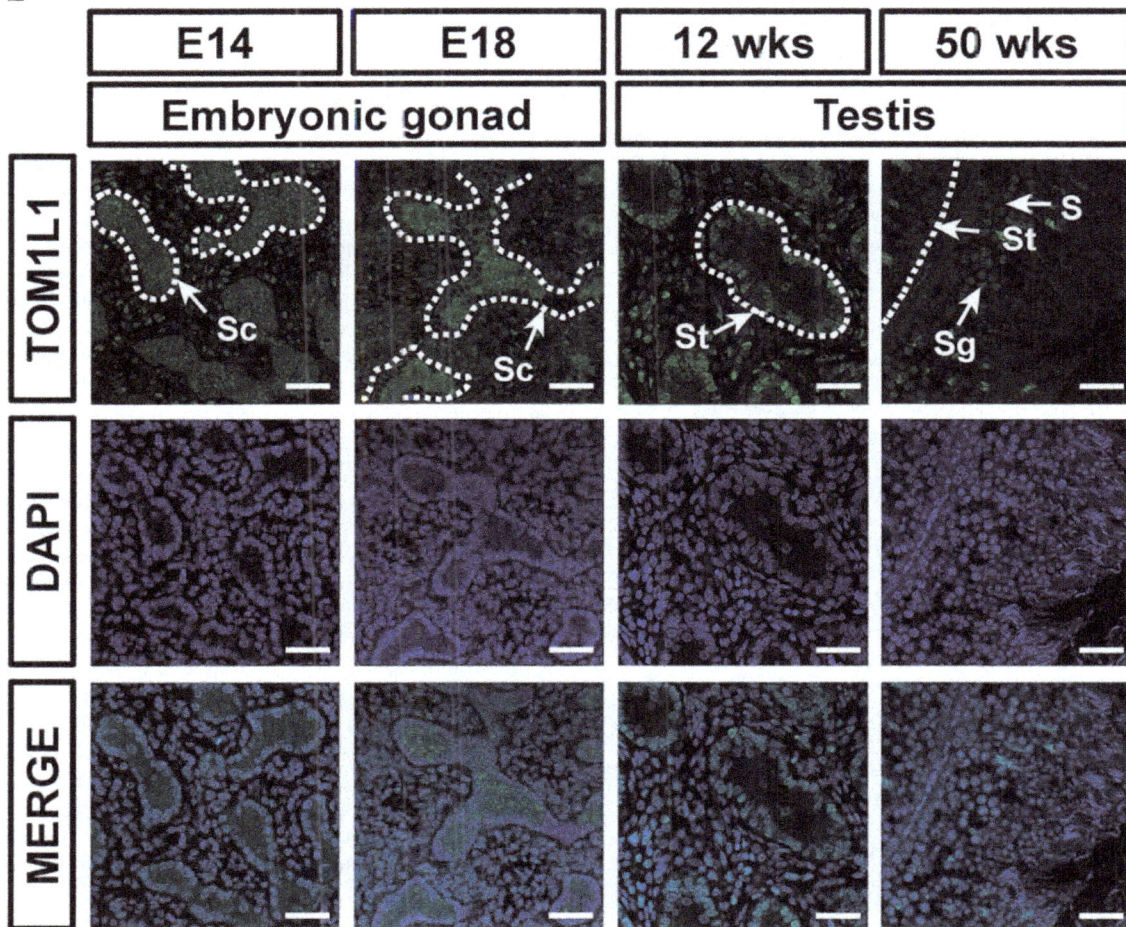

Figure 6. Cell-specific localization of mRNA and protein for TOM1L1 in male reproductive tracts during their development. Localization of TOM1L1 expression was analyzed in the male reproductive tract of chickens during their development by *in situ* hybridization (A) and immunofluorescence analyses (B). Cell nuclei were stained with DAPI (blue). Legend: S, Sertoli cell; Sc, seminiferous cord; Sg, spermatogonia; St, seminiferous tubule. Scale bar represents 100 μm and 20 μm for first and second horizontal panels of (A) and 50 μm for (B). See *Materials and Methods* for a complete description of the methods.

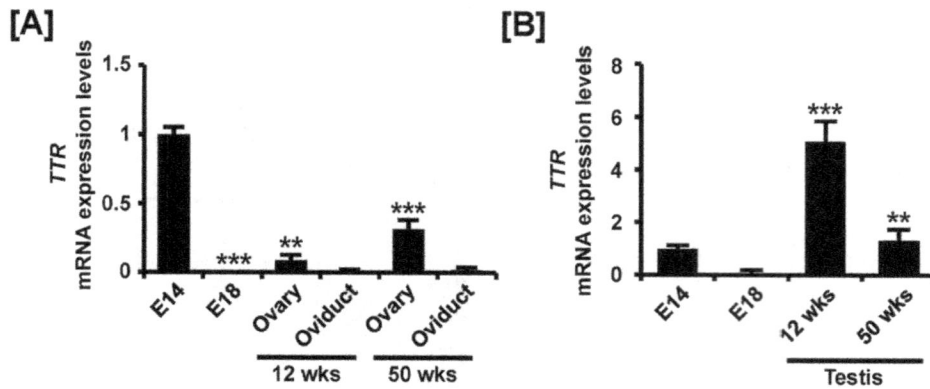

Figure 7. Quantitative analysis of *TTR* mRNA expression in female and male reproductive tracts during their development. Quantitative RT-PCR was conducted using cDNA templates from female (A) and male (B) gonads at embryonic days 14 and 18, 12-week-old ovary and oviduct and 50-week-old ovary and oviduct. The asterisks denote statistically significant differences (*** $p<0.001$ and ** $p<0.01$).

synuclein. The SNCA gene has 7 exons (5 protein-coding) and is expressed mainly in the brain, particularly in the hippocampus, caudate nucleus, amygdala, substantia nigra and thalamus in adult humans [19]. The SCNA gene has been specifically related to several neurodegenerative diseases such as Parkinson's disease via three point mutations (A53T, A30P and E46K) [20,21] and Alzheimer's disease through accumulation of the gene product in humans [22]. In addition, testosterone increases *SNCA* mRNA expression in the brain of chipping sparrows to affect their song system [23]. In the chicken, there is expression of SNCA in the majority of neurons in brain and spinal cord during embryogenesis [24]. However, there are no published results on SNCA and development of reproductive organs of any animal or human model. We reported that SNCA is expressed in both male and female embryonic gonads in chickens. Therefore, we investigated expression of SNCA during development of the reproductive organs from embryonic gonads to adult male and female reproductive tracts. SNCA was detected predominantly in ovary and testis of 12-week-old chickens which suggests that SNCA might have an important role in morphogenesis of ovary and testis.

TOM1L1 is also known as Src-activating and signaling molecule (SRCASM). This gene is an activator and substrate for Src family tyrosine kinases (SFKs) that include nine members that have significant roles in mitogenesis and morphological alterations via induction of growth factors [25]. TOM1L1 is tyrosine-phosphorylated in response to EGFR ligand as a SFK substrate downstream of EGFR. Increased expression of TOM1L1 activates endogenous SFKs preferably for phosphorylating Fyn and Src. Therefore, TOM1L1 links with EGFR and SFK-dependent signaling in differentiation of keratinocytes [26,27]. In addition, TOM1L1 has a role as a regulatory adaptor bridging activated EGFR in endocytosis by EGF stimulus [28]. In the present study, we determined that TOM1L1 was expressed strongly in luminal epithelium of the immature oviduct and follicles of adult ovaries in female chickens and in testes of 12-week-old male chickens. These results show that TOM1L1 likely has a role in regulating development of the immature oviduct, ovarian folliculogenesis and seminiferous tubules in chickens.

TTR (also called prealbumin) is one of the transporters of thyroid hormones and cooperates with retinol-binding protein (RBP) and vitamin A (retinol). TTR directly binds the thyroid hormones (T_3 and T_4) in the central channel constituted by tetrameric assembly of the monomers [29,30,31], and it indirectly provides vitamin A as retinol bound to RBP [32]. TTR has a well-established role in regulating spermatogenesis through effects on retinol metabolism in the adult testis of rats. Circulating retinol binds to a complex of RBP and TTR which is present in the peritubular cells associated with intracellular CRBP which has a high affinity for binding retinol [33]. The peritubular cells secrete retinol as a complex form to the Sertoli cells that oxidize retinol into retinoic acid which stimulates differentiation of germ cells [34]. Therefore, TTR assists in development of germ cells within the developing seminiferous tubules. In females, it is not known if TTR regulates oogenesis or embryogenesis. However, thyroid hormones transported by TTR from serum into the oocyte play a crucial role in embryogenesis in various species, especially as it is accumulates in the yolk of oocytes during oogenesis in chickens [35]. In our study, TTR was highly expressed during testes development, predominantly in the seminiferous tubules of immature testes. Otherwise, in the female reproductive tract, expression of TTR was weak in immature and mature ovaries. These results indicate that the TTR gene might have an important role in development and maturation of the postnatal testis in chickens.

ZEB1 (also known as EF1, TCF8, AREB6, Nil-2-a) is a transcription factor binding to DNA via two zinc finger clusters, one at the N-terminus and one at the C-terminus, and it can modulate transcription of target genes by binding directly to 5′-CACCT sequences in their promoter regions [36]. ZEB1 plays an important role in development, cell proliferation, differentiation, migration and reproduction [37,38]. In addition, ZEB1 protein induces cell migration during development and cancer progression by repressing expression of E-cadherin in epithelial cells [39,40,41,42]. ZEB1 is regulated by steroid hormones, estrogen [43], progesterone [44] and androgen [45]. In chickens, estrogen induces proliferation and differentiation of tubular gland cells associated with production of egg white proteins and stimulates ZEB1 expression leading to activation of transcription of downstream targets in the chick oviduct [43,46]. In this study, ZEB1 was highly expressed in both the immature and mature female reproductive tract as compared to the embryonic gonads suggesting a key role in development of the oviduct of adult female chickens. Moreover, in male chickens, ZEB1 expression increased gradually with progressive development of the testes from an E18 gonad to an adult testis. Thus, ZEB1 may play a crucial role in egg

[A]

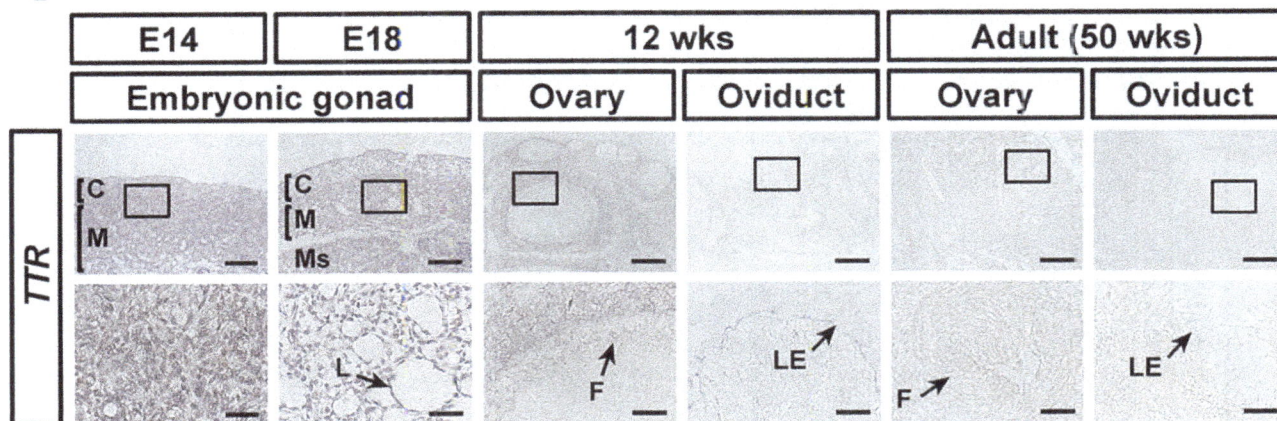

[B]

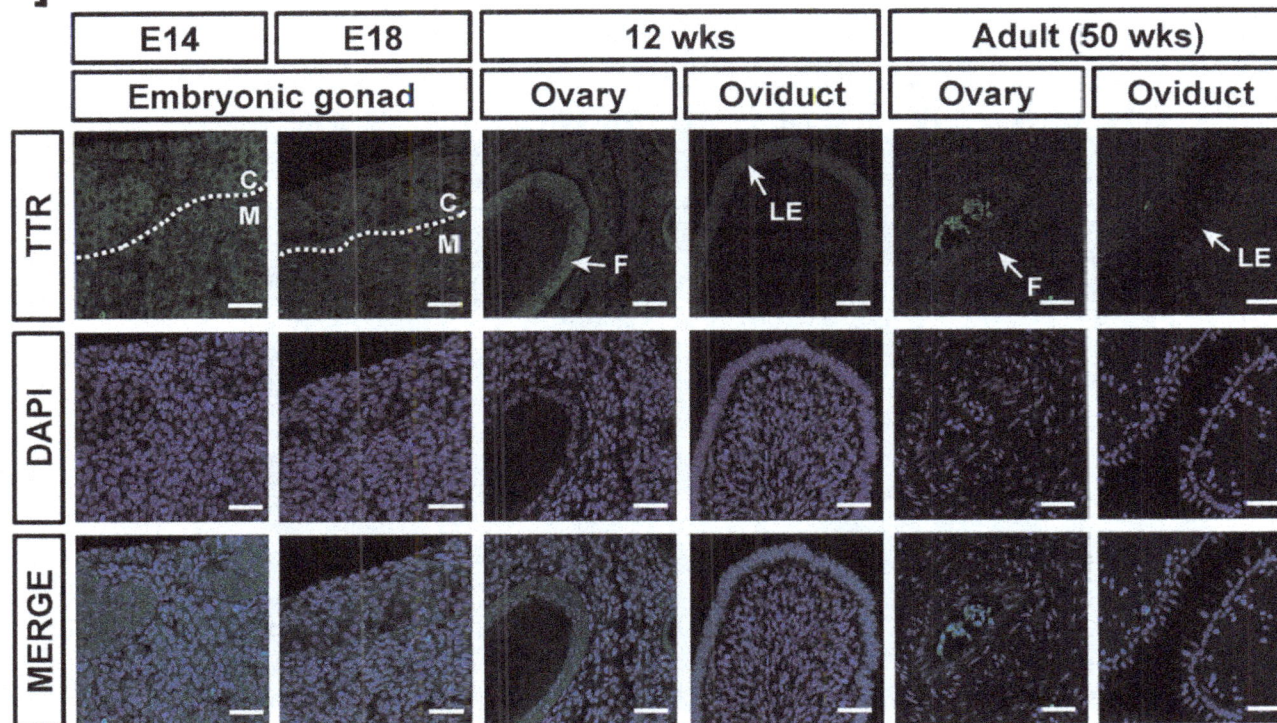

Figure 8. Cell-specific localization of mRNA and protein for TTR in female reproductive tracts during their development. Cell-specific expression of *TTR* mRNA and protein in development of the female reproductive tract was demonstrated by *in situ* hybridization (A) and immunofluorescence analyses (B). Cell nuclei were stained with DAPI (blue). Legend: C, cortex; F, follicle; L, lacunae; LE, luminal epithelium; M, medullar; Ms, mesonephros; S, stroma. Scale bar represents 100 µm and 20 µm for first and second horizontal panels of (A) and 50 µm for (B). See *Materials and Methods* for a complete description of the methods.

production through effects on development of the oviduct, as well as all stages of development of the testis in male chickens.

Based on results from validation of gene expression during development of chicken reproductive organs, we next investigated whether target genes undergo post-transcriptional regulation by specific microRNAs. MicroRNAs (miRNAs) are small non-coding single stranded RNAs of 18–23 nucleotides that play a role as post-transcriptional regulators and transformers of cell fate through modulation of target-mRNA translation in various cells and

tissues. In other words, miRNAs have crucial regulatory effects in a variety of biological events including growth, development, differentiation and control of cell cycle by modulating gene expression [47,48,49]. For example, expression of miRNAs during gonadal development in chickens and mammals has been reported [50,51,52]. In addition, several miRNAs regulate mechanisms required for development and differentiation of the oviduct and ovarian cancer in female chickens [14,16,53,54]. Moreover, *miR-34c* down-regulates genes related to germ cell differentiation and

[A]

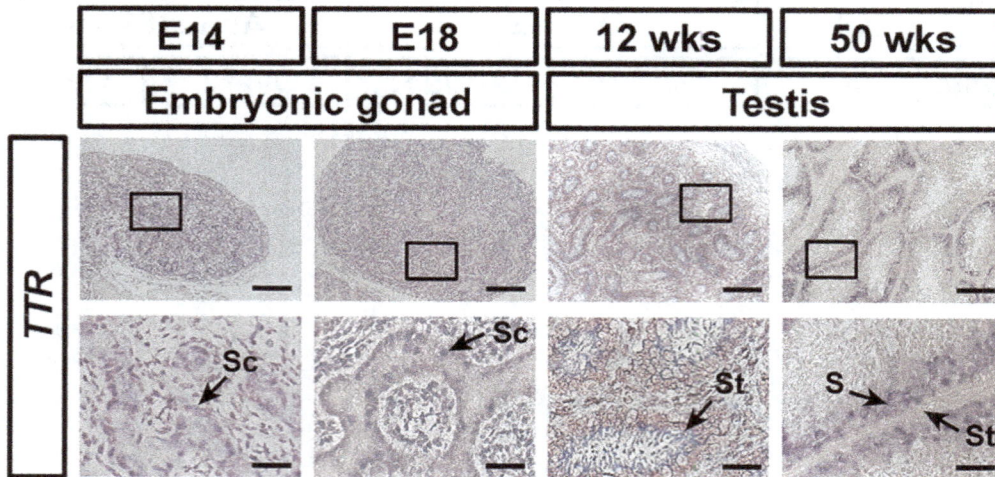

[B]

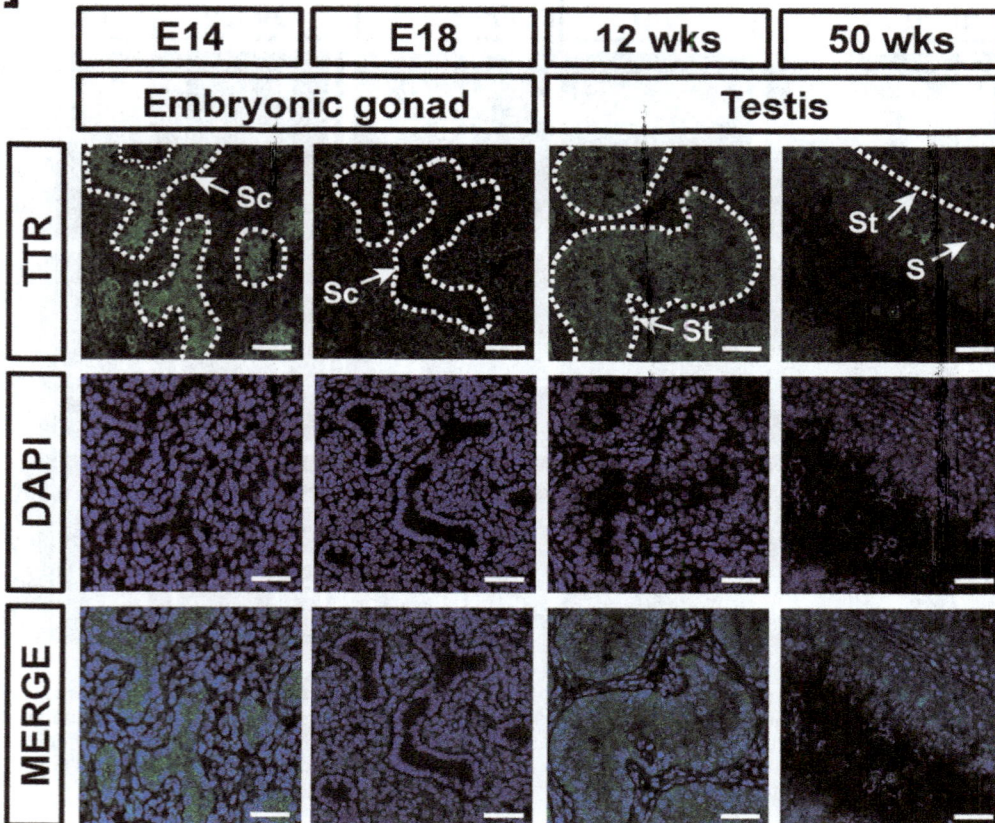

Figure 9. Cell-specific localization of mRNA and protein for TTR in male reproductive tracts during their development. Localization of TTR expression was analyzed in the male reproductive tract of chickens during their development by *in situ* hybridization (A) and immunofluorescence analyses (B). Cell nuclei were stained with DAPI (blue). Legend: S, Sertoli cell; Sc, seminiferous cord; St, seminiferous tubule. Scale bar represents 100 μm and 20 μm for first and second horizontal panels of (A) and 50 μm for (B). See *Materials and Methods* for a complete description of the methods.

[A] **[B]**

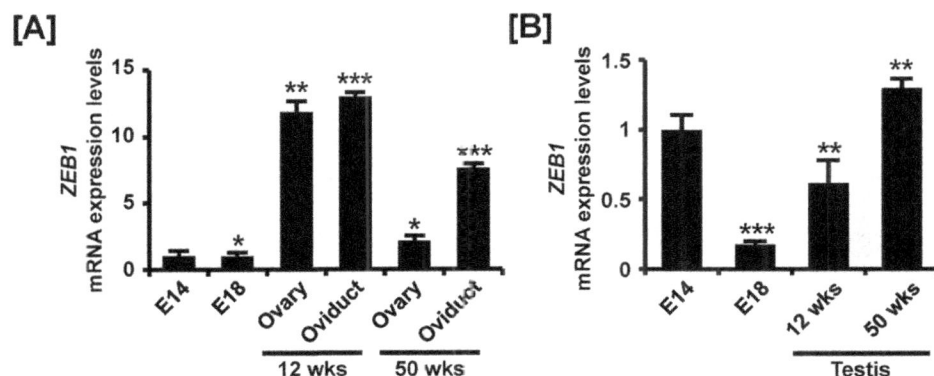

Figure 10. Quantitative analysis of *ZEB1* mRNA expression in female and male reproductive tracts during their development. Quantitative RT-PCR was conducted using cDNA templates from female (A) and male (B) gonads at embryonic days 14 and 18, 12-week-old ovary and oviduct and 50-week-old ovary and oviduct. The asterisks denote statistically significant differences (*** $p<0.001$, ** $p<0.01$ and * $p<0.05$).

its expression was detected mainly in the later stages of meiosis in spermatogenesis in chickens [55]. Based on previous reports, miRNAs might play a role during morphogenesis of the ovary, oviduct and testis in chickens. However, few miRNAs have been investigated with respect to their regulation of target genes and mechanisms whereby they act remain unknown. In this study, we performed an *in vitro* target assay of miRNAs to determine if *SNCA*, *TTR* and *ZEB1* transcripts are regulated at the post-transcriptional level by target miRNAs. As illustrated in Figures 13, 14, and 15, specific target miRNAs of chickens attenuate intensity of GFP-*SNCA*, *-TTR* and *-ZEB1* expressing cells. These results indicate that at least one to two miRNAs directly bind to the developmental-regulatory genes of reproductive organs and post-transcriptionally regulate their expression during development of the male and female reproductive tracts of chickens.

In conclusion, our results provide evidence for temporal and spatial expression of five genes that influence development of reproductive organs of chickens from the embryonic stage to the immature and mature stages of development. Expression of SNCA, TTR and ZEB1 are modulated via post-transcriptional regulation by specific target miRNAs which warrant further study. These results suggest roles for four important genes that likely regulate development of reproductive organs in chickens.

Materials and Methods

Experimental Animals and Animal Care

The experimental use of chickens for this study was approved by the Animal Care and Use Committee of Korea University. All chickens were exposed to a light regimen of 15 h light and 9 h dark with *ad libitum* access to feed and water, and subjected to standard poultry husbandry guidelines.

Tissue Samples

The left and right gonads were collected separately from the mesonephric kidney of chicken embryos at E14 and E18 in a 1.5 ml tube containing diethylpyrocarbonate treated PBS (DEPC-PBS). Then we centrifuged the sample at 1,080 x g for 5 min to allow collection of each gonad from the bottom of the tubes. After removal of the DEPC-PBS, the gonads were stored at −80°C until RNA was extracted. Also we collected whole embryos and fixed them in freshly prepared 4% paraformaldehyde in PBS (pH 7.4). Tissue samples were collected from ovary, oviduct and testis of 12- and 50-week-old females (n = 4) and males (n = 4). The collected

samples were either stored at −80°C until RNA was extracted or fixed immediately upon collection in freshly prepared 4% paraformaldehyde in PBS (pH 7.4). After 24 h, the samples fixed in 4% paraformaldehyde were changed to 70% ethanol for 24 h and then dehydrated in a graded series of increasing concentrations of ethanol. Embryos were then incubated in xylene for 3h and embedded in Paraplast-Plus. Paraffin-embedded tissues were sectioned at 5 μm.

RNA Isolation

Total cellular RNA was isolated from frozen tissues using Trizol reagent according to manufacturer's recommendations. The quantity and quality of total RNA was determined by spectrometry and denaturing agarose gel electrophoresis, respectively.

Quantitative RT-PCR Analysis

Total RNA was extracted from gonads on embryonic day 14 and 18 from both sexes and ovaries, oviducts and testes from 12- and 50-week-old females and males using TRIzol and purified using an RNeasy Mini Kit. Complementary DNA was synthesized using a Superscript III First-Strand Synthesis System. Gene expression levels were measured using SYBR Green and a StepOnePlus Real-Time PCR System. The *glyceraldehydes 3-phosphate dehydrogenase (GAPDH)* gene was analyzed simultaneously as a control and used for normalization of data. *GAPDH* expression is most stable among other housekeeping genes and it is used commonly for normalizing for variations in loading. Each target gene and *GAPDH* were analyzed in triplicate. Using the standard curve method, we determined expression of the examined genes using the standard curves and Ct values, and normalized them using *GAPDH* expression. The PCR conditions were 95°C for 3 min, followed by 40 cycles at 95°C for 20 sec, 60°C for 40 sec, and 72°C for 1 min using a melting curve program (increasing the temperature from 55°C to 95°C at 0.5°C per 10 sec) and continuous fluorescence measurement. ROX dye was used as a negative control for the fluorescence measurements. Sequence-specific products were identified by generating a melting curve in which the Ct value represented the cycle number at which a fluorescent signal was statistically greater than background, and relative gene expression was quantified using the $2^{-\Delta\Delta Ct}$ method [56]. For the control, the relative quantification of gene expression was normalized to the Ct value for the control oviduct.

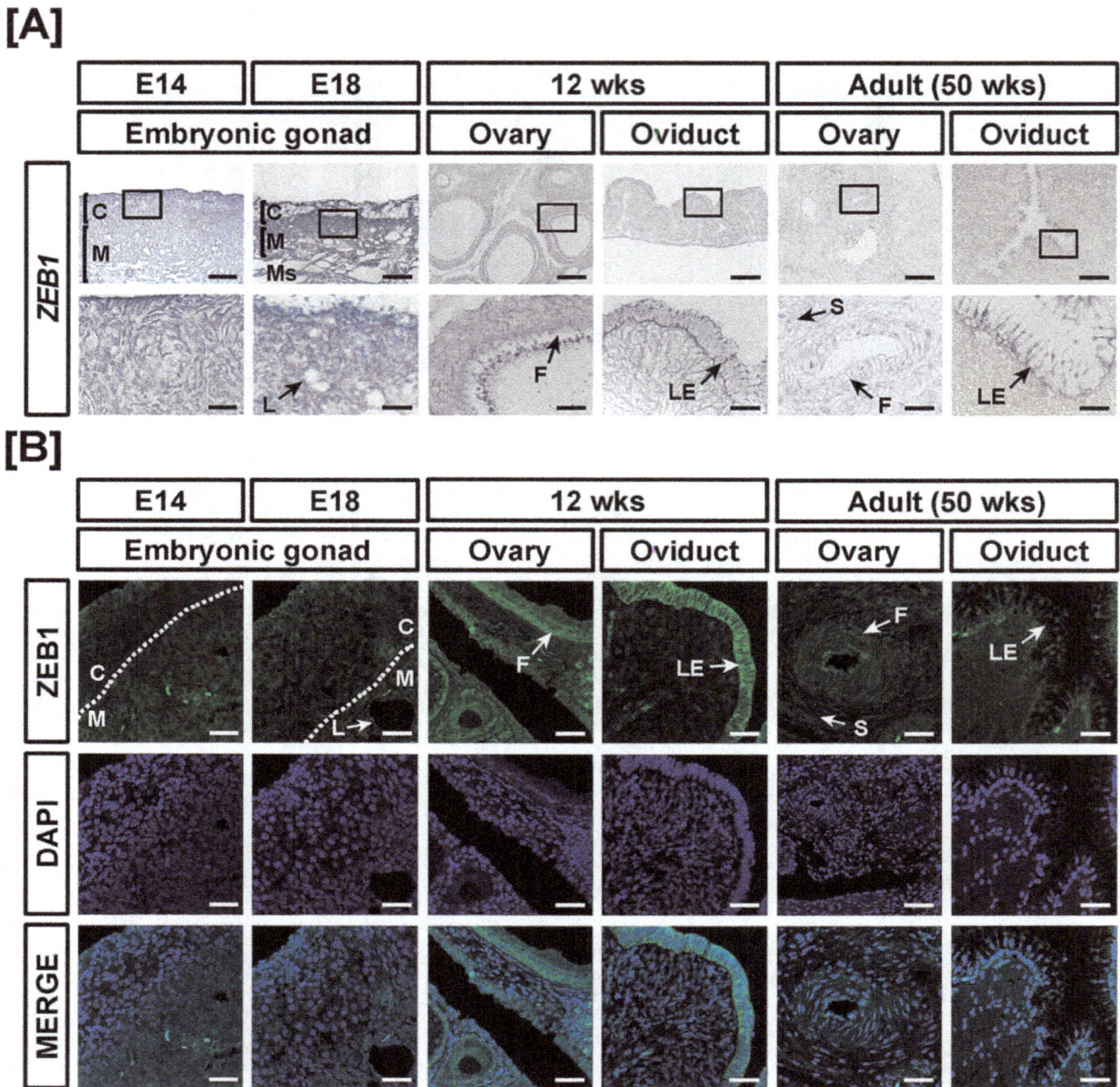

Figure 11. Cell-specific localization of mRNA and protein for ZEB1 in female reproductive tracts during their development. Cell-specific expression of *ZEB1* mRNA and protein in development of the female reproductive tract was demonstrated by *in situ* hybridization (A) and immunofluorescence analyses (B). Cell nuclei were stained with DAPI (blue). Legend: C, cortex; F, follicle; L, lacunae; LE, luminal epithelium; M, medullar; Ms, mesonephros; S, stroma. Scale bar represents 100 μm and 20 μm for first and second horizontal panels of (A) and 50 μm for (B). See *Materials and Methods* for a complete description of the methods.

Information on the primer sets was provided previously (Lim and Song, 2014, in submission).

In Situ Hybridization Analysis

For hybridization probes, PCR products were generated from cDNA with the primers used for RT-PCR analysis. The products were extracted from the gel and cloned into TOPO TA cloning vector. After verification of the sequences, plasmids containing gene sequences were linearized and transcribed using a DIG RNA labeling kit with T7 or SP6 polymerase. Information on the probes has been published (Lim and Song, 2014, in submission). Tissues were collected and fixed in freshly prepared 4% paraformaldehyde, embedded in paraffin and sectioned at 5 μm on APES-treated (silanized) slides. The sections were then deparaffinized in xylene and rehydrated to diethylpyrocarbonate (DEPC)-treated water through a graded series of alcohol. The sections were treated with 1% Triton X-100 in PBS for 20 min and washed two times in DEPC-treated PBS. After washing in DEPC-treated PBS, the

[A]

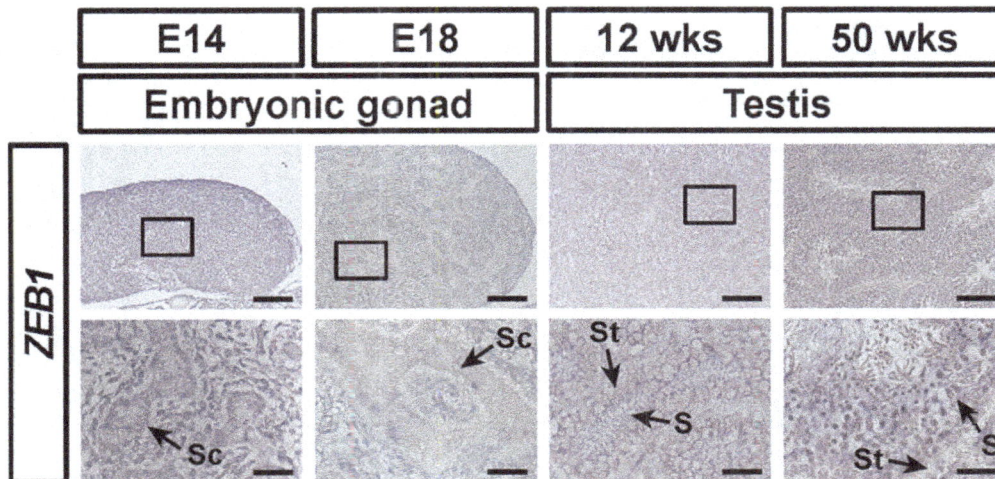

[B]

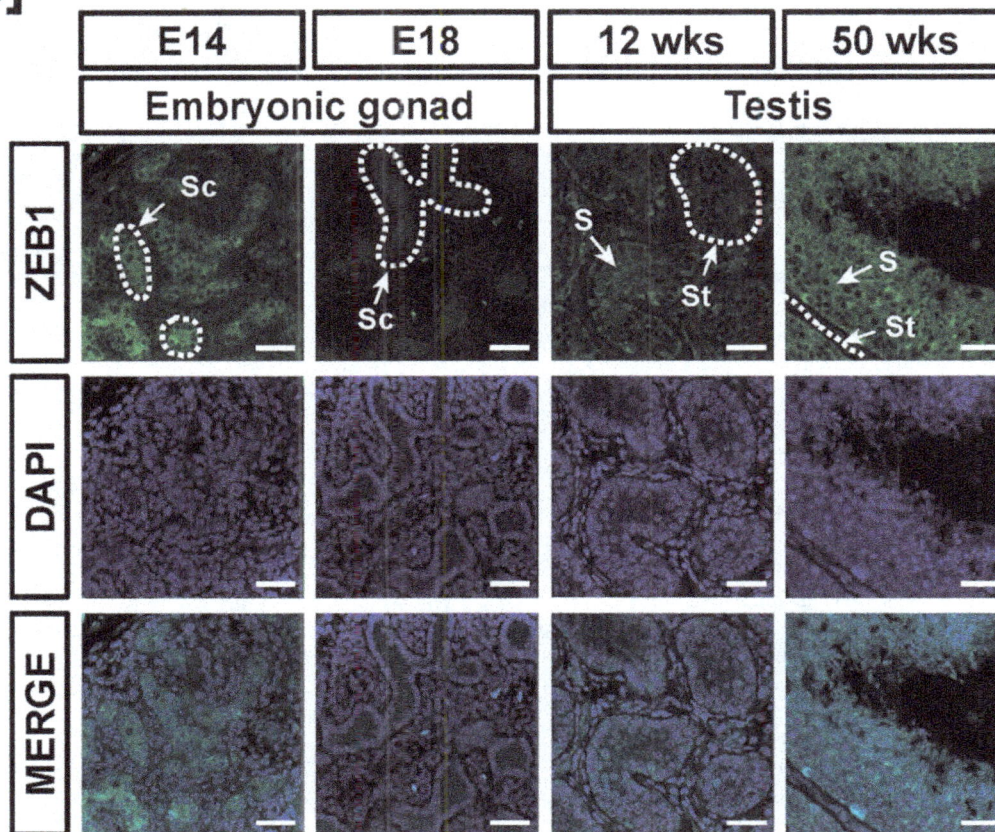

Figure 12. Cell-specific localization of mRNA and protein for ZEB1 in male reproductive tracts during their development. Localization of ZEB1 expression was analyzed in the male reproductive tract of chickens during their development by *in situ* hybridization (A) and immunofluorescence analyses (B). Cell nuclei were stained with DAPI (blue). Legend: S, Sertoli cell; Sc, seminiferous cord; St, seminiferous tubule. Scale bar represents 100 μm and 20 μm for first and second horizontal panels of (A) and 50 μm for (B). See *Materials and Methods* for a complete description of the methods.

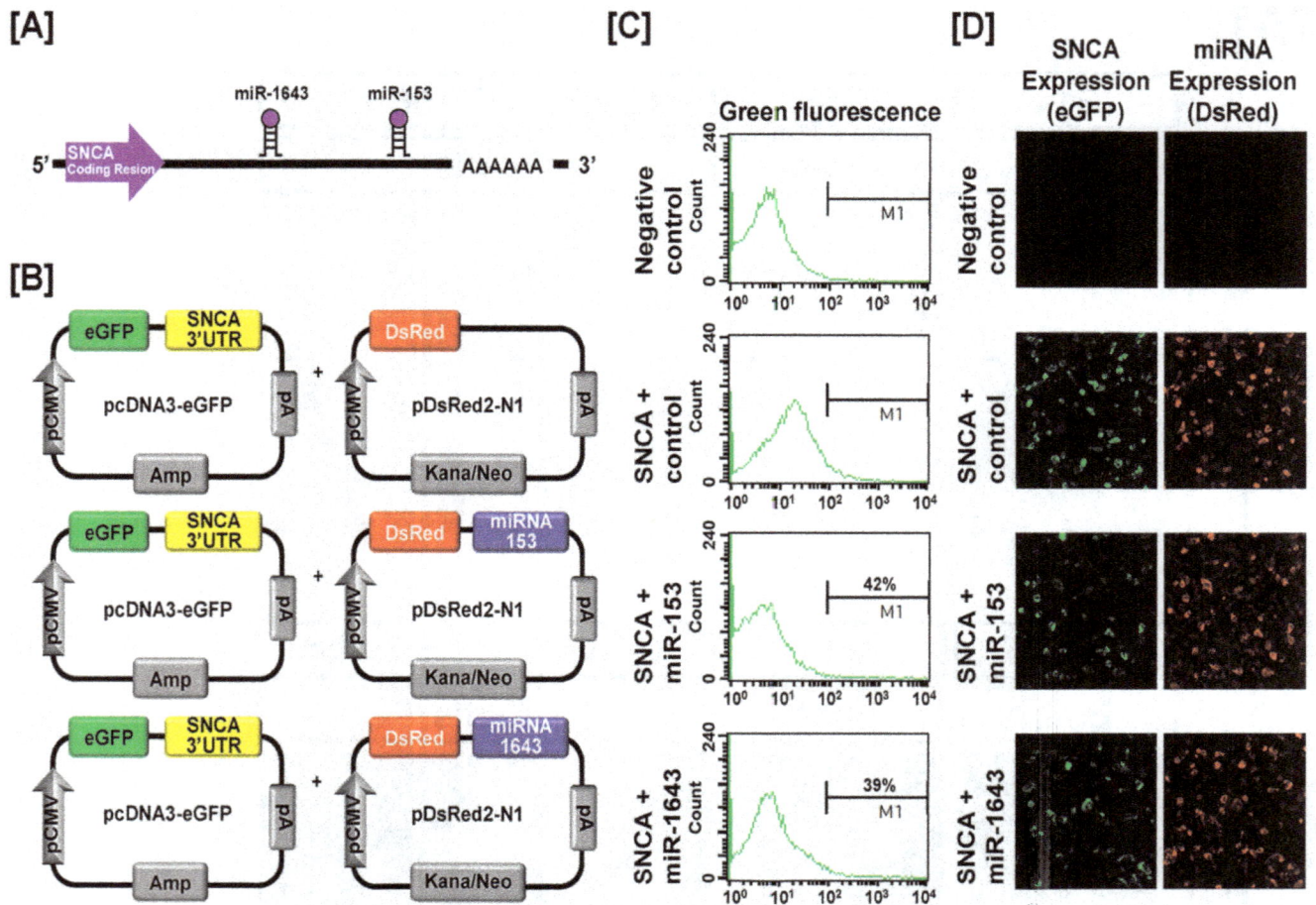

Figure 13. *In vitro* **target assay for** *miR-153* **and** *miR-1643* **on the SNCA transcript.** (A) Diagram showing *miR-153* and *miR-1643* binding sites in SNCA 3′-UTR. (B) Schematic expression of vector maps for eGFP with SNCA 3′-UTR and DsRed with each miRNA. (C and D) The fluorescence signals of GFP and DsRed were detected using FACS (C) and fluorescent microscopy (D) after co-transfection of pcDNA-eGFP-3′-UTR for the *SNCA* transcript and pcDNA-DsRed-miRNA for the *miR-153* and *miR-1643*.

sections were digested with 5 μg/ml Proteinase K in TE buffer (100 mM Tris-HCl, 50 mM EDTA, pH 8.0) at 37°C. After postfixation in 4% paraformaldehyde, sections were incubated twice for 5 min each in DEPC-treated PBS and incubated in TEA buffer (0.1M triethanolamine) containing 0.25% (v/v) acetic anhydride. The sections were incubated in a prehybridization mixture containing 50% formamide and 4X standard saline citrate (SSC) for at least 10 min at room temperature. After prehybridization, the sections were incubated overnight at 42°C in a humidified chamber in a hybridization mixture containing 40% formamide, 4X SSC, 10% dextran sulfate sodium salt, 10mM DTT, 1 mg/ml yeast tRNA, 1mg/ml salmon sperm DNA, 0.02% Ficoll, 0.02% polyvinylpyrrolidone, 0.2mg/ml RNase-free bovine serum albumin and denatured DIG-labeled cRNA probe. After hybridization, sections were washed for 15 min in 2X SSC at 37°C, 15min in 1X SSC at 37°C, 30 min in NTE buffer (10mM Tris, 500mM NaCl and 1mM EDTA) at 37°C and 30 min in 0.1X SSC at 37°C. After blocking with 2% normal sheep serum, the sections were incubated overnight with sheep anti-DIG antibody conjugated to alkaline phosphatase. The signal was visualized following exposure to a solution containing 0.4 mM 5-bromo-4-chloro-3-indolyl phosphate, 0.4 mM nitroblue tetrazolium, and 2 mM levamisole.

Immunofluorescence Analysis

The localization of four proteins in the reproductive tract of both sexes during their development was evaluated by immunofluorescence (IF) using anti-human SNCA polyclonal antibody (ab21975), anti-human TOM1L1 polyclonal antibody (ab126972), anti-human TTR polyclonal antibody (ab9015) and anti-human ZEB1 polyclonal antibody (ab81972). Antigen retrieval was performed using boiling 10mM sodium citrate buffer pH 6.0 for 10 min after which the slides were cooled on the bench top for 20 min. After antigen retrieval the slides were washed three times in 1X PBS for 5 min. Slides were incubated in blocking buffer (10% normal serum from the same species as the secondary antibody in 1X PBS) for 1 h. After the blocking solution was aspirated, slides were incubated overnight at 4°C with primary antibody. The slides were then rinsed three times in 1X PBS for 5 min each. Slides were then incubated with Alexa Fluor 488 rabbit anti-goat IgG secondary antibody for ZEB1, goat anti-rabbit IgG secondary antibody for TOM1L1 and donkey anti-sheep IgG secondary antibody for SNCA and TTR at a 1:200 dilution for 1 h at room temperature in the dark. Slides were then washed and overlaid with Prolong Gold Antifade with DAPI. For primary antibody, images were captured using a Zeiss confocal

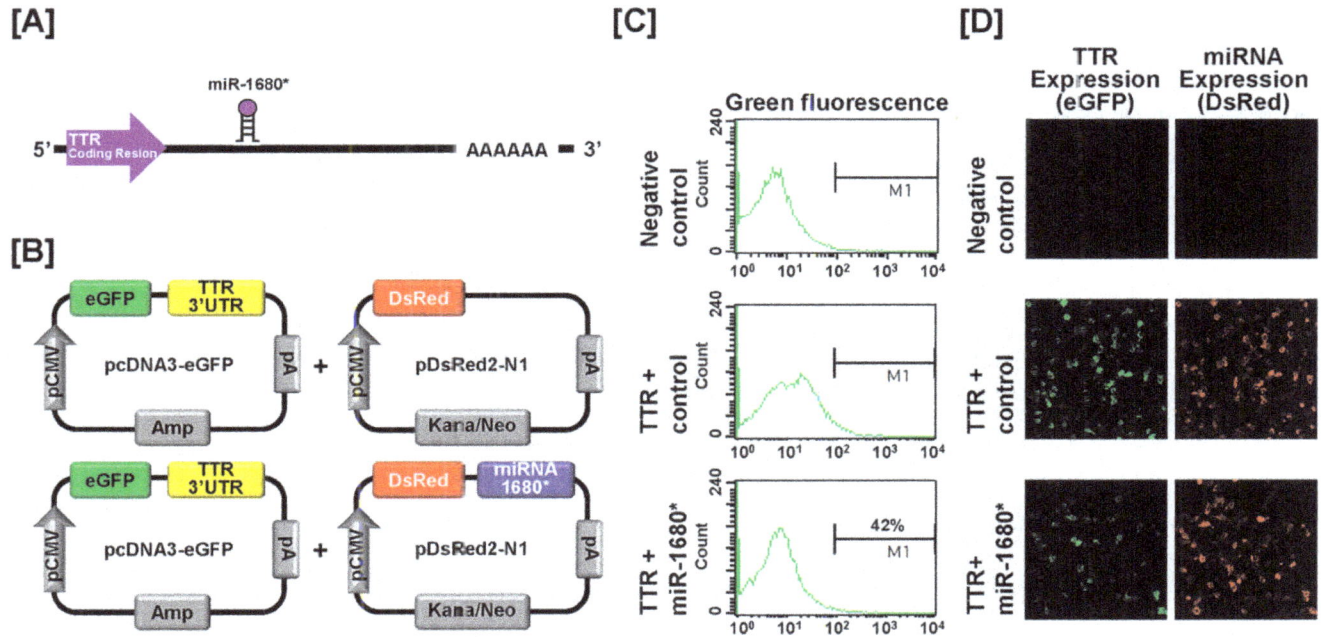

Figure 14. *In vitro* **target assay for** *miR-1680** **on the TTR transcript.** (A) Diagram showing *miR-1680** binding sites in TTR 3'-UTR. (B) Schematic expression of vector maps for eGFP with TTR 3'-UTR and DsRed with each miRNA. (C and D) The fluorescence signals of GFP and DsRed were detected using FACS (C) and fluorescent microscopy (D) after co-transfection of pcDNA-eGFP-3'-UTR for the *TTR* transcript and pcDNA-DsRed-miRNA for *miR-1680**.

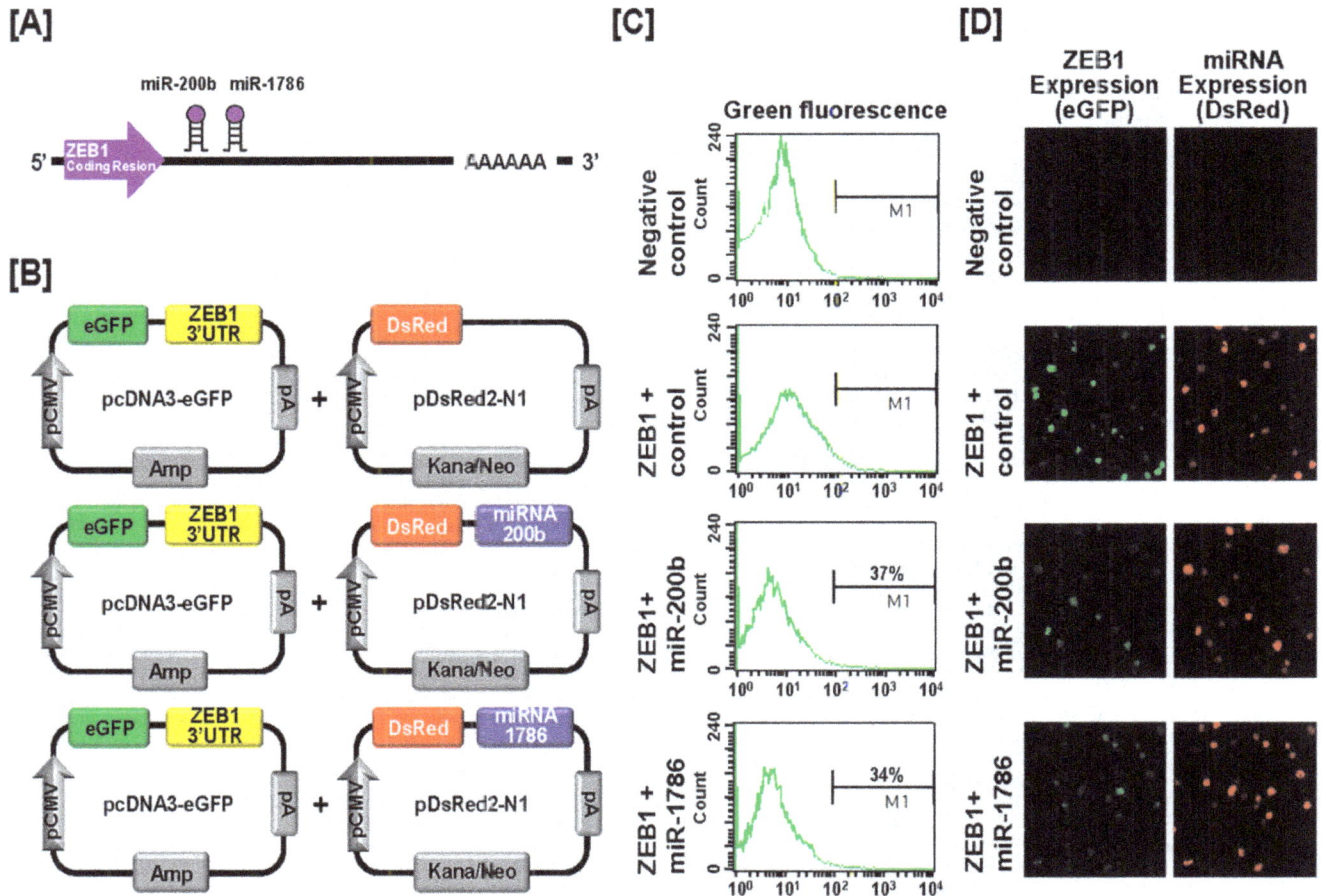

Figure 15. *In vitro* **target assay for** *miR-200b* **and** *miR-1786* **on the ZEB1 transcript.** (A) Diagram showing *miR-200b* and *miR-1786* binding sites in ZEB1 3'-UTR. (B) Schematic expression of vector maps for eGFP with ZEB1 3'-UTR and DsRed with each miRNA. (C and D) The fluorescence signals of GFP and DsRed were detected using FACS (C) and fluorescent microscopy (D) after co-transfection of pcDNA-eGFP-3'-UTR for the *ZEB1* transcript and pcDNA-DsRed-miRNA for the *miR-200b* and *miR-1786*.

microscope LSM710 fitted with a digital microscope camera AxioCam using Zen 2009 software.

MicroRNA Target Validation Assay

The 3'-UTR of SNCA, TTR and ZEB1 were cloned and confirmed by sequencing. Each 3'-UTR was subcloned between the eGFP gene and the bovine growth hormone (bGH) poly-A tail in pcDNA3eGFP to generate the eGFP-miRNA target 3'-UTR (pcDNA-eGFP-3'UTR) fusion constructs. For the dual fluorescence reporter assay, the fusion constructs containing the DsRed gene and target miRNAs were designed to be co-expressed under control of the CMV promoter (pcDNA-DsRed-miRNA). The pcDNA-eGFP-3'UTR and pcDNA-DsRed-miRNA (4µg) were co-transfected into 293FT cells using the calcium phosphate method. When the DsRed-miRNA is expressed and binds to the target site of the 3'-UTR downstream of the GFP transcript, green fluorescence intensity decreases due to degradation of the GFP transcript. At 48 h post-transfection, dual fluorescence was detected by fluorescence microscopy and calculated by FACSCalibur flow cytometry. For flow cytometry, the cells were fixed in freshly prepared 4% paraformaldehyde and analyzed using FlowJo software.

Statistical Analyses

All quantitative data were subjected to analysis of variance (ANOVA) according to the general linear model (PROC-GLM) of the SAS program. All tests of significance were performed using the appropriate error terms according to the expectation of the mean square for error. Data are presented as mean ± SEM unless otherwise stated. Differences with a probability value of $P<0.05$ were considered statistically significant.

Acknowledgments

We appreciate Dr. Fuller W. Bazer (Texas A&M University, USA) for thoughtful editing and comments on our paper and Ms. Wooyoung Jeong (Seoul National University, Korea) for experimental assistance.

Author Contributions

Conceived and designed the experiments: WL GS. Performed the experiments: WL. Analyzed the data: WL GS. Contributed reagents/materials/analysis tools: GS. Wrote the paper: WL GS.

References

1. Fujimoto T, Ukeshima A, Kiyofuji R (1976) The origin, migration and morphology of the primordial germ cells in the chick embryo. Anat Rec 185: 139–145.

2. Carlon N, Stahl A (1985) Origin of the Somatic Components in Chick Embryonic Gonads. Archives D Anatomie Microscopique Et De Morphologie Experimentale 74: 52–59.

3. Smith CA, Sinclair AH (2004) Sex determination: insights from the chicken. Bioessays 26: 120–132.

4. Ukeshima A, Fujimoto T (1991) A fine morphological study of germ cells in asymmetrically developing right and left ovaries of the chick. Anat Rec 230: 378–386.

5. Hughes GC (1963) The Population of Germ Cells in the Developing Female Chick. J Embryol Exp Morphol 11: 513–536.

6. Callebau.M (1967) Premeiosis and Premeiotic DNA Synthesis in Left Ovary of Female Chick Embryo. Journal of Embryology and Experimental Morphology 18: 299-&.

7. Diaz FJ, Anthony K, Halfhill AN (2011) Early Avian Follicular Development is Characterized by Changes in Transcripts Involved in Steroidogenesis, Paracrine Signaling and Transcription. Molecular Reproduction and Development 78: 212–223.

8. Ocon-Grove OM, Poole DH, Johnson AL (2012) Bone morphogenetic protein 6 promotes FSH receptor and anti-Mullerian hormone mRNA expression in granulosa cells from hen prehierarchal follicles. Reproduction 143: 825–833.

9. Whittow GC (2000) Sturkie's avian physiology. San Diego: Academic Press. xiii, 685 p. p.

10. Ha Y, Tsukada A, Saito N, Zadworny D, Shimada K (2008) Identification of differentially expressed genes involved in the regression and development of the chicken Mullerian duct. International Journal of Developmental Biology 52: 1135–1141.

11. Palmiter RD, Wrenn JT (1971) Interaction of Estrogen and Progesterone in Chick Oviduct Development.3. Tubular Gland Cell Cytodifferentiation. Journal of Cell Biology 50: 598-&.

12. Lim W, Ahn SE, Jeong W, Kim JH, Kim J, et al. (2012) Tissue specific expression and estrogen regulation of SERPINB3 in the chicken oviduct. General and Comparative Endocrinology 175: 65–73.

13. Lim W, Kim JH, Ahn SE, Jeong W, Kim J, et al. (2011) Avian SERPINB11 Gene: Characteristics, Tissue-Specific Expression, and Regulation of Expression by Estrogen. Biology of Reproduction 85: 1260–1268.

14. Jeong W, Kim J, Ahn SE, Lee SI, Bazer FW, et al. (2012) AHCYL1 Is Mediated by Estrogen-Induced ERK1/2 MAPK Cell Signaling and MicroRNA Regulation to Effect Functional Aspects of the Avian Oviduct. Plos One 7.

15. Lim W, Jeong W, Kim JH, Lee JY, Kim J, et al. (2011) Differential expression of alpha 2 macroglobulin in response to dietylstilbestrol and in ovarian carcinomas in chickens. Reproductive Biology and Endocrinology 9.

16. Lee JY, Jeong W, Lim W, Kim J, Bazer FW, et al. (2012) Chicken Pleiotrophin: Regulation of Tissue Specific Expression by Estrogen in the Oviduct and Distinct Expression Pattern in the Ovarian Carcinomas. Plos One 7.

17. Gonzalez-Moran MG, Guerra-Araiza C, Campos MG, Camacho-Arroyo I (2008) Histological and sex steroid hormone receptor changes in testes of immature, mature, and aged chickens. Domestic Animal Endocrinology 35: 371–379.

18. Chue J, Smith CA (2011) Sex determination and sexual differentiation in the avian model. Febs Journal 278: 1027–1034.

19. Lavedan C (1998) The synuclein family. Genome Res 8: 871–880.

20. Li J, Uversky VN, Fink AL (2001) Effect of familial Parkinson's disease point mutations A30P and A53T on the structural properties, aggregation, and fibrillation of human alpha-synuclein. Biochemistry 40: 11604–11613.

21. Zarranz JJ, Alegre J, Gomez-Esteban JC, Lezcano E, Ros R, et al. (2004) The new mutation, E46K, of alpha-synuclein causes Parkinson and Lewy body dementia. Annals of Neurology 55: 164–173.

22. Masliah E, Iwai A, Mallory M, Ueda K, Saitoh T (1996) Altered presynaptic protein NACP is associated with plaque formation and neurodegeneration in Alzheimer's disease. Am J Pathol 148: 201–210.

23. Hartman VN, Miller MA, Clayton DF, Liu WC, Kroodsma DE, et al. (2001) Testosterone regulates alpha-synuclein mRNA in the avian song system. Neuroreport 12: 943–946.

24. Tiunova AA, Anokhin KV, Saha AR, Schmidt O, Hanger DP, et al. (2000) Chicken synucleins: cloning and expression in the developing embryo. Mechanisms of Development 99: 195–198.

25. Franco M, Furstoss O, Simon V, Benistant C, Hong WJ, et al. (2006) The adaptor protein Tom1L1 is a negative regulator of Src mitogenic signaling induced by growth factors. Molecular and Cellular Biology 26: 1932–1947.

26. Li WJ, Marshall C, Mei LJ, Schmults C, Dans M, et al. (2005) Srcasm modulates EGF and Src-kinase signaling in keratinocytes. Journal of Biological Chemistry 280: 6036–6046.

27. Elmarghani A, Abuabaid H, Kjellen P (2009) TOM1L Is Involved in a Novel Signaling Pathway Important for the IL-2 Production in Jurkat T Cells Stimulated by CD3/CD28 CoLigation. Mediators of Inflammation.

28. Liu NS, Loo LS, Loh E, Seet LF, Hong WJ (2009) Participation of Tom1L1 in EGF-stimulated endocytosis of EGF receptor. Embo Journal 28: 3485–3499.

29. Duan W, Achen MG, Richardson SJ, Lawrence MC, Wettenhall RE, et al. (1991) Isolation, characterization, cDNA cloning and gene expression of an avian transthyretin. Implications for the evolution of structure and function of transthyretin in vertebrates. Eur J Biochem 200: 679–687.

30. Blake CC, Geisow MJ, Oatley SJ, Rerat B, Rerat C (1978) Structure of prealbumin: secondary, tertiary and quaternary interactions determined by Fourier refinement at 1.8 A. J Mol Biol 121: 339–356.

31. Ferguson RN, Edelhoch H, Saroff HA, Robbins J, Cahnmann HJ (1975) Negative cooperativity in the binding of thyroxine to human serum prealbumin. Preparation of tritium-labeled 8-anilino-1-naphthalenesulfonic acid. Biochemistry 14: 282–289.

32. Kanai M, Raz A, Goodman DS (1968) Retinol-binding protein: the transport protein for vitamin A in human plasma. J Clin Invest 47: 2025–2044.

33. Blaner WS, Galdieri M, Goodman DS (1987) Distribution and levels of cellular retinol- and cellular retinoic acid-binding proteins in various types of rat testis cells. Biol Reprod 36: 130–137.

34. Livera G, Rouiller-Fabre V, Pairault C, Levacher C, Habert R (2002) Regulation and perturbation of testicular functions by vitamin A. Reproduction 124: 173–180.

35. Vieira AV, Sanders EJ, Schneider WJ (1995) Transport of Serum Transthyretin into Chicken Oocytes - a Receptor-Mediated Mechanism. Journal of Biological Chemistry 270: 2952–2956.

36. Funahashi J, Sekido R, Murai K, Kamachi Y, Kondoh H (1993) Delta-crystallin enhancer binding protein delta EF1 is a zinc finger-homeodomain protein implicated in postgastrulation embryogenesis. Development 119: 433–446.

37. Sekido R, Takagi T, Okanami M, Moribe H, Yamamura M, et al. (1996) Organization of the gene encoding transcriptional repressor deltaEF1 and cross-species conservation of its domains. Gene 173: 227–232.

38. Vandewalle C, Van Roy F, Berx G (2009) The role of the ZEB family of transcription factors in development and disease. Cell Mol Life Sci 66: 773–787.

39. Eger A, Aigner K, Sonderegger S, Dampier B, Oehler S, et al. (2005) DeltaEF1 is a transcriptional repressor of E-cadherin and regulates epithelial plasticity in breast cancer cells. Oncogene 24: 2375–2385.

40. Guaita S, Puig I, Franci C, Garrido M, Dominguez D, et al. (2002) Snail induction of epithelial to mesenchymal transition in tumor cells is accompanied by MUC1 repression and ZEB1 expression. J Biol Chem 277: 39209–39216.

41. Spoelstra NS, Manning NG, Higashi Y, Darling D, Singh M, et al. (2006) The transcription factor ZEB1 is aberrantly expressed in aggressive uterine cancers. Cancer Res 66: 3893–3902.

42. Comijn J, Berx G, Vermassen P, Verschueren K, van Grunsven L, et al. (2001) The two-handed E box binding zinc finger protein SIP1 downregulates E-cadherin and induces invasion. Mol Cell 7: 1267–1278.

43. Chamberlain EM, Sanders MM (1999) Identification of the novel player deltaEF1 in estrogen transcriptional cascades. Mol Cell Biol 19: 3600–3606.

44. Richer JK, Jacobsen BM, Manning NG, Abel MG, Wolf DM, et al. (2002) Differential gene regulation by the two progesterone receptor isoforms in human breast cancer cells. J Biol Chem 277: 5209–5218.

45. Anose BM, Sanders MM (2011) Androgen Receptor Regulates Transcription of the ZEB1 Transcription Factor. Int J Endocrinol 2011: 903918.

46. Schutz G, Nguyen-Huu MC, Giesecke K, Hynes NE, Groner B, et al. (1978) Hormonal control of egg white protein messenger RNA synthesis in the chicken oviduct. Cold Spring Harb Symp Quant Biol 42 Pt 2: 617–624.

47. Bartel DP (2009) MicroRNAs: Target Recognition and Regulatory Functions. Cell 136: 215–233.

48. Garzon R, Fabbri M, Cimmino A, Calin GA, Croce CM (2006) MicroRNA expression and function in cancer. Trends in Molecular Medicine 12: 580–587.

49. Gregory RI, Chendrimada TP, Cooch N, Shiekhattar R (2005) Human RISC couples microRNA biogenesis and posttranscriptional gene silencing. Cell 123: 631–640.

50. Huang P, Gong YZ, Peng XL, Li SJ, Yang Y, et al. (2010) Cloning, identification, and expression analysis at the stage of gonadal sex differentiation of chicken miR-363 and 363*. Acta Biochimica Et Biophysica Sinica 42: 522–529.

51. Tripurani SK, Xiao CD, Salem M, Yao JB (2010) Cloning and analysis of fetal ovary microRNAs in cattle. Animal Reproduction Science 120: 16–22.

52. Torley KJ, da Silveira JC, Smith P, Anthony RV, Veeramachaneni DNR, et al. (2011) Expression of miRNAs in ovine fetal gonads: potential role in gonadal differentiation. Reproductive Biology and Endocrinology 9.

53. Lim W, Kim HS, Jeong W, Ahn SE, Kim J, et al. (2012) SERPINB3 in the Chicken Model of Ovarian Cancer: A Prognostic Factor for Platinum Resistance and Survival in Patients with Epithelial Ovarian Cancer. Plos One 7.

54. Lim W, Jeong W, Kim J, Ka H, Bazer FW, et al. (2012) Differential expression of secreted phosphoprotein 1 in response to estradiol-17 beta and in ovarian tumors in chickens. Biochemical and Biophysical Research Communications 422: 494–500.

55. Bouhallier F, Allioli N, Lavial F, Chalmel F, Perrard MH, et al. (2010) Role of miR-34c microRNA in the late steps of spermatogenesis. Rna-a Publication of the Rna Society 16: 720–731.

56. Livak KJ, Schmittgen TD (2001) Analysis of relative gene expression data using real-time quantitative PCR and the 2(T)(-Delta Delta C) method. Methods 25: 402–408.

Permissions

The contributors of this book come from diverse backgrounds, making this book a truly international effort. This book will bring forth new frontiers with its revolutionizing research information and detailed analysis of the nascent developments around the world.

We would like to thank all the contributing authors for lending their expertise to make the book truly unique. They have played a crucial role in the development of this book. Without their invaluable contributions this book wouldn't have been possible. They have made vital efforts to compile up to date information on the varied aspects of this subject to make this book a valuable addition to the collection of many professionals and students.

This book was conceptualized with the vision of imparting up-to-date information and advanced data in this field. To ensure the same, a matchless editorial board was set up. Every individual on the board went through rigorous rounds of assessment to prove their worth. After which they invested a large part of their time researching and compiling the most relevant data for our readers.

The editorial board has been involved in producing this book since its inception. They have spent rigorous hours researching and exploring the diverse topics which have resulted in the successful publishing of this book. They have passed on their knowledge of decades through this book. To expedite this challenging task, the publisher supported the team at every step. A small team of assistant editors was also appointed to further simplify the editing procedure and attain best results for the readers.

Apart from the editorial board, the designing team has also invested a significant amount of their time in understanding the subject and creating the most relevant covers. They scrutinized every image to scout for the most suitable representation of the subject and create an appropriate cover for the book.

The publishing team has been an ardent support to the editorial, designing and production team. Their endless efforts to recruit the best for this project, has resulted in the accomplishment of this book. They are a veteran in the field of academics and their pool of knowledge is as vast as their experience in printing. Their expertise and guidance has proved useful at every step. Their uncompromising quality standards have made this book an exceptional effort. Their encouragement from time to time has been an inspiration for everyone.

The publisher and the editorial board hope that this book will prove to be a valuable piece of knowledge for researchers, students, practitioners and scholars across the globe.

List of Contributors

Rui Hao and Xiaoxiang Hu
State Key Laboratory for Agrobiotechnology, China Agricultural University, Beijing, P. R. China

Changxin Wu
College of Animal Science and Technology, China Agricultural University, Beijing, P. R. China

Ning Li
State Key Laboratory for Agrobiotechnology, China Agricultural University, Beijing, P. R. China
College of Animal Science, Yunnan Agricultural University, Kunming, P. R. China

Anna C. Davies and Christine J. Nicol
School of Clinical Veterinary Science, University of Bristol, Bristol, United Kingdom

Mia E. Persson
Avian Behavioural Genomics and Physiology Group, Linköping University Linköping, Sweden

Andrew N. Radford
School of Biological Sciences, University of Bristol, Bristol, United Kingdom

Taiki Kawagoshi and Yoshinobu Uno
Laboratory of Animal Genetics, Department of Applied Molecular Biosciences, Graduate School of Bioagricultural Sciences, Nagoya University, Nagoya, Japan

Chizuko Nishida
Department of Natural History Sciences, Faculty of Science, Hokkaido University, Sapporo, Japan

Yoichi Matsuda
Laboratory of Animal Genetics, Department of Applied Molecular Biosciences, Graduate School of Bioagricultural Sciences, Nagoya University, Nagoya, Japan
Avian Bioscience Research Center, Graduate School of Bioagricultural Sciences, Nagoya University, Nagoya, Japan

Luke S. Lambeth and Thomas Ohnesorg
Murdoch Childrens Research Institute, Royal Children's Hospital, Melbourne, VIC, Australia
Poultry Cooperative Research Centre, Armidale, NSW, Australia

Andrew H. Sinclair
Murdoch Childrens Research Institute, Royal Children's Hospital, Melbourne, VIC, Australia
Department of Paediatrics, The University of Melbourne, Melbourne, VIC, Australia
Poultry Cooperative Research Centre, Armidale, NSW, Australia

Craig A. Smith
Murdoch Childrens Research Institute, Royal Children's Hospital, Melbourne, VIC, Australia
Department of Paediatrics, The University of Melbourne, Melbourne, VIC, Australia
Poultry Cooperative Research Centre, Armidale, NSW, Australia

David M. Cummins
CSIRO Animal, Food and Health Sciences, Australian Animal Health Laboratory, Geelong, VIC, Australia

Alistair Bailey
Institute for Life Sciences, University of Southampton, Southampton, United Kingdom
Cancer Sciences Unit, Faculty of Medicine, University of Southampton, Southampton, United Kingdom

Andy van Hateren
Institute for Life Sciences, University of Southampton, Southampton, United Kingdom
Cancer Sciences Unit, Faculty of Medicine, University of Southampton, Southampton, United Kingdom

Tim Elliott
Institute for Life Sciences, University of Southampton, Southampton, United Kingdom
Cancer Sciences Unit, Faculty of Medicine, University of Southampton, Southampton, United Kingdom

Jörn M. Werner
Institute for Life Sciences, University of Southampton, Southampton, United Kingdom
Centre for Biological Sciences, Faculty of Natural & Environmental Sciences, University of Southampton, Southampton, United Kingdom

Laëtitia Trapp-Fragnet, Djihad Bencherit, Daniéle Chabanne-Vautherot, Sylvie Remy, Jean-François Vautherot and Caroline Denesvre
INRA, UMR1282 Infectiologie et Santé Publique, Equipe Biologie des Virus Aviaires, Nouzilly, France

Yves Le Vern
INRA, UMR1282 Infectiologieet Santé Publique, Laboratoire de Cytométrie, Nouzilly France

Elisa Boutet-Robinet and Gladys Mirey
INRA, UMR 1331, Toxalim, Research Centre in Food Toxicology, Toulouse, France
University of Toulouse, UPS, UMR1331, Toxalim, Toulouse, France

Nowlan H. Freese, Allison Scott and Susan C. Chapman
Department of Biological Sciences, Clemson University, Clemson, South Carolina, United States of America

Brianna A. Lam and Meg Staton
Department of Entomology and Plant Pathology, University of Tennessee, Knoxville, Tennessee, United States of America

Shaoying Min, Yueling Zhang, Mingqi Zhong, Jinsong Cao and Jiehui Chen
Department of Biology and Guangdong Provincial Key Laboratory of Marine Biotechnology, Shantou University, Shantou, China

Fang Yan
Department of Biology and Guangdong Provincial Key Laboratory of Marine Biotechnology, Shantou University, Shantou, China
Mariculture Institute of Shandong Province, Qingdao, China

Haiying Zou and Xiangqun Ye
Medical College, Shantou University, Shantou, China

Alissa Piekarski, Stephanie Khaldi, Elizabeth Greene, Kentu Lassiter, James G. Mason
Nicholas Anthony, Walter Bottje and Sami Dridi
Center of Excellence for Poultry Science, University of Arkansas, Fayetteville, Arkansas, United States of America

Shotaro Wani, Yosuke Fujiwara, Masaya Yamamoto, Yoshiaki Ohkuma and Yutaka Hirose
Laboratory of Gene Regulation, Graduate School of Medicine and Pharmaceutical Sciences, University of Toyama, Sugitani, Toyama, Japan

Masamichi Yuda and Fumio Harada
Department of Molecular and Cellular Biology, Cancer Research Institute, Kanazawa University, Kakuma-machi, Kanazawa, Japan

Gahee Jo, Whasun Lim, Seung-Min Bae and Gwonhwa Song
Department of Biotechnology, College of Life Sciences and Biotechnology, Korea University, Seoul, Republic of Korea

Fuller W. Bazer
Center for Animal Biotechnology and Genomics and Department of Animal Science, Texas A & M University, College Station, Texas, United States of America

Huiqing Shang and Jun Ji
College of Animal Science, South China Agricultural University, Guangzhou, P R China

Hongxin Li, Baoli Sun, Hongmei Li and Qingmei Xie
College of Animal Science, South China Agricultural University, Guangzhou, P R China
Key Laboratory of Chicken Genetics, Breeding and Reproduction, Ministry of Agriculture Guangzhou, P R China

Dingming Shu
State Key Laboratory of Livestock and Poultry Breeding, Guangzhou, P R China

Huanmin Zhang
United States Department of Agriculture (USDA), Agriculture Research Service, Avian Disease and Oncology Laboratory, East Lansing, Michigan, United States of America

Robert E. Poelmann
Department of Anatomy and Embryology, Leiden University Medical Center, Leiden, The Netherlands
Department of Cardiology, Leiden University Medical Center, Leiden, The Netherlands
Institute of Biology Leiden (IBL), Leiden University, Sylvius Laboratory, Leiden, The Netherlands

Rebecca Vicente-Steijn, Lambertus J. Wisse, Margot M. Bartelings and Sonja Everts
Department of Anatomy and Embryology, Leiden University Medical Center, Leiden, The Netherlands

Adriana C. Gittenberger-de Groot
Department of Cardiology, Leiden University Medical Center, Leiden, The Netherlands

Boudewijn P. T. Kruithof
Department of Molecular Cell Biology, Leiden University Medical Center, Leiden, The Netherlands

Bjarke Jensen
Department of Anatomy, Embryology and Physiology, AMC Amsterdam, Amsterdam, The Netherlands
Department of Bioscience-Zoophysiology, Aarhus University, Aarhus, Denmark

Paul W. de Bruin
Department of Radiology, Leiden University Medical Center, Leiden, The Netherlands

Tatsuya Hirasawa and Shigeru Kuratani
Laboratory for Evolutionary Morphology, RIKEN Center for Developmental Biology, Kobe, Japan

Freek Vonk
Naturalis Biodiversity Center, Darwinweg , Leiden, The Netherlands

Jeanne M. M. S. van de Put, Merijn A. de Bakker, Michael K. Richardson and Tamara Hoppenbrouwers
Institute of Biology Leiden (IBL), Leiden University, Sylvius Laboratory, Leiden, The Netherlands

Jason J. Lavinder and Yariv Wine
Department of Chemical Engineering, University of Texas at Austin, Austin, Texas, United States of America
Institute for Cellular and Molecular Biology, University of Texas at Austin, Austin, Texas, United States of America

George Georgiou
Department of Chemical Engineering, University of Texas at Austin, Austin, Texas, United States of America
Institute for Cellular and Molecular Biology, University of Texas at Austin, Austin, Texas, United States of America
Department of Biomedical Engineering, University of Texas at Austin, Austin, Texas, United States of America
Section of Molecular Genetics and Microbiology, University of Texas at Austin, Austin, Texas, United States of America

Kam Hon Hoi
Department of Biomedical Engineering, University of Texas at Austin, Austin, Texas, United States of America

Sai T. Reddy
Department of Chemical Engineering, University of Texas at Austin, Austin, Texas, United States of America
Department of Biomedical Engineering, University of Texas at Austin, Austin, Texas, United States of America

Zhenjie Zhang, Chengtai Ma, Peng Zhao, Luntao Duan, Wenqing Chen, Fushou Zhang and Zhizhong Cui
College of Veterinary Medicine, Shandong Agricultural University, Taian, China
Animal Disease Prevention Technology and Research Center of Shandong Province, Taian, China

Nelson Henrique de Almeida Curi, Marcelo Passamani and Guilherme Ramos Demétrio
Postgraduate program in Applied Ecology, Department of Biology, Federal University of Lavras, Lavras, Brazil

Ana Maria de Oliveira Paschoal and Rodrigo Lima Massara
Postgraduate program in Ecology, Conservation and Management of Wildlife, Department of Biology, Institute of Biological Sciences, Federal University of Minas Gerais, Belo Horizonte, Brazil

Pain Marcelino and Andreza Adriana Aparecida Ribeiro
Laboratory of Leishmaniasis, Ezequiel Dias Foundation-FUNED, Belo Horizonte, Brazil

Adriano Garcia Chiarello
Department of Biology, University of São Paulo, Ribeirão Preto, Brazil

Sophie Blanc, Florence Ruggiero, Anne-Marie Birot, Jacques Samarut and Anne Mey
Institut de Génomique Fonctionnelle de Lyon, Université de Lyon, Université Lyon 1, CNRS UMR 5242, INRA USC 1370, Ecole Normale Supérieure de Lyon, Lyon, France

Hervé Acloque
Institut de Génomique Fonctionnelle de Lyon, Université de Lyon, Université Lyon 1, CNRS UMR 5242, INRA USC 1370, Ecole Normale Supérieure de Lyon, Lyon, France
Laboratoire de Génétique Cellulaire-INRA, ENVT, Castanet Tolosan, France

Didier Décimo and Thé ophile Ohlmann
CIRI, International Center for Infectiology Research, Université de Lyon, INSERM U1111, Ecole Normale Supérieure de Lyon, Lyon, France

Emmanuelle Lerat
Université de Lyon, Lyon, France; Université Lyon 1, Villeurbanne, France; CNRS, UMR5558, Laboratoire de Biométrie et Biologie Evolutive, Villeurbanne, France

Ana Rita Amândio, Pedro Gaspar, Jessica L. Whited and Florence Janody
Instituto Gulbenkian de Ciência, Oeiras, Portugal

Whasun Lim and Gwonhwa Song
Department of Biotechnology, College of Life Sciences and Biotechnology, Korea University, Seoul, Republic of Korea

Index